# Engineering Optimization

# Engineering Optimization

## Methods and Applications

**G. V. REKLAITIS**

Chemical Engineering
Purdue University

**A. RAVINDRAN**

Industrial Engineering
University of Oklahoma

**K. M. RAGSDELL**

Aerospace and Mechanical Engineering
University of Arizona

**A Wiley-Interscience Publication**
**JOHN WILEY AND SONS**
New York  •  Chichester  •  Brisbane  •  Toronto  •  Singapore

*Library of Congress Cataloging in Publication Data:*

Reklaitis, G. V., 1942–
  Engineering optimization

  "A Wiley-Interscience publication."
  Bibliography: p.
  Includes index.
  1. Engineering—Mathematical models.  2. Mathematical
optimization.  I. Ravindran, A., 1944– .  II. Ragsdell,
K. M.  III. Title.
TA342.R44 1983          620′.0042          83-3545
ISBN 0-471-05579-4

Printed in the United States of America

15

To Janine, Bhuvana, and Janet

# Preface

This is a text on the practical aspects of optimization methodology, with a major focus on the techniques and stratagems relevant to engineering applications arising in design, operations, and analysis. Attention is given primarily to techniques applicable to problems in continuous variables that involve real-valued constraint functions and a single real-valued objective function. In short, we treat the methodology often categorized as nonlinear programming. Within this framework a broad survey is given of all important families of optimization methods, ranging from those applicable to the minimization of a single-variable function to those most suitable for large-scale nonlinear constrained problems. Included are discussions not only of classical methods, important for historical reasons and for their role in motivating subsequent developments in this field, but also of promising new techniques such as those of successive quadratic programming.

Our development is aimed at imparting an understanding of the logic of the methods, of the key assumptions that underlie them, and of the comparative merits of the methods. Proofs and mathematical derivations are given only if they serve to explain key steps or properties of algorithms. Generally, we simply cite the original literature source for proofs and use the pages of this book to motivate and explain the key steps underlying the mathematical constructions. Thus, our aim is to inform the engineer–user of optimization methodology rather than to prepare the software specialist who will develop computer implementations of the algorithms. In keeping with this aim, we have given considerable attention to practical issues such as model formulation, implementation, preparation for solution, starting point generation, and the selection of execution strategies. A major chapter (Chapter 13) is devoted to strategies for carrying out optimization studies; another (Chapter 12) reviews the results of existing comparative studies involving state-of-the-art optimization software; and a third (Chapter 14) discusses three significant engineering case studies. In addition, a considerable fraction of each chapter is allocated to engineering examples drawn from the chemical, industrial, and mechanical engineering backgrounds of the authors. While a number of excellent books are available that deal in detail with the rich theoretical and numerical analysis issues that are relevant to nonlinear programming, this book is unique in the features outlined above: broad treatment of up-to-date methods; conceptual, rather than formal, presentation; and focus on issues relevant to engineering studies.

The book was developed over a period of eight years in which various drafts were used in a one-semester interdisciplinary engineering optimization course, team-taught by the authors to senior undergraduate and first-year graduate students at Purdue University. For these students, this course was typically the first systematic exposure to optimization methods. The students' mathematical preparation consisted of the calculus and linear algebra coursework typical of BS engineering curricula; hence, that is all that is required of the reader of this book. The organization of the manuscript has also benefited considerably from the authors' experience in teaching a television course on engineering optimization broadcast to regional campuses, nonresident MS, and professional engineering audiences. We are therefore confident that the book can serve as a text for conventional classroom lectures, for television courses, and for industrial short courses, as well as for self-study.

Two different course organizations have been followed in teaching from this text: an all-lecture, one-semester course involving 45 fifty-minute lectures; and a lecture-recitation format involving 30 lectures and 15 recitation–discussion sessions. In the former case, the entire contents of the book, except for Chapter 14, can be covered in numerical chapter sequence. In the latter case, Chapters 1, 13, and 14, as well as additional case studies and examples, are discussed in the recitation sessions, while the methodology chapters (Chapters 2–10) and Chapter 12 are covered in the lectures. In this format, Chapter 11 and Sections 8.3 and 9.4 typically must be omitted because of the limited lecture hours. Homework problems in both formats should include selections from the manual and computer exercises given at the end of the chapters. Student computer solutions can be carried out using the OPTLIB, OPT, and BIAS programs referenced in Chapter 12.

The development and evolution of this book have benefited substantially from the advice and counsel, both conscious and unintentional, of our colleagues and mentors in our respective engineering disciplines, who are too numerous to acknowledge individually. Professor Don Philips, while he was a colleague at Purdue University, contributed materially to the conception and initial development of notes that were early precursors to this book. The course and the notes have enjoyed the strong support of the past heads of our respective schools, Lowell B. Koppel, Wilbur Meier, R. W. Fox, and A. H. Lefebvre. Professors F. Kayihan and C. Knopf; Drs. F. W. Ahrens, G. A. Gabriele, R. Root, E. Sandgren, P. V. L. N. Sarma, Brad Overturf, and Mohinder Sood; and Messrs. P. K. Eswaran, J. Posey, and B. Nelson have offered corrections and improvements. We are also indebted for the numerous questions and pleas for clarification raised by our students in engineering optimization, who have persevered through various revisions of the manuscript. Their persistent, often biting, and usually justifiable criticism, overt and sub rosa, has been a key driving force in the revision process. Finally, we wish to thank our Wiley editor, Frank Cerra, and his predecessor, Meyer Kutz, for their perseverance and patience. We are grateful to our wives and children for allowing us the time to write and for supporting us through the period of production.

G. V. REKLAITIS
A. RAVINDRAN
K. M. RAGSDELL

*West Lafayette, Indiana*
*Summer 1983*

# Contents

**Chapter 1.** **Introduction to Optimization, 1**

**Chapter 2.** **Functions of a Single Variable, 25**

## Chapter 3. Functions of Several Variables, 68

## Chapter 4. Linear Programming, 137

*Very Important*

## Chapter 5. Constrained Optimality Criteria, 184

## Chapter 6. Transformation Methods, 216

*only*

## Chapter 7. Constrained Direct Search, 261

**Chapter 10.   Quadratic Approximation Methods for Constrained Problems, 438**

**Chapter 11.   Structured Problems and Algorithms, 468**

# Engineering Optimization

# Chapter 1

# Introduction to Optimization

This text is an introduction to optimization theory and its application to problems arising in engineering. In the most general terms, optimization theory is a body of mathematical results and numerical methods for finding and identifying the best candidate from a collection of alternatives without having to explicitly enumerate and evaluate all possible alternatives. The process of optimization lies at the root of engineering, since the classical function of the engineer is to design new, better, more efficient, and less expensive systems as well as to devise plans and procedures for the improved operation of existing systems.

The power of optimization methods to determine the best case without actually testing all possible cases comes through the use of a modest level of mathematics and at the cost of performing iterative numerical calculations using clearly defined logical procedures or algorithms implemented on computing machines. The development of optimization methodology will therefore require some facility with basic vector-matrix manipulations, a bit of linear algebra and calculus, and some elements of real analysis. We use mathematical concepts and constructions not simply to add rigor to the proceedings but because they are the language in terms of which calculation procedures are best developed, defined, and understood.

Because of the scope of most engineering applications and the tedium of the numerical calculations involved in optimization algorithms, the techniques of optimization are intended primarily for computer implementation. However, although the methodology is developed with computers in mind, we do not delve into the details of program design and coding. Instead, our emphasis is on the ideas and logic underlying the methods, on the factors involved in selecting the appropriate techniques, and on the considerations important to successful engineering application.

## 1.1 REQUIREMENTS FOR THE APPLICATION OF OPTIMIZATION METHODS

In order to apply the mathematical results and numerical techniques of optimization theory to concrete engineering problems it is necessary to clearly delineate the boundaries of the engineering system to be optimized, to define the quantitative criterion on the basis of which candidates will be ranked to determine the "best," to select the system variables that will be used to characterize or identify candidates, and to define a model that will express the manner in which the variables are related. This composite activity constitutes the process of *formulating* the engineering optimization problem. Good problem formulation is the key to the success of an optimization study and is to a large degree an art. It is learned through practice and the study of successful applications and is based on the knowledge of the strengths, weaknesses, and peculiarities of the techniques provided by optimization theory. For these reasons, this text is liberally laced with engineering applications drawn from the literature and the experience of the authors. Moreover, along with presenting the techniques, we attempt to elucidate their relative advantages and disadvantages, wherever possible by presenting or citing the results of actual computational tests.

In the next several subsections we discuss the elements of problem formulation in a bit more detail. In Section 1.2 we follow up this discussion by examining a few application formulations.

### 1.1.1  Defining the System Boundaries

Before undertaking any optimization study, it is important to clearly define the boundaries of the system under investigation. In this context a system is the restricted portion of the universe under consideration. The system boundaries are simply the limits that separate the system from the remainder of the universe. They serve to isolate the system from its surroundings, because, for purposes of analysis, all interactions between the system and its surroundings are assumed to be frozen at selected representative levels. Nonetheless, since interactions always exist, the act of defining the system boundaries is the first step in the process of approximating the real system.

In many situations it may turn out that the initial choice of boundary is too restrictive. To fully analyze a given engineering system, it may be necessary to expand the system boundaries to include other subsystems that strongly affect the operation of the system under study. For instance, suppose a manufacturing operation has a paint shop in which finished parts are mounted on an assembly line and painted in different colors. In an initial study of the paint shop, we may consider it in isolation from the rest of the plant. However, we may find that the optimal batch size and color sequence we deduce for this system are strongly influenced by the operation of the fabrication department that produces the finished parts. A decision thus has to be made whether to expand the system boundaries to include the fabrication system. An expansion of the system boundaries certainly increases the size and complexity of the composite system and thus may make the study much more difficult. Clearly, in order to make our work as engineers more manageable, we would prefer as much as possible to break down large complex systems into smaller subsystems that can be dealt with individually. However, we must recognize that such a decomposition may constitute a potentially misleading simplification of reality.

### 1.1.2  The Performance Criterion

Given that we have selected the system of interest and have defined its boundaries, we next need to select a criterion on the basis of which the performance or design of the system can be evaluated so that the "best" design or set of operating conditions can be identified. In many engineering applications, an economic criterion is selected. However, there is a considerable choice in the precise definition of such a criterion: total capital cost, annual cost, annual net profit, return on investment, cost-benefit ratio, or net present worth. In other applications a criterion may involve some technological factors —for instance, minimum production time, maximum production rate, minimum energy utilization, maximum torque, maximum weight, and so on. Regardless of the criterion selected, in the context of optimization the "best" will always mean the candidate system with either the *minimum* or *maximum* value of the performance index.

It is important to note that within the context of the optimization methods discussed in this book, only *one* criterion or performance measure can be used to define the optimum. It is not possible to find a solution that, say, simultaneously minimizes cost and maximizes reliability and minimizes energy utilization. This again is an important simplification of reality, because in many practical situations it would be desirable to achieve a solution that is "best" with respect to a number of different criteria.

One way of treating multiple competing objectives is to select one criterion as primary and the remaining criteria as secondary. The primary criterion is then used as an optimization performance measure, while the secondary criteria are assigned acceptable minimum or maximum values and are treated as problem constraints. For instance, in the case of the paint shop study, the following criteria may well be selected by different groups in the company:

1. The shop foreman may seek a design that will involve long production runs with a minimum of color and part changes. This will maximize the number of parts painted per unit time.
2. The sales department would prefer a design that maximizes the inventory of parts of every type and color. This will minimize the time between customer order and order dispatch.
3. The company financial officer would prefer a design that will minimize inventories so as to reduce the amount of capital tied up in parts inventory.

These are clearly conflicting performance criteria that cannot all be optimized simultaneously. A suitable compromise would be to select as the primary performance index the minimum annual cost, but then to require as secondary conditions that the inventory of each part not be allowed to fall below or rise above agreed-upon limits and that production runs involve no more than some maximum acceptable number of part and color changes per week.

In summary, for purposes of applying the methods discussed in this text, it is necessary to formulate the optimization problem with a single performance criterion. Advanced techniques do exist for treating certain types of multi-criteria optimization problems. However, this new and growing body of techniques is quite beyond the scope of this book. The interested reader is directed to recent specialized texts [1, 2].

### 1.1.3  The Independent Variables

The third key element in formulating a problem for optimization is the selection of the independent variables that are adequate to characterize the possible candidate designs or operating conditions of the system. There are several factors to be considered in selecting the independent variables.

First, it is necessary to distinguish between variables whose values are amenable to change and variables whose values are fixed by external factors, lying outside the boundaries selected for the system in question. For instance, in the case of the paint shop, the types of parts and the colors to be used are clearly fixed by product specifications or customer orders. These are specified system parameters. On the other hand, the order in which the colors are sequenced is, within constraints imposed by the types of parts available and inventory requirements, an independent variable that can be varied in establishing a production plan.

Furthermore, it is important to differentiate between system parameters that can be treated as fixed and those that are subject to fluctuations influenced by external and uncontrollable factors. For instance, in the case of the paint shop, equipment breakdown and worker absenteeism may be sufficiently high to seriously influence the shop operations. Clearly, variations in these key system parameters must be taken into account in the formulation of the production planning problem if the resulting optimal plan is to be realistic and operable.

Second, it is important to include in the formulation all the important variables that influence the operation of the system or affect the design definition. For instance, if in the design of a gas storage system we include the height, diameter, and wall thickness of a cylindrical tank as independent variables but exclude the possibility of using a compressor to raise the storage pressure, we may well obtain a very poor design. For the selected fixed pressure, we would certainly find the least-cost tank dimensions. However, by including the storage pressure as an independent variable and adding the compressor cost to our performance criteria we could obtain a design with a lower overall cost because of a reduction in the required tank volume. Thus, the independent variables must be selected so that all important alternatives are included in the formulation. In general, the exclusion of possible alternatives will lead to suboptimal solutions.

Finally, another consideration in the selection of variables is the level of detail to which the system is considered. While it is important to treat all key independent variables, it is equally important not to obscure the problem by the inclusion of a large number of fine details of subordinate importance. For instance, in the preliminary design of a process involving a number of different pieces of equipment—pressure vessels, towers, pumps, compressors, and heat exchangers—one would normally not explicitly consider all the fine details of the design of each individual unit. A heat exchanger may well be characterized by a heat-transfer surface area as well as shell-side and tube-side pressure drops. Detailed design variables such as number and size of tubes, number of tube and shell passes, baffle spacing, header type, and shell dimensions would normally be considered in a separate design study involving that unit by itself. In selecting the independent variables, a good rule is to include only those variables that have a significant impact on the composite system performance criterion.

### 1.1.4   The System Model

Once the performance criterion and the independent variables have been selected, the next step in problem formulation is to assemble the model that describes the manner in which the problem variables are related and the way in which the performance criterion is influenced by the independent variables. In principle, optimization studies may be performed by experimenting directly with the system. Thus, the independent variables of the system or process may be set to selected values, the system operated under those conditions, and the system performance index evaluated using the observed performance. The optimization methodology would then be used to predict improved choices of the independent variable values, and the experiments continued in this fashion. In practice, most optimization studies are carried out with the help of a simplified mathematical representation of the real system, called a *model*. Models are used because it is too expensive or time-consuming or risky to use the real system to carry out the study. Models are typically used in engineering design because they offer the cheapest and fastest way of studying the effects of changes in key design variables on system performance.

In general, the model will be composed of the basic material and energy balance equations, engineering design relations, and physical property equations that describe the physical phenomena taking place in the system. These equations will normally be supplemented by inequalities that define allowable operating ranges, specify minimum or maximum performance requirements, or set bounds on resource availabilities. In sum, the model consists of all elements that normally must be considered in calculating a design or in predicting the performance of an engineering system. Quite clearly, the assembly of a model is a very time-consuming activity and one that requires a thorough understanding of the system being considered. In later chapters we will have occasion to discuss the mechanics of model development in more detail. For now, we simply observe that a model is a collection of equations and inequalities that define how the system variables are related and that constrain the variables to take on acceptable values.

From the preceding discussion, we observe that a problem suitable for the application of optimization methodology consists of a performance measure, a set of independent variables, and a model relating the variables. Given these rather general and abstract requirements, it is evident that the methods of optimization can be applied to a very wide variety of applications. In fact, the methods we will discuss have been applied to problems that include the optimum design of process and structures, the planning of investment policies, the layout of warehouse networks, the determination of optimal trucking routes, the planning of health care systems, the deployment of military forces, and the design of mechanical components, to name but a few. In this text our focus will be on engineering applications. Some of these applications and their formulations are discussed in the next section.

## 1.2  APPLICATIONS OF OPTIMIZATION IN ENGINEERING

Optimization theory finds ready application in all branches of engineering in four primary areas:

1. Design of components or entire systems
2. Planning and analysis of existing operations
3. Engineering analysis and data reduction
4. Control of dynamic systems

In this section we briefly consider representative applications from each of the first three areas. The control of dynamic systems is an important area to which the methodology discussed in this book is applicable but which requires the consideration of specialized topics quite beyond the scope of this book.

In considering the application of optimization methods in design and operations, keep in mind that the optimization step is but one step in the overall process of arriving at an optimal design or an efficient operation. Generally, that overall process will, as shown in Figure 1.1, consist of an iterative cycle involving synthesis or definition of the structure of the system, model formulation, model parameter optimization, and analysis of the resulting solution. The final optimal design or new operating plan will be obtained only after solving a series of optimization problems, the solution to each of which will serve to generate new ideas for further system structures. In the interests of brevity, the examples in this section show only one pass of this iterative cycle and deal mainly with preparations for the optimization step.

This focus should not be interpreted as an indication of the dominant role of optimization methods in the engineering design and systems analysis process. Optimization theory is a very powerful tool, but to be effective it must be used skillfully and intelligently by an engineer who thoroughly understands the system under study. The primary objective of the following examples is simply to illustrate the wide variety but common form of the optimization problems that arise in the process of design and analysis.

### 1.2.1　Design Applications

Applications in engineering design range from the design of individual structural members to the design of separate pieces of equipment to the preliminary design of entire production facilities. For purposes of optimization, the shape or structure of the system is assumed to be known, and the optimization problem reduces to that of selecting values of the unit dimensions and

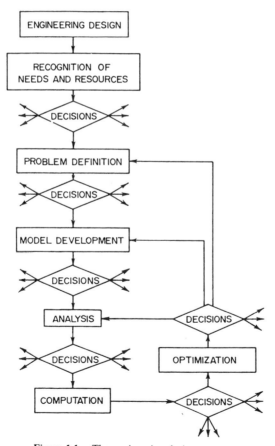

**Figure 1.1.**　The engineering design process.

operating variables that will yield the best value of the selected performance
criterion.

### *Example 1.1   Design of an Oxygen Supply System*

**Description.**  The basic oxygen furnace (BOF) used in the production of steel
is a large fed-batch chemical reactor that employs pure oxygen. The furnace is
operated in a cyclical fashion. Ore and flux are charged to the unit, treated for
a specified time period, and then discharged. This cyclical operation gives rise
to a cyclically varying demand rate for oxygen. As shown in Figure 1.2, over
each cycle there is a time interval of length $t_1$ of low demand rate $D_0$, and a
time interval $(t_2 - t_1)$ of high demand rate $D_1$. The oxygen used in the BOF is
produced in an oxygen plant, in a standard process in which oxygen is
separated from air by using a combination of refrigeration and distillation.
Oxygen plants are highly automated and are designed to deliver oxygen at a
fixed rate. In order to mesh the continuous oxygen plant with the cyclically
operating BOF, a simple inventory system (Figure 1.3), consisting of a compressor and a storage tank, must be designed. A number of design possibilities
can be considered. In the simplest case, the oxygen plant capacity could be
selected to be equal to $D_1$, the high demand rate. During the low-demand
interval the excess oxygen could just be vented to the air. At the other extreme,
the oxygen plant capacity could be chosen to be just enough to produce the
amount of oxygen required by the BOF over a cycle. During the low-demand
interval, the excess oxygen produced would then be compressed and stored for
use during the high-demand interval of the cycle. Intermediate designs could
use some combination of venting and storage of oxygen. The problem is to
select the optimal design.

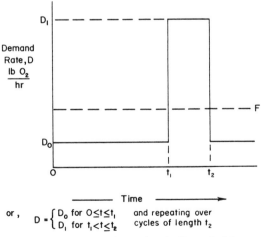

or ,   $D = \begin{cases} D_0 \text{ for } 0 \le t \le t_1 & \text{and repeating over} \\ D_1 \text{ for } t_1 < t \le t_2 & \text{cycles of length } t_2 \end{cases}$

**Figure 1.2.**  Oxygen demand cycle, Example 1.1.

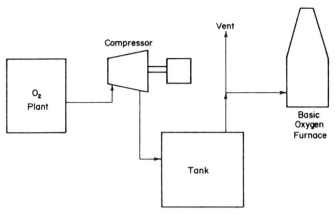

**Figure 1.3.** Design of oxygen production system, Example 1.1.

**Formulation.** The system of concern will consist of the $O_2$ plant, the compressor, and the storage tank. The BOF and its demand cycle are assumed fixed by external factors. A reasonable performance index for the design is the total annual cost, which consists of the oxygen production cost (fixed and variable), the compressor operating cost, and the fixed costs of the compressor and storage vessel. The key independent variables are the oxygen plant production rate $F$ (lb $O_2$/hr), the compressor and storage tank design capacities, $H$ (hp) and $V$ (ft³), respectively, and the maximum tank pressure, $p$ (psia). Presumably the oxygen plant design is standard so that the production rate fully characterizes the plant. Similarly, we assume that the storage tank will be of a standard design approved for $O_2$ service.

The model will consist of the basic design equations that relate the key independent variables.

If $I_{max}$ is the maximum amount of oxygen that must be stored, then using the corrected gas law we have

$$V = \frac{I_{max}}{M} \frac{RT}{p} z \qquad (1.1)$$

where $R$ is the gas constant

$T$ is the gas temperature (assume fixed)

$z$ is the compressibility factor

$M$ is the molecular weight of $O_2$

From Figure 1.2, the maximum amount of oxygen that must be stored is equal to the area under the demand curve between $t_1$ and $t_2$ and $D_1$ and $F$. Thus,

$$I_{max} = (D_1 - F)(t_2 - t_1) \qquad (1.2)$$

Substituting (1.2) into (1.1), we obtain

$$V = \frac{(D_1 - F)(t_2 - t_1)}{M} \frac{RT}{p} z \tag{1.3}$$

The compressor must be designed to handle a gas flowrate of $(D_1 - F)(t_2 - t_1)/t_1$ and to compress the gas to the maximum pressure $p$. Assuming isothermal ideal gas compression [3],

$$H = \frac{(D_1 - F)(t_2 - t_1)}{t_1} \frac{RT}{k_1 k_2} \ln\left(\frac{p}{p_0}\right) \tag{1.4}$$

where $k_1$ is a unit conversion factor

$k_2$ is the compressor efficiency

$p_0$ is the $O_2$ delivery pressure

In addition to (1.3) and (1.4), the $O_2$ plant rate $F$ must be adequate to supply the total oxygen demand, or

$$F \geqslant \frac{D_0 t_1 + D_1(t_2 - t_1)}{t_2} \tag{1.5}$$

Moreover, the maximum tank pressure must be greater than the $O_2$ delivery pressure,

$$p \geqslant p_0 \tag{1.6}$$

The performance criterion will consist of the oxygen plant annual cost,

$$C_1 \, (\$/\text{yr}) = a_1 + a_2 F \tag{1.7}$$

where $a_1$ and $a_2$ are empirical constants for plants of this general type and include fuel, water, and labor costs.

The capital cost of storage vessels is given by a power law correlation,

$$C_2 \, (\$) = b_1 V^{b_2} \tag{1.8}$$

where $b_1$ and $b_2$ are empirical constants appropriate for vessels of a specific construction.

The capital cost of compressors is similarly obtained from a correlation:

$$C_3 \, (\$) = b_3 H^{b_4} \tag{1.9}$$

The compressor power cost will, as an approximation, be given by $b_5 t_1 H$ where $b_5$ is the cost of power. The total cost function will thus be of the form,

$$\text{Annual cost} = a_1 + a_2 F + d\{b_1 V^{b_2} + b_3 H^{b_4}\} + N b_5 t_1 H \tag{1.10}$$

where $N$ is the number of cycles per year and $d$ is an appropriate annual cost factor.

The complete design optimization problem thus consists of the problem of minimizing (1.10) by the appropriate choice of $F$, $V$, $H$, and $p$, subject to Eqs. (1.3) and (1.4) as well as inequalities (1.5) and (1.6).

The solution of this problem will clearly be affected by the choice of the cycle parameters ($N$, $D_0$, $D_1$, $t_1$, and $t_2$), the cost parameters ($a_1$, $a_2$, $b_1$–$b_5$, and d), and the physical parameters ($T$, $p_0$, $k_2$, $z$, and $M$).

In principle we could solve this problem by eliminating $V$ and $H$ from (1.10) using (1.3) and (1.4), thus obtaining a two-variable problem. We could then plot the contours of the cost function (1.10) in the plane of the two variables $F$ and $p$, impose the inequalities (1.5) and (1.6), and determine the minimum point from the plot. However, the methods discussed in subsequent chapters allow us to obtain the solution with much less work. For further details and a study of solutions for various parameter values, the reader is invited to consult Jen et al. [4].

Example 1.1 presented a preliminary design problem formulation for a system consisting of several pieces of equipment. The next example illustrates a detailed design of a single structural element.

### *Example 1.2   Design of a Welded Beam*

**Description.**   A beam A is to be welded to a rigid support member B. The welded beam is to consist of 1010 steel and is to support a force $F$ of 6000 lb. The dimensions of the beam are to be selected so that the system cost is minimized. A schematic of the system is shown in Figure 1.4.

**Formulation.**   The appropriate system boundaries are quite self-evident. The system consists of the beam A and the weld required to secure it to B. The independent or design variables in this case are the dimensions $h$, $l$, $t$, and $b$, as shown in Figure 1.4. The length $L$ is assumed to be specified at 14 in. For notational convenience we redefine these four variables in terms of the vector of unknowns **x**:

$$\mathbf{x} = [x_1, x_2, x_3, x_4,]^T = [h, l, t, b]^T$$

The performance index appropriate to this design is the cost of a weld assembly. The major cost components of such an assembly are (1) setup labor

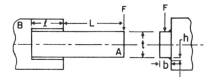

Figure 1.4.   A welded beam, Example 1.2.

cost, (2) welding labor cost, and (3) material cost.

$$F(x) = c_0 + c_1 + c_2 \tag{1.10}$$

where $F(x)$ = cost function

$c_0$ = setup cost

$c_1$ = welding labor cost

$c_2$ = material cost

*Setup Cost $c_0$.* The company has chosen to make this component a weldment, because of the existence of a welding assembly line. Furthermore, assume that fixtures for setup and holding of the bar during welding are readily available. The cost $c_0$ can therefore be ignored in this particular total-cost model.

*Welding Labor Cost $c_1$.* Assume that the welding will be done by machine at a total cost of \$10/hr (including operating and maintenance expense). Furthermore, suppose that the machine can lay down a cubic inch of weld in 6 min. The labor cost is then

$$c_1 = \left(10\frac{\$}{hr}\right)\left(\frac{1}{60}\frac{hr}{min}\right)\left(6\frac{min}{in.^3}\right)V_w = 1\left(\frac{\$}{in.^3}\right)V_w$$

where $V_w$ = weld volume, in.$^3$.

*Material Cost $c_2$*

$$c_2 = c_3 V_w + c_4 V_B$$

where $c_3$ = \$/volume of weld material = $(0.37)(0.283)(\$/in.^3)$

$c_4$ = \$/volume of bar stock = $(0.17)(0.283)(\$/in.^3)$

$V_B$ = volume of bar A, in.$^3$

From the geometry,

$$V_w = 2(\tfrac{1}{2}h^2 l) = h^2 l$$

and

$$V_B = tb(L + l)$$

so

$$c_2 = c_3 h^2 l + c_4 tb(L + l)$$

Therefore, the cost function becomes

$$F(x) = h^2 l + c_3 h^2 l + c_4 tb(L + l) \tag{1.11}$$

or, in terms of the $x$ variables,

$$F(x) = (1 + c_3)x_1^2 x_2 + c_4 x_3 x_4 (L + x_2) \tag{1.12}$$

Note all combinations of $x_1$, $x_2$, $x_3$ and $x_4$ can be allowed if the structure is to support the load required. Several functional relationships between the design variables that delimit the region of feasibility must certainly be defined. These relationships, expressed in the form of inequalities, represent the design model. Let us first define the inequalities and then discuss their interpretation. The inequalities are

$$g_1(x) = \tau_d - \tau(x) \geqslant 0 \tag{1.13}$$

$$g_2(x) = \sigma_d - \sigma(x) \geqslant 0 \tag{1.14}$$

$$g_3(x) = x_4 - x_1 \geqslant 0 \tag{1.15}$$

$$g_4(x) = x_2 \geqslant 0 \tag{1.16}$$

$$g_5(x) = x_3 \geqslant 0 \tag{1.17}$$

$$g_6(x) = P_c(x) - F \geqslant 0 \tag{1.18}$$

$$g_7(x) = x_1 - 0.125 \geqslant 0 \tag{1.19}$$

$$g_8(x) = 0.25 - \delta(x) \geqslant 0 \tag{1.20}$$

where    $\tau_d$ = design shear stress of weld

       $\tau(x)$ = maximum shear stress in weld; a function of $x$

       $\sigma_d$ = design normal stress for beam material

       $\sigma(x)$ = maximum normal stress in beam; a function of $x$

       $P_c(x)$ = bar buckling load; a function of $x$

       $\delta(x)$ = bar end deflection; a function of $x$

In order to complete the model it is necessary to define the important stress states.

*Weld Stress*: $\tau(x)$. After Shigley [5], the weld shear stress has two components, $\tau'$ and $\tau''$, where $\tau'$ is the primary stress acting over the weld throat area and $\tau''$ is a secondary torsional stress.

$$\tau' = \frac{F}{\sqrt{2}\, x_1 x_2} \quad \text{and} \quad \tau'' = \frac{MR}{J}$$

with

$$M = F[L + (x_2/2)]$$

$$R = \left\{ \frac{x_2^2}{4} + \left[ \frac{x_3 + x_1}{2} \right]^2 \right\}^{1/2}$$

$$J = 2\left\{ .707x_1x_2 \left[ \frac{x_2^2}{12} + \left( \frac{x_3 + x_1}{2} \right)^2 \right] \right\}$$

where  $M$ = moment of $F$ about the center of gravity of the weld group
    $J$ = polar moment of inertia of the weld group

Therefore, the weld stress $\tau$ becomes

$$\tau(x) = \left[ (\tau')^2 + 2\tau'\tau'' \cos \theta + (\tau'')^2 \right]^{1/2}$$

where

$$\cos \theta = x_2/2R.$$

*Bar Bending Stress $\sigma(x)$.*   The maximum bending stress can be shown to be equal to

$$\sigma(x) = \frac{6FL}{x_4 x_3^2}$$

*Bar Buckling Load $P_c(x)$.*   If the ratio $t/b = x_3/x_4$ grows large, there is a tendency for the bar to buckle. Those combinations of $x_3$ and $x_4$ that will cause this buckling to occur must be disallowed. It has been shown [6] that for narrow rectangular bars, a good approximation to the buckling load is

$$P_c(x) = \frac{4.013\sqrt{EI\alpha}}{L^2} \left[ 1 - \frac{x_3}{2L} \sqrt{\frac{EI}{\alpha}} \right]$$

where  $E$ = Young's modulus = $30 \times 10^6$ psi
    $I = \frac{1}{12}x_3 x_4^3$
    $\alpha = \frac{1}{3}Gx_3 x_4^3$
    $G$ = shearing modulus = $12 \times 10^6$ psi

*Bar Deflection $\delta(x)$.*   To calculate the deflection, assume the bar to be a cantilever of length $L$. Thus,

$$\delta(x) = \frac{4FL^3}{Ex_3^3 x_4}$$

The remaining inequalities are interpreted as follows: $g_3$ states that it is not practical to have the weld thickness greater than the bar thickness. $g_4$ and $g_5$ are nonnegativity restrictions on $x_2$ and $x_3$. Note that the nonnegativity of $x_1$ and $x_4$ are implied by $g_3$ and $g_7$. Constraint $g_6$ ensures that the buckling load is not exceeded. Inequality $g_7$ specifies that it is not physically possible to produce an extremely small weld.

Finally, the two parameters $\tau_d$ and $\sigma_d$ in $g_1$ and $g_2$ depend on the material of construction. For 1010 steel, $\tau_d = 13{,}600$ psi and $\sigma_d = 30{,}000$ psi are appropriate.

The complete design optimization problem thus consists of the cost function (1.12) and the complex system of inequalities that results when the stress formulas are substituted into (1.13)–(1.20). All of these functions are expressed in terms of four independent variables.

This problem is sufficiently complex that graphical solution is patently infeasible. However, the optimum design can readily be obtained numerically by using the methods of subsequent chapters.

For a further discussion of this problem and its solution, see reference 7.

### 1.2.2  Operations and Planning Applications

The second major area of engineering application of optimization is found in the tuning of existing operations and development of production plans for multiproduct processes. Typically an operations analysis problem arises when an existing production facility designed under one set of conditions must be adapted to operate under different conditions. The reasons for doing this might be to

1.   Accommodate increased production throughout
2.   Adapt to different feedstocks or a different product slate
3.   Modify the operations because the initial design is itself inadequate or unreliable

The solution to such problems might require the selection of new temperature, pressure, or flow conditions; the addition of further equipment; or the definition of new operating procedures. Production planning applications arise from the need to schedule the joint production of several products in a given plant or to coordinate the production plans of a network of production facilities. Since in such applications the capital equipment is already in place, only the variable costs need to be considered. Thus, this type of application can often be formulated in terms of linear or nearly linear models. We will illustrate this class of applications using a refinery planning problem.

### Example 1.3  Refinery Production Planning

**Description.** A refinery processes crude oils to produce a number of raw gasoline intermediates, which must subsequently be blended to make two grades of motor fuel, regular and premium. Each raw gasoline has a known performance rating, a maximum availability, and a fixed unit cost. The two motor fuels have a specified minimum performance rating and selling price, and their blending is achieved at a known unit cost. Contractual obligations impose minimum production requirements of both fuels. However, all excess fuel production or unused raw gasoline amounts can be sold in the open market at known prices. The optimal refinery production plan is to be determined over the next specified planning period.

**Formulation.** The system in question consists of the raw gasoline intermediates, the blending operation, and the fluid motor fuels as shown schematically in Figure 1.5. Excluded from consideration are the refinery processes involved in the production of the raw gasoline intermediates, as well as the inventory and distribution subsystems for crudes, intermediates, and products. Since equipment required to carry out the blending operations is in place, only variable costs will be considered.

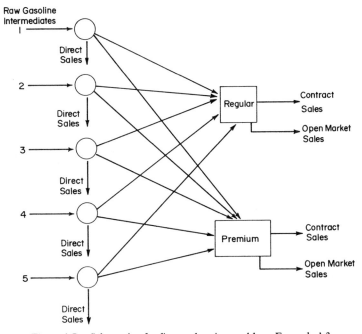

**Figure 1.5.**  Schematic of refinery planning problem, Example 1.3.

The performance index in this case will be the net profit over the planning period. The net profit will be composed of motor fuel and intermediate sales minus blending costs minus the charged costs of the intermediates. The independent variables will simply be the flows depicted as directed arcs in Figure 1.5. Thus, each intermediate will have associated with it a variable that represents the amount of that intermediate allocated to the production of regular grade gasoline, another that represents the amount used to make premium, and a third that represents the amount sold directly.

Thus, for each intermediate $i$,

$$x_i = \text{amount used for regular, bbl/period}$$

$$y_i = \text{amount used for premium, bbl/period}$$

$$z_i = \text{amount sold directly, bbl/period}$$

Each product will have two variables associated with it: one to represent the contracted sales and one to represent the open-market sales.

Thus, for each product $j$,

$$u_j = \text{amount allocated to contracts, bbl/period}$$

$$v_j = \text{amount sold in open market, bbl/period}$$

The model will consist of material balances on each intermediate and product, blending constraints that ensure that product performance ratings are met, and bounds on the contract sales.

1.  Material balances on each intermediate $i$:

$$x_i + y_i + z_i \leqslant \alpha_i \tag{1.21}$$

   where $\alpha_i$ is the availability of intermediate $i$ over the period, in bbl/period.

2.  Material balances on each product:

$$\sum_i x_i = u_1 + v_1 \qquad \sum_i y_i = u_2 + v_2 \tag{1.22}$$

3.  Blending constraints on each product:

$$\sum_i \beta_i x_i \geqslant \gamma_1(u_1 + v_1) \qquad \sum_i \beta_i y_i \geqslant \gamma_2(u_2 + v_2) \tag{1.23}$$

   where $\beta_i$ is the performance rating of intermediate $i$, and $\gamma_j$ is the minimum performance rating of product $j$.

**4.** Contract sales restrictions for each product $j$.

$$u_j \geq \delta_j \tag{1.24}$$

where $\delta_j$ is the minimum contracted production, in bbl/period.

The performance criterion (net profit) is given by

$$\sum c_j^{(1)} u_j + \sum c_j^{(2)} v_j + \sum_i c_i^{(3)} z_i - \sum_i c_i^{(4)}(x_i + y_i + z_i) - \sum_i c_i^{(5)}(x_i + y_i)$$

where $c_j^{(1)}$ = unit selling price for contract sales of $j$
$\quad\quad c_j^{(2)}$ = unit selling price for open market sales of $j$
$\quad\quad c_i^{(3)}$ = unit selling price of direct sales of intermediate $i$
$\quad\quad c_i^{(4)}$ = unit charged cost of intermediate $i$
$\quad\quad c_i^{(5)}$ = blending cost of intermediate $i$

Using the data given in Table 1.1, the planning problem reduces to

$$\text{Maximize } 40u_1 + 55u_2 + 46v_1 + 60v_2 + 6z_1 + 8z_2 + 7.50z_3$$

$$+ 7.50z_4 + 20z_5 - 25(x_1 + y_1) - 28(x_2 + y_2) - 29.50(x_3 + y_3)$$

$$- 35.50(x_4 + y_4) - 41.50(x_5 + y_5)$$

**Table 1.1   Data for Example 1.3**

| Raw Gasoline Intermediate | Availability, $\alpha_i$ (bbl/period) | Performance Rating, $\beta_i$ | Selling Price $c_i^{(3)}$ | Charged Cost $c_i^{(4)}$ | Blending Cost $c_i^{(5)}$ |
|---|---|---|---|---|---|
| 1 | $2 \times 10^5$ | 70 | 30.00 | 24.00 | 1.00 |
| 1 | $4 \times 10^5$ | 80 | 35.00 | 27.00 | 1.00 |
| 3 | $4 \times 10^5$ | 85 | 36.00 | 28.50 | 1.00 |
| 4 | $5 \times 10^5$ | 90 | 42.00 | 34.50 | 1.00 |
| 5 | $5 \times 10^5$ | 99 | 60.00 | 40.00 | 1.50 |

| Product Type | Minimum Contract Sales $\delta_j$ | Minimum Performance Rating | Selling Price ($/bbl) Contract $c_j^{(1)}$ | Open Market $c_j^{(2)}$ |
|---|---|---|---|---|
| Regular (1) | $5 \times 10^5$ | 85 | $40.00 | $46.00 |
| Premium (2) | $4 \times 10^5$ | 95 | $55.00 | $60.00 |

Subject to:
Constraints of type (1.21):

$$x_1 + y_1 + z_1 \leqslant 2 \times 10^5$$

$$x_2 + y_2 + z_2 \leqslant 4 \times 10^5$$

$$x_3 + y_3 + z_3 \leqslant 4 \times 10^5$$

$$x_4 + y_4 + z_4 \leqslant 5 \times 10^5$$

$$x_5 + y_5 + z_5 \leqslant 5 \times 10^5$$

Constraints of type (1.22):

$$x_1 + x_2 + x_3 + x_4 + x_5 = u_1 + v_1$$

$$y_1 + y_2 + y_3 + y_4 + y_5 = u_2 + v_2$$

Constraints of type (1.23):

$$70x_1 + 80x_2 + 85x_3 + 90x_4 + 99x_5 \geqslant 85(u_1 + v_1)$$

$$70y_1 + 80y_2 + 85y_3 + 90y_4 + 99y_5 \geqslant 95(u_2 + v_2)$$

Constraints of type (1.24):

$$u_1 \geqslant 5 \times 10^5 \qquad u_2 \geqslant 4 \times 10^5$$

In addition, all variables must be greater than or equal to zero to be physically realizable. The composite optimization problem involves 19 variables and 11 constraints plus the nonnegativity conditions. Note that all model functions are linear in the independent variables.

In general, refineries will involve many more intermediate streams and products than were considered in this example. Moreover, in practice it may be important to include further variables, reflecting the material in inventory, as well as to expand the model to cover several consecutive planning periods. In the latter case, a second subscript could be added to the set of variables, for example,

$x_{ik}$ = amount of intermediate $i$ used for regular grade in planning period $k$

The resulting production planning model can then become very large. In practice, models of this type with over a thousand variables are solved quite routinely.

### 1.2.3  Analysis and Data Reduction Applications

A further fertile area for the application of optimization techniques in engineering can be found in nonlinear regression problems as well as in many analysis problems arising in engineering science. A very common problem arising in engineering model development is the need to determine the parameters of some semitheoretical model given a set of experimental data. This data reduction or regression problem inherently transforms to an optimization problem, because the model parameters must be selected so that the model fits the data as closely as possible.

Suppose some variable $y$ is assumed to be dependent upon an independent variable $x$ and related to $x$ through a postulated equation $y = f(x, \theta_1, \theta_2)$, which depends upon two parameters $\theta_1$ and $\theta_2$. To establish the appropriate values of $\theta_1$ and $\theta_2$, we run a series of experiments in which we adjust the independent variable $x$ and measure the resulting $y$. As a result of a series of $N$ experiments covering the range of $x$ of interest, a set of $y$ and $x$ values $(y_i, x_i)$, $i = 1, \ldots, N$, is available. Using these data we now try to "fit" our function to the data by adjusting $\theta_1$ and $\theta_2$ until we get a "good fit." The most commonly used measure of a "good fit" is the *least squares criterion*,

$$L(\theta_i, \theta_2) = \sum_{i=1}^{N} [y_i - f(x_i, \theta_1, \theta_2)]^2 \qquad (1.25)$$

The difference $y_i - f(x_i, \theta_1, \theta_2)$ between the experimental value $y_i$ and the predicted value $f(x_i, \theta_1, \theta_2)$ measures how close our model prediction is to the data and is called the *residual*. The sum of the squares of the residuals at all the experimental points gives an indication of goodness of fit. Clearly, if $L(\theta_1, \theta_2)$ is equal to zero, then the choice of $\theta_1, \theta_2$ has led to a perfect fit; the data points fall exactly on the predicted curve. The data-fitting problem can thus be viewed as an optimization problem in which $L(\theta_1, \theta_2)$ is minimized by appropriate choice of $\theta_1$ and $\theta_2$.

*Example 1.4   Nonlinear Curve Fitting*

**Description.**  The pressure–molar volume–temperature relationship of real gases is known to deviate from that predicted by the ideal gas relationship,

$$Pv = RT$$

where  $P$ = pressure (atm)
   $v$ = molar volume (cm$^3$/g · mol)
   $T$ = temperature (K)
   $R$ = gas constant (82.06 atm · cm$^3$/g · mole · K)

The semiempirical Redlich-Kwong equation [3],

$$P = \frac{RT}{v - b} - \frac{a}{T^{1/2}v(v + b)} \qquad (1.26)$$

is intended to correct for the departure from ideality but involves two empirical constants $a$ and $b$ whose values are best determined from experimental data. A series of $PvT$ measurements, listed in Table 1.2, are made for $CO_2$, from which $a$ and $b$ are to be estimated using nonlinear regression.

**Formulation.** Parameters $a$ and $b$ will be determined by minimizing the least squares function (1.25). In the present case, the function will take the form

$$\sum_{i=1}^{8}\left[P_i - \frac{RT_i}{v_i - b} + \frac{a}{T_i^{1/2}v_i(v_i + b)}\right]^2 \qquad (1.27)$$

where $P_i$ is the experimental value at experiment $i$, and the remaining two terms correspond to the value of $P$ predicted from Eq. (1.26) for the conditions of experiment $i$ for some selected value of the parameters $a$ and $b$. For instance, the term corresponding to the first experimental point will be

$$\left(33 - \frac{82.06(273)}{500 - b} + \frac{a}{(273)^{1/2}(500)(500 + b)}\right)^2$$

Function (1.27) is thus a two-variable function whose value is to be minimized by appropriate choice of the independent variables $a$ and $b$. If the Redlich-Kwong equation were to precisely match the data, then at the optimum the function (1.27) would be exactly equal to zero. In general, because of experimental error and because the equation is too simple to accurately model

Table 1.2  PVT Data for $CO_2$

| Experiment No. | $P$, atm | $v$, cm$^3$/g · mol | $T$, K |
|---|---|---|---|
| 1 | 33 | 500 | 273 |
| 2 | 43 | 500 | 323 |
| 3 | 45 | 600 | 373 |
| 4 | 26 | 700 | 273 |
| 5 | 37 | 600 | 323 |
| 6 | 39 | 700 | 373 |
| 7 | 38 | 400 | 273 |
| 8 | 63.6 | 400 | 373 |

the $CO_2$ nonidealities, Eq. (1.27) will not be equal to zero at the optimum. For instance, the optimal values of $a = 6.377 \times 10^7$ and $b = 29.7$ still yield a squared residual of $9.7 \times 10^{-2}$.

In addition to regression applications, a number of problems arise in engineering science that can be solved by posing them as optimization problems. One rather classical application is the determination of the equilibrium composition of a chemical mixture [3]. It is known that the equilibrium composition of a closed system at fixed temperature and pressure with specified initial composition will be that composition which minimizes the Gibbs free energy of the system. As shown by White et al. [8], the determination of the equilibrium composition can thus be posed as the problem of minimizing a nonlinear function subject to a set of linear equations in nonnegative variables.

Another classical engineering problem that can be posed and solved as an optimization problem is the determination of the steady-state current flows in an electrical resistance network [9]. Given a network with specified arc resistances and a specified overall current flow, the arc current flows can be determined as the solution of the problem of minimizing the total $I^2R$ power loss subject to a set of linear constraints that ensure that Kirchhoff's current law is satisfied at each arc junction in the network.

## 1.3  STRUCTURE OF OPTIMIZATION PROBLEMS

Although the application problems discussed in the previous section originate from radically different sources and involve different systems, at root they have a remarkably similar form. All four can be expressed as problems requiring the minimization of a real-valued function $f(x)$ of an $N$-component vector argument $x = (x_1, x_2, \ldots, x_N)$ whose values are restricted to satisfy a number of real-valued equations $h_k(x) = 0$, a set of inequalities $g_j(x) \geqslant 0$, and the variable bounds $x_i^{(U)} \geqslant x_i \geqslant x_i^{(L)}$. In subsequent discussions we will refer to the function $f(x)$ as the *objective function*, to the equations $h_k(x) = 0$ as the *equality constraints*, and to the inequalities $g_j(x) \geqslant 0$ as the *inequality constraints*. For our purposes, these problem functions will always be assumed to be real valued, and their number will always be finite.

The general problem,

$$
\begin{aligned}
&\text{Minimize} && f(x) \\
&\text{Subject to} && h_k(x) = 0 && k = 1, \ldots, K \\
& && g_j(x) \geqslant 0 && j = 1, \ldots, J \\
& && x_i^{(U)} \geqslant x_i \geqslant x_i^{(L)} && i = 1, \ldots, N
\end{aligned}
$$

is called the *constrained* optimization problem. For instance, Examples 1.1, 1.2, and 1.3 are all constrained problems. The problem in which there are no constraints, that is,

$$J = K = 0$$

and

$$x_i^{(U)} = -x_i^{(L)} = \infty \qquad i = 1, \ldots, N$$

is called the *unconstrained* optimization problem. Example 1.4 is an unconstrained problem.

Optimization problems can be classified further based on the structure of the functions $f$, $h_k$, and $g_j$ and on the dimensionality of $x$. Unconstrained problems in which $x$ is a one-component vector are called *single-variable* problems and form the simplest, but nonetheless a very important, subclass. Constrained problems in which the function $h_k$ and $g_j$ are all linear are called *linearly constrained* problems. This subclass can further be subdivided into those with a linear objective function $f$ and those in which $f$ is nonlinear. The category in which all problem functions are linear in $x$ includes problems with continuous variables, which are called *linear programs*, and problems in integer variables, which are called *integer programs*. Example 1.3 (Section 1.2) is a linear programming problem.

Problems with nonlinear objective and linear constraints are sometimes called *linearly constrained nonlinear programs*. This class can further be subdivided according to the particular structure of the nonlinear objective function. If $f(x)$ is quadratic, the problem is a *quadratic program*; if it is a ratio of linear functions, it is called a *fractional* linear program; and so on. Subdivision into these various classes is worthwhile, because the special structure of these problems can be efficiently exploited in devising solution techniques. We will consider techniques applicable to most of these problem structures in subsequent chapters.

## 1.4  SCOPE OF THIS BOOK

In this text we study the methodology applicable to constrained and unconstrained optimization problems. Our primary focus is on general-purpose techniques applicable to problems in continuous variables, involving real-valued constraint functions and a single real-valued objective function. Problems posed in terms of integer or discrete variables are considered only briefly. Moreover, we exclude from consideration optimization problems involving functional equations, non-steady-state models, or stochastic elements. These very interesting but more complex elements are the appropriate subject matter for more advanced, specialized texts. While we make every effort to be precise in stating mathematical results, we do not normally present detailed proofs of these results unless such a proof serves to explain subsequent algorithmic constructions. Generally we simply cite the original literature source for proofs and use the pages of this book to motivate and explain the key concepts underlying the mathematical constructions.

One of the goals of this text is to demonstrate the applicability of optimization methodology to engineering problems. Hence, a considerable portion is devoted to engineering examples, to the discussion of formulation alternatives, and to consideration of computational devices that expedite the solution of applications problems. In addition, we review and evaluate available computational evidence to help elucidate why particular methods are preferred under certain conditions.

In Chapter 2, we begin with a discussion of the simplest problem, the single-variable unconstrained problem. This is followed by an extensive treatment of the multivariable unconstrained case. In Chapter 4, the important linear programming problem is analyzed. With Chapter 5 we initiate the study of nonlinear constrained optimization by considering tests for determining optimality. Chapters 6 through 10 focus on solution methods for constrained problems. Chapter 6 considers strategies for transforming constrained problems into unconstrained problems, while Chapter 7 discusses direct search methods. Chapters 8 and 9 develop the important linearization-based techniques, and Chapter 10 discusses methods based on quadratic approximations. Then, in Chapter 11 we summarize some of the methods available for specially structured problems. Next, in Chapter 12 we review the results of available comparative computational studies. The text concludes with a survey of strategies for executing optimization studies (Chapter 13) and a discussion of three engineering case studies (Chapter 14).

## REFERENCES

1. Zeleny, Milan, *Multiple Criteria Decision Making*, McGraw-Hill, New York, 1982.

2. Vincent, T. L., and W. J. Grantham, *Optimality in Parametric Systems*, Wiley, New York, 1981.

3. Bett, K. E., J. S. Rowlinson, and G. Saville, *Thermodynamics for Chemical Engineers*, MIT Press, Cambridge, MA, 1975.

4. Jen, F. C., C. C. Pegels, and T. M. Dupuis, "Optimal Capacities of Production Facilities," *Management Sci.* **14B**, 570–580 (1968).

5. Shigley, J. E., *Mechanical Engineering Design*, McGraw-Hill, New York, 1973, p. 271.

6. Timoshenko S. and J. Gere, *Theory of Elastic Stability*, McGraw-Hill, New York, 1961, p. 257

7. Ragsdell, K. M., and Phillips, D. T., "Optimal Design of a Class of Welded Structures using Geometric Programming," *ASME J. Eng. Ind. Ser. B*, **98**, (3), 1021–1025 (1975).

8. White, W. B., S. M. Johnson, and G. B. Dantzig, "Chemical Equilibrium in Complex Mixtures," *J. Chem. Phys.*, **28**, 251–255 (1959).

9. Hayt, W. H., and J. E. Kemmerly, *Engineering Circuit Analysis*, McGraw-Hill, New York, 1971, Chap. 2.

# Chapter 2

# Functions of a Single Variable

The optimization problem in which the performance measure is a function of one variable is the most elementary type of optimization problem. Yet, it is of central importance to optimization theory and practice, not only because it is a type of problem that the engineer commonly encounters in practice, but also because single-variable optimization often arises as a subproblem within the iterative procedures for solving multivariable optimization problems. Because of the central importance of the single-variable problem, it is not surprising that a large number of algorithms have been devised for its solution. Basically, these methods can be categorized according to the nature of the function $f(x)$ and the variable $x$ involved as well as the type of information that is available about $f(x)$.

## 2.1 PROPERTIES OF SINGLE-VARIABLE FUNCTIONS

A *function* $f(x)$, in the most elementary sense, is a rule that assigns to every choice of $x$ a unique value $y = f(x)$. In this case, $x$ is called the *independent variable* and $y$ is called the *dependent variable*. Mathematically, consider a set $S \subset R$, where $R$ is the set of all real numbers. We can define a correspondence or a transformation that assigns a single numerical value to every point $x \in S$. Such a correspondence is called a *scalar function f* defined on the set $S$.

When the set $S = R$, we have an *unconstrained function* of one variable. When $S$ is a proper subset of $R$, we have a function defined on a *constrained region*. For example,

$$f(x) = x^3 + 2x^2 - x + 3 \qquad \text{for all } x \in R$$

is an unconstrained function, while

$$f(x) = x^3 + 2x^2 - x + 3 \qquad \text{for all } x \in S = \{x \mid -5 \leqslant x \leqslant 5\}$$

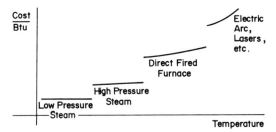

**Figure 2.1.** A discontinuous function.

is a constrained function. In engineering optimization, we call $f$ the *objective function* and $S$ the *feasible region*, the *constraint set*, or the *domain* of interest of $x$.

Most physical processes can be modeled or described by *continuous functions*, that is, functions that are continuous at every point $x$ in their domain. However, it is not at all unusual in engineering applications to encounter *discontinuous* functions. For instance, if we construct a function that measures the cost of heating per Btu at different delivery temperatures, we quite naturally arrive at the broken curve shown in Figure 2.1. The cost is a discontinuous function of the delivery temperature; however, the temperature itself can assume all values over the range of, say, 200–3000 F.

Of course, it is not necessary that the domain of the independent variable $x$ assume all real values over the range in question. It is quite possible that the variable could assume only *discrete* values. For instance, if we construct a function that gives the cost of commercial pipe per foot of length at different pipe diameters, we quite naturally arrive at the sequence of discrete points shown in Figure 2.2. Pipe is made only in a finite number of sizes.

*Note:* It is important to remember the following properties of continuous functions:

1.  A sums or products of a continuous function is continuous.
2.  The ratio of two continuous functions is continuous at all points where the denominator does not vanish.

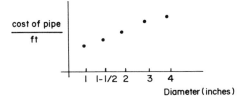

**Figure 2.2.** A discrete function.

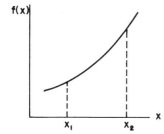

**Figure 2.3.** Monotonic increasing function.

Clearly, depending upon whether the function to be optimized is continuous or discontinuous, or depending upon the nature of its domain, different methods will have to be employed to conduct the search for the optimum. A method that is effective for continuous functions will probably not be effective for discontinuous functions, although the converse may be possible.

In addition to these properties, functions can also be classified according to their shape or topology within the range of interest.

**Monotonic Functions.** A function $f(x)$ is *monotonic* (either increasing or decreasing) if for any two points $x_1$ and $x_2$, with $x_1 \leqslant x_2$, it follows that

$$f(x_1) \leqslant f(x_2) \quad \text{(monotonically increasing)}$$

$$f(x_1) \geqslant f(x_2) \quad \text{(monotonically decreasing)}$$

Figure 2.3 illustrates a monotonically increasing function, and Figure 2.4 illustrates a monotonically decreasing function. Note that a function does not have to be continuous to be monotonic.

Figure 2.5 illustrates a function that is monotonically decreasing for $x \leqslant 0$ and increasing for $x \geqslant 0$. The function attains its minimum at $x = x^*$ (origin), and it is monotonic on either side of the minimum point. Such a function is called a *unimodal function*.

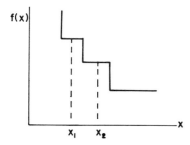

**Figure 2.4.** Monotonic decreasing function.

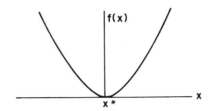

**Figure 2.5.**   Unimodal function.

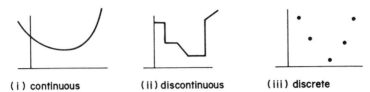

( i ) continuous        ( ii ) discontinuous        ( iii ) discrete

**Figure 2.6.**   Unimodal functions.

### *Definition*

A function $f(x)$ is *unimodal* on the interval $a \leqslant x \leqslant b$ if and only if it is monotonic on either side of the single optimal point $x^*$ in the interval. In other words, if $x^*$ is the single minimum point of $f(x)$ in the range $a \leqslant x \leqslant b$, then $f(x)$ is unimodal on the interval if and only if for any two points $x_1$ and $x_2$,

$$x^* \leqslant x_1 \leqslant x_2 \text{ implies that } f(x^*) \leqslant f(x_1) \leqslant f(x_2)$$

and

$$x^* \geqslant x_1 \geqslant x_2 \text{ implies that } f(x^*) \leqslant f(x_1) \leqslant f(x_2)$$

As shown in Figure 2.6, a function does not have to be continuous to be unimodal. *Unimodality* is an extremely important functional property used in optimization. We shall discuss its usefulness in Section 2.3.

## 2.2   OPTIMALITY CRITERIA

In considering optimization problems, two questions generally must be addressed:

1. *The static question*: How can one determine whether a given point $x^*$ is the optimal solution?
2. *The dynamic question*: If $x^*$ is not the optimal point, then how does one go about finding a solution that is optimal?

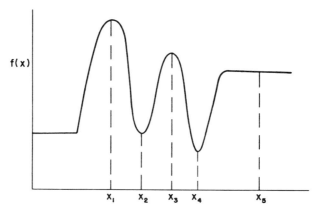

**Figure 2.7.** Local and global optima.

In this section we are concerned primarily with the static question, namely, developing a set of *optimality criteria* for determining whether a given solution is optimal.

### Definitions

A function $f(x)$ defined on a set $S$ attains its *global minimum* at a point $x^{**} \in S$ if and only if

$$f(x^{**}) \leqslant f(x) \qquad \text{for all } x \in S.$$

A function $f(x)$ defined on $S$ has a *local minimum* (*relative minimum*) at a point $x^* \in S$ if and only if

$$f(x^*) \leqslant f(x) \qquad \text{for all } x \text{ within a distance } \varepsilon \text{ from } x^*$$

that is, there exists an $\varepsilon > 0$ such that for all $x$ satisfying $|x - x^*| < \varepsilon$, $f(x^*) \leqslant f(x)$.

### Remarks

1. By reversing the directions of inequality, we can get the equivalent definitions of *global maximum* and *local maximum*.
2. Under the assumption of unimodality, the local minimum automatically becomes the global minimum.
3. When the function is not unimodal, multiple local optima are possible and the global minimum can be found only by locating all local optima and selecting the best one.

In Figure 2.7, $x_1$ is the global maximum, $x_2$ is a local minimum, $x_3$ is a local maximum, $x_4$ is the global minimum, and $x_5$ may be considered as both local minimum and local maximum points.

**Identification of Single-Variable Optima.** Suppose $f(x)$ is a function of a single variable $x$ defined on an open interval $(a, b)$ and $f$ is differentiable to the $n$th order over the interval. If $x^*$ is a point within that interval, then Taylor's theorem allows us to write the change in the value of $f$ from $x^*$ to $(x^* + \varepsilon)$ as follows:

$$f(x^* + \varepsilon) - f(x^*) = (\varepsilon) \left.\frac{df}{dx}\right|_{x=x^*} + \frac{(\varepsilon^2)}{2!} \left.\frac{d^2f}{dx^2}\right|_{x=x^*} + \cdots$$

$$+ \frac{(\varepsilon^n)}{n!} \left.\frac{d^nf}{dx^n}\right|_{x=x^*} + O_{n+1}(\varepsilon) \tag{2.1}$$

where $O_{n+1}(\varepsilon)$ indicates terms of $(n + 1)$st order or higher in $\varepsilon$. If $x^*$ is a local minimum of $f$ on $(a, b)$, then from the definition there must be an $\varepsilon$ neighborhood of $x^*$ such that for all $x$ within a distance $\varepsilon$,

$$f(x) \geqslant f(x^*) \tag{2.2}$$

Inequality (2.2) implies that

$$\varepsilon \left.\frac{df}{dx}\right|_{x=x^*} + \frac{(\varepsilon^2)}{2!} \left.\frac{d^2f}{dx^2}\right|_{x=x^*} + \cdots + \frac{(\varepsilon^n)}{n!} \left.\frac{d^nf}{dx^n}\right|_{x=x^*} + O_{n+1}(\varepsilon) \geqslant 0$$

$$\tag{2.3}$$

For $\varepsilon$ sufficiently small, the first term will dominate the others, and since $\varepsilon$ can be chosen both positive and negative, it follows that inequality (2.3) will hold only when

$$\left.\frac{df}{dx}\right|_{x=x^*} = 0 \tag{2.4}$$

Continuing this argument further, inequality (2.3) will be true only when

$$\left.\frac{d^2f}{dx^2}\right|_{x=x^*} \geqslant 0 \tag{2.5}$$

The same construction applies in the case of a local maximum but with inequality (2.2) reversed, and we obtain the following general result:

### Theorem 2.1

Necessary conditions for $x^*$ to be a local minimum (maximum) of $f$ on the open interval $(a, b)$, providing that $f$ is twice differentiable, are that

1. $\dfrac{df}{dx}\bigg|_{x=x^*} = 0$

2. $\dfrac{d^2f}{dx^2}\bigg|_{x=x^*} \geq 0 \ (\leq 0)$

These are necessary conditions, which means that if they are *not* satisfied then $x^*$ is *not* a local minimum (maximum). On the other hand, if they *are* satisfied, we still have no guarantee that $x^*$ is a local minimum (maximum). For example, consider the function $f(x) = x^3$, shown in Figure 2.8. It satisfies the necessary conditions for both local minimum and local maximum at the origin, but the function does not achieve a minimum or a maximum at $x^* = 0$.

### Definitions

A *stationary point* is a point $x^*$ at which

$$\frac{df}{dx}\bigg|_{x=x^*} = 0$$

An *inflection point or saddle point* is a stationary point that does not correspond to a local optimum (minimum or maximum).

In order to distinguish whether a stationary point corresponds to a local minimum, a local maximum, or an inflection point, we need the *sufficient* conditions of optimality.

### Theorem 2.2

Suppose at a point $x^*$ the first derivative is zero and the first nonzero higher order derivative is denoted by $n$.

(i)  If $n$ is odd, then $x^*$ is a point of inflection.
(ii) If $n$ is even, then $x^*$ is a local optimum. Moreover,

   (a) If that derivative is positive, then the point $x^*$ is a local minimum.
   (b) If that derivative is negative, then the point $x^*$ is a local maximum.

### Proof

This result is easily verified by recourse to the Taylor series expansion given in Eq. (2.1). Since the first nonvanishing higher order derivative is $n$, Eq. (2.1)

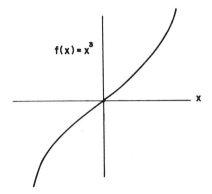

$f(x) = x^3$

**Figure 2.8.**   Illustration of inflection point.

reduces to

$$f(x^* + \varepsilon) - f(x^*) = \frac{\varepsilon^n}{n!} \left. \frac{d^n f}{dx^n} \right|_{x=x^*} + O_{n+1}(\varepsilon) \qquad (2.6)$$

If $n$ is odd, then the right-hand side of Eq. (2.6) can be made positive or negative by choosing $\varepsilon$ positive or negative. This implies that depending on the sign of $\varepsilon$, $f(x^* + \varepsilon) - f(x^*)$ could be positive or negative. Hence the function does not attain a minimum or a maximum at $x^*$, and $x^*$ is an inflection point.

Now, consider the case when $n$ is even. Then the term $\varepsilon^n$ is always positive, and for all $\varepsilon$ sufficiently small the sign of Eq. (2.6) will be dominated by the first term. Hence, if $(d^n f/dx^n)|_{x=x^*}$ is positive, $f(x^* + \varepsilon) - f(x^*) > 0$, and $x^*$ corresponds to a local minimum. A similar argument can be applied in the case of a local maximum.

Applying Theorem 2.2 to the function $f(x) = x^3$ shown in Figure 2.8, we note that

$$\left. \frac{df}{dx} \right|_{x=0} = 0, \qquad \left. \frac{d^2 f}{dx^2} \right|_{x=0} = 0; \qquad \left. \frac{d^3 f}{dx^3} \right|_{x=0} = 6.$$

Thus the first nonvanishing derivative is 3 (odd), and $x = 0$ is an inflection point.

### Remark

In the above discussion, we have assumed that the function is always differentiable, in other words, that continuous first derivatives always exist. However, if the function is not differentiable at all points, then even the necessary condition for an unconstrained optimum, that the point is a stationary point, may not hold. For example, consider the piecewise linear function given by

$$f(x) = \begin{cases} x & \text{for } x \leqslant 2 \\ 4 - x & \text{for } x \geqslant 2 \end{cases}$$

The function is continuous at *all* points. But it is not differentiable at the point $x = 2$, and the function attains its maximum at $x = 2$, which is not a stationary point by our definition.

**Example 2.5**

Consider the function

$$f(x) = 5x^6 - 36x^5 + \tfrac{165}{2}x^4 - 60x^3 + 36$$

defined on the real line. The first derivative of this function is

$$\frac{df}{dx} = 30x^5 - 180x^4 + 330x^3 - 180x^2 = 30x^2(x - 1)(x - 2)(x - 3)$$

Clearly, the first derivative vanishes at $x = 0, 1, 2, 3$, and hence these points can be classified as stationary points. The second derivative of $f$ is

$$\frac{d^2f}{dx^2} = 150x^4 - 720x^3 + 990x^2 - 360x$$

Evaluating this derivative at the four candidate points $x = 0, 1, 2, 3$, we obtain

| $x$ | $f(x)$ | $d^2f/dx^2$ |
|---|---|---|
| 0 | 36 | 0 |
| 1 | 27.5 | 60 |
| 2 | 44 | − 120 |
| 3 | 5.5 | 540 |

We conclude that $x = 1, 3$ are local minima and $x = 2$ is a local maximum. In order to characterize $x = 0$, the next higher derivative must be evaluated:

$$\frac{d^3f}{dx^3} = 600x^3 - 2160x^2 + 1980x - 360 = -360$$

Since this is an odd-order derivative and is nonzero, the point $x = 0$ is not an optimum point but an inflection point.

The next question is how to determine the global maximum or minimum of a function of one variable. Since the global optimum has to be a local optimum, a simple approach is to compute all local optima and choose the best. An algorithm based on this is given below:

Maximize   $f(x)$

Subject to   $a \leqslant x \leqslant b$

where $a$ and $b$ are practical limits on the values of the variable $x$.

Once a function is bounded in an interval, you should notice that in addition to the *stationary points*, the *boundary points* can also qualify for the local optimum.

**Step 1.** Set $df/dx = 0$, and compute all stationary points.

**Step 2.** Select all stationary points that belong to the interval $[a, b]$. Call them $x_1, x_2, \ldots, x_N$. These points, along with $a$ and $b$, are the only points that can qualify for a local optimum.

**Step 3.** Find the largest value of $f(x)$ out of $f(a), f(b), f(x_1), \ldots, f(x_N)$. This value becomes the global maximum point.

*Note:* We did not try to classify the stationary points as local minimum, local maximum, or inflection points, which would involve the calculation of higher order derivatives. Instead it is easier to just compute their functional values and discard them.

### *Example 2.6*

Maximize $f(x) = -x^3 + 3x^2 + 9x + 10$ in the interval $-2 \leqslant x \leqslant 4$. Set

$$\frac{df}{dx} = -3x^2 + 6x + 9 = 0$$

Solving this equation, we get $x = 3$ and $x = -1$ as the two stationary points, and both are in the search interval.

To find the global maximum, evaluate $f(x)$ at $x = 3, -1, -2,$ and 4.

$$f(3) = 37 \qquad f(-1) = 5$$

$$f(-2) = 12 \qquad f(4) = 30$$

Hence $x = 3$ maximizes $f$ over the interval $(-2, 4)$.

Instead of evaluating all the stationary points and their functional values, we could use certain properties of the function to determine the global optimum faster. At the end of Section 2.1, we introduced unimodal functions, for which a local optimum is the global optimum. Unfortunately, the definition of unimodal functions does not suggest a simple test for determining whether or not a function is unimodal. However there exists an important class of unimodal functions in optimization theory, known as *convex* and *concave* functions, which can be identified with some simple tests. A review of convex and concave functions and their properties is given in Appendix B.

**Example 2.7**

Let us examine the properties of the function

$$f(x) = (2x + 1)^2(x - 4)$$
$$f'(x) = (2x + 1)(6x - 15)$$
$$f''(x) = 24(x - 1)$$

For all $x \leqslant 1$, $f''(x) \leqslant 0$, and the function is concave in this region. For all $x \geqslant 1$, $f''(x) \geqslant 0$, and the function is convex in the region.

Note that the function has two stationary points, $x = -\frac{1}{2}$ and $x = \frac{5}{2}$. $f''(-\frac{1}{2}) < 0$, and the function has a local maximum at $x = -\frac{1}{2}$. At $x = \frac{5}{2}$, $f''(\frac{5}{2}) > 0$, and the function attains a local minimum at this point. If we restricted the region of interest to $x \leqslant 1$, then $f(x)$ attains the global maximum at $x = -\frac{1}{2}$ since $f(x)$ is concave in this region and $x = -\frac{1}{2}$ is a local maximum. Similarly, if we restricted the region of interest to $x \geqslant 1$, then $f(x)$ attains the global minimum at $x = \frac{5}{2}$. However, over the entire range of $x$ from $-\infty$ to $+\infty$, $f(x)$ has no finite global maximum or minimum.

**Example 2.8  Inventory Control**

Many firms maintain an inventory of goods to meet future demand. Among the reasons for holding inventory is to avoid the time and cost of constant replenishment. On the other hand, to replenish only infrequently would imply large inventories that would tie up unnecessary capital and incur huge storage costs. Determining the optimal size of the inventory is a classic optimization problem, and a frequently used model in inventory control is the *economic order quantity* (EOQ) model.

The EOQ model assumes that there is a steady demand of $\lambda$ units per year for the product. Frequent replenishment is penalized by assuming that there is a setup or ordering cost of \$$K$ each time an order is placed, irrespective of how many units are ordered. The acquisition cost of each unit is \$$c$. Excessive inventory is penalized by assuming that each unit will cost \$$h$ to store for one year. To keep things simple, we will assume that all demand must be met immediately (i.e., no back orders are allowed) and that replenishment occurs instantaneously as soon as an order is sent.

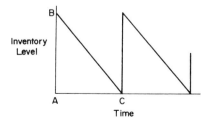

Figure 2.9.  An inventory problem.

Figure 2.9 graphically illustrates the change in the inventory level with respect to time. Starting at any time $A$ with an inventory of $B$, the inventory level will decrease at the rate of $\lambda$ units per unit time until it becomes zero at time $C$, when a fresh order is placed.

The triangle $ABC$ represents one *inventory cycle* which will be repeated throughout the year. The problem is to determine the optimal order quantity $B$, denoted by variable $Q$ and the common length of time $C - A$, denoted by $T$, between reorders.

Since $T$ is just the length of time required to deplete $Q$ units at rate $\lambda$, we get

$$T = \frac{Q}{\lambda}$$

The only remaining problem is to determine $Q$. Note that when $Q$ is small, $T$ will be small, implying more frequent reorders during the year. This will result in higher reorder costs but lower inventory holding cost. On the other hand, a large inventory (large $Q$) will result in a higher inventory cost but lower reorder costs. The basic inventory control problem is to determine the optimal value of $Q$ that will minimize the sum of the inventory cost and reorder cost in a year.

We shall now develop the necessary mathematical expression to optimize the yearly cost (cost/cycle × number of cycles/year).

$$\text{Number of cycles (reorders)/year} = \frac{1}{T} = \frac{\lambda}{Q}$$

$$\text{Cost per cycle} = \text{reorder cost} + \text{inventory cost}$$

$$= (K + cQ) + \left( \frac{Q}{2} hT \right)$$

$$= K + cQ + \left( \frac{hQ^2}{2\lambda} \right)$$

*Note:* The inventory cost per cycle is simply the cost of holding an average inventory of $Q/2$ for a length of time $T$.

Thus, the yearly cost to be minimized is

$$f(Q) = \frac{\lambda K}{Q} + \lambda c + \frac{hQ}{2}$$

$$f'(Q) = \frac{-\lambda K}{Q^2} + \frac{h}{2}$$

$$f''(Q) = \frac{2\lambda K}{Q^3} > 0 \qquad \text{for all } Q > 0$$

Hence $f(Q)$ is a convex function, and if there exists a positive $Q^*$ such that $f'(Q^*) = 0$, then $Q^*$ minimizes $f(Q)$.

Solving $f'(Q) = 0$, we get

$$Q^* = \sqrt{\frac{2\lambda K}{h}} > 0$$

Thus, the optimal order quantity is

$$Q^* = \sqrt{\frac{2\lambda K}{h}}$$

and

$$T^* = \text{time between reorders} = \sqrt{\frac{2K}{h\lambda}}$$

$Q^*$ is the famous *economic order quantity* or EOQ used frequently in inventory control.

## 2.3 REGION-ELIMINATION METHODS

In Section 2.2, we addressed the static question, namely, how to determine whether or not a given solution is optimal. For this we developed a set of necessary and sufficient optimality conditions that a solution must satisfy in order to be optimal. Now we shall address the *dynamic* question, namely, how to determine the optimal or candidate optimal solutions. For this we shall develop a number of single-variable *search methods* for locating the optimal point in a given interval. Search methods that locate a single-variable optimum by successively eliminating subintervals so as to reduce the remaining interval of search are called *region-elimination methods*.

In Section 2.1, we introduced the definition of *unimodal functions*. Unimodality is an extremely important functional property and virtually all the single-variable search methods currently in use require *at least* the assumption that within the domain of interest the function is unimodal. The usefulness of this property lies in the fact that if $f(x)$ is unimodal, then it is only necessary to compare $f(x)$ at two different points in order to predict in which of the subintervals defined by those two points the optimum does *not* lie.

### Theorem 2.3

Suppose $f$ is strictly unimodal' on the interval $a \leq x \leq b$ with a minimum at $x^*$. Let $x_1$ and $x_2$ be two points in the interval such that $a < x_1 < x_2 < b$.

---

†A function is strictly unimodal if it is unimodal and has no intervals of finite length in which the function is of constant value.

Comparing the functional values at $x_1$ and $x_2$, we can conclude:

(i)  If $f(x_1) > f(x_2)$, then the minimum of $f(x)$ does not lie in the interval $(a, x_1)$. In other words, $x^* \in (x_1, b)$ (see Figure 2.10).

(ii)  If $f(x_1) < f(x_2)$, then the minimum does not lie in the interval $(x_2, b)$ or $x^* \in (a, x_2)$. (See Figure 2.10).

*Proof*

Consider case (i), where $f(x_1) > f(x_2)$. Suppose we assume the contrary that $a \leqslant x^* \leqslant x_1$. Since $x^*$ is the minimum point, by definition we have $f(x^*) \leqslant f(x)$ for all $x \in (a, b)$. This implies that

$$f(x^*) \leqslant f(x_1) > f(x_2) \qquad \text{with } x^* < x_1 < x_2$$

But this is impossible, because the function has to be monotonic on either side of $x^*$ by the unimodality of $f(x)$. Thus, the theorem is proved by contradiction. A similar argument holds for case (ii).

*Note:*  When $f(x_1) = f(x_2)$, we could eliminate both ends, $(a, x_1)$ and $(x_2, b)$, and the minimum must occur in the interval $(x_1, x_2)$ providing $f(x)$ is strictly unimodal.

By means of Theorem 2.3, sometimes called the *elimination property*, one can organize a search in which the optimum is found by recursively eliminating sections of the initial bounded interval. When the remaining subinterval is reduced to a sufficiently small length, the search is terminated. Note that without the elimination property, nothing less than an exhaustive search would suffice. The greatest advantage of these search methods is that they require only functional evaluations. The optimization functions need not be differentiable. As a matter of fact, the functions need not be in a mathematical or analytical form. All that is required is that given a point $x$ the value of the function $f(x)$ can be determined by direct evaluation or by a simulation experiment. Generally, these search methods can be broken down into two

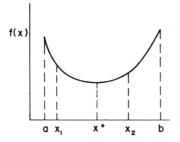 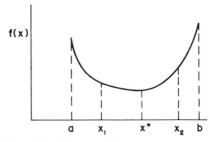

**Figure 2.10.**   Case (i) and case (ii) of Theorem 2.3.

phases:

*Bounding Phase*: An initial coarse search that will bound or bracket the optimum.

*Interval Refinement Phase*: A finite sequence of interval reductions or refinements to reduce the initial search interval to desired accuracy.

### 2.3.1 Bounding Phase

In the initial phase, starting at some selected trial point, the optimum is roughly bracketed within a finite interval by using the elimination property. Typically, this bounding search is conducted using some heuristic expanding pattern, although extrapolation methods have also been devised. An example of an expanding pattern is Swann's method [1], in which the $(k + 1)$st test point is generated using the recursion

$$x_{k+1} = x_k + 2^k \Delta \qquad \text{for } k = 0, 1, 2, \ldots .$$

where $x_0$ is an arbitrarily selected starting point and $\Delta$ is a step-size parameter of suitably chosen magnitude. The sign of $\Delta$ is determined by comparing $f(x_0)$, $f(x_0 + |\Delta|)$, and $f(x_0 - |\Delta|)$. If

$$f(x_0 - |\Delta|) \geqslant f(x_0) \geqslant f(x_0 + |\Delta|)$$

then, because of the unimodality assumption, the minimum must lie to the right of $x_0$, and $\Delta$ is chosen to be positive. If the inequalities are reversed, $\Delta$ is chosen to be negative; if

$$f(x_0 - |\Delta|) \geqslant f(x_0) \leqslant f(x_0 + |\Delta|)$$

the minimum has been bracketed between $x_0 - |\Delta|$ and $x_0 + |\Delta|$ and the bounding search can be terminated. The remaining case,

$$f(x_0 - |\Delta|) \leqslant f(x_0) \geqslant f(x_0 + |\Delta|)$$

is ruled out by the unimodality assumption. However, occurrence of the above condition indicates that the given function is not unimodal.

*Example 2.9*

Consider the problem of minimizing $f(x) = (100 - x)^2$ given the starting point $x_0 = 30$ and a step size $|\Delta| = 5$.

The sign of $\Delta$ is determined by comparing

$$f(x_0) \qquad\quad = f(30) = 4900$$

$$f(x_0 + |\Delta|) = f(35) = 4225$$

$$f(x_0 - |\Delta|) = f(25) = 5625$$

Since

$$f(x_0 - |\Delta|) \geqslant f(x_0) \geqslant f(x_0 + |\Delta|)$$

$\Delta$ must be positive, and the minimum point $x^*$ must be greater than 30. Thus, $x_1 = x^0 + \Delta = 35$.

Next,

$$x_2 = x_1 + 2\Delta = 45$$

$$f(45) = 3025 < f(x_1)$$

therefore, $x^* > 35$.

$$x_3 = x_2 + 2^2\Delta = 65$$

$$f(65) = 1225 < f(x_2)$$

therefore, $x^* > 45$.

$$x_4 = x_3 + 2^3\Delta = 105$$

$$f(105) = 25 < f(x_3)$$

therefore, $x^* > 65$.

$$x_5 = x_4 + 2^4\Delta = 185$$

$$f(185) = 7225 > f(x_4)$$

therefore, $x^* < 185$. Consequently, in six evaluations $x^*$ has been bracketed within the interval

$$65 \leqslant x^* \leqslant 185$$

Note that the effectiveness of the bounding search depends directly on the step size $\Delta$. If $\Delta$ is large, a poor bracket, that is, a large initial interval, is obtained. On the other hand, if $\Delta$ is small, many evaluations may be necessary before a bound can be established.

### 2.3.2   Interval Refinement Phase

Once a bracket has been established around the optimum, then more sophisticated interval reduction schemes can be applied to obtain a refined estimate of the optimum point. The amount of subinterval eliminated at each step depends on the location of the trial points $x_1$ and $x_2$ within the search interval. Since we have no prior knowledge of the location of the optimum, it is reasonable to expect that the location of the trial points ought to be such that regardless of the outcome the interval should be reduced the same amount. Moreover, in the interest of efficiency, that same amount should be as large as possible. This is sometimes called the minimax criterion of search strategy.

**Interval Halving.** This method deletes *exactly* one-half the interval at each stage. This is also called a *three-point equal-interval search* since it works with three equally spaced trial points in the search interval. The basic steps of the search procedure for finding the minimum of a function $f(x)$ over the interval $(a, b)$ are as follows:

**Step 1.**   Let $x_m = (a + b)/2$ and $L = b - a$. Compute $f(x_m)$.

**Step 2.**   Set $x_1 = a + L/4$ and $x_2 = b - L/4$.

Note that the points $x_1$, $x_m$, and $x_2$ are all equally spaced at one-fourth the interval. Compute $f(x_1)$ and $f(x_2)$.

**Step 3.**   Compare $f(x_1)$ and $f(x_m)$.
　　　　　(i)   If $f(x_1) < f(x_m)$, then drop the interval $(x_m, b)$ by setting $b = x_m$. The midpoint of the new search interval will now be $x_1$. Hence, set $x_m = x_1$. Go to step 5.
　　　　　(ii)   If $f(x_1) \geqslant f(x_m)$, go to step 4.

**Step 4.**   Compare $f(x_2)$ and $f(x_m)$.
　　　　　(i)   If $f(x_2) < f(x_m)$, drop the interval $(a, x_m)$ by setting $a = x_m$. Since the midpoint of the new interval will now be $x_2$, set $x_m = x_2$. Go to step 5.
　　　　　(ii)   If $f(x_2) \geqslant f(x_m)$, drop the interval $(a, x_1)$ and $(x_2, b)$. Set $a = x_1$ and $b = x_2$. Note that $x_m$ continues to be the midpoint of the new interval. Go to step 5.

**Step 5.**   Compute $L = b - a$. If $|L|$ is small, terminate. Otherwise return to step 2.

*Remarks*

1.  At each stage of the algorithm, exactly half the length of the search interval is deleted.
2.  The midpoint of subsequent intervals is always equal to one of the previous trial points $x_1$, $x_2$, or $x_m$. Hence, at most two functional evaluations are necessary at each subsequent step.
3.  After $n$ functional evaluations, the initial search interval will be reduced to $(\frac{1}{2})^{n/2}$.
4.  It has been shown by Kiefer [2] that out of all equal-interval searches (two-point, three-point, four-point, etc.), the three-point search or interval halving is the most efficient.

*Example 2.10   Interval Halving*

Minimize $f(x) = (100 - x)^2$ over the interval $60 \leqslant x \leqslant 150$. Here $a = 60$, $b = 150$, and $L = 150 - 60 = 90$.

$$x_m = \frac{60 + 150}{2} = 105$$

**Stage 1**

$$x_1 = a + \frac{L}{4} = 60 + \frac{90}{4} = 82.5$$

$$x_2 = b - \frac{L}{4} = 150 - \frac{90}{4} = 127.5$$

$$f(82.5) = 306.25 > f(105) = 25$$

$$f(127.5) = 756.25 > f(105)$$

Hence, drop the intervals (60, 82.5) and (127.5, 150). The length of the search interval is reduced from 90 to 45.

**Stage 2**

$$a = 82.5 \qquad b = 127.5 \qquad x_m = 105$$

$$L = 127.5 - 82.5 = 45$$

$$x_1 = 82.5 + \tfrac{45}{4} = 93.75$$

$$x_2 = 127.5 - \tfrac{45}{4} = 116.25$$

$$f(93.75) = 39.06 > f(105) = 25$$

$$f(116.2\mathfrak{5}) = 264.06 > f(105)$$

Hence, the interval of uncertainty is (93.75, 116.25).

**Stage 3**

$$a = 93.75 \qquad b = 116.25 \qquad x_m = 105$$

$$L = 116.25 - 93.75 = 22.5$$

$$x_1 = 99.375$$

$$x_2 = 110.625$$

$$f(x_1) = .39 < f(105) = 25$$

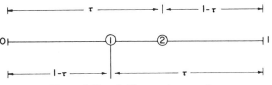

**Figure 2.11.** Golden section search.

Hence, delete the interval (105, 116.25). The new interval of uncertainty is now (93.75, 105), and its midpoint is 99.375 (old $x_1$). Thus, in three stages (six functional evaluations), the initial search interval of length 90 has been reduced exactly to $(90)(\frac{1}{2})^3 = 11.25$.

**Golden Section Search.** In our discussion of region-elimination methods and minimax search strategies, we have observed the following:

1. If only two trials are available, then it is best to locate them equi-distant from the center of the interval.
2. In general, the minimax strategy suggests that trials be placed symmetrically in the interval so that the subinterval eliminated will be of the same length regardless of the outcome of the trials.
3. The search scheme should require the evaluation of *only one* new point at each step.

With this goal in mind, consider the symmetric placement of two trial points shown in Figure 2.11. Starting with a unit interval, (purely for convenience) two trials are located a fraction $\tau$ from either end. With this symmetric placement, regardless of which of the corresponding function values is smaller, the length of the remaining interval is always $\tau$. Suppose that the right-hand subinterval is eliminated. It is apparent from Figure 2.12 that the remaining subinterval of length $\tau$ has the one old trial located interior to it at a distance $1 - \tau$ from the left endpoint.

In order to retain the symmetry of the search pattern, the distance $1 - \tau$ should correspond to a fraction $\tau$ of the length of interval (which itself is of length $\tau$). With this choice of $\tau$, the next evaluation can be located at a fraction $\tau$ of the length of the interval from the right-hand endpoint (Figure 2.13).

Hence, with the choice of $\tau$ satisfying $1 - \tau = \tau^2$, the symmetric search pattern of Figure 2.11 is retained in the reduced interval of Figure 2.13. The

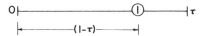

**Figure 2.12.** Golden section intervals.

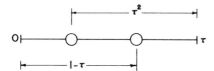

**Figure 2.13.** Golden section symmetry.

solution of the quadratic equation is

$$\tau = \frac{-1 \pm \sqrt{5}}{2}$$

the positive component of which is $\tau = 0.61803\ldots$ . The search scheme for locating trial points based upon this ratio is known as the *golden section search*. Note that after the first two evaluations, each succeeding evaluation will eliminate $1 - \tau$ of the remaining interval. Hence, the interval remaining after $N$ evaluations, assuming the initial interval was of unit length, will be of length $\tau^{N-1}$. It can be shown that this is the asymptotic form of the optimum minimax search.

### *Example 2.11    Golden Section Search*

Consider again the problem of Example 2.10:

Minimize $f(x) = (100 - x)^2$ over the bounded interval $60 \leqslant x \leqslant 150$

For convenience, we rescale the interval to unit length by defining

$$w = \frac{x - 60}{90}$$

Thus, our problem becomes

Minimize    $f(w) = (40 - 90w)^2$

Subject to   $0 \leqslant w \leqslant 1$

**Step 1.**  $I_1 = (0, 1)$; $L_1 = 1$. The first two evaluations are at

$$w_1 = \tau = 0.618 \qquad\qquad \text{with } f(w_1) = 244.0$$

$$w_2 = 1 - \tau = \tau^2 = 0.382 \qquad \text{with } f(w_2) = 31.6$$

Since $f(w_2) < f(w_1)$ and $w_2 < w_1$, the interval $w \geqslant w_1$ is eliminated.

**Step 2.**  $I_2 = (0, .618)$; $L_2 = 0.618 = \tau$. The next evaluation is located at

$$w_3 = \tau - \tau^2 = \tau(1 - \tau) = \tau^3 = 0.236 \qquad \text{with } f(w_3) = 352$$

Since $f(w_3) > f(w_2)$ and $w_3 < w_2$, the interval $w \leqslant w_3$ is eliminated.

**Step 3.** $I_3 = (.236, .618)$, $L_3 = 0.382 = \tau^2$. The next evaluation is at a distance $\tau$ times the length of the remaining interval from the current left endpoint or, equivalently, a distance $1 - \tau$ times the interval length from the current right endpoint. Thus,

$$w_4 = 0.618 - (1 - \tau)L_3 = 0.618 - \tau^2 L_3$$

$$= 0.618 - \tau^2(\tau^2) = 0.618 - \tau^4 = 0.472$$

with

$$f(w_4) = 6.15$$

Since $f(w_4) < f(w_2)$ and $w_4 > w_2$, the interval $w \leqslant w_2$ is eliminated.

At this point the remaining interval of uncertainty is $0.382 \leqslant w \leqslant 0.618$ in the $w$ variable, or $94.4 \leqslant x \leqslant 115.6$ in the original variable.

If the search is continued through six evaluations, then the final interval in $w$ will be

$$\tau^{N-1} = \tau^5 = 0.09$$

which corresponds to an interval of 8.1 in variable $x$. This is a substantial improvement over the 11.25 achieved with interval halving.

In general, if the left and right endpoints of the remaining interval of uncertainty (indicated by XR and XL) are saved, then all subsequent golden section trials can be calculated as either

$$w = XR - \tau^n \quad \text{or} \quad w = XL + \tau^n$$

depending upon whether the left or right subinterval was last eliminated. In the above expressions, $\tau^n$ indicates $\tau$ raised to the $n$th power, where $n$ is the evaluation index.

The golden section iterations can be terminated either by specifying a limit on the number of evaluations (and hence on the accuracy in the variable) or by specifying a relative accuracy in the function value. Preferably both tests should be applied.

### 2.3.3 Comparison of Region-Elimination Methods

Let us now compare the relative efficiencies of the region-elimination methods we have discussed so far. Let the original interval of uncertainty be denoted $L_1$ and the final interval of uncertainty after $N$ functional evaluations be $L_N$. Suppose we consider the *fractional reduction* (FR) of the original interval as a measure of merit of region-elimination methods. Then

$$FR(N) = \frac{L_N}{L_1}$$

Recall that for the interval-halving and golden section search methods, the length of the final interval is given by $L_1(0.5)^{N/2}$ and $L_1(0.618)^{N-1}$, respectively. Hence, the fractional reduction after $N$ functional evaluations becomes

$$\text{FR}(N) = \begin{cases} (0.5)^{N/2} & \text{for interval halving} \\ (0.618)^{N-1} & \text{for golden section} \end{cases}$$

For the purpose of comparison, let us also include an exhaustive (brute force) search, wherein the function is evaluated at $N$ equally distant points [i.e., $L_1$ is divided into $(N + 1)$ equal intervals of length $L_1/(N + 1)$]. Let $x^*$ be the point where the minimum of $f(x)$ is observed. Then the true minimum of $f(x)$ is bracketed in the interval

$$\left[ \left( x^* - \frac{L_1}{N + 1} \right), \left( x^* + \frac{L_1}{N + 1} \right) \right]$$

In other words, $L_N = 2L_1/(N + 1)$. Hence, for an exhaustive search,

$$\text{FR}(N) = \frac{2}{N + 1}$$

Table 2.1 illustrates the values of $\text{FR}(N)$ for the three search methods for selected values of $N$. From the table it is clear that the golden section search achieves the most fractional reduction of the original interval for the same number of functional evaluations. Looking at it another way, we can also compare the number of functional evaluations required for the same fractional reduction or accuracy. For a given $\text{FR}(N) = E$, the value of $N$ may be computed as follows:

For interval-halving:

$$N = \frac{2\ln(E)}{\ln(0.5)}$$

**Table 2.1  Fractional Reduction Achieved**

| Search Method | Number of Functional Evaluations | | | | |
|---|---|---|---|---|---|
|  | $N = 2$ | $N = 5$ | $N = 10$ | $N = 15$ | $N = 20$ |
| Interval halving | 0.5 | 0.177 | 0.031 | .006 | .0009 |
| Golden section | 0.618 | 0.146 | 0.013 | .001 | .0001 |
| Exhaustive | 0.667 | 0.333 | 0.182 | .125 | .095 |

For golden section:

$$N = 1 + \frac{\ln(E)}{\ln(0.618)}$$

For exhaustive search:

$$N = \frac{2}{E} - 1$$

Table 2.2 gives the number of functional evaluations required for a desired accuracy in the determination of the minimum point. Once again, it is clear that the golden section search out-performs the other two methods by requiring the least number of functional evaluations for the same desired accuracy.

**Table 2.2    Required Number of Functional Evaluations**

| Search Method | Accuracy Desired | | | |
| --- | --- | --- | --- | --- |
| | $E = 0.1$ | $E = 0.05$ | $E = .01$ | $E = .001$ |
| Interval halving | 7 | 9 | 14 | 20 |
| Golden section | 6 | 8 | 11 | 16 |
| Exhaustive | 19 | 39 | 199 | 1999 |

## 2.4  POLYNOMIAL APPROXIMATION OR POINT-ESTIMATION METHODS

The region-elimination methods discussed in the previous section require only that the function be unimodal. Hence, they are applicable to both continuous and discontinuous functions as well as to discrete variables. The entire search logic in these methods is based upon a simple comparison of function values at two trial points. Moreover, this comparison only takes account of the ordering of the function values and in no way involves the magnitude of the difference between the function values. The search methods discussed in this section do take into account the relative magnitudes of the function values, and as a consequence often perform better than the region-elimination methods. However, this gain in efficiency is achieved at the price of having to require that the functions optimized be sufficiently smooth.

The basic motivating idea of the techniques to be presented here is that if the function is smooth then it can be approximated by a polynomial, and the approximating polynomial can then be used to predict the location of the optimum. For this strategy to be effective it is only necessary for the function

in question to be both unimodal and continuous. The Weierstrass approximation theorem [3] guarantees that if the function is continuous on the interval
then it can be approximated as closely as desired by polynomials of sufficiently
high order. Consequently, if the function is unimodal and a reasonably good
approximating polynomial is at hand, then the function optimum can be
reasonably well predicted by the polynomial optimum. As suggested by the
Weierstrass theorem, improvements in the estimated optimum obtained through
approximating polynomials can be achieved in two ways: by using a higher
order polynomial or by reducing the interval over which the function is to be
approximated. Of these, the second alternative is generally to be preferred,
because beyond the third order the polynomial-fitting algebra becomes more
complicated, and as a result of the unimodality assumption, interval reduction
is rather easily accomplished.

### Quadratic Estimation Methods

The simplest polynomial interpolation is a quadratic approximation. It is based
upon the observation that if a function takes on its minimum in the interior of
an interval then it must be at least quadratic. If it is linear, it will assume its
optimal value at one or the other of the endpoints of the interval. Thus, a
quadratic estimation scheme assumes that within the bounded interval the
function can be approximated by a quadratic and this approximation will
improve as the points used to construct the approximation approach the actual
minimum.

Given three consecutive points $x_1$, $x_2$, $x_3$ and their corresponding function
values $f_1$, $f_2$, $f_3$, we seek to determine three constants $a_0$, $a_1$, and $a_2$ such that
the quadratic function

$$q(x) = a_0 + a_1(x - x_1) + a_2(x - x_1)(x - x_2)$$

agrees with $f(x)$ at these three points. We proceed by evaluating $q(x)$ at each
of the three given points. First of all, since

$$f_1 = f(x_1) = q(x_1) = a_0$$

we have

$$a_0 = f_1$$

Next, since

$$f_2 = f(x_2) = q(x_2) = f_1 + a_1(x_2 - x_1)$$

we have

$$a_1 = \frac{f_2 - f_1}{x_2 - x_1}$$

Finally, at $x = x_3$,

$$f_3 = f(x_3) = q(x_3) = f_1 + \frac{f_2 - f_1}{x_2 - x_1}(x_3 - x_1) + a_2(x_3 - x_1)(x_3 - x_2)$$

Solving for $a_2$, we obtain

$$a_2 = \frac{1}{x_3 - x_2}\left(\frac{f_3 - f_1}{x_3 - x_1} - \frac{f_2 - f_1}{x_2 - x_1}\right)$$

Hence, given three points and their function values, the quadratic estimate can be constructed by simply evaluating the expressions for $a_0, a_1, a_2$ given above.

Now following the proposed strategy, if the approximating quadratic is a good approximation to the function to be optimized over the interval $x_1$ to $x_3$, then it can be used to predict the location of the optimum. Recall that the stationary points of a single-variable function can be determined by setting its first derivative to zero and solving for the roots of the resulting equation. In the case of our quadratic approximating function,

$$\frac{dq}{dx} = a_1 + a_2(x - x_2) + a_2(x - x_1) = 0$$

can be solved to yield the estimate

$$\bar{x} = \frac{x_2 + x_1}{2} - \frac{a_1}{2a_2}$$

Since the function $f(x)$ is unimodal over the interval in question and since the approximating quadratic is also a unimodal function, it is reasonable to expect that $\bar{x}$ will be a good estimate of the desired exact optimum $x^*$.

### Example 2.12   Quadratic Search

Consider the estimation of the minimum of

$$f(x) = 2x^2 + \frac{16}{x}$$

on the interval $1 \leqslant x \leqslant 5$. Let $x_1 = 1$, $x_3 = 5$, and choose as $x_2$ the midpoint $x_2 = 3$. Evaluating the function, we obtain

$$f_1 = 18 \qquad f_2 = 23.33 \qquad f_3 = 53.2$$

In order to calculate the estimate $\bar{x}$, the constants $a_1$ and $a_2$ of the approximating function must be evaluated. Thus,

$$a_1 = \frac{23.33 - 18}{3 - 1} = \frac{8}{3}$$

$$a_2 = \frac{1}{5 - 3}\left(\frac{53.2 - 18}{5 - 1} - \frac{8}{3}\right) = \frac{46}{15}$$

Substituting into the expression for $\bar{x}$,

$$\bar{x} = \frac{3+1}{2} - \frac{8/3}{2(46/15)} = 1.565$$

The exact minimum is $x^* = 1.5874$.

### Successive Quadratic Estimation Method

Developed by Powell [4], this method uses successive quadratic estimation. The algorithm is outlined as follows. Let $x_1$ be the initial point and $\Delta x$ the selected step size.

**Step 1.**  Compute $x_2 = x_1 + \Delta x$.

**Step 2.**  Evaluate $f(x_1)$ and $f(x_2)$.

**Step 3.**  If $f(x_1) > f(x_2)$, let $x_3 = x_1 + 2\Delta x$.

        If $f(x_1) \leqslant f(x_2)$, let $x_3 = x_1 - \Delta x$.

**Step 4.**  Evaluate $f(x_3)$ and determine

$$F_{\min} = \min\{f_1, f_2, f_3\}$$

$$X_{\min} = \text{point } x_i \text{ corresponding to } F_{\min}$$

**Step 5.**  Use the points $x_1, x_2, x_3$ to calculate $\bar{x}$ using the quadratic estimation formula.

**Step 6.**  Check for termination.
(a) Is $F_{\min} - f(\bar{x})$ small enough?
(b) Is $X_{\min} - \bar{x}$ small enough?
If both are satisfied, terminate. Otherwise, go to step 7.

**Step 7.**  Save the currently best point ($X_{\min}$ or $\bar{x}$), the two points bracketing it, or the two closest to it. Relabel them and go to step 4.

Note that with the first pass through step 5 a bound of the minimum may not yet have been established. Consequently, the $\bar{x}$ obtained may be an extrapolation beyond $x_3$. To ensure against a very large extrapolation step, a test should be added after step 5 that will replace $\bar{x}$ with a point calculated using a preset step size if $\bar{x}$ is very far from $x_3$.

### *Example 2.13   Powell's Method*

Consider the problem of Example 2.12:

$$\text{Minimize } f(x) = 2x^2 + \frac{16}{x}$$

with initial point $x_1 = 1$ and step size $\Delta x = 1$. For convergence parameters use

$$\left| \frac{\text{Difference in } x}{x} \right| \leqslant 3 \times 10^{-2} \qquad \left| \frac{\text{Difference in } F}{F} \right| \leqslant 3 \times 10^{-3}$$

### Iteration 1

**Step 1.** $x_2 = x_1 + \Delta x = 2$

**Step 2.** $f(x_1) = 18 \qquad f(x_2) = 16$

**Step 3.** $f(x_1) > f(x_2)$, therefore set $x_3 = 1 + 2 = 3$.

**Step 4.** $f(x_3) = 23.33$

$F_{\min} = 16$

$X_{\min} = x_2$

**Step 5.**
$$a_1 = \frac{16 - 18}{2 - 1} = -2$$

$$a_2 = \frac{1}{3 - 2} \left\{ \frac{23.33 - 18}{3 - 1} - a_1 \right\} = \frac{5.33}{2} + 2 = 4.665$$

$$\bar{x} = \frac{1 + 2}{2} - \frac{(-2)}{2(4.665)} = 1.5 + \frac{1}{4.665} = 1.714$$

$$f(\bar{x}) = 15.210$$

**Step 6.** Test for termination:

(i) $\left| \dfrac{16 - 15.210}{15.210} \right| = 0.0519 > .003$

Therefore continue.

**Step 7.** Save $\bar{x}$, the currently best point, and $x_1$ and $x_2$, the two points that bound it. Relabel the points in order and go to iteration 2, starting with step 4.

### Iteration 2

**Step 4.** $x_1 = 1 \qquad f_1 = 18$

$x_2 = 1.714 \qquad f_2 = 15.210 = F_{\min}$ and $X_{\min} = x_2$

$x_3 = 2 \qquad f_3 = 16$

**Step 5.**
$$a_1 = \frac{15.210 - 18}{1.714 - 1} = -3.908$$

$$a_2 = \frac{1}{2 - 1.714} \left\{ \frac{16 - 18}{2 - 1} - (-3.908) \right\} = \frac{1.908}{0.286} = 6.671$$

$$\bar{x} = \frac{2.714}{2} - \frac{-3.908}{2(6.671)} = 1.357 + 0.293 = 1.650$$

$$f(\bar{x}) = 15.142$$

**Step 6.** Test for termination:

$$\text{(i)} \quad \left| \frac{15.210 - 15.142}{15.142} \right| = 0.0045 > .003 \qquad \text{not satisfied}$$

**Step 7.** Save $\bar{x}$, the currently best point, and $x_1 = 1$ and $x_2 = 1.714$, the two points that bracket it.

### *Iteration 3*

**Step 4.** $\quad x_1 = 1 \qquad\qquad f_1 = 18$

$\qquad\qquad x_2 = 1.65 \qquad\quad f_2 = 15.142 = F_{\min} \quad \text{and} \quad X_{\min} = x_2$

$\qquad\qquad x_3 = 1.714 \qquad f_3 = 15.210$

**Step 5.** $\quad a_1 = \dfrac{15.142 - 18}{1.65 - 1} = -4.397$

$$a_2 = \frac{1}{1.714 - 1.650} \left\{ \frac{15.210 - 18}{1.714 - 1} - (-4.397) \right\} = 7.647$$

$$\bar{x} = \frac{2.65}{2} - \frac{(-4.397)}{2(7.647)} = 1.6125$$

$$f(\bar{x}) = 15.123$$

**Step 6.** Test for termination:

$$\text{(i)} \quad \left| \frac{15.142 - 15.123}{15.123} \right| = 0.0013 < .003$$

$$\text{(ii)} \quad \left| \frac{1.65 - 1.6125}{1.6125} \right| = 0.023 < 0.03$$

Therefore, terminate iterations.

## 2.5  METHODS REQUIRING DERIVATIVES

The search methods discussed in the preceding sections required the assumptions of unimodality and, in some cases, continuity of the performance index being optimized. It seems reasonable that, in addition to continuity, if it is assumed that the function is differentiable then further efficiencies in the search could be achieved. Recall from Section 2.2 that the necessary condition for a point $z$ to be a local minimum is that the derivative at $z$ vanish, that is, $f'(z) = df/dx|_{x=z} = 0$.

When $f(x)$ is a function of third degree or higher terms involving $x$, direct analytic solution of $f'(x) = 0$ would be difficult. Hence, a search method that successively approximates the stationary point of $f$ is required. We first describe a classical search scheme for finding the root of a nonlinear equation that was originally developed by Newton and later refined by Raphson [5].

### 2.5.1  Newton-Raphson Method

The Newton-Raphson scheme requires that the function $f$ be twice differentiable. It begins with a point $x_1$ that is the initial estimate or approximation to the

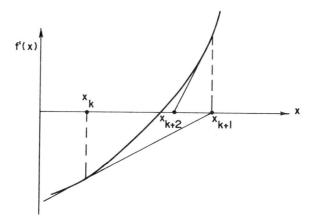

**Figure 2.14.** Newton-Raphson method (convergence).

stationary point or root of the equation $f'(x) = 0$. A linear approximation of the function $f'(x)$ at the point $x_1$ is constructed, and the point at which the linear approximation vanishes is taken as the next approximation. Formally, given the point $x_k$ to be the current approximation of the stationary point, the linear approximation of the function $f'(x)$ at $x_k$ is given by

$$\tilde{f}'(x; x_k) = f'(x_k) + f''(x_k)(x - x_k) \qquad (2.7)$$

Setting Eq. (2.7) to be zero, we get the next approximation point as

$$x_{k+1} = x_k - \frac{f'(x_k)}{f''(x_k)}$$

Figure 2.14 illustrates the general steps of Newton's method. Unfortunately, depending on the starting point and the nature of the function, it is quite possible for Newton's method to diverge rather than converge to the true stationary point. Figure 2.15 illustrates this difficulty. If we start at a point to the right of $x_0$, the successive approximations will be moving away from the stationary point $z$.

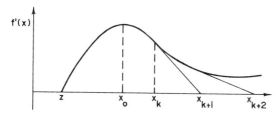

**Figure 2.15.** Newton-Raphson method (divergence).

*Example 2.14   Newton-Raphson Method*

Consider the problem

$$\text{Minimize } f(x) = 2x^2 + \frac{16}{x}$$

Suppose we use the Newton-Raphson method to determine a stationary point of $f(x)$ starting at the point $x_1 = 1$:

$$f'(x) = 4x - \frac{16}{x^2} \qquad f''(x) = 4 + \frac{32}{x^3}$$

**Step 1.** $x_1 = 1 \qquad f'(x_1) = -12 \qquad f''(x_1) = 36$

$\qquad\quad x_2 = 1 - \dfrac{-12}{36} = 1.33$

**Step 2.** $x_2 = 1.33 \qquad f'(x_2) = -3.73 \qquad f''(x_2) = 17.6$

$\qquad\quad x_3 = 1.33 - \dfrac{-3.73}{17.6} = 1.54$

We continue until $|f'(x_k)| < \varepsilon$, where $\varepsilon$ is a prespecified tolerance.

### 2.5.2   Bisection Method

If the function $f(x)$ is unimodal over a given search interval, then the optimal point will be the one where $f'(x) = 0$. If both the function value and the derivative of the function are available, then an efficient region-elimination search can be conducted using just a single point rather than a pair of points to identify a point at which $f'(x) = 0$. For example, if at a point $z$, $f'(z) < 0$, then assuming that the function is unimodal, the minimum cannot lie to the left of $z$. In other words, the interval $x \leqslant z$ can be eliminated. On the other hand, if $f'(z) > 0$, then the minimum cannot lie to the right of $z$, and the interval $x \geqslant z$ can be eliminated. Based on these observations, the *bisection method* (sometimes called the *Bolzano search*) can be constructed.

Determine two points $L$ and $R$ such that $f'(L) < 0$ and $f'(R) > 0$. The stationary point is between the points $L$ and $R$. We determine the derivative of the function at the midpoint,

$$z = \frac{L + R}{2}$$

If $f'(z) > 0$, then the interval $(z, R)$ can be eliminated from the search. On the other hand, if $f'(z) < 0$, then the interval $(L, z)$ can be eliminated. We shall now state the formal steps of the algorithm.

Given a bounded interval $a \leqslant x \leqslant b$ and a termination criterion $\varepsilon$.

**Step 1.**  Set $R = b$, $L = a$; assume $f'(a) < 0$ and $f'(b) > 0$.

**Step 2.**  Calculate $z = (R + L)/2$, and evaluate $f'(z)$.

**Step 3.**  If $|f'(z)| \leqslant \varepsilon$, terminate. Otherwise, if $f'(z) < 0$, set $L = z$ and go to step 2. If $f'(z) > 0$, set $R = z$ and go to step 2.

Note that the search logic of this region-elimination method is based purely on the sign of the derivative and does not use its magnitude. A method that uses both is the secant method, to be discussed next.

### 2.5.3  Secant Method

The *secant method* combines Newton's method with a region-elimination scheme for finding a root of the equation $f'(x) = 0$ in the interval $(a, b)$ if it exists. Suppose we are interested in finding the stationary point of $f(x)$ and we have two points $L$ and $R$ in $(a, b)$ such that their derivatives are opposite in sign. The secant algorithm then approximates the function $f'(x)$ as a "secant line" (a straight line between the two points) and determines the next point where the secant line of $f'(x)$ is zero (see Figure 2.16). Thus, the next approximation to the stationary point $x^*$ is given by

$$z = R - \frac{f'(R)}{(f'(R) - f'(L))/(R - L)}$$

If $|f'(z)| \leqslant \varepsilon$, we terminate the algorithm. Otherwise, we select $z$ and one of the points $L$ or $R$ such that their derivatives are opposite in sign and repeat the secant step. For example, in Figure 2.16, we would have selected $z$ and $R$ as the next two points. It is easy to see that, unlike the bisection search, the secant method uses both the magnitude and sign of the derivatives and hence can eliminate more than half the interval in some instances (see Figure 2.16).

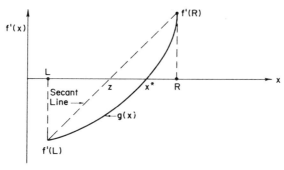

**Figure 2.16.**  Secant method.

*Example 2.15  Secant Method*

Consider again the problem of Example 2.14:

$$\text{Minimize } f(x) = 2x^2 + \frac{16}{x} \quad \text{over the interval} \quad 1 \leqslant x \leqslant 5$$

$$f'(x) = \frac{df(x)}{dx} = 4x - \frac{16}{x^2}$$

*Iteration 1*

**Step 1.** $R = 5 \qquad L = 1 \qquad f'(R) = 19.36 \qquad f'(L) = -12$

**Step 2.** $z = 5 - \dfrac{19.36}{(19.36 + 12)/4} = 2.53$

**Step 3.** $f'(z) = 7.62 > 0$; set $R = 2.53$.

*Iteration 2*

**Step 2.** $z = 2.53 - \dfrac{7.62}{(7.62 + 12)/1.53} = 1.94$

**Step 3.** $f'(z) = 3.51 > 0$; set $R = 1.94$.

Continue until $|f'(z)| \leqslant \varepsilon$.

### 2.5.4  Cubic Search Method

This is a polynomial approximation method in which a given function $f$ to be minimized is approximated by a third-order polynomial. The basic logic is similar to the quadratic approximation scheme. However, in this instance, because both the function value and the derivative value are available at each point, the approximating polynomial can be constructed using fewer points.

The cubic search starts with an arbitrary point $x_1$ and finds another point $x_2$ by a bounding search such that the derivatives $f'(x_1)$ and $f'(x_2)$ are of opposite sign. In other words, the stationary point $\bar{x}$ where $f'(x) = 0$ is bracketed between $x_1$ and $x_2$. A cubic approximation function of the form

$$\bar{f}(x) = a_0 + a_1(x - x_1) + a_2(x - x_1)(x - x_2) + a_3(x - x_1)^2(x - x_2)$$

$$(2.8)$$

is fitted such that Eq. (2.8) agrees with $f(x)$ at the two points $x_1$ and $x_2$. The first derivative of $\bar{f}(x)$ is given by

$$\frac{d\bar{f}(x)}{dx} = a_1 + a_2(x - x_1) + a_2(x - x_2)$$

$$+ a_3(x - x_1)^2 + 2a_3(x - x_1)(x - x_2) \qquad (2.9)$$

The coefficients $a_0$, $a_1$, $a_2$, and $a_3$ of Eq. (2.8) can now be determined using the values of $f(x_1)$, $f(x_2)$, $f'(x_1)$, and $f'(x_2)$ by solving the following linear equations:

$$f_1 = f(x_1) = a_0$$

$$f_2 = f(x_2) = a_0 + a_1(x_2 - x_1)$$

$$f_1' = f'(x_1) = a_1 + a_2(x_1 - x_2)$$

$$f_2' = f'(x_2) = a_1 + a_2(x_2 - x_1) + a_3(x_2 - x_1)^2$$

Note that the above system can be solved very easily in a recursive manner. Then, as in the quadratic case discussed earlier, given these coefficients, an estimate of the stationary point of $f$ can be obtained from the approximating cubic of Eq. (2.8). In this case, when we set the derivative of $\bar{f}(x)$ given by Eq. (2.9) to zero, we get a quadratic equation. By applying the formula for the root of the quadratic equation, a closed-form solution to the stationary point $\bar{x}$ of the approximating cubic is obtained as follows:

$$\bar{x} = \begin{cases} x_2 & \text{if } \mu < 0 \\ x_2 - \mu(x_2 - x_1) & \text{if } 0 \leqslant \mu \leqslant 1 \\ x_1 & \text{if } \mu > 1 \end{cases} \qquad (2.10)$$

where

$$\mu = \frac{f_2' + w - z}{f_2' - f_1' + 2w}$$

$$z = \left( \frac{3(f_1 - f_2)}{x_2 - x_1} \right) + f_1' + f_2'$$

$$w = \begin{cases} (z^2 - f_1' f_2')^{1/2} & \text{if } x_1 < x_2 \\ -(z^2 - f_1' f_2')^{1/2} & \text{if } x_1 > x_2 \end{cases}$$

The definition of $w$ in Eq. (2.10) ensures that the proper root of the quadratic equation is selected, while forcing $\mu$ to lie between 0 and 1 ensures that the predicted point $\bar{x}$ lies between the bounds $x_1$ and $x_2$. Once again we select the next two points for the cubic approximation as $\bar{x}$ and one of the points $x_1$ or $x_2$ such that the derivatives of the two points are of opposite sign, and repeat the cubic approximation.

Given an initial point $x_0$, positive step size $\Delta$, and termination parameters $\varepsilon_1$ and $\varepsilon_2$, the formal steps of the cubic search algorithm are as follows:

**Step 1.**  Compute $f'(x_0)$.
  If $f'(x_0) < 0$, compute $x_{K+1} = x_K + 2^K \Delta$ for $K = 0, 1, \ldots$ .
  If $f'(x_0) > 0$, compute $x_{K+1} = x_K - 2^K \Delta$ for $K = 0, 1, 2, \ldots$ .

**Step 2.**  Evaluate $f'(x)$ for the points $x_{K+1}$ for $K = 0, 1, 2, \ldots$ until a point $x_M$ is reached at which $f'(x_{M-1})f'(x_M) \leqslant 0$.
  Then set $x_1 = x_{M-1}$, $x_2 = x_M$.
  Compute $f_1, f_2, f_1', $ and $f_2'$.

**Step 3.**  Calculate the stationary point $\bar{x}$ of the cubic approximating function using Eq. (2.10).

**Step 4.**  If $f(\bar{x}) < f(x_1)$, go to step 5. Otherwise, set $\bar{x} = \bar{x} + \frac{1}{2}(\bar{x} - x_1)$ until $f(\bar{x}) \leqslant f(x_1)$ is achieved.

**Step 5.**  Termination check:
  If $|f'(\bar{x})| \leqslant \varepsilon_1$ and $\left| \dfrac{\bar{x} - x_1}{\bar{x}} \right| \leqslant \varepsilon_2$, stop. Otherwise, set either
  (i)   $x_2 = x_1$ and $x_1 = \bar{x}$     if $f'(\bar{x})f'(x_1) < 0$
  or
  (ii)  $x_1 = \bar{x}$     if  $f'(\bar{x})f'(x_2) < 0$
  In either case, continue with step 3.

Note that steps 1 and 2 constitute a bounding search proceeding along an expanding pattern in which the sign of the derivative is used to determine when the optimum has been overshot. Step 3 represents the calculation of the optimum of the approximating cubic. Step 4 is merely a safety check to ensure that the generated estimate is in fact an improvement. Provided that direct values of the derivative are available, this search is unquestionably the most efficient search method currently available. However, if derivatives must be obtained by differences, then the quadratic based search of Powell is preferred.

*Example 2.16   Cubic Search with Derivatives*

Consider again,

$$\text{Minimize } f(x) = 2x^2 + \frac{16}{x}$$

with the initial point $x_0 = 1$ and step size $\Delta = 1$. For convergence parameters use

$$\varepsilon_1 = 10^{-2} \qquad \varepsilon_2 = 3 \times 10^{-2}$$

$$f'(x) = \frac{df}{dx} = 4x - \frac{16}{x^2}$$

### Iteration 1

**Step 1.** $f'(1) = -12 < 0$. Hence, $x_1 = 1 + 1 = 2$.

**Step 2.** $f'(2) = 4$

Since $f'(1)f'(2) = -48 < 0$, a stationary point has been bracketed between 1 and 2. Set $x_1 = 1, x_2 = 2$. Then, $f_1 = 18, f_2 = 16, f'_1 = -12$, $f'_2 = 4$.

**Step 3.** $z = \dfrac{3}{1}(18 - 16) + (-12) + 4 = -2$

$w = +[4 - (-12)(4)]^{1/2} = (52)^{1/2} = 7.211$

$\mu = \dfrac{4 + 7.211 - (-2)}{4 - (-12) + 2(7.211)} = 0.4343$

$\bar{x} = 2 - 0.4343(2 - 1) = 1.5657$

**Step 4.** $f(1.5657) = 15.1219 < f(x_1) = 18$

Therefore, continue.

**Step 5.** Termination check:

$$f'(1.5657) = -0.2640 \qquad \text{clearly not terminated.}$$

Since

$$f'(\bar{x})f'(x_2) = (-0.2640)(4) < 0$$

set $x_1 = \bar{x} = 1.5657$.

### Iteration 2

**Step 3.** $z = \dfrac{3}{0.4343}(15.1219 - 16) + (-0.2640) + 4 = -2.3296$

$w = +[(2.3296)^2 - (-0.2640)(4)]^{1/2} = 2.5462$

$\mu = \dfrac{4 + 2.5462 - (-2.3296)}{4 - (-0.2640) + 2(2.5462)} = 0.9486$

$\bar{x} = 2 - 0.9486(2 - 1.5657) = 1.5880$

**Step 4.** $f(1.5880) = 15.1191 < f(x_1) = 15.1219$

Therefore, continue.

**Step 5.** Termination check:

$f'(1.5880) = 0.0072 < 10^{-2}$

$\left| \dfrac{1.5880 - 1.5657}{1.5880} \right| = 0.0140 < 3 \times 10^{-2}$

Therefore, terminate.

Note that using the same two points, $x_1 = 1$ and $x_2 = 2$, the quadratic search of Example 2.13 returned an estimate $\bar{x} = 1.714$, whereas the cubic search produced 1.5657. The exact minimum is 1.5874, indicating the clear superiority of the higher order polynomial fit.

## 2.6  COMPARISON OF METHODS

From a theoretical point of view, it can be shown that point estimation methods such as Powell's quadratic search or the cubic search with derivatives are intrinsically superior to region-elimination methods such as golden section. In particular, it has been shown that the quadratic scheme exhibits superlinear asymptotic convergence to the true minimum; that is, the deviation of the $(k + 1)$st estimate from the true minimum is proportional to the deviation of the $k$th estimate raised to the $\alpha$ power, where $\alpha > 1$ ($\alpha = 1.3$ has been claimed [6]). By contrast, if in the case of the golden section search, the $k$th estimate of the true minimum is taken to be the midpoint of the interval remaining after $k$ evaluations, then the deviation of this estimate from the actual minimum decreases linearly from the $k$th to the $(k + 1)$st iterations. Assuming comparable regions of convergence for both methods, this indicates that the quadratic search method will converge more rapidly than any region-elimination scheme. Of course, this comparison is based on assuming comparable regions of convergence and on well-behaved unimodal functions.

The limited numerical comparisons available in the literature do not indicate any overwhelming advantage of either the derivative-based or the quadratic or region-elimination search method over the others. If functional evaluations are lengthy and costly in terms of computer time, Box et al. [7] claim superiority of a search strategy based on a modification of Powell's method. This claim seems to be supported by the limited computational experiments presented by Himmelblau [8], who compares this strategy with one based on the golden section search and shows that a modified Powell's search requires fewer functional evaluations to attain a specified accuracy in the estimate of the minimizing point. Certainly, if very high accuracy is desired, the polynomial approximation methods are quite superior. On the other hand, for strongly skewed or possibly multimodal functions, Powell's search method has been known to converge at a much slower rate than region-elimination schemes. Hence, if reliability is of prime importance, then golden section is the ideal choice. Because of these considerations, it is recommended that Powell-type search methods generally be used along with a golden section search to which the program can default if it encounters difficulties in the course of iterations. Readers may also refer to Brent [9] for a good discussion of the relative efficiencies of different search methods.

In a small study conducted as a class project for the optimization course at Purdue University, the following point-estimation methods were compared on the function $f(x) = \sin^k x$ for different values of $k$:

Bisection method

Powell's method

Cubic search

In addition, the best region-elimination method, namely the golden section search, was also included for comparison.

Three criteria—solution time, solution accuracy, and sensitivity to convergence parameter—were used to evaluate these methods. The first two criteria were examined by varying the value of the power $k$. Odd powers between 1 and 79 were chosen for $k$. Note that the minimum point is always $x^* = 4.71239$ and $f(x^*) = -1.0$ for all values of $k$. However, as $k$ increases, the function becomes less smooth and has narrow valleys near the minimum. This would make the point-estimation methods slow and less accurate.

The sensitivity criterion was tested by varying the convergence parameter. As expected, results showed an increase in solution times as $k$ increased. The bisection method showed the greatest increase in solution times with respect to $k$, due to rapid increase in the values of the gradient close to the minimum. Golden section search was unaffected by the increase in power $k$. Similarly, solution accuracy, measured in terms of percentage error to the true minimum, became worse as $k$ increased for all three methods—bisection, Powell, and cubic. However, golden section search was unaffected by changes in the steepness of the function. Sensitivity to change in the convergence criterion was minimal for all four methods tested.

## 2.7  SUMMARY

In this chapter, we discussed necessary and sufficient optimality conditions for functions of a single variable. We showed that a necessary condition for the optimum is that the point be a stationary point, that is, that the first derivative be zero. Second-order and higher order derivatives were then used to test whether a point corresponds to a minimum, a maximum, or an inflection point. Next we addressed the question of identifying candidate optima. We developed a number of search methods called region-elimination methods for locating an optimal point in a given interval. We showed that the golden section search is generally the preferred algorithm due to its computational efficiency and ease of implementation.

The region-elimination method required a simple comparison of function values at two trial points and hence took into account the ordering of the function values only. In order to involve the magnitude of the difference between the function values as well, we developed point-estimation methods. These involved a quadratic or cubic approximation of the function to determine the optimum. We pointed out that, assuming comparable regions of convergence, point-estimation methods converge more rapidly than region-elimination methods as a class for well-behaved unimodal functions. However, for strongly skewed or multimodal functions, the golden section search is more reliable. We concluded with a recommendation that Powell-type successive quadratic estimation search be generally used along with a golden section search to which the program can default if it encounters difficulties in the course of iterations.

## REFERENCES

1. Swann, W. H., "Report on the Development of a Direct Search Method of Optimization," ICI Ltd., *Central Instr. Res. Lab*, *Res. Note*, **64** / 3, London, 1964.

2. Kiefer, J., "Optimum Sequential Search and Approximation Methods Under Minimum Regularity Assumptions," *J. Soc. Ind. Appl. Math.*, **5**(3), 105–125 (1957).

3. Bartle, R., *Elements of Real Analysis*, Wiley, New York, 1976.

4. Powell, M. J. D., "An Efficient Method for Finding the Minimum of a Function of Several Variables Without Calculating Derivatives," *Computer J.*, **7**, 155–162 (1964).

5. Raphson, J., *History of Fluxion*, London, 1725.

6. Kowalik, J., and M. R. Osborne, *Methods for Unconstrained Optimization Problems*, American Elsevier, New York, 1968.

7. Box, M. J., D. Davies, and W. H. Swann, *Nonlinear Optimization Techniques*, I.C.I. Monograph, Oliver and Boyd, Edinburgh, 1969.

8. Himmelblau, D. M., *Applied Nonlinear Programming*, McGraw-Hill, New York, 1972.

9. Brent, R. P., *Algorithms for Minimization Without Derivatives*, Prentice-Hall, Englewood Cliffs, NJ, 1973.

10. Phillips, D. T., A. Ravindran, and J. J. Solberg, *Operations Research: Principles and Practice*, Wiley, New York, 1976.

## PROBLEMS

**2.1.** What is an inflection point and how do you identify it?

**2.2.** How do you test a function to be convex or concave?

**2.3.** What is the unimodal property and what is its significance in single variable optimization?

**2.4.** Suppose a point satisfies sufficiency conditions for a local minimum. How do you establish that it is a global minimum?

**2.5.** Cite a condition under which a search method based on polynomial interpolation may fail.

**2.6.** Are region elimination methods as a class more efficient than point estimation methods? Why or why not?

**2.7.** In terminating search methods, it is recommended that both the difference in variable values and the difference in the function values be

**Figure 2.17.** Problem 2.9 schematic.

tested. Is it possible for one test alone to indicate convergence to a minimum while the point reached is really not a minimum? Illustrate graphically.

**2.8.** Given the following functions of one variable:

(a) $f(x) = x^5 + x^4 - \dfrac{x^3}{3} + 2$

(b) $f(x) = (2x + 1)^2(x - 4)$

Determine, for each of the above functions, the following:

(i) Region(s) where the function is increasing; decreasing.

(ii) Inflexion points, if any.

(iii) Region(s) where the function is concave; convex.

(iv) Local and global maxima, if any.

(v) Local and global minima, if any.

**2.9.** A cross-channel ferry is constructed to transport a fixed number of tons ($L$) across each way per day (see Figure 2.17). If the cost of construction of the ferry without the engines varies as the load ($l$), and the cost of the engines varies as the product of the load and the cube of the speed ($v$), show that the total cost of construction is least when twice as much money is spent on the ferry as on the engine. (You may neglect the time of loading and unloading, and assume the ferry runs continuously.)

**2.10.** A forest fire is burning down a narrow valley of width 2 miles at a velocity of 32 fpm. (Figure 2.18). A fire can be contained by cutting a fire break through the forest across the width of the valley. A man can clear 2 ft of the fire break in a minute. It costs $20 to transport each man to the scene of fire and back, and each man is paid $6 per hour while there. The value of timber is $2000 per square mile. How many men should be sent to fight the fire so as to minimize the total costs?

**2.11.** Consider the unconstrained optimization of a single-variable function $f(x)$. Given the following information about the derivatives of orders $1, 2, 3, 4$ at the point $x_i$ ($i = 1, 2, \ldots, 10$), identify the nature of the test point $x_i$ (i.e., whether it is maximum, minimum, inflection point,

**Figure 2.18.** Problem 2.10 schematic.

nonoptimal, no conclusion possible, etc.):

| $x_i$ | $f'(x_i)$ | $f''(x_i)$ | $f'''(x_i)$ | $f''''(x_i)$ |
|-------|-----------|------------|-------------|--------------|
| $x_1$ | 0 | + | anything | anything |
| $x_2$ | 0 | 0 | + | anything |
| $x_3$ | 0 | − | anything | anything |
| $x_4$ | − | − | anything | anything |
| $x_5$ | 0 | 0 | − | anything |
| $x_6$ | 0 | 0 | 0 | − |
| $x_7$ | 0 | 0 | 0 | 0 |
| $x_8$ | 0 | 0 | 0 | + |
| $x_9$ | + | + | anything | anything |
| $x_{10}$ | 0 | − | + | − |

**2.12.** State whether each of the following functions is convex, concave, or neither.

(a) $f(x) = e^x$

(b) $f(x) = e^{-x}$

(c) $f(x) = \dfrac{1}{x^2}$

(d) $f(x) = x + \log x$      for   $x > 0$

(e) $f(x) = |x|$

(f) $f(x) = x \log x$      for   $x > 0$

(g) $f(x) = x^{2k}$      where $k$ is an integer

(h) $f(x) = x^{2k+1}$      where $k$ is an integer

**2.13.** Consider the function

$$f(x) = x^3 - 12x + 3 \qquad \text{over the region} \quad -4 \leqslant x \leqslant 4$$

Determine the local minima, local maxima, global minimum, and global maximum of $f$ over the given region.

**2.14.** Identify the regions over which the following function is convex and where it is concave:

$$f(x) = e^{-x^2}$$

Determine its global maximum and global minimum.

**2.15.** Consider a single-period inventory model for perishable goods as follows:

Demand is a random variable with density $f$; i.e., $P(\text{demand} \leqslant x) = \int_0^x f(x)\, dx$.

All stockouts are lost sales.

$C$ is the unit cost of each item.

$p$ is the loss due to each stockout (includes loss of revenue and goodwill).

$r$ is the selling price per unit.

$l$ is the salvage value of each unsold item at the end of the period.

The problem is to determine the optimal order quantity $Q$ that will maximize the expected net revenue for the season. The expected net revenue, denoted by $\Pi(Q)$, is given by

$$\Pi(q) = r\mu + l\int_0^Q (Q - x)f(x)\, dx$$

$$- (p + r)\int_Q^\infty (x - Q)f(x)\, dx - CQ$$

where $\mu$ = expected demand = $\int_0^\infty xf(x)\, dx$.

(a) Show that $\Pi(Q)$ is a concave function in $Q(\geqslant 0)$.

(b) Explain how you will get an optimal ordering policy using (a).

(c) Compute the optimal order policy for the example given below:

$$C = \$2.50 \qquad r = \$5.00 \qquad l = 0 \qquad p = \$2.50$$

$$f(x) = \frac{1}{400} \qquad \text{for} \quad 100 \leqslant x \leqslant 500$$

$$= 0 \qquad \text{otherwise}$$

*Hint:* Use the Leibniz rule for differentiation under the integral sign.

**2.16.** Suppose we conduct a single-variable search using (a) the golden section method, and (b) the bisection method with derivatives estimated numerically by differences. Which is likely to be more efficient? Why?

**2.17.** Carry out a single-variable search to minimize the function

$$f(x) = 3x^2 + \frac{12}{x^3} - 5 \qquad \text{on the interval} \quad \tfrac{1}{2} \leqslant x \leqslant \tfrac{5}{2},$$

using (a) golden section, (b) interval halving, (c) quadratic based search, and (d) cubic derivative based search. Each search method is to use four functional evaluations only.

Compare the final search intervals obtained by the above methods.

**2.18.** Determine the minimum of

$$f(x) = (10x^3 + 3x^2 + x + 5)^2$$

starting at $x = 2$ and using a step size $\Delta = 0.5$,

(a) Using region elimination: expanding pattern bounding plus six steps of golden section.

(b) Using quadratic point estimation: three iterations of Powell's method.

**2.19.** Determine the real roots of the equation (within one decimal point accuracy)

$$f(x) = 3000 - 100x^2 - 4x^5 - 6x^6 = 0$$

using (a) the Newton-Raphson method; (b) the bisection method; (c) the secant method. You may find it helpful to write a small-computer program to these search methods.

**2.20.** (a) Explain how problem 2.19 can be converted to an unconstrained nonlinear optimization problem in one variable.

(b) Write a computer program for the golden section search, and solve the optimization problem formulated in part (a). Compare the solution with those obtained in problem 19.

**2.21.** An experimenter has obtained the following equation used to describe the trajectory of a space capsule (Phillips, Ravindran, and Solberg [10]):

$$f(x) = 4x^3 + 2x - 3x^2 + e^{x/2}$$

Determine a root of the above equation using any of the derivative-based methods.

**2.22.** Minimize the following functions using any univariate search method up to one-decimal-point accuracy (Phillips, Ravindran, and Solberg [10]).

(a) Minimize $f(x) = 3x^4 + (x - 1)^2$ over the range $[0, 4]$.

(b) Minimize $f(x) = (4x)(\sin x)$ over the range $[0, \Pi]$.

(c) Minimize $f(x) = 2(x - 3)^2 + e^{0.5x^2}$ over the range $(0, 100)$.

**2.23.** In a chemical plant, the cost of pipes, their fittings, and pumping are important investment costs. Consider the design of a pipeline $L$ feet long that should carry fluid at the rate of $Q$ gpm. The selection of economic pipe diameter $D$ (in.) is based on minimizing the annual cost of pipe, pump, and pumping. Suppose the annual cost of a pipeline with a standard carbon steel pipe and a motor-driven centrifugal pump can be expressed as

$$f = 0.45L + 0.245LD^{1.5} + 325(hp)^{1/2} + 61.6(hp)^{0.925} + 102$$

where

$$hp = 4.4 \times 10^{-8}\frac{LQ^3}{D^5} + 1.92 \times 10^{-9}\frac{LQ^{2.68}}{d^{4.68}}$$

Formulate the appropriate single-variable optimization problem for designing a pipe of length 1000 ft with a fluid rate of 20 gpm. The diameter of the pipe should be between 0.25 to 6 in. Solve using the golden section search.

# Chapter 3

---

# Functions of Several Variables

In this chapter we examine fundamental concepts and useful methods for finding minima of unconstrained functions of several variables. We build and expand upon the material given in Chapter 2, recognizing the value of the methods considered there in the discussion to follow.

With the aid of a few necessary mathematical notational conventions and results from linear analysis and the calculus (see Appendix A), we address the *static question*. That is, we examine conditions that allow us (in the most general circumstances possible) to characterize optima. We use these conditions to test candidate points to see whether they are (or are not) minima, maxima, saddlepoints, or none of the above. In doing so, we are not interested in the generation of the candidate points, but in determining if they solve the unconstrained multivariable problem:

$$\text{Minimize} \quad f(x) \qquad x \in R^N \tag{3.1}$$

in the absence of constraints on $x$, where $x$ is a vector of *design variables* of dimension $N$, and $f$ is a scalar *objective function*. We typically assume that $x_i$ (for all $i = 1, 2, 3, \ldots, N$) can take on any value, even though in many practical applications $x$ must be selected from a discrete set. In addition, we shall often find it convenient to assume that $f$ and its derivatives exist and are continuous everywhere, even though we know that optima may occur at points of discontinuity in $f$ or its *gradient*,

$$\nabla f = \left[ \frac{\partial f}{\partial x_1}, \frac{\partial f}{\partial x_2}, \frac{\partial f}{\partial x_3}, \ldots, \frac{\partial f}{\partial x_N} \right]^T \tag{3.2}$$

It is well to remember that $f$ may take on a minimum value at a point $\bar{x}$, where $f$ is discontinuous or $\nabla f$ is discontinuous or does not exist. We must at least temporarily discard these complicating factors in order to develop useful optimality criteria. Finally, we must often be satisfied to identify *local* optima, since the nonlinear objective $f$ will typically not be convex (see Appendix B) and therefore will be multimodal as in Figure 3.1, which is a contour plot of Himmelblau's function:

$$f(x) = \left[x_1^2 + x_2 - 11\right]^2 + \left[x_1 + x_2^2 - 7\right]^2 \qquad (3.3)$$

We see that this function has four distinct minima.

Next, we are led to examine the *dynamic question*, given $x^{(0)}$, a point that does *not* satisfy the above-mentioned optimality criteria, what is a "good" new estimate $x^{(1)}$ of the solution $x^*$? This leads naturally to a discussion of a number of methods, which in fact constitutes the greater part of the chapter. We classify the methods according to their need for derivative information. The chapter closes with a discussion of the relative merits of the various methods with supportive numerical results.

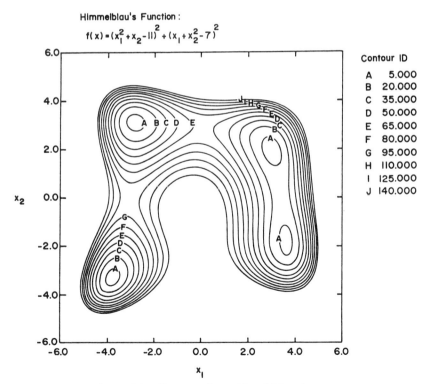

**Figure 3.1.** Contours of a multimodal function.

## 3.1  OPTIMALITY CRITERIA

Here we develop conditions or tests that will allow us to characterize (that is, classify) points in the design space. We examine optimality criteria for basically two reasons, (1) because they are necessary to recognize solutions, and (2) because they provide motivation for most of the useful methods. Consider the Taylor expansion of a function of several variables:

$$f(x) = f(\bar{x}) + \nabla f(\bar{x})^T \Delta x + \frac{1}{2} \Delta x^T \nabla^2 f(\bar{x}) \Delta x + O_3(\Delta x) \quad (3.4)$$

where         $\bar{x} =$ the current or expansion point in $R^N$

$\Delta x = x - \bar{x} =$ the change in x

$\nabla f(\bar{x}) =$ the $N$-component column vector of first derivatives of $f(x)$ evaluated at $\bar{x}$

$\nabla^2 f(\bar{x}) = \mathbf{H}_f(\bar{x}) =$ the $N \times N$ symmetric matrix of second partial derivatives of $f(x)$ evaluated at $\bar{x}$, often called the *Hessian* matrix. The element in the $i$th row and $j$th column is $\partial^2 f / \partial x_i\, \partial x_j$.

$O_3(\Delta x) =$ all terms of order greater than 2 in $\Delta x$

Let us now ignore the higher order terms [that is, drop $O_3(\Delta x)$], and examine the change in the magnitude of the objective $f(x)$ corresponding to arbitrary changes in $x$.

$$\Delta f(x) = f(x) - f(\bar{x}) = \nabla f(\bar{x})^T \Delta x + \frac{1}{2} \Delta x^T \nabla^2 f(\bar{x}) \Delta x \quad (3.5)$$

Recall that by definition a minimum is a point such that all other points in a neighborhood produce a greater objective value. That is,

$$\Delta f = f(x) - f(\bar{x}) \geqslant 0 \quad (3.6)$$

The point $\bar{x}$ is a *global* minimum if (3.6) holds for all $x \in R^N$, and we give this point the symbol $x^{**}$. When (3.6) holds for some δ-neighborhood that is, for all $x$ such that $\|x - \bar{x}\| \leqslant δ$ for some $δ > 0$, then $\bar{x}$ is a *local* minimum or $x^*$. When

$$\Delta f = f(x) - f(\bar{x}) \leqslant 0 \quad (3.7)$$

then $\bar{x}$ is a maximum, either local or global as before. Removal of the equality in (3.6) and (3.7) produces *strict* minimum and maximum points. When $\Delta f$ is either positive, negative, or zero depending on the choice of nearby points in a δ-neighborhood, then $\bar{x}$ is a *saddlepoint*.

Let us return to reason on (3.5), and recall that we assume that $f(x)$, $\nabla f(x)$, and $\nabla^2 f(x)$ exist and are continuous for all $x \in R^N$. Therefore, from (3.5), in order that the sign of $\Delta f$ be known for arbitrary values of $\Delta x$, $\nabla f(\bar{x})$ must be zero; that is, $\bar{x}$ must be a *stationary point*. Otherwise, we could force $\Delta f$ to be plus or minus depending on the sign of $\nabla f(\bar{x})$ and $\Delta x$. Accordingly, $\bar{x}$ must satisfy the *stationary conditions*:

$$\nabla f(\bar{x}) = 0 \tag{3.8}$$

so (3.5) becomes

$$\Delta f(x) = \tfrac{1}{2}\Delta x^T \nabla^2 f(\bar{x})\, \Delta x \tag{3.9}$$

Clearly, then, the sign of $\Delta f(x)$ depends on the nature of the quadratic form;

$$Q(x) = \Delta x^T \nabla^2 f(\bar{x})\, \Delta x \tag{3.10}$$

or

$$Q(z) = z^T A z$$

We know from linear algebra (see Appendix A) that:

| | |
|---|---|
| $A$ is *positive definite* if | for all $z$, $Q(z) > 0$ |
| $A$ is *positive semidefinite* if | for all $z$, $Q(z) \geqslant 0$ |
| $A$ is *negative definite* if | for all $z$, $Q(z) < 0$ |
| $A$ is *negative semidefinite* if | for all $z$, $Q(z) \leqslant 0$ |
| $A$ is *indefinite* if | for some $z$, $Q(z) > 0$ |
| and | for other $z$, $Q(z) < 0$ |

(3.11)

Therefore from (3.11) the stationary point $\bar{x}$ is a

minimum, if $\nabla^2 f(\bar{x})$ is positive definite

maximum, if $\nabla^2 f(\bar{x})$ is negative definite                    (3.12)

saddlepoint, if $\nabla^2 f(\bar{x}) \gtrless 0$ is indefinite

In addition, it may be helpful to examine the nature of the stationary point $\bar{x}$ in a slightly different light. Consider the stationary point $\bar{x}$, a surrounding $\delta$-neighborhood, and directions emanating from $\bar{x}$ (Figure 3.2) such that

$$\tilde{x} = \bar{x} + \alpha s(\bar{x}) \tag{3.13}$$

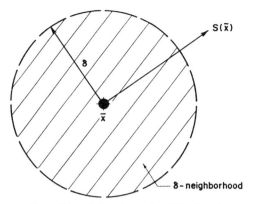

**Figure 3.2.** A "stationary" neighborhood.

We can generate all points in the neighborhood $\tilde{x}$ by appropriate choice of $\alpha$ and $s$. Substitution of (3.13) into (3.9) produces

$$\Delta f(x) = \frac{\alpha^2}{2} s^T \nabla^2 f(\bar{x}) s \qquad (3.14)$$

Now we can classify $s(\bar{x})$ as a descent direction, ascent direction, or neither, using the previous reasoning in (3.11) and (3.12). Hence $\bar{x}$ becomes $x^*$, a local minimum when a descent direction *cannot* be found, and this corresponds to the case when $\nabla^2 f(\bar{x})$ is positive semidefinite.

We are now in a position to state the necessary and sufficient conditions for the existence of a minimum of a function of several variables.

### Theorem 3.1   Necessary Conditions

For $x^*$ to be a local minimum, it is necessary that

$$\nabla f(x^*) = 0 \qquad (3.15a)$$

and

$$\nabla^2 f(x^*) \text{ is positive semidefinite} \qquad (3.15b)$$

### Theorem 3.2   Sufficient Conditions

If

$$\nabla f(x^*) = 0 \qquad (3.16a)$$

and

$$\nabla^2 f(x^*) \text{ is positive definite} \qquad (3.16b)$$

then $x^*$ is an isolated local minimum of $f(x)$.

The proofs of these theorems follow directly from the previous discussion and are left as exercises for the reader. Typically we must be satisfied to find a local minimum, but when we can show that $x^T \nabla^2 f(x) x \geqslant 0$ for all $x$ we say that $f(x)$ is a *convex* function and a local minimum is a global minimum.

### Example 3.1   Optimality Criteria

Consider the function

$$f(x) = 2x_1^2 + 4x_1 x_2^3 - 10x_1 x_2 + x_2^2$$

as shown in Figure 3.3, and the point $\bar{x} = [0, 0]^T$. Classify $\bar{x}$.

**Solution**

$$\frac{\partial f}{\partial x_1} = 4x_1 + 4x_2^3 - 10x_2$$

$$\frac{\partial f}{\partial x_2} = 12x_1 x_2^2 - 10x_1 + 2x_2$$

$$\nabla f(\bar{x}) = [0, 0]^T$$

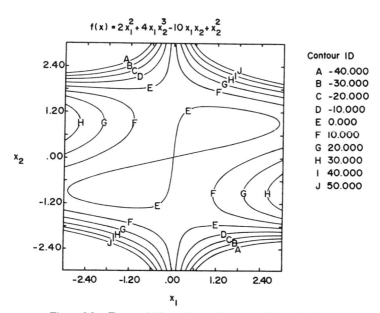

Figure 3.3.   Two-variable nonlinear function of Example 3.1.

Therefore, $\bar{x}$ is a stationary point.

$$\frac{\partial^2 f}{\partial x_1^2} = 4$$

$$\frac{\partial^2 f}{\partial x_2^2} = 24 x_1 x_2 + 2 = +2$$

$$\frac{\partial^2 f}{\partial x_1 \partial x_2} = 12 x_2^2 - 10 = -10$$

Therefore,

$$\nabla^2 f(\bar{x}) = H_f(\bar{x}) = \begin{bmatrix} +4 & -10 \\ -10 & +2 \end{bmatrix}$$

$\nabla^2 f(\bar{x})$ is indefinite, since the quadratic form $z^T H_f z > 0$ for $z = (0, 1)$ and $< 0$ for $z = (1, 1)$ (see Appendix A). Accordingly, $\bar{x}$ is a *saddlepoint*, as the contour map (Figure 3.3) suggests.

## 3.2  DIRECT SEARCH METHODS

In this and the following sections of this chapter we consider the *dynamic question* for functions of several variables. That is, we examine methods or algorithms that iteratively produce estimates of $x^*$, that set of design variables that causes $f(x)$ to take on its minimum value. The methods work equally well for maximization by letting the objectives be $-f(x)$. The methods that have been devised for the solution of this problem can be classified into three broad categories based on the type of information that must be supplied by the user:

1.  Direct search methods, which use only function values
2.  Gradient methods, which require accurate values of the first derivative of $f(x)$
3.  Second-order methods, which, in addition to the above, also use the second derivative of $f(x)$

We will consider examples from each class, since no one method or class of methods can be expected to uniformly solve all problems with equal efficiency. For instance, in some applications available computer storage is limited; in others, function evaluations are very time-consuming; in still others, great accuracy in the final solution is desired. In some applications it is either impossible or else very time-consuming to obtain analytic expressions for derivatives. Consequently, difference approximations must be employed if gradient-based techniques are to be used. This in turn may mean considerable experimentation to determine step sizes that strike the proper balance between roundoff and truncation errors. Clearly it behooves the engineer to tailor the method used to the characteristics of the problem at hand.

Methods for the unconstrained minimization problem are relatively well developed compared to the other areas of nonlinear programming. In fact, excellent surveys of the most powerful method do exist. The book by Murray [1] is a good example. In this section we treat *direct methods*, which are methods that require only values of the objective to proceed, and in the following section we discuss gradient and second-order methods. We assume here that $f(x)$ is continuous, and $\nabla f(x)$ may or may not exist but certainly is not available. In addition, we recognize that these direct methods *can* be used on problems where $\nabla f$ does exist and often are when $\nabla f$ is a complex vector function of the design variables. Finally, in this and the following sections we assume that $f(x)$ has a single minimum in the domain of interest. When these methods are applied to multimodal functions, we must be content to locate local minima.

Multivariable methods that employ only function values to guide the search for the optimum can be classified broadly into heuristic techniques and theoretically based techniques. The heuristic techniques, as the name implies, are search methods constructed from geometric intuition for which no performance guarantees other than empirical results can be stated. The theoretically based techniques, on the other hand, do have a mathematical foundation that allows performance guarantees, such as convergence, to be established, at least under restricted conditions. We will examine in particular detail three direct search techniques:

1. The simplex search or $S^2$ method
2. The Hooke-Jeeves pattern search method
3. Powell's conjugate direction method

The first two of these are heuristic techniques and exhibit fundamentally different strategies. The $S^2$ method employs a regular pattern of points in the design space sequentially, whereas the Hooke-Jeeves method employs a fixed set of directions (the coordinate directions) in a recursive manner. Powell's method is a theoretically based method that was devised assuming a quadratic objective, and for such functions will converge in a finite number of iterations. The common characteristic of the three methods is that computationally they are relatively uncomplicated, hence easy to implement and quick to debug. On the other hand, they can be, and often are, slower than the derivative-based methods. We give little attention to methods based upon the region-elimination concepts of Chapter 2, such as given by Box et al. [2] and Krolak and Cooper [3], since they tend to be vastly inferior to other available methods.

### 3.2.1  The $S^2$ or Simplex Search Method

The earliest attempts to solve unconstrained problems by direct search produced methods modeled essentially on single-variable methods. Typically, the domain of the performance index was subdivided into a grid of points, and then various strategies for shrinking the area in which the solution lay were applied. Often this led to nearly exhaustive enumeration and hence proved unsatisfactory for all but two-dimensional problems. A more useful idea was to pick a base point and, rather than attempting to cover the entire range of the variables, evaluate the performance index in some pattern about the base point. For example, in two dimensions, a square pattern such as in Figure 3.4 is located. Then the best of the five points is selected as the next base point around which to locate the next pattern of points. If none of the corner points is better than the base point, the scale of the grid is reduced and the search continues.

This type of *evolutionary optimization* was advocated by Box [4] and others to optimize the performance of operating plants when there is error present in the measured response of the plant to imposed process-variable changes. In higher dimensions this corresponds to evaluating the performance index at each vertex as well as the centroid of a hypercube and is called a *factorial design pattern*. If the number of variables, that is, the dimension of the search space, is $N$, then a factorial search pattern will require $2^N + 1$ function evaluations per pattern. The number of evaluations, even for problems of modest dimension, rapidly becomes overwhelming. Hence, even though the logic here is simple, more efficient methods must be investigated if we are to use these direct-search methods on problems of reasonable size.

One particularly ingenious search strategy is the *simplex search method* of Spendley, Hext, and Himsworth [5]. This method and others in this class have *no* relationship to the simplex method of linear programming. The similarity in name is indeed unfortunate. The simplex search method of Spendley, Hext, and Himsworth is based upon the observation that the first-order experimental design requiring the smallest number of points is the regular simplex. In $N$ dimensions, a regular simplex is a polyhedron composed of $N + 1$ equidistant points, which form its vertices. For example, an equilateral triangle is a simplex in two dimensions; a tetrahedron is a simplex in three dimensions. The main

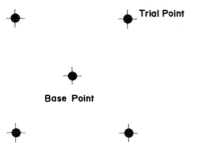

Figure 3.4.  Factorial design.

property of a simplex employed by their algorithm is that a new simplex can be generated on any face of the old one by projecting any chosen vertex a suitable distance through the centroid of the remaining vertices of the old simplex. The new simplex is then formed by replacing the old vertex by the newly generated projected point. In this way each new simplex is generated with a single evaluation of the objective. This process is demonstrated for two dimensions in Figure 3.5.

The method begins by setting up a regular simplex in the space of the independent variables and evaluating the function at each vertex. The vertex with highest functional value is located. This "worst" vertex is then reflected through the centroid to generate a new point, which is used to complete the next simplex. As long as the performance index decreases smoothly, the iterations move along crabwise until either the minimum is straddled or the iterations begin to cycle between two or more simplexes. These situations are resolved using the following three rules:

### Rule 1  Minimum "Straddled"

If the selected "worse" vertex was generated in the previous iteration, then choose instead the vertex with the next highest function value.

### Rule 2  Cycling

If a given vertex remains unchanged for more than $M$ iterations, reduce the size of the simplex by some factor. Set up a new simplex with the currently lowest point as the base point. Spendley et al. suggest that $M$ be predicted via

$$M = 1.65N + 0.05N^2$$

where $N$ is the problem dimension and $M$ is rounded to the nearest integer. This rule requires the specification of a reduction factor.

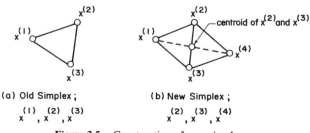

(a) Old Simplex ;
$$x^{(1)}, x^{(2)}, x^{(3)}$$

(b) New Simplex ;
$$x^{(2)}, x^{(3)}, x^{(4)}$$

Figure 3.5.  Construction of new simplex.

**FUNCTIONS OF SEVERAL VARIABLES**

### Rule 3   *The Termination Criterion*

The search is terminated when the simplex gets small enough or else if the standard deviation of the function values at the vertices gets small enough. This rule requires the specification of a termination parameter.

The implementation of this algorithm requires only two types of calculations: (1) generation of a regular simplex given a base point and appropriate scale factor, and (2) calculation of the reflected point. The first of these calculations is readily carried out, since it can be shown from elementary geometry that given an $N$-dimensional starting or base point $x^{(0)}$ and a scale factor $\alpha$, then the other $N$ vertices of the simplex in $N$ dimensions are given by

$$
x_j^{(i)} = \begin{cases} x_j^{(0)} + \delta_1 & \text{if } j = i \\ x_j^{(0)} + \delta_2 & \text{if } j \neq i \end{cases} \tag{3.17}
$$

for $i$ and $j = 1, 2, 3, \ldots, N$.

The increments $\delta_1$ and $\delta_2$, which depend only on $N$ and the selected scale factor $\alpha$, are calculated from

$$
\delta_1 = \left[ \frac{(N+1)^{1/2} + N - 1}{N\sqrt{2}} \right] \alpha \tag{3.18}
$$

$$
\delta_2 = \left[ \frac{(N+1)^{1/2} - 1}{N\sqrt{2}} \right] \alpha \tag{3.19}
$$

Note that the scale factor $\alpha$ is chosen by the user to suit the problem at hand. The choice $\alpha = 1$ leads to a regular simplex with sides of unit length.

The second calculation, reflection through the centroid, is equally straightforward. Suppose $x^{(j)}$ is the point to be reflected. Then the centroid of the remaining $N$ points is

$$
x_c = \frac{1}{N} \sum_{\substack{i=0 \\ i \neq j}}^{N} x^{(i)} \tag{3.20}
$$

All points on the line from $x^{(j)}$ through $x_c$ are given by

$$
x = x^{(j)} + \lambda \left( x_c - x^{(j)} \right) \tag{3.21}
$$

The choice $\lambda = 0$ yields the original point $x^{(j)}$, while the choice $\lambda = 1$ corresponds to the centroid $x_c$. In order to retain the regularity of the simplex, the reflection should be symmetric. Hence, $\lambda = 2$ will yield the desired new vertex point. Thus,

$$x_{new}^{(j)} = 2x_c - x_{old}^{(j)} \tag{3.22}$$

Both calculations are illustrated in the following example.

### Example 3.2  Simplex Search Calculations

$$\text{Minimize} \quad f(x) = (1 - x_1)^2 + (2 - x_2)^2$$

**Solution.** The construction of the initial simplex requires the specification of an initial point and a scale factor. Suppose $x^{(0)} = [0, 0]^T$ and $\alpha = 2$ are selected. Then

$$\delta_1 = \left[ \frac{\sqrt{3} + 1}{2\sqrt{2}} \right] \alpha = 1.9318$$

and

$$\delta_2 = \left[ \frac{\sqrt{3} - 1}{2\sqrt{2}} \right] \alpha = 0.5176$$

With these two parameters, the other two vertices are calculated as

$$x^{(2)} = [0 + 0.5176, 0 + 1.9318]^T = [0.5176, 1.9318]^T$$

$$x^{(1)} = [0 + 1.9318, 0 + 0.5176]^T = [1.9318, 0.5176]^T$$

with function values $f(x^{(1)}) = 3.0658$ and $f(x^{(2)}) = 0.2374$. Since $f(x^{(0)}) = 5$, $x^{(0)}$ is the point to be reflected to form the new simplex. The replacement $x^{(3)}$ is calculated as follows:

$$x_c = \frac{1}{2} \sum_{i=1}^{2} x^{(i)} = \frac{1}{2} (x^{(1)} + x^{(2)})$$

Therefore, using (3.22) we get

$$x^{(3)} = x^{(1)} + x^{(2)} - x^{(0)}$$

$$x^{(3)} = [2.4494, 2.4494]^T$$

At the new point $f(x^{(3)}) = 2.3027$, an improvement. The new simplex is composed of points $x^{(1)}$, $x^{(2)}$, and $x^{(3)}$. The algorithm would now continue by reflecting out the point with the highest function value, $x^{(1)}$. The iteration proceeds as before except when situations are encountered that require rules 1, 2, or 3 given before.

The $S^2$ algorithm as stated above has several obvious advantages:

1. The calculations are simple and the logic uncomplicated, hence a program will be short.
2. Storage requirements are relatively low: essentially an array of dimension $(N + 1, N + 2)$.
3. Comparatively few adjustable parameters need to be supplied: the scale factor $\alpha$, a factor by which to reduce $\alpha$ if rule 2 is encountered, and termination parameters.
4. The algorithm is effective when evaluation errors are significant because it operates on the worst rather than the best point.

These factors make it quite useful for on-line performance improvement calculations.

The algorithm also has several serious disadvantages:

1. It can encounter scaling problems because all coordinates are subjected to the same scale factor $\alpha$. To mitigate this, the user should in any given application scale all variables to have comparable magnitude.
2. The algorithm is slow, since it does not use past information to generate acceleration moves.
3. There is no simple way of expanding the simplex without recalculating the entire pattern. Thus, once $\alpha$ is reduced (because, for example, a region of narrow contours is encountered), the search must continue with this reduced step size.

To partially eliminate some of these disadvantages a modified simplex procedure has been developed by Nelder and Mead [6]. They observed that although it is convenient to use the formula for a regular simplex when generating the initial pattern, there is no real incentive for maintaining the regularity of the simplex as the search proceeds. Consequently, they allow both expansion and contraction in the course of the reflection calculation. Their modifications require consideration of the point with highest current function value, $x^{(h)}$; the next highest point, $x^{(g)}$; the point with the lowest current function value, $x^{(l)}$; and the corresponding function values, $f^{(h)}$, $f^{(g)}$, and $f^{(l)}$. Recall that the reflection step is described by the line

$$x = x^{(h)} + \lambda\left(x_c - x^{(h)}\right)$$

or

$$x = x^{(h)} + (1 + \theta)\left(x_c - x^{(h)}\right) \qquad (3.23)$$

If we let $\theta = 1$ we get the normal simplex reflection, such that $x_{new}$ is located a distance $\|x_c - x^{(j)}\|$ from $x_c$. When $-1 \leqslant \theta < 1$ we produce a shortened reflection or a *contraction*, while the choice $\theta > 1$ will generate a lengthened reflection step or an *expansion* of the simplex. The situations are depicted in Figure 3.6. The three values of $\theta$ used for normal reflection, contraction, and expansion are denoted respectively $\alpha$, $\beta$, and $\gamma$. The method proceeds from the initial simplex by determination of $x^{(h)}$, $x^{(g)}$, $x^{(l)}$, and $x_c$. A normal reflection of the simplex function values is checked to see if termination is appropriate. If not, a normal reflection, expansion, or contraction is taken, using the tests outlines in Figure 3.6. The iterations continue until the simplex function values do not vary significantly. Nelder and Mead recommend that the values $\alpha = 1$, $\beta = .5$, and $\gamma = 2$ be employed.

Some limited numerical comparisons indicate that the Nelder-Mead variant is very reliable in the presence of noise or error in the objective and is reasonably efficient. In 1969, Box and Draper [7] stated that this algorithm is the "most efficient of all current sequential techniques." A 1972 study by Parkinson and Hutchinson [8] investigated the effects of the choice of $\alpha$, $\beta$, $\gamma$, and the manner of construction of the initial simplex. They determined that the shape of the initial simplex was not important but its orientation was. The best overall choice of parameters was $(\alpha, \beta, \gamma) = (2, 0.25, 2.5)$ and this choice worked well if repeated successive expansions were allowed.

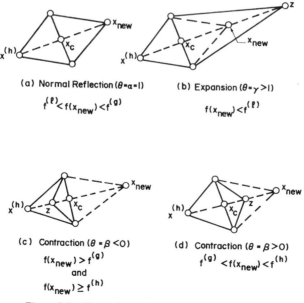

(a) Normal Reflection $(\theta = \alpha = 1)$

$f^{(l)} < f(x_{new}) < f^{(g)}$

(b) Expansion $(\theta = \gamma > 1)$

$f(x_{new}) < f^{(l)}$

(c) Contraction $(\theta = \beta < 0)$

$f(x_{new}) > f^{(g)}$

and

$f(x_{new}) \geqslant f^{(h)}$

(d) Contraction $(\theta = \beta > 0)$

$f^{(g)} < f(x_{new}) < f^{(h)}$

**Figure 3.6.** Expansion and contraction of a simplex.

### 3.2.2. The Hooke-Jeeves Pattern Search Method

The preceding search techniques were based on the systematic disposition and manipulation of a pattern of trial points. Yet despite our preoccupation with the geometric configuration in which the trial points are located, it is apparent that the main role of the set of points is to generate a direction of search. The exact placement of the points really only influences how sensitive the generated search direction is to local variations in the function topology. In particular, the equation for the reflection step,

$$x_{\text{new}} = x^{(j)} + \lambda \left( x_c - x^{(j)} \right)$$

clearly reveals that the entire set of points is condensed into a single vector difference that defines a direction. The remaining search logic is merely concerned with adjusting the step size $\lambda$ so that reasonable improvement is attained. If the generation of search directions is, then, the primary motivation for manipulating the pattern of trial points, then a reasonable improvement over the simplex strategy might be to operate directly with a set of direction vectors using them to guide the search. In the simplest approach, a fixed arbitrary set of directions could be selected and searched recursively for improvement. Alternatively, one could devise strategies for modifying one or more of the search directions with each iteration so as to better align them with the gross topology of the function. In either case, to ensure that the complete domain of the function is searched, it is reasonable to insist that the set of search directions be independent and span the entire domain of $f(x)$. For example, it clearly would be inadequate to search for the optimum of a three-variable function by alternating use of two search directions. Hence, all the direct-search methods discussed here will require at least $N$ independent search directions, where $N$ is the dimension of $x$.

An elementary example of a method that recursively uses a fixed set of search directions is the "one-variable-at-a-time" or *sectioning* technique. In this approach the set of directions is chosen to be the coordinate directions in the space of the problem variables. Each of the coordinate directions is sequentially searched using a single-variable optimization scheme until no further improvement is possible. If the objective function is spherically symmetrical, then this search will succeed admirably. However, if the function contours are distorted or elongated, as is usually the case, then the iteration can degenerate into an infinite sequence of shorter and shorter steps, and the search becomes quite inefficient. In fact, it has been shown by Powell [9] that the *cyclic* use of coordinate searches (or in fact of any fixed set of search directions) can not only prove inefficient but actually fail to converge to a local optimum even if an infinite number of iterations are employed.

In the course of attempting to improve the efficiency of this method, it was noted that considerable acceleration in convergence could be achieved by periodically searching in the direction $d^{(i)} = x^{(i)} - x^{(i-N)}$. This simple observation forms the basis of a modified sectioning method devised by Hooke and Jeeves [10], one of the first algorithms to incorporate the past history of a sequence of iterations into the generation of a new search direction. Basically, the Hooke-Jeeves (HJ) procedure is a combination of "exploratory" moves of the one-variable-at-a-time kind with "pattern" or acceleration moves regulated by some heuristic rules. The exploratory moves examine the local behavior of the function and seek to locate the direction of any sloping valleys that might be present. The pattern moves utilize the information generated in the exploration to step rapidly along the valleys.

**Exploratory Moves.** Given a specified step size, which may be different for each coordinate direction and change during the search, the exploration proceeds from an initial point by the specified step size in each coordinate direction. If the function value does not increase, the step is considered successful. Otherwise, the step is retracted and replaced by a step in the opposite direction, which in turn is retained depending upon whether it succeeds or fails. When all $N$ coordinates have been investigated, the exploratory move is completed. The resulting point is termed a *base point*.

**Pattern Move.** A pattern move consists of a single step from the present base point along the line from the previous to the current base point. Thus, a new pattern point is calculated as

$$x_p^{(k+1)} = x^{(k)} + \left( x^{(k)} - x^{(k-1)} \right) \qquad (3.24)$$

Now, recognizing that this move may not result in an improvement, the point $x_p^{(k+1)}$ is accepted only temporarily. It becomes the temporary base point for a new exploratory move. If the result of this exploratory move is a better point than the previous base point $x^{(k)}$, then this point is accepted as the new base point $x^{(k+1)}$. On the other hand, if the exploratory move does not produce improvement, then the pattern move is discarded and the search returns to $x^{(k)}$, where an exploratory search is undertaken to find a new pattern. Eventually a situation is reached when even this exploratory search fails. In this case the step sizes are reduced by some factor and the exploration resumed. The search is terminated when the step size becomes sufficiently small. In order to summarize the logic of the method, we employ the following notation:

$$x^{(k)} \qquad \text{current base point}$$
$$x^{(k-1)} \qquad \text{previous base point}$$
$$x_p^{(k+1)} \qquad \text{pattern move point}$$
$$x^{(k+1)} \qquad \text{next (or new) base point}$$

Using this notation we outline the pattern search method of Hooke and Jeeves.

## The Pattern Search Method of Hooke and Jeeves

**Step 1.** Define:

The starting point $x^{(0)}$

The increments $\Delta_i$ for $i = 1, 2, 3, \ldots, N$

The step reduction factor, $\alpha > 1$

A termination parameter, $\varepsilon > 0$

**Step 2.** Perform exploratory search.

**Step 3.** Was exploratory search successful (i.e., was a lower point found)?

Yes: Go to 5.

No: Continue.

**Step 4.** Check for termination.

Is $\|\Delta\| < \varepsilon$?

Yes: Stop; current point approximates $x^*$.

No: Reduce the increments:

$$\Delta_i = \frac{\Delta_i}{\alpha} \qquad i = 1, 2, 3, \ldots, N$$

Go to 2.

**Step 5.** Perform pattern move:

$$x_p^{(k+1)} = x^{(k)} + \left( x^{(k)} - x^{(k-1)} \right)$$

**Step 6.** Perform exploratory research using $x_p^{(k+1)}$ as the base point; let the result be $x^{(k+1)}$.

**Step 7.** Is $f(x^{(k+1)}) < f(x^{(k)})$?

Yes: Set $x^{(k-1)} = x^{(k)}$; $x^{(k)} = x^{(k+1)}$.

Go to 5.

No: Go to 4.

### *Example 3.3 Hooke-Jeeves Pattern Search*

Find the minimum of

$$f(x) = 8x_1^2 + 4x_1x_2 + 5x_2^2$$

from $x^{(0)} = [-4, -4]^T$.

**Solution.** In order to use the HJ direct-search method we must choose

$$\Delta = \text{step-size increments} = \begin{bmatrix} 1, & 1 \end{bmatrix}^T$$

$$\alpha = \text{step reduction factor} = 2$$

$$\varepsilon = \text{termination parameter} = 10^{-4}$$

Begin the iteration with an exploration about $x^{(0)}$, with $f(x^{(0)}) = 272$. With $x_2$ fixed, we increment $x_1$:
$x_2 = -4$:

$$x_1 = -4 + 1 \rightarrow f(-3, -4) = 200 < f(x^{(0)}) \rightarrow \text{success}$$

Therefore, we fix $x_1$ at $-3$ and increment $x_2$:
$x_1 = -3$:

$$x_2 = -4 + 1 \rightarrow f(-3, -3) = 153 < 200 \rightarrow \text{success}$$

Therefore the result of the first exploration is

$$x^{(1)} = [-3, -3]^T \qquad f(x^{(1)}) = 153$$

Since the exploration was a success, we go directly to the pattern move:

$$x_p^{(2)} = x^{(1)} + (x^{(1)} - x^{(0)}) = [-2, -2]^T$$

$$f(x_p^{(2)}) = 68$$

Now we perform an exploratory search about $x_p^{(2)}$. We find that positive increments on $x_1$ and $x_2$ produce success. The result is the point

$$x^{(2)} = [-1, -1]^T \qquad f(x^{(2)}) = 17$$

Since $f(x^{(2)}) < f(x^{(1)})$, the pattern move is deemed a success, and $x^{(2)}$ becomes the new base point for an additional pattern move. The method continues until step reduction causes termination near the solution, $x^* = [0, 0]^T$. Progress of the method is displayed in Figure 3.7.

As can be seen from Example 3.3, the search strategy is quite simple, the calculations uninvolved, and the storage requirements less than even that associated with the simplex search method. Because of these factors, the HJ algorithm has enjoyed wide application in all areas of engineering, especially when used with penalty functions (to be discussed in Chapter 6). Because of its reliance on coordinate steps, the algorithm can, however, terminate prematurely, and in the presence of severe nonlinearities will degenerate to a sequence of exploratory moves without benefit of pattern acceleration.

Numerous modifications to the basic Hooke-Jeeves method have been reported since it was introduced. For instance, Bandler and McDonald [11] modified the basic HJ procedure by inserting rules for expanding and contracting the exploratory increments as well as allowing a contracted pattern step if the normal pattern step fails. Our own experiments suggest the addition of another phase of the algorithm, which we call pattern exploitation. That is,

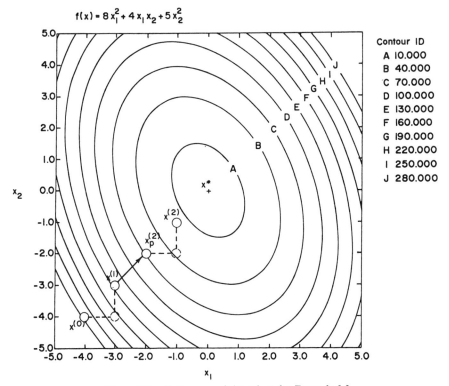

**Figure 3.7.** Pattern search iterations for Example 3.3.

when a pattern move is a success, why not *completely* exploit the direction by conducting a line search along the pattern direction, or at least trying pattern steps of increasing magnitude? This can often significantly accelerate the convergence of the method. Emery and O'Hagan [12] altered the exploratory phase by using a set of orthogonal search directions whose orientation was redirected randomly after each iteration. Rosenbrock [13] developed a method that, like HJ, generated a new search direction based on the accumulated progress of recent iterations. However, unlike HJ, Rosenbrock's approach continually changed the set of direction vectors used in the exploratory phase by a process of orthogonalizations. Another technique, reported by Swann [14] and sometimes called the DSC multivariate search method, used a strategy similar to Rosenbrock's. Rather than relying on only simple increments for each direction, a complete line search is performed in each direction. Each of the many variants claims to offer advantages over its class siblings in selected applications. However, it is questionable whether the added complications are justifiable. If a more sophisticated algorithm is in fact to be used, then a better algorithm than these heuristic methods is available, namely, that due to Powell.

### 3.2.3  Powell's Conjugate Direction Method

To date the most successful of the direct search algorithms is the method due to Powell [15], especially with the modifications suggested by Zangwill [16] and Brent [17]. This algorithm effectively uses the history of the iterations to build up directions for acceleration and at the same time avoids degenerating to a sequence of coordinate searches. It is based upon the model of a quadratic objection, and thus has a theoretical basis.

There are two reasons for choosing a quadratic model:

1. It is the simplest type of nonlinear function to minimize (linear functions do not take on interior optima), and hence any general technique must work well on a quadratic if it is to have any success with a general function.

2. Near the optimum, all nonlinear functions can be approximated by a quadratic (since in the Taylor expansion, the linear part must vanish). Hence, the behavior of the algorithm on the quadratic will give some indication of how the algorithm will converge for general functions.

The motivation for the algorithm stems from the observation that if a quadratic function in $N$ variables can be transformed so that it is just the sum of perfect squares, then the optimum can be found after exactly $N$ single-variable searches, one with respect to each of the transformed variables.

The process of transforming a quadratic function

$$q(x) = a + b^T x + \tfrac{1}{2} x^T C x \qquad (3.25)$$

into a sum of perfect squares is equivalent to finding a transformation matrix $T$ such that the quadratic term is reduced to diagonal form. Thus, given the quadratic form

$$Q(x) = x^T C x \qquad (3.26)$$

the desired transformation

$$x = Tz \qquad (3.27)$$

will yield

$$Q(x) = z^T T^T C T z = z^T D z \qquad (3.28)$$

where $D$ is a diagonal matrix, that is, its elements are nonzero only if $i = j$.

Let $t_j$ be the $j$th column of $T$. Then the transformation in (3.27) expresses the fact that we are rewriting each vector $x$ as a linear combination of the column vectors $t_j$:

$$x = \mathbf{T}z = t_1 z_1 + t_2 z_2 + \cdots + t_N z_N \qquad (3.29)$$

In other words, instead of writing $x$ in terms of the standard coordinate system represented by the set of vectors $e^{(i)}$, we are expressing it in terms of a new coordinate system given by the set of vectors $t_j$. Moreover, since they diagonalize the quadratic, the set of vectors $t_j$ correspond to the principal axes of the quadratic form. Graphically, this corresponds to taking a general quadratic function with cross terms, as shown in Figure 3.8, and realigning the new coordinate axes to coincide with the major and minor axes of the quadratic, as shown in Figure 3.9.

To summarize, by taking the quadratic through the transformation we are really choosing a new coordinate system for the quadratic that coincides with the principal axes of the quadratic. Consequently, single-variable searches for the optimum performed in the space of the transformed or $z$ variables simply correspond to single-variable searches along each of the principal axes of the quadratic. Since the principal axes are the same as the vectors $t_j$, the single-

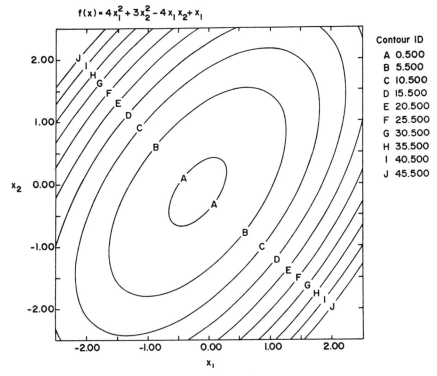

**Figure 3.8.** A quadratic with cross terms.

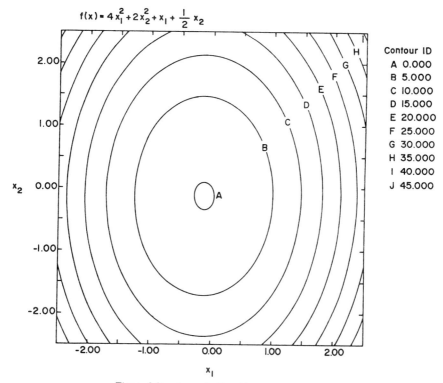

$$f(x) = 4x_1^2 + 2x_2^2 + x_1 + \frac{1}{2}x_2$$

| Contour ID | |
| --- | --- |
| A | 0.000 |
| B | 5.000 |
| C | 10.000 |
| D | 15.000 |
| E | 20.000 |
| F | 25.000 |
| G | 30.000 |
| H | 35.000 |
| I | 40.000 |
| J | 45.000 |

**Figure 3.9.** A quadratic without cross terms.

variable searches are actually performed along each of these vectors. Let us illustrate with an example.

### *Example 3.4   Sum-of-Squares Transformation*

Consider the function

$$f(x) = 4x_1^2 + 3x_2^2 - 4x_1x_2 + x_1$$

and the transformation

$$x_1 = z_1 + \tfrac{1}{2}z_2 \qquad x_2 = z_2$$

or

$$x = \begin{bmatrix} 1 & \frac{1}{2} \\ 0 & 1 \end{bmatrix} z$$

The transformed quadratic becomes

$$f(z) = 4z_1^2 + 2x_2^2 + z_1 + \tfrac{1}{2}z_2$$

Note that this transformation is not unique, since the $t_j$'s are not scaled to be orthonormal. For instance, you can readily verify that the transformation

$$x = \begin{bmatrix} 1 & 0 \\ \tfrac{2}{3} & 1 \end{bmatrix} z$$

will also diagonalize the quadratic function. Given $x^{(0)} = [0,0]^T$ and the two columns of the transformation matrix;

$$t_1 = [1,0]^T \qquad t_2 = [\tfrac{1}{2},1]^T$$

the optimum $[-\tfrac{3}{16}, -\tfrac{1}{8}]^T$ should be found by two successive single-variable searches in the $t_1$ and $t_2$ directions. Searching along $t_1$ first,

$$x^{(1)} = x^{(0)} + \lambda t_1$$

we find $\lambda = -\tfrac{1}{8}$ and $x^{(1)} = [-\tfrac{1}{8}, \ 0]^T$. Now perform the second search along $t_2$ from $x^{(1)}$. Again, $\lambda = -\tfrac{1}{8}$ and $x^{(2)} = [-\tfrac{3}{16}, -\tfrac{1}{8}]^T$, the solution.

Generalizing from the above example and the preceding analysis, if a suitable set of transforming vectors $t_j$; $j = 1,\ldots, N$, conventionally called *conjugate directions*, can be obtained, then the optimum of a quadratic function can be found by exactly $N$ single-variable searches, one along each of the $N$ directions $t_j$; $j = 1,\ldots, N$. The question thus remains how to calculate a set of such vectors. Clearly, if an estimate of the matrix $\mathbf{C}$ were at hand, then the transformation $\mathbf{T}$ could be obtained by Gaussian elimination (as explained in Appendix A) followed by a matrix inversion. Gaussian elimination will yield the factorization

$$\mathbf{C} = \mathbf{P}^T \mathbf{D} \mathbf{P} \qquad\qquad (3.30)$$

hence,

$$(\mathbf{P}^{-1})^T \mathbf{C}(\mathbf{P}^{-1}) = \mathbf{D} \quad \text{and} \quad \mathbf{T} = \mathbf{P}^{-1} \qquad\qquad (3.31)$$

would satisfy our needs. However, an estimate of $\mathbf{C}$ is not available in our case, because we are seeking to develop a method for optimizing $f(x)$ that uses only function values, not first derivatives and certainly not second derivatives. Fortunately, a set of conjugate directions can nonetheless be obtained by using only function values on the basis of the following elementary property of quadratic functions.

### *Parallel Subspace Property*

Given a quadratic function $q(x)$, two arbitrary but distinct points $x^{(1)}$ and $x^{(2)}$, and a direction $d$; if $y^{(1)}$ is the solution to min $q(x^{(1)} + \lambda d)$, and $y^{(2)}$ is the solution to min $q(x^{(2)} + \lambda d)$, then the direction $(y^{(2)} - y^{(1)})$ is $\mathbf{C}$ conjugate to $d$.

In two dimensions, this property is illustrated in Figure 3.10, where we see that a single-variable search from $y^{(1)}$ or $y^{(2)}$ along the direction $(y^{(2)} - y^{(1)})$ will produce the minimum. Thus, it can be concluded that in two dimensions by performing three single-variable searches, a set of conjugate directions can be generated, and in addition the optimum of the quadratic can be obtained. Before proceeding to exploit this construction further, we pause to develop the proof of the above very important property.

First, recall that by definition a set of $\mathbf{C}$ conjugate directions is a set of column vectors with the property that the matrix $\mathbf{T}$ formed from these columns diagonalizes $\mathbf{C}$:

$$\mathbf{T}^T \mathbf{C} \mathbf{T} = \mathbf{D} \tag{3.32}$$

Alternatively, since all off-diagonal elements of $\mathbf{D}$ are zero, this means that

$$t_j^T \mathbf{C} t_k = \begin{cases} d_{kk} & \text{if } j = k \\ 0 & \text{if } j \neq k \end{cases} \tag{3.33}$$

where $t_i$ is the $i$th column of $\mathbf{T}$. Hence, we are able now to give the more conventional, equivalent, and possibly more useful, definition of conjugacy.

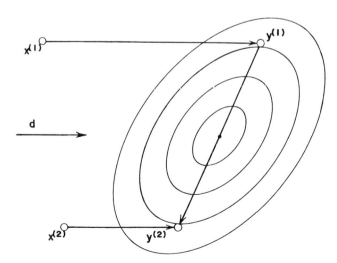

**Figure 3.10.** Conjugacy in two dimensions.

### Conjugate Directions

Given an $N \times N$ symmetric matrix $C$, the directions $s^{(1)}, s^{(2)}, s^{(3)}, \ldots, s^{(r)}$; $r \leqslant N$ are said to be $C$ conjugate if the directions are linearly independent, and

$$s^{(i)T} C s^{(j)} = 0 \qquad \text{for all } i \neq j \qquad (3.34)$$

Consider once again the general quadratic function given earlier:

$$q(x) = a + b^T x + \tfrac{1}{2} x^T C x$$

The points along the direction $d$ from $x^{(1)}$ are

$$x = x^{(1)} + \lambda d$$

and the minimum of $q(x)$ along $d$ is obtained by finding $\lambda^*$ such that $\partial q / \partial \lambda = 0$. The derivative is calculated using the chain rule:

$$\frac{\partial q}{\partial \lambda} = \frac{\partial q}{\partial x} \frac{\partial x}{\partial \lambda} = (b^T + x^T C)\, d \qquad (3.35)$$

By hypothesis the minimum occurs at $y^{(1)}$; hence,

$$\left[ (y^{(1)})^T C + b^T \right] d = 0 \qquad (3.36)$$

Similarly, since the minimum of $q(x)$ along $d$ from $x^{(2)}$ is attained at $y^{(2)}$, we have

$$\left[ (y^{(2)})^T C + b^T \right] d = 0 \qquad (3.37)$$

Subtracting Eq. (3.36) from Eq. (3.37), the result is

$$\left( y^{(2)} - y^{(1)} \right)^T C d = 0 \qquad (3.38)$$

Accordingly, by definition, the directions $d$ and $(y^{(2)} - y^{(1)})$ are $C$ conjugate, and the *parallel subspace property* of quadratic functions has been verified.

### Example 3.5   Minimization via the Parallel Subspace Property

We illustrate the above property by considering once again the quadratic

$$q(x) = 4x_1^2 + 3x_2^2 - 4x_1 x_2 + x_1$$

Suppose the two points $x^{(1)} = [0, 0]^T$, $x^{(2)} = [1, 0]^T$ and the direction $d = [1, 1]^T$

are given. The first search follows the line

$$x = [0,0]^T + \lambda[1,1]^T$$

and yields the point ($\lambda^* = -\frac{1}{6}$): $y^{(1)} = [-\frac{1}{6}, -\frac{1}{6}]^T$. The second search follows the line

$$x = [1,0]^T + \lambda[1,1]^T$$

and yields the point ($\lambda^* = -\frac{5}{6}$): $y^{(2)} = [\frac{1}{6}, -\frac{5}{6}]^T$. According to the parallel subspace property,

$$y^{(2)} - y^{(1)} = [\frac{1}{6}, -\frac{5}{6}]^T - [-\frac{1}{6}, -\frac{1}{6}]^T = [\frac{1}{3}, -\frac{2}{3}]^T$$

is conjugate to $d = [1,1]^T$. By definition, this means that $[1,1]C[\frac{1}{3}, -\frac{2}{3}]^T = 0$; which it is.

Note that according to our previous observation, since we are dealing with a two-variable function, if a single-variable search is now performed along the line $(y^{(2)} - y^{(1)})$ the optimum of $q(x)$ will be found. This is readily verified, since the minimum along the line

$$x = [-\frac{1}{6}, -\frac{1}{6}]^T + \lambda[\frac{1}{3}, -\frac{2}{3}]^T$$

is ($\lambda^* = -\frac{1}{16}$): $x^* = [-\frac{3}{16} - \frac{1}{8}]^T$, which is the solution previously obtained.

In the previous construction, we require the initial specification of two points and a direction in order to produce a single conjugate direction. This is clumsy from a computational point of view, and we would prefer to generate the conjugate directions from a single starting point. This is easily accomplished by employing the coordinate unit vectors $e^{(1)}, e^{(2)}, e^{(3)}, \ldots, e^{(N)}$, as follows. (Here we demonstrate the procedure in two dimensions; the extension to $N$ dimensions should be obvious.) Let $e^{(1)} = [1,0]^T$ and $e^{(2)} = [0,1]^T$. Given a starting point $x^{(0)}$, calculate $\lambda^{(0)}$ such that $f(x^{(0)} + \lambda^{(0)}e^{(1)})$ is minimized. Let

$$x^{(1)} = x^{(0)} + \lambda^{(0)}e^{(1)}$$

and calculate $\lambda^{(1)}$ so that $f(x^{(1)} + \lambda^{(1)}e^{(2)})$ is minimized, and let

$$x^{(2)} = x^{(1)} + \lambda^{(1)}e^{(2)}$$

Next calculate $\lambda^{(2)}$ so that $f(x^{(2)} + \lambda^{(2)}e^{(1)})$ is minimized, and let

$$x^{(3)} = x^{(2)} + \lambda^{(2)}e^{(1)}$$

Then, the directions $(x^{(3)} - x^{(1)})$ and $e^{(1)}$ will be conjugate. To see this, consider Figure 3.11. Note that the point $x^{(1)}$ is found by searching along $e^{(1)}$

from $x^{(0)}$, and $x^{(3)}$ is found by searching along $e^{(1)}$ from $x^{(2)}$. Consequently, by the parallel subspace property, the directions $e^{(1)}$ and $(x^{(3)} - x^{(1)})$ are conjugate. Furthermore, if we continue the iteration with a search along the line $(x^{(3)} - x^{(1)})$, we will have searched along two conjugate directions, and since we assume that $f(x)$ is a two-dimensional quadratic, the solution $x^*$ will be found.

The preceding construction and the conjugacy property upon which it is based were stated in terms of only two conjugate directions. However, both can be readily extended to more than two conjugate directions and hence to higher dimensions. In particular, it is easy to show that if from $x^{(1)}$ the point $y^{(1)}$ is found after searches along each of $M(< N)$ conjugate directions; and, similarly, if from $x^{(2)}$ the point $y^{(2)}$ is found after searches along the same $M$ conjugate directions, $s^{(1)}, s^{(2)}, s^{(3)}, \ldots, s^{(M)}$; then the vector $(y^{(2)} - y^{(1)})$ will be conjugate to *all* of the $M$ previous directions. This is known as the *extended parallel subspace property*. Using this extension, the construction shown in Figure 3.11 can immediately be generalized to higher dimensions. A three-dimensional case is shown in Figure 3.12.

In the construction shown, the search begins with the three coordinate directions $e^{(1)}$, $e^{(2)}$, and $e^{(3)}$, and these are replaced in turn by newly generated conjugate directions. From $x^{(0)}$ a cycle of line searches is made along $e^{(3)}$, $e^{(1)}$, $e^{(2)}$, and again $e^{(3)}$. At the conclusion of this cycle the directions $e^{(3)}$ and $(x^{(4)} - x^{(1)})$ will be conjugate. The new search direction designated ④ is Figure 3.12 now replaces $e^{(1)}$. A new cycle of line searches is executed using

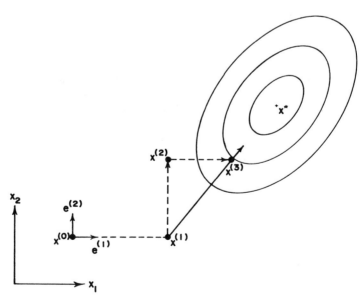

**Figure 3.11.** Conjugacy from a single point.

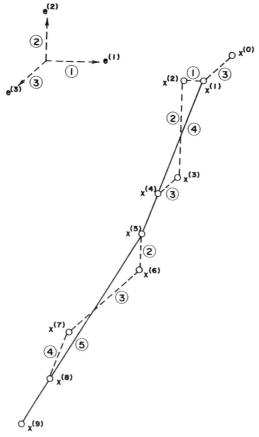

**Figure 3.12.** Construction of conjugate directions in three dimensions.

directions ④, $e^{(2)}$, $e^{(3)}$, and again ④. Once again, by the extended parallel subspace property, the new direction $(x^{(8)} \quad x^{(5)})$, designated ⑤ in the figure, will be conjugate not only to ④ but also to $e^{(3)}$. Hence, the set of directions $e^{(3)}$, $(x^{(4)} - x^{(1)})$, and $(x^{(8)} - x^{(5)})$ are mutually conjugate. Therefore, if one additional line search is executed from $x^{(8)}$ along $(x^{(8)} - x^{(5)})$—that is, ⑤ in the figure—the point $x^{(9)}$ is found, which must be the optimum if $f(x)$ is a three-dimensional quadratic, since we have searched three mutually conjugate directions in turn. Thus we see that in three dimensions, nine line searches using only function values are required to determine exactly (assuming exact arithmetic, of course) the optimum of a quadratic function. The construction is easily generalized and will in $N$ dimensions require $N^2$ line searches to optimize a quadratic function. The generalization to $N$ dimensions is outlined below.

## Powell's Conjugate Direction Method

**Step 1.** Define $x^{(0)}$, the starting point, and a set of $N$ linearly independent directions; possibly $s^{(i)} = e^{(i)}$, $i = 1, 2, 3, \ldots, N$.

**Step 2.** Minimize along the $N + 1$ directions, using the previous minimum to begin the next search and letting $s^{(N)}$ be the first and last searched.

**Step 3.** Form the new conjugate direction using the extended parallel subspace property.

**Step 4.** Delete $s^{(1)}$ and replace it with $s^{(2)}$, and so on. Place the new conjugate direction in $s^{(N)}$. Go to step 2.

In order for the method (as described here) to be of practical use, a convergence check must be inserted as well as a test to ensure linear independence of the direction set. The independence check is particularly important when $f(x)$ is not quadratic. [16, 17]

It follows from the way the algorithm is constructed that if the function is quadratic and has a minimum, then this minimum will be reached in exactly $N$ loops through steps 2 through 4, where $N$ is the number of variables. If the function is not quadratic, then of course more than $N$ loops will be required. However, it can be proven rigorously under reasonable hypothesis that the method will converge to a local minimum and will do this at a *superlinear* rate (as defined below).

**Rate of Convergence.** The methods considered in this text generate a sequence of iterates $x^{(k)}$ that approximate the solution $x^*$. We say a method is *convergent* if

$$\frac{\|\varepsilon^{(k+1)}\|}{\|\varepsilon^{(k)}\|} \leqslant 1 \tag{3.39}$$

holds at each iteration, where

$$\varepsilon^{(k)} = x^{(k)} - x^* \tag{3.40}$$

Since we are performing calculations using finite arithmetic, even the best algorithms are doomed to generate an infinite sequence of iterates. Accordingly, we are interested in the asymptotic convergence properties of these methods. Therefore, by convention we say (see Gill et al. [18] or Fletcher [19] for more complete discussions) that an algorithm has $r$ order (or rate) of convergence if

$$\lim_{k \to \infty} \frac{\|\varepsilon^{(k+1)}\|}{\|\varepsilon^{(k)}\|^r} = C \tag{3.41}$$

where $C$ is the error constant. From the above we see that $C \leqslant 1$ when $r = 1$. When $r = 1$ we say the algorithm exhibits a *linear convergence rate*, and when $r = 2$ we say the rate is *quadratic*. When $r = 1$ and $C = 0$, the algorithm has a *superlinear* rate of convergence.

*Example 3.6  Powell's Conjugate Direction Method.*   Find the minimum of

$$f(x) = 2x_1^3 + 4x_1x_2^3 - 10x_1x_2 + x_2^2$$

from the starting point $x^{(0)} = [5, 2]^T$, where $f(x^{(0)}) = 314$.

**Step 1.**  $s^{(1)} = [1, 0]^T$    $s^{(2)} = [0, 1]^T$

**Step 2.**  (a) Find $\lambda$ such that $f(x^{(0)} + \lambda s^{(2)}) \to$ min. The result is $\lambda^* = -0.81$, so

$$x^{(1)} = [5, 2]^T - .81[0, 1]^T = [5, 1.19]^T \qquad f(x^{(1)}) = 250$$

(b) Find $\lambda$ such that $f(x^{(1)} + \lambda s^{(1)}) \to$ min.

$$\lambda^* = -3.26 \qquad x^{(2)} = [1.74, 1.19]^T \qquad f(x^{(2)}) = 1.10$$

(c) Find $\lambda$ such that $f(x^{(2)} + \lambda s^{(2)}) \to$ min.

$$\lambda^* = -0.098 \qquad x^{(3)} = [1.74, 1.092] \qquad f(x^{(3)}) = 0.72$$

**Step 3.**  Let $s^{(3)} = x^{(3)} - x^{(1)} = [-3.26, -.098]^T$. Normalize.

$$s^{(3)} = \frac{s^{(3)}}{\|s^{(3)}\|} = [-0.99955, -0.03]^T$$

Let $s^{(1)} = s^{(2)}$; $s^{(2)} = s^{(3)}$, and go to step 2.

**Step 4.**  Find $\lambda$ such that $f(x^{(3)} + \lambda s^{(2)}) \to$ min.

$$\lambda^* = 0.734: \qquad x^{(4)} = [1.006, 1.070]^T \qquad f(x^{(4)}) = -2.86$$

*Note:*  If $f(x)$ were a quadratic, the problem would be solved at this point (ignoring round-off error); since it is not, the iteration continues to the solution.

The search directions and iterates are shown in Figure 3.13.

All available computational evidence indicates that Powell's method (with the linear dependence check added) is at least as reliable as any other direct search method and is usually much more efficient. Consequently, if a sophisticated algorithm is in order, this method is clearly to be recommended.

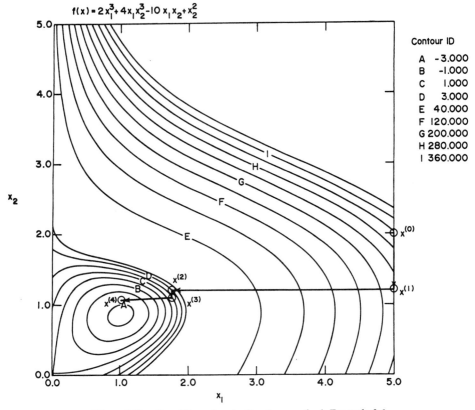

$$f(x) = 2x_1^3 + 4x_1 x_2^3 - 10 x_1 x_2 + x_2^2$$

**Figure 3.13.** Powell's conjugate directions method, Example 3.6.

This completes our discussion of direct search methods for unconstrained functions. In the next section we go on to consider methods that employ derivative information.

## 3.3   GRADIENT-BASED METHODS

In the previous section we examined several methods that require only objective function values to proceed toward the solution. These direct methods are important, because very often in a practical engineering problem this is the only reliable information available. On the other hand, even the best direct methods can require an excessive number of function evaluations to locate the solution. This, combined with the quite natural desire to seek stationary points [that is, points that satisfy the first necessary condition, (3.15a)], motivates us to consider methods that employ gradient information. The methods will be iterative, since the elements of the gradient will in general be nonlinear functions of the design variables.

We shall assume throughout that $f(x)$, $\nabla f(x)$, and $\nabla^2 f(x)$ exist and are continuous. We briefly consider methods that require both first and second derivatives, but primarily as a point of reference and as motivation for the more useful methods. In particular, we will examine in detail *conjugate gradient* methods, which exploit the conjugacy concept previously discussed but by use of gradient information, and *quasi-Newton* methods, which mimic Newton's method but once again require only first-order information. We assume that the elements of the gradient are available in closed form or can be reliably approximated numerically. In addition, we give suggestions for making numerical approximations to the gradient.

All of the methods considered here employ a similar iteration procedure:

$$x^{(k+1)} = x^{(k)} + \alpha^{(k)} s(x^{(k)}) \tag{3.42}$$

where $x^{(k)}$ = current estimate of $x^*$, the solution

$\alpha^{(k)}$ = step-length parameter

$s(x^{(k)}) = s^{(k)}$ = search direction in the $N$ space of the design variables $x_i$;
$\quad i = 1, 2, 3, \ldots, N$

The manner in which $s(x)$ and $\alpha$ are determined at each iteration specifies a particular method. Usually $\alpha^{(k)}$ is selected so as to minimize $f(x)$ in the $s(x^{(k)})$ direction. Therefore, we will require efficient single-variable minimization algorithms to implement these methods.

### 3.3.1 Cauchy's Method [20]

Let us assume for the moment that we are stationed at the point $\bar{x}$ in the design space and wish to determine the direction of "most local descent." As before, consider the Taylor expansion of the objective about $\bar{x}$:

$$f(x) = f(\bar{x}) + \nabla f(\bar{x})^T \Delta x + \cdots \tag{3.43}$$

where we ignore terms of degree 2 and higher. Clearly the second term will dictate the local decrease in the objective value, since $f(\bar{x})$ is fixed. The greatest decrease in $f$ is associated with the choice of direction in (3.42) that produces the greatest *negative* scalar product in the second term. The greatest negative scalar product results from the choice

$$s(\bar{x}) = -\nabla f(\bar{x}) \tag{3.44}$$

clearly from the projection property of scalar products. The second term then becomes

$$-\alpha \nabla f(\bar{x})^T \nabla f(\bar{x})$$

producing the most local descent. This is the motivation for the *simple gradient method*:

$$x^{(k+1)} = x^{(k)} - \alpha \nabla f(x^{(k)}) \qquad (3.45)$$

where $\alpha$ is a fixed positive parameter. The method has two disadvantages: the need to make an appropriate choice for $\alpha$, and inherent sluggishness near the minimum due to the corrections vanishing as $\nabla f$ goes to zero.

Accordingly, we are led to adjust $\alpha$ at each iteration:

$$x^{(k+1)} = x^{(k)} - \alpha^{(k)} \nabla f(x^{(k)}) \qquad (3.46)$$

We determine $\alpha^{(k)}$ such that $f(x^{(k+1)})$ is a minimum along $\nabla f(x^{(k)})$, using an appropriate single-variable searching method. We call this the method of steepest descent or *Cauchy's method*, since Cauchy first employed a similar technique to solve systems of linear equations.

The line search implied in (3.46) will make the method more reliable than the simple gradient method, but still the rate of convergence will be intolerably slow for most practical problems. It seems reasonable that the method will approach the minimum slowly, since the change in the variables is directly related to the magnitude of the gradient, which is itself going to zero. There is, then, no mechanism to produce an acceleration toward the minimum in the final iterations. One major advantage of Cauchy's method is its surefootedness. The method possesses a *descent property*, since by the very nature of the iteration we are assured that

$$f(x^{(k+1)}) \leqslant f(x^{(k)}) \qquad (3.47)$$

With this in mind we note that the method typically produces good reduction in the objective from points far from the minimum. It is then a logical starting procedure for all gradient-based methods. Finally, the method does demonstrate many of the essential ingredients of all methods in this class.

### *Example 3.7   Cauchy's Method*

Consider the function

$$f(x) = 8x_1^2 + 4x_1 x_2 + 5x_2^2$$

to which we wish to apply Cauchy's method.

**Solution.**   To begin, calculate the elements of the gradient;

$$\frac{\partial f}{\partial x_1} = 16x_1 + 4x_2 \qquad \frac{\partial f}{\partial x_2} = 10x_2 + 4x_1$$

## Table 3.1   Cauchy Method

| $k$ | $x_1^{(k)}$ | $x_2^{(k)}$ | $f(x^{(k)})$ |
|---|---|---|---|
| 1 | $-1.2403$ | 2.1181 | 24.2300 |
| 2 | 0.1441 | 0.1447 | 0.3540 |
| 3 | $-0.0181$ | 0.0309 | 0.0052 |
| 4 | 0.0021 | 0.0021 | 0.0000 |

Now to use the method of steepest descent, we estimate the solution,

$$x^{(0)} = [10, 10]^T$$

and generate a new estimate using expression (3.46):

$$x^{(1)} = x^{(0)} - \alpha^{(0)} \nabla f(x^{(0)})$$

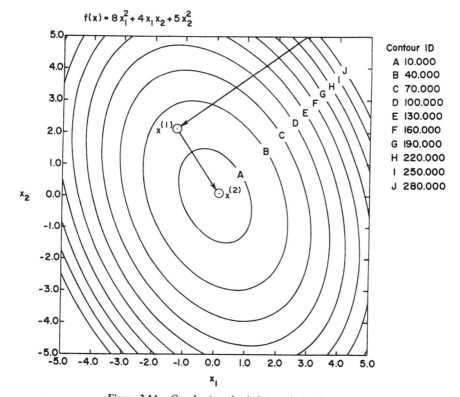

Figure 3.14.   Cauchy/quadratic interpolation iterates.

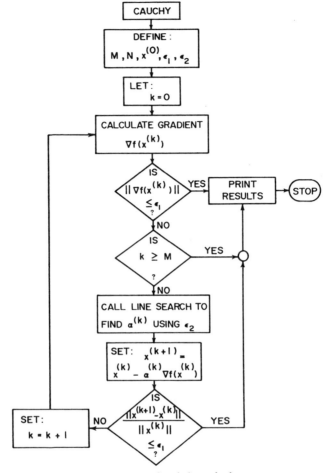

**Figure 3.15.** Cauchy's method.

Now we choose $\alpha^{(0)}$ such that $f(x^{(1)}) \rightarrow$ min; $\alpha^{(0)} = 0.056$. Therefore, $x^{(1)} = [-1.20, 2.16]^T$. Next we calculate

$$x^{(2)} = x^{(1)} - \alpha^{(1)} \nabla f\left(x^{(1)}\right)$$

by evaluating the gradient at $x^{(1)}$ and performing the line search. If we perform the line search using the quadratic interpolation method [21, 22], we get the iterates given in Table 3.1 and displayed in Figure 3.14.

Cauchy's method, even though not of great practical value, exhibits the major steps of most gradient-based methods. A flowchart of the algorithm is given in Figure 3.15. Note that termination tests include the length of the gradient and $\Delta x$ vectors.

### 3.3.2 Newton's Method [23]

We have seen that Cauchy's method is the "best" local gradient-based strategy, but observe that the negative gradient points directly toward the minimum only when the contours of $f$ are circular, and therefore the negative gradient is *not* a good *global* direction (in general) for nonlinear functions. Cauchy's method employs successive linear approximations to the objective and requires values of the objective and gradient at each iteration. We are then led to consider the use of higher order information, namely second derivatives, in an effort to construct a more global strategy.

Consider once again the Taylor expansion of the objective:

$$f(x) = f(x^{(k)}) + \nabla f(x^{(k)})^T \Delta x + \tfrac{1}{2} \Delta x^T \nabla^2 f(x^{(k)}) \, \Delta x + O(\Delta x^3)$$

We form a quadratic approximation to $f(x)$ by dropping terms of order 3 and above:

$$\tilde{f}(x; x^{(k)}) = f(x^{(k)}) + \nabla f(x^{(k)})^T \Delta x + \tfrac{1}{2} \Delta x^T \nabla^2 f(x^{(k)}) \, \Delta x \quad (3.48)$$

where we use $\tilde{f}(x; x^{(k)})$ to denote an *approximating function* constructed at $x^{(k)}$, which is itself a function of $x$. Now let us use this quadratic approximation of $f(x)$ to form an iteration sequence by forcing $x^{(k+1)}$, the next point in the sequence, to be a point where the gradient of the *approximation* is zero. Therefore,

$$\nabla \tilde{f}(x; x^{(k)}) = \nabla f(x^{(k)}) + \nabla^2 f(x^{(k)}) \, \Delta x = 0 \quad (3.49)$$

so

$$\Delta x = - \nabla^2 f(x^{(k)})^{-1} \nabla f(x^{(k)}) \quad (3.50)$$

Accordingly, this successive quadratic approximation scheme produces *Newton's optimization method*:

$$x^{(k+1)} = x^{(k)} - \nabla^2 f(x^{(k)})^{-1} \nabla f(x^{(k)}) \quad (3.51)$$

which we demonstrate in the following example.

### *Example 3.8 Newton's Method*

Consider once again the function given in the previous example:

$$f(x) = 8x_1^2 + 4x_1 x_2 + 5x_2^2$$

where

$$\nabla f(x) = [16x_1 + 4x_2, 10x_2 + 4x_1]^T$$

and

$$\nabla^2 f(x) = H = \begin{bmatrix} 16 & 4 \\ 4 & 10 \end{bmatrix}$$

Using (3.51), with $x^{(0)} = [10, 10]^T$ as before, we get

$$x^{(1)} = [10, 10]^T - \left(\frac{1}{144}\right)\begin{bmatrix} 10 & -4 \\ -4 & 16 \end{bmatrix}[200, 140]^T$$

and therefore

$$x^{(1)} = [10, 10]^T - \left(\frac{1}{144}\right)[1440, 1440]^T = [0, 0]^T$$

which is the exact solution.

We see, as we might expect from the previous example, that Newton's method will minimize a quadratic function (from any starting point) in exactly one step.

**Convergence Properties.** It is worthwhile to consider the convergence properties of Newton's method. Mangasarian [24] has shown, under rather mild regularity conditions on $f(x)$, that Newton's method exhibits a quadratic rate of convergence:

$$\|\varepsilon^{(k+1)}\| \leq C\|\varepsilon^{(k)}\|^2 \tag{3.52}$$

where the error constant $C$ is related to the condition of the Hessian $\nabla^2 f$. Clearly, Newton's method will converge whenever $x^{(0)}$ is chosen such that

$$\|\varepsilon(x^{(0)})\| < \frac{1}{C} \tag{3.53}$$

We might have expected the quadratic convergence rate, since the method is based on a quadratic approximation. For arbitrary functions it is entirely feasible to assume that we might choose a point such that $\|\varepsilon(x^{(0)})\| > 1/C$, and the method may perform poorly. Note that the algorithm does not possess a descent property. To see this more clearly, assume the current iterate $\bar{x}$ is not a stationary point (i.e., $\nabla f(\bar{x}) \neq 0$), and examine the projection of the Newton direction onto the local gradient. By definition, the search direction is a descent direction if

$$\nabla f(\bar{x})^T s(\bar{x}) < 0 \tag{3.54}$$

Therefore, from (3.51) this leads to

$$-\nabla f(\bar{x})^T \nabla^2 f(\bar{x})^{-1} \nabla f(\bar{x}) < 0 \tag{3.55}$$

Clearly this condition is satisfied, and the Newton direction is "downhill" if $\nabla^2 f(\bar{x})$ is positive definite. If we encounter a point where $\nabla^2 f(\bar{x})$ is negative definite, then the direction points "uphill"; and when the Hessian is indefinite, there is no assurance of ascent or descent.

### 3.3.3 Modified Newton's Method

Experience has shown that Newton's method can be unreliable for non-quadratic functions. That is, the Newton step will often be large when $x^{(0)}$ is far from $x^*$, and there is the real possibility of divergence. It is possible to modify the method in a logical and simple way to ensure descent by adding a line search as in Cauchy's method. That is, we form the sequence of iterates:

$$x^{(k+1)} = x^{(k)} - \alpha^{(k)} \nabla^2 f(x^{(k)})^{-1} \nabla f(x^{(k)}) \tag{3.56}$$

by choosing $\alpha^{(k)}$ such that

$$f(x^{(k+1)}) \to \min$$

which ensures that

$$f(x^{(k+1)}) \leqslant f(x^{(k)})$$

This is the *modified Newton's method*, and we find it reliable and efficient when first and second derivatives are accurately and inexpensively calculated. The major difficulty is the need to calculate and solve the linear equation involving the Hessian $\nabla^2 f(\bar{x})$ at each iteration.

### 3.3.4 Marquardt's Method

This method combines Cauchy's and Newton's methods in a convenient manner that exploits the strengths of both but *does* require second-order information. We have previously noted that the gradient is the direction of most local ascent and typically provides a good reduction in the objective when $x^{(0)}$ is far from $x^{(*)}$. On the other hand, Newton's method generates ideal search directions near the solution. It is this simple idea that motivates the method that Marquardt introduced in 1963 [25]. He specified the search direction to be

$$s(x^{(k)}) = -[\mathbf{H}^{(k)} + \lambda^{(k)}\mathbf{I}]^{-1} \nabla f(x^{(k)}) \tag{3.57}$$

and set $\alpha^{(k)} = +1$ in (3.42), since $\lambda$ is used to control both the direction of search and the length of the step. $I$ is the identity matrix, that is, it contains zeros except for the diagonal elements, which are $+1$. To begin the search, let

$\lambda^{(0)}$ be a large constant, say $10^4$, such that

$$[\mathbf{H}^{(0)} + \lambda^{(0)}\mathbf{I}]^{-1} = [\lambda^{(0)}\mathbf{I}]^{-1} = \left(\frac{1}{\lambda^{(0)}}\right)\mathbf{I} \qquad (3.58)$$

Therefore, for $\lambda^{(0)}$ sufficiently large, $s(x^{(0)}) \rightarrow -\nabla f(x^{(0)})$. We notice from (3.57) that as $\lambda$ decreases from a large value to zero, $s(x)$ goes from the gradient to the Newton direction. Accordingly, we test descent after the first step, and when the step is a success [that is, $f(x^{(1)}) < f(x^{(0)})$], we let $\lambda^{(1)} < \lambda^{(0)}$ and step again; otherwise we set $\lambda^{(0)} = \beta\lambda^{(0)}$, where $\beta > 1$, and perform the previous step again. The algorithm is summarized below.

### Marquardt's Compromise

**Step 1.**  Define $x^{(0)}$ = initial estimate of $x^*$

$M$ = maximum number of iterations allowed

$\varepsilon$ = convergence criterion

**Step 2.**  Set $k = 0$. $\lambda^{(0)} = 10^4$.

**Step 3.**  Calculate $\nabla f(x^{(k)})$.

**Step 4.**  Is $\|\nabla f(x^{(k)})\| < \varepsilon$?

Yes:  Go to step 11.

No:   Continue.

**Step 5.**  Is $k \geqslant M$?

Yes:  Go to step 11.

No:   Continue.

**Step 6.**  Calculate $s(x^{(k)}) = -[\mathbf{H}^{(k)} + \lambda^{(k)}\mathbf{I}]^{-1}\nabla f(x^{(k)})$.

**Step 7.**  Set $x^{(k+1)} = x^{(k)} + s(x^{(k)})$.

**Step 8.**  Is $f(x^{(k+1)}) < f(x^{(k)})$?

Yes:  Go to step 9.

No:   Go to step 10.

**Step 9.**  Set $\lambda^{(k+1)} = \frac{1}{2}\lambda^{(k)}$ and $k = k + 1$. Go to step 3.

**Step 10.**  Set $\lambda^{(k)} = 2\lambda^{(k)}$. Go to step 6.

**Step 11.**  Print results and stop.

The major advantages of the method are its simplicity, the descent property, the excellent convergence rate near $x^*$, and the absence of a line search. The major disadvantage is the need to calculate $\mathbf{H}^{(k)}$ and then solve the linear equation set corresponding to (3.57). This method has been used quite extensively with problems where $f(x)$ is a sum of squares, that is,

$$f(x) = f_1^2(x) + f_2^2(x) + \cdots + f_m^2(x) \qquad (3.59)$$

(In fact this is the problem originally considered by Marquardt.) Powell [26] and Bard [27] give numerical results that indicate that the method is particularly attractive for this application.

### 3.3.5 Conjugate Gradient Methods

In the previous sections we have discussed Cauchy's and Newton's methods. We have seen that Cauchy's method tends to be effective far from the minimum but becomes less so as $x^*$ is approached, whereas Newton's method can be unreliable far from $x^*$ but is very efficient as $x^{(k)}$ approaches the minimum. In this and the following sections we discuss methods that tend to exhibit the positive characteristics of the Cauchy and Newton methods, using only first-order information. That is, the methods to be considered here will be reliable far from $x^*$ and will accelerate as the sequence of iterates approaches the minimum.

We call these methods *quadratically convergent* if they terminate in approximately $N$ steps when applied to a quadratic function and exact arithmetic is employed. One such class of methods are those that employ *conjugate directions*. We have previously given the conjugacy conditions for a set of directions $s^{(k)}$: $k = 1, 2, 3, \ldots, r \leqslant N$; and an $N \times N$ symmetric matrix $\mathbf{C}$. In addition, we have examined these directions as transformations into the form of a sum of perfect squares for an arbitrary quadratic function. In this way we concluded that searching in each of $N$ $\mathbf{C}$-conjugate directions in turn would produce the minimum for an $N$-dimensional quadratic. Furthermore, we generated the set of conjugate directions using values of the objective only. Here a quadratic approximation to $f(x)$ and gradient information are employed to generate conjugate directions. In addition, we shall insist upon descent at each iteration.

Consider any two distinct points in the design space $x^{(0)}$ and $x^{(1)}$ and the gradient of the quadratic objective at these points:

$$\nabla f(x) = \nabla q(x) = \mathbf{C}x + b = g(x) \qquad (3.60)$$

where for convenience we use the symbol $g(x)$ for the gradient. Therefore,

$$g(x^{(0)}) = \mathbf{C}x^{(0)} + b$$

$$g(x^{(1)}) = \mathbf{C}x^{(1)} + b$$

Now form the change in the gradient from $x^{(0)}$ to $x^{(1)}$:

$$\Delta g(x) = g(x^{(1)}) - g(x^{(0)}) = C(x^{(1)} - x^{(0)})$$

$$\Delta g(x) = C \Delta x \qquad (3.61)$$

which is the property of quadratic functions we shall exploit.

In 1952, Hestenes and Stiefel [28] published an efficient iterative technique for solving systems of linear equations, which is essentially the method of conjugate gradients. They viewed the set of linear equations as elements of the gradient vector of a quadratic to which they sought the minimum. Later, Fletcher and Reeves [29] proved the quadratic convergence and extended the method to nonquadratic functions. Fried and Metzler [30] have demonstrated the utility of the method when it is applied to linear sets resulting from finite element discretizations, where the coefficient matrix is sparse (see Appendix for definition) but not in an orderly way. They stress the ease of implementing the method as compared to other more commonly used techniques. This is an attractive feature to us as well.

We assume in the development of the method that the objective is quadratic:

$$f(x) = q(x) = a + b^T x + \tfrac{1}{2} x^T C x$$

and the iteration prescription in (3.42) is employed; namely,

$$x^{(k+1)} = x^{(k)} + \alpha^{(k)} s(x^{(k)})$$

Let us form the search directions at each iteration, using

$$s^{(k)} = -g^{(k)} + \sum_{i=0}^{k-1} \gamma^{(i)} s^{(i)} \qquad (3.62)$$

with

$$s^{(0)} = -g^{(0)} \qquad (3.63)$$

where $g^{(k)} = \nabla f(x^{(k)})$. Since we shall perform line searches along each direction in turn as they are formed, it is useful to recall that the following condition holds as a result of termination of the line search:

$$\nabla f(x^{(k+1)})^T s^{(k)} = 0 \qquad (3.64)$$

Now we wish to choose $\gamma^{(i)}$, $i = 1, 2, 3, \ldots, k - 1$, such that $s^{(k)}$ is C-conjugate to all previous search directions. Consider the first direction:

$$s^{(1)} = -g^{(1)} + \gamma^{(0)} s^{(0)} = -g^{(1)} - \gamma^{(0)} g^{(0)}$$

and force conjugacy with $s^{(0)}$ only:

$$s^{(1)T}\mathbf{C}s^{(0)} = 0$$

$$\left[g^{(1)} + \gamma^{(0)}g^{(0)}\right]^T \mathbf{C}s^{(0)} = 0$$

From the previous iteration, we know that

$$s^{(0)} = \frac{\Delta x}{\alpha^{(0)}}$$

and therefore

$$\left[g^{(1)} + \gamma^{(0)}g^{(0)}\right]^T \mathbf{C}\left[\frac{\Delta x}{\alpha^{(0)}}\right] = 0$$

From the quadratic property (3.61) we get

$$\left[g^{(1)} + \gamma^{(0)}g^{(0)}\right]^T \Delta g = 0 \tag{3.65}$$

or

$$\gamma^{(0)} = \frac{-\Delta g^T g^{(1)}}{\Delta g^T g^{(0)}} \tag{3.66}$$

Expansion of (3.65) produces

$$g^{(1)T}g^{(1)} + \gamma^{(0)}g^{(0)T}g^{(1)} - g^{(1)T}g^{(0)} - \gamma^{(0)}g^{(0)T}g^{(0)} = 0$$

but by choice of $\alpha^{(0)}$ and (3.64) we know that

$$g^{(1)T}g^{(0)} = 0$$

so that

$$\gamma^{(0)} = \frac{\|g^{(1)}\|^2}{\|g^{(0)}\|^2} \tag{3.67}$$

Continuing the process, we form the next search direction:

$$s^{(2)} = -g^{(2)} + \gamma^{(0)}s^{(0)} + \gamma^{(1)}s^{(1)}$$

and choose $\gamma^{(0)}$ and $\gamma^{(1)}$ such that

$$s^{(2)T}\mathbf{C}s^{(0)} = 0 \quad \text{and} \quad s^{(2)T}\mathbf{C}s^{(1)} = 0$$

$s^{(2)}$ is C-conjugate to $s^{(0)}$ and $s^{(1)}$. We leave it as an exercise to show that $\gamma^{(0)} = 0$ here and $\gamma^{(i)} = 0$ for $i = 0, 1, 2, \ldots, k - 2$ for all $k$ in general, using the previous reasoning and conditions (3.61) and (3.64). This leads to the general iteration:

$$s^{(k)} = -g^{(k)} + \left[ \frac{\|g^{(k)}\|^2}{\|g^{(k-1)}\|^2} \right] s^{(k-1)} \qquad (3.68)$$

which is the form suggested by Fletcher and Reeves. When $f(x)$ is a quadratic, $N - 1$ such directions and $N$ line searches will be required (in the absence of round-off error) to locate the minimum. When $f(x)$ is not quadratic, additional directions and line searches are required.

Several investigators and some computational experience suggest it is wise to restart [that is, set $s(x) = -g(x)$] every $N$ or $N + 1$ steps. This remains a subject of research interest, and for general functions restarting at the $N$th stage may actually slow convergence. On the other hand, regular restarts do seem to provide a safer procedure, since the generation of a linearly dependent direction is less likely. The generation of a direction that is linearly dependent upon one or more of the previous directions may cause the method to fail, since a subspace of $R^N$ is searched thereafter. Quite happily, this seldom occurs in practice. In fact the method has performed quite well in practice and is especially appealing because of the simplicity of the one-term update required and the small amount of storage needed to advance the search. As the number of variables grows large, the method and its several variants become increasingly attractive due to the small relative storage requirement.

### Example 3.9   The Fletcher-Reeves Method

Find the minimum of the function

$$f(x) = 4x_1^2 + 3x_2^2 - 4x_1x_2 + x_1$$

from $x^{(0)} = [0, 0]^T$.

### Solution

**Step 1.**  $\nabla f(x) = [8x_1 - 4x_2 + 1, \quad 6x_2 - 4x_1]^T$
$s^{(0)} = -\nabla f(x^{(0)}) = -[1, 0]^T$

**Step 2.**  Line search:

$$x^{(1)} = x^{(0)} - \alpha^{(0)} \nabla f(x^{(0)}) \rightarrow \alpha^{(0)} = \tfrac{1}{8}$$

$$x^{(1)} = [0, 0]^T - \tfrac{1}{8}[1, 0] = \left[-\tfrac{1}{8}, 0\right]^T$$

**Step 3.**  $k = 1$

$$s^{(1)} = -\left[0, \tfrac{1}{2}\right]^T - (\tfrac{1}{4})[1, 0]^T = -\left[\tfrac{1}{4}, \tfrac{1}{2}\right]^T$$

**Step 4.** Line search:

$$x^{(2)} = x^{(1)} + \alpha^{(1)}s^{(1)} \rightarrow \alpha^{(1)} = \tfrac{1}{4}$$

$$x^{(2)} = \left[-\tfrac{1}{8}, 0\right]^T - \tfrac{1}{4}\left[\tfrac{1}{4}, \tfrac{1}{2}\right]^T = \left[-\tfrac{3}{16}, -\tfrac{1}{8}\right]^T$$

$$\nabla f(x^{(2)}) = [0,0]^T$$

Therefore, $x^{(2)} = x^*$, and the solution is achieved in two line searches, because they were performed exactly and the objective is quadratic.

Miele and Cantrell [31] have given an extension of the Fletcher-Reeves approach:

$$x^{(k+1)} = x^{(k)} + \alpha^{(k)}\{-\nabla f(x^{(k)}) + \gamma^{(k)}s(x^{(k-1)})\} \qquad (3.69)$$

where $\alpha^{(k)}$ and $\gamma^{(k)}$ are sought directly at each iteration. The method, known as the *memory gradient method*, is remarkably efficient in terms of number of iterations but requires more function and gradient evaluations than Fletcher-Reeves. Accordingly, the method seems useful only when objective and gradient evaluations are very inexpensive.

Recall that in the development of the Fletcher-Reeves method both a quadratic objective and exact line searches were assumed. A host of methods have been proposed that generate conjugate directions while relaxing one or both of these conditions. Possibly the most noteworthy was given by Polak and Ribiere [32] in 1969. This method demands exact line searches but assumes a more general objective model, to give

$$\gamma^{(k)} = \frac{\Delta g(x^{(k)})^T g(x^{(k)})}{\|g(x^{(k-1)})\|^2} \qquad (3.70)$$

where

$$\Delta g(x^{(k)}) = g(x^{(k)}) - g(x^{(k-1)}) \qquad (3.71)$$

as before. Using $\alpha$ supplied by the line search and $\gamma$ in (3.70), we have the *Polak and Ribiere conjugate gradient method*:

$$x^{(k+1)} = x^{(k)} + \alpha^{(k)}s(x^{(k)})$$

$$s(x^{(k)}) = -\nabla f(x^{(k)}) + \gamma^{(k)}s(x^{(k-1)}) \qquad (3.72)$$

Quite obviously, the only difference between the Polak-Ribiere and Fletcher-Reeves methods is the choice of $\gamma$.

Other similar methods have been proposed that assume exact line searches but employ a more general (than quadratic) objective model. See, for instance, Davison and Wong [33] and Boland et al. [34, 35]. In 1972, Crowder and Wolfe [36] and later Powell [37] proved that conjugate gradient methods were doomed to a linear rate of convergence in the absence of periodic restarts. A restart is defined as a special procedure that interrupts the normal generation of search directions, like that which is required for $s(x^{(0)})$. As we have said, there are several reasons one might expect an algorithm to require a restart procedure in order to be robust. The most obvious is to avoid the generation of a dependent direction. Powell [38] proved that the Polak-Ribiere method also gives a linear rate of convergence without restart but is superior to the Fletcher-Reeves method for general functions and exhibits less sensitivity to inexact line searches.

The formulation of efficient restart procedures and methods with greater tolerance of inexact line searches remain active areas of research. Beale [39] has given a conjugate gradient procedure similar to the standard Fletcher-Reeves method that allows restarting without the use of the gradient direction. He shows how the direction just prior to restart might be used to reduce the overall computation effort for functions that require many restarts. Powell [38] carefully examines Beale's and other restarting strategies and suggests that a restart should occur every $N$ steps, or when

$$|g(x^{(k)})g(x^{(k-1)})| \geqslant 0.2\|g(x^{(k)})\|^2 \qquad (3.73)$$

He demonstrates that Beale's restart strategy (using his criterion (3.73)) works equally well with the Fletcher-Reeves or Polak-Ribiere direction formulas and gives numerical results that suggest the superiority of the Polak-Ribiere (with restart) method. Shanno [40] has studied the relation of inexact line searches and restart strategies on the efficiency of conjugate gradient methods. He shows that Beale's restart strategy (along with the associated two-term update formula) with Powell's restart criterion allows a significant decrease in the required precision of the line searches and therefore a significant increase in the overall computational efficiency of the conjugate gradient method. Shanno gives numerical results that indicate the superiority of the Polak-Ribiere method with restart and inexact line searches. Gabriele and Ragsdell [41] recently demonstrated the value of these techniques for large nonlinear programs, where their primary value remains.

### 3.3.6  Quasi-Newton Methods

These methods are similar to those of Section 3.3.5 in that they are based primarily upon properties of quadratic functions. Whereas the previous methods drew their strength from conjugate search directions, the current methods are designed to more directly mimic the positive characteristics of Newton's method using only first-order information. All methods of this class generate directions to be used in (3.42) of the form

$$s(x^{(k)}) = -A^{(k)}\nabla f(x^{(k)}) \tag{3.74}$$

where $A^{(k)}$ is an $N \times N$ matrix called the *metric*. Methods that employ directions of this form are often called *variable metric* methods, because $A$ changes at each iteration. To be precise, a variable metric method is a *quasi-Newton* method if it is designed so that the iterates satisfy the following quadratic property:

$$\Delta x = C^{-1}\Delta g \tag{3.75}$$

Unfortunately, the literature is not precise or consistent in the use of these terms [42]; therefore, we shall use them interchangeably. This is appropriate, since both expressions are of equal importance in the design and execution of these methods.

Let us assume a recursion for the estimate to the inverse of the Hessian:

$$A^{(k+1)} = A^{(k)} + A_c^{(k)} \tag{3.76}$$

where $A_c^{(k)}$ is a correction to the current metric. We will use $A^{(k)}$ in (3.74) and (3.42). The plan is to form $A_c^{(k)}$ such that the sequence $A^{(0)}, A^{(1)}, A^{(2)}, \ldots, A^{(k+1)}$ approaches $H^{-1} = \nabla^2 f(x^*)^{-1}$, for in so doing one additional line search will produce $x^*$ if $f(x)$ is quadratic. As we have said several times before, we hope that success with quadratic functions will lead to success with general nonlinear functions.

Recall the important property of a quadratic function given in (3.75), and assume our approximation of $C^{-1}$ is of the form

$$C^{-1} = \beta A^{(k)} \tag{3.77}$$

where $\beta$ is a scalar. We would like the approximation to satisfy (3.75); that is,

$$\Delta x^{(k)} = A^{(k)}\Delta g^{(k)} \tag{3.78}$$

but this is clearly impossible, since we need $A^{(k)}$ to find $\Delta g^{(k)}$, where

$$\Delta x^{(k)} = x^{(k+1)} - x^{(k)} \tag{3.79}$$

$$\Delta g^{(k)} = g(x^{(k+1)}) - g(x^{(k)}) \tag{3.80}$$

On the other hand, we can require that the *new* approximation satisfy (3.75):

$$\Delta x^{(k)} = \beta A^{(k+1)}\Delta g^{(k)} \tag{3.81}$$

Combining (3.76) and (3.81) leads to

$$A_c^{(k)}\Delta g^{(k)} = \left(\frac{1}{\beta}\right)\Delta x^{(k)} - A^{(k)}\Delta g^{(k)} \tag{3.82}$$

A solution is not obvious, but one can verify by direct substitution that

$$A_c^{(k)} = \frac{1}{\beta}\left(\frac{\Delta x^{(k)}y^T}{y^T\Delta g^{(k)}}\right) - \frac{A^{(k)}\Delta g^{(k)}z^T}{z^T\Delta g^{(k)}} \tag{3.83}$$

is a solution. $y$ and $z$ are arbitrary vectors, so this is really a family of solutions. If we let

$$y = \Delta x^{(k)} \quad \text{and} \quad z = A^{(k)}\Delta g^{(k)} \tag{3.84}$$

we get the well-known and widely used *Davidon-Fletcher-Powell (DFP) method* [43, 44]:

$$A^{(k)} = A^{(k-1)} + \frac{\Delta x^{(k-1)}\Delta x^{(k-1)T}}{\Delta x^{(k-1)T}\Delta g^{(k-1)}} - \frac{A^{(k-1)}\Delta A^{(k-1)}\Delta g^{(k-1)T}A^{(k-1)}}{\Delta g^{(k-1)T}A^{(k-1)}\Delta g^{(k-1)}} \tag{3.85}$$

It can be shown that this updating formula preserves symmetry and positive definiteness, so that if $A^{(0)}$ is any symmetric positive-definite matrix, $A^{(1)}, A^{(2)}, \ldots$ will also be symmetric and positive definite in the absence of round-off error. It is convenient to let $A^{(0)} = I$.

Recalling the first variation of $f(x)$, we write

$$\Delta f(x) = \nabla f(x^{(k)})^T \Delta x \tag{3.86}$$

Now using (3.42) and (3.74) with this expression, we get

$$\Delta f(x) = -\nabla f(x^{(k)})^T \alpha^{(k)} A^{(k)} \nabla f(x^{(k)}) \tag{3.87}$$

Therefore,

$$\Delta f(x) = -\alpha^{(k)}\nabla f(x^{(k)})^T A^{(k)} \nabla f(x^{(k)}) \tag{3.88}$$

assures us that $f(x^{(k+1)}) < f(x^{(k)})$ for all $\alpha^{(k)} > 0$ and $A^{(k)}$ positive definite. So the algorithm has a descent property. The DFP method has been and continues to be a very widely used gradient-based technique. The method tends to be robust; that is, it will typically do well on a wide variety of practical problems. The major disadvantage associated with methods in this class is the need to store the $N \times N$ matrix $A$.

***Example 3.10*** *The Davidon-Fletcher-Powell Method*

Find the minimum of

$$f(x) = 4x_1^2 + 3x_2^2 - 4x_1x_2 + x_1$$

from $x^{(0)} = [0,0]^T$, using the DFP method.

**Solution**

1. Let $s^{(0)} = -\nabla f(x^{(0)}) = -[1,0]^T$.
2. Line search, same as in Example 3.9.

$$x^{(1)} = \left[-\tfrac{1}{8}, 0\right]^T.$$

3. $k = 1;\ \mathbf{A}^{(1)} = \mathbf{A}^{(0)} + \mathbf{A}_c^{(0)}$

$$\mathbf{A}^{(0)} = \begin{bmatrix} 1 & 0 \\ 0 & 1 \end{bmatrix} = \mathbf{I}$$

$$\mathbf{A}_c^{(0)} = \frac{\Delta x^{(0)}\Delta x^{(0)T}}{\Delta x^{(0)T}\Delta g^{(0)}} - \frac{\mathbf{A}^{(0)}\Delta g^{(0)}\Delta g^{(0)T}\mathbf{A}^{(0)}}{\Delta g^{(0)T}\mathbf{A}^{(0)}\Delta g^{(0)}}$$

$$\Delta x^{(0)} = \left[-\tfrac{1}{8}, 0\right]^T - [0,0]^T = \left[-\tfrac{1}{8}, 0\right]^T$$

$$\Delta g^{(0)} = \left[0, \tfrac{1}{2}\right]^T - [1,0]^T = \left[-1, \tfrac{1}{2}\right]^T$$

$$\mathbf{A}_c^{(0)} = \frac{\left[-\tfrac{1}{8},0\right]^T\left[-\tfrac{1}{8},0\right]}{\left[-\tfrac{1}{8},0\right]\left[-1,\tfrac{1}{2}\right]^T} - \frac{\begin{bmatrix} 1 & 0 \\ 0 & 1 \end{bmatrix}\left[-1,\tfrac{1}{2}\right]^T\left[-1,\tfrac{1}{2}\right]\begin{bmatrix} 1 & 0 \\ 0 & 1 \end{bmatrix}}{\left[-1,\tfrac{1}{2}\right]\begin{bmatrix} 1 & 0 \\ 0 & 1 \end{bmatrix}\left[-1,\tfrac{1}{2}\right]^T}$$

$$\mathbf{A}_c^{(0)} = \begin{bmatrix} \tfrac{1}{8} & 0 \\ 0 & 0 \end{bmatrix} - \begin{bmatrix} \tfrac{4}{5} & -\tfrac{2}{5} \\ -\tfrac{2}{5} & \tfrac{1}{5} \end{bmatrix}$$

$$\mathbf{A}^{(1)} = \begin{bmatrix} .325 & \tfrac{2}{5} \\ \tfrac{2}{5} & \tfrac{4}{5} \end{bmatrix}$$

$$s^{(1)} = -\mathbf{A}^{(1)}\nabla f(x^{(1)}) = -\left[\tfrac{1}{5}, \tfrac{2}{5}\right]^T$$

4. Line search:

$$x^{(2)} = x^{(1)} + \alpha^{(1)}s^{(1)} \rightarrow \alpha^{(1)} = \tfrac{5}{16}$$

$$x^{(2)} = \left[-\tfrac{1}{8}, 0\right]^T - \tfrac{5}{16}\left[\tfrac{1}{5}, \tfrac{2}{5}\right]^T = \left[-\tfrac{3}{16}, -\tfrac{1}{8}\right]^T$$

which is the solution, as before.

In Examples 3.9 and 3.10, it is noteworthy that the Fletcher-Reeves and DFP methods generated exactly the same search directions, and they are in fact **H**-conjugate:

$$s^{(1)T}\mathbf{H}s^{(0)} = \left[\tfrac{1}{4}, \tfrac{1}{2}\right]\begin{bmatrix} 8 & -4 \\ -4 & 6 \end{bmatrix}[1,0]^T = 0$$

We also notice that the metric $\mathbf{A}^{(k)}$ does converge to $\mathbf{H}^{-1}$:

$$\mathbf{H} = \begin{bmatrix} 8 & -4 \\ -4 & 6 \end{bmatrix} \rightarrow \mathbf{H}^{-1} = \begin{bmatrix} \tfrac{3}{16} & \tfrac{1}{8} \\ \tfrac{1}{8} & \tfrac{1}{4} \end{bmatrix}$$

$$\Delta x = x^{(2)} - x^{(1)} = \left[-\tfrac{3}{16}, -\tfrac{1}{8}\right]^T - \left[-\tfrac{1}{8}, 0\right]^T = \left[-\tfrac{1}{16}, -\tfrac{1}{8}\right]^T$$

$$\Delta g = g^{(2)} - g^{(1)} = [0,0]^T - \left[0, \tfrac{1}{2}\right]^T = \left[0, -\tfrac{1}{2}\right]^T$$

$$\mathbf{A}_c^{(1)} = \frac{\left[-\tfrac{1}{16}, -\tfrac{1}{8}\right]^T\left[-\tfrac{1}{16}, -\tfrac{1}{8}\right]}{\left[-\tfrac{1}{16}, -\tfrac{1}{8}\right]\left[0, -\tfrac{1}{2}\right]^T} - \frac{\begin{bmatrix} .325 & .4 \\ .4 & .8 \end{bmatrix}\left[0, -\tfrac{1}{2}\right]^T\left[0, -\tfrac{1}{2}\right]\begin{bmatrix} .325 & .4 \\ .4 & .8 \end{bmatrix}}{\left[0, -\tfrac{1}{2}\right]\begin{bmatrix} .325 & .4 \\ .4 & .8 \end{bmatrix}\left[0, -\tfrac{1}{2}\right]^T}$$

$$\mathbf{A}^{(2)} = \begin{bmatrix} .325 & .4 \\ .4 & .8 \end{bmatrix} + \begin{bmatrix} \tfrac{1}{16} & \tfrac{1}{8} \\ \tfrac{1}{8} & \tfrac{1}{4} \end{bmatrix} - \begin{bmatrix} .2 & .4 \\ .4 & .8 \end{bmatrix} = \begin{bmatrix} \tfrac{3}{16} & \tfrac{1}{8} \\ \tfrac{1}{8} & \tfrac{1}{4} \end{bmatrix}$$

So, $\mathbf{A}^{(2)} = \mathbf{H}^{-1}(x^*)$. Since the publication of the Davidon updating formula, there have been various other variable metric methods proposed that result from different choices for $\beta$, $y$, and $z$ in (3.83). A common practical difficulty is the tendency of $\mathbf{A}^{(k+1)}$ to be ill-conditioned (see Appendix C.3 for definition), causing increased dependence on a restarting procedure.

Another method proposed almost simultaneously by Broyden [45], Fletcher [46], and Shanno [47] has received rather wide acclaim and has in fact been strongly recommended by Powell [48]. The updating formula for the *Broyden-Fletcher-Shanno (BFS) method* is

$$\mathbf{A}^{(k+1)} = \left[\mathbf{I} - \frac{\Delta x^{(k)}\Delta g^{(k)T}}{\Delta x^{(k)T}\Delta g^{(k)}}\right]\mathbf{A}^{(k)}\left[\mathbf{I} - \frac{\Delta x^{(k)}\Delta g^{(k)T}}{\Delta x^{(k)T}\Delta g^{(k)}}\right] + \frac{\Delta x^{(k)}\Delta x^{(k)T}}{\Delta x^{(k)T}\Delta g^{(k)}} \quad (3.89)$$

The major advantage of the BFS method is the decreased need for restart, and the method seems significantly less dependent upon exact line searching.

There is a definite similarity between current efforts to improve variable metric methods and those to improve conjugate gradient methods. Therefore, our remarks on other conjugate gradient methods hold to some degree for variable metric methods as well. Davidon [49] has proposed a class of updates that permit inexact line searches. Powell [50] later established the quadratic termination property of these methods in the absence of line searches. It would appear that a variable metric method with quadratic termination but without the need of expensive line searches would be both robust and fast on general functions. Goldfarb [51] has also investigated this very promising possibility. In general, we restart the DFP method (that is, set $\mathbf{A} = \mathbf{I}$) after $N$ updates. This is usually a very conservative measure, since many more steps may often be taken before a restart is required. We sense the need to restart as the condition number of the update worsens. Be aware that even though restarting does provide a degree of safety (or robustness) with variable metric methods, it does typically slow progress to the solution, since a second-order estimate is being discarded.

McCormick [52], Shanno [40], and Shanno and Phua [53] have extensively investigated the relationship between conjugate gradient and variable metric methods. Shanno views conjugate gradient methods as "memoryless" variable metric methods. Most would now agree that these two classes of methods share much more than was originally thought. In addition, it is clear that the many variable metric variants have very much in common in actual practice [54, 55], which causes one to weigh carefully the additional computational overhead of the more complex methods. Shanno and Phua [56, 57] give extensive numerical results that favor the relatively simple BFS update.

Excellent Fortran implementations of the best methods are available in the open literature [53, 58]. In addition, variable metric methods have been used extensively in the development of techniques for constrained problems. LaFrance et al. [59] have given a combined generalized reduced gradient and DFP method, whereas Root [60] used DFP with his implementation of the method of multipliers. Of course, the latest work of Han [61] and Powell [62] on constrained problems exploits the efficiency of variable metric methods.

Finally, there is a significant amount of work currently under way to make the variable metric methods more attractive for large problems. As previously noted, conjugate gradient methods are currently the best methods for truly large problems. But progress is being made on problems where the Hessian is sparse [63–66]. One difficulty with these methods is that we often have knowledge of the sparsity of some estimate of $\mathbf{H}^{-1}$ instead of $\mathbf{H}$.

### 3.3.7 A Gradient-Based Algorithm

The similarities of the conjugate gradient and quasi-Newton methods suggest that we might devise a general algorithm including all of the methods previously discussed. Such an algorithm is the *gradient method*

### Gradient Method

**Step 1.** Define $\quad M$ = maximum number of allowable iterations

$\qquad\qquad\quad N$ = number of variables

$\qquad\qquad x^{(0)}$ = initial estimate of $x^*$

$\qquad\qquad\quad \varepsilon_1$ = overall convergence criteria

$\qquad\qquad\quad \varepsilon_2$ = line search convergence criteria

**Step 2.** Set $k = 0$.

**Step 3.** Calculate $\nabla f(x^{(k)})$.

**Step 4.** Is $\|\nabla f(x^{(k)})\| \leqslant \varepsilon_1$?
    Yes:  Print "convergence: gradient", go to 13.
    No:   Continue.

**Step 5.** Is $k > M$?
    Yes:  Print "termination: $k = M$", go to 13.
    No:   Continue.

**Step 6.** Calculate $s(x^{(k)})$.

**Step 7.** Is $\nabla f(x^{(k)})s(x^{(k)}) < 0$?
    Yes:  Go to 9.
    No:   Set $s(x^{(k)}) = -\nabla f(x^{(k)})$. Print "restart: bad direction". Go
           to 9.

**Step 8.** Find $\alpha^{(k)}$ such that $f(x^{(k)} + \alpha^{(k)}s(x^{(k)})) \rightarrow$ minimum, using $\varepsilon_2$.

**Step 9.** Set $x^{(k+1)} = x^{(k)} + \alpha^{(k)}s(x^{(k)})$.

**Step 10.** Is $f(x^{(k+1)}) < f(x^{(k)})$?
    Yes:  Go to 11.
    No:   Print "termination: no descent", go to 13.

**Step 11.** Is $\|\Delta x\|/\|x^{(k)}\| \leqslant \varepsilon_1$?
    Yes:  Print "termination: no progress", go to 13.
    No:   Go to 12.

**Step 12.** Set $k = k + 1$. Go to 3.

**Step 13.** Stop.

We produce the various methods by defining the appropriate search direction in step 6 of the algorithm. The required calculation for each method has been given previously. Admittedly, some of the calculations (for instance, line 7) would not be appropriate for the Cauchy method, but we have discussed an algorithm for this method before. A purist might insist upon a restart feature for the FR method, but the checks included in the algorithm would catch any difficulties associated with the lack of restart for a conjugate gradient method.

We might insert additional tests in line 7 similar to those suggested by Beale and Powell for restart. An efficient implementation would certainly contain additional tests in line 8 along the lines suggested by Davidon, Powell, and Shanno. Certainly, exact line searches should be avoided whenever possible. Our tests indicate that line searching is typically the most time-consuming phase of these methods.

### 3.3.8 Numerical Gradient Approximations

To this point in our discussion of gradient methods we have assumed that the gradient $\nabla f(x)$ and the Hessian matrix $\nabla^2 f(x)$ are explicitly available. Modern methods seek to mimic the positive characteristics of Newton's method, such as rate of convergence near the minimum, without use of second-order information ($\nabla^2 f(x)$). In many practical problems the gradient is not explicitly available or is difficult and inconvenient to formulate. An example is when $f(x)$ is the result of a simulation. In addition, even when the analytical expressions can be obtained, they are prone to error. Therefore, if for no other reason than to check the analytical expressions, it seems prudent to have a procedure to approximate the gradient numerically. We have used with considerable success the simple *forward difference* approximation:

$$\frac{\partial f(x)}{\partial x_i}\bigg|_{x=\bar{x}} = \frac{f(\bar{x} + \varepsilon e^{(i)}) - f(\bar{x})}{\varepsilon} \qquad (3.90)$$

This approximation follows directly from the definition of a partial derivative and is a good estimate for appropriately small values of $\varepsilon$. The proper choice for $\varepsilon$ depends on $f(x)$ and $\bar{x}$, and the accuracy or word length of the computing machine. We recall that theoretically the approximation is exact in the limit as $\varepsilon$ goes to zero. This is obviously an unacceptable choice for $\varepsilon$, and in fact $\varepsilon$ must be large enough so that the difference in the numerator is finite. When $\varepsilon$ is less than the precision of the machine, the numerator will be zero. In addition, we must not choose $\varepsilon$ too large or lose contact with any semblance of a limiting process. In other words, for $\varepsilon$ too large we get good numbers with regard to machine precision, but poor estimates of the derivative.

We can, at the expense of an additional function evaluation, increase the inherent accuracy of the approximation by using the central difference formula:

$$\frac{\partial f(x)}{\partial x_i}\bigg|_{x=\bar{x}} = \frac{f(\bar{x} + \varepsilon e^{(i)}) - f(\bar{x} - \varepsilon e^{(i)})}{2\varepsilon} \qquad (3.91)$$

For a given machine, $f(x)$, $\bar{x}$, and $\varepsilon$, the central difference approximation will be more accurate but requires an additional function value. In many cases the increased accuracy does not warrant the additional expense.

Stewart [67] proposed a procedure for calculating $\varepsilon$ at each iteration of the DFP method. His method is based upon estimates of the inherent error of the forward difference approximation (3.90) and the error introduced by subtracting nearly equal numbers in the numerator when $\varepsilon$ becomes very small. The method is attractive because it requires no additional function values. Gill et al. [68] have shown that the method may produce poor values for $\varepsilon$ and suggest a different approach. Their method requires $2N$ function values to calculate the required $N\varepsilon$ values, but needs to be performed only one at $x^{(0)}$. Their technique is based upon better estimates of the errors involved, and numerical results suggest superiority of the Gill approach.

## 3.4  COMPARISON OF METHODS AND NUMERICAL RESULTS

Only recently has it been possible to say much from a theoretical viewpoint concerning the convergence or rate of convergence of the methods discussed in the previous sections. Accordingly, much of what is known about the efficiency of these methods when applied to general functions has come from numerical experiments. Very often the publication of a new method is associated with very sparse comparative numerical results. Often the test problems are chosen to highlight the outstanding characteristics of the new method. This manner of presentation limits to a large degree the validity of the results.

Himmelblau (see chap. 5 of [69]) gives numerical results of quite a different nature. His results have increased significance since he is not a personal advocate of any single method. He characterizes each method (direct and gradient) according to robustness, number of function evaluations, and computer time to termination when applied to a family of test problems. Robustness is a measure of success in obtaining an optimal solution of given precision for a wide range of problems. Himmelblau concludes that the BFS, DFP, and Powell's direct-search method are the superior methods. The reader is referred directly to Himmelblau for complete conclusions.

A similar but less complete study is given by Sargent and Sebastian [70]. They give results for gradient-based methods, including the BFS, DFP, and Fletcher-Reeves algorithms. They test the effect of line-search termination criteria, restart frequency, positive definiteness for quasi-Newton methods, and gradient accuracy. Their results once again indicate the superiority of the quasi-Newton methods for general functions (in particular BFS). They also note that computer word length affects the quasi-Newton methods more than the conjugate gradient methods. This leads to the possibility that the Fletcher-Reeves method might be superior on a small computer such as is sometimes used in a process control environment.

Carpenter and Smith [71] have reported on the relative computational efficiency of Powell's direct-search method, DFP, and Newton's method for structural applications. They conclude that for the test problems considered, DFP and Newton's method are superior to Powell's direct search method, and DFP was superior to Newton's method for large problems. More recently, rather extensive tests have been conducted by Shanno and Phua [40, 56, 57]. They report on numerical experiments with a variety of conjugate gradient and variable metric methods. Their results are difficult to summarize briefly (the reader is referred directly to the cited papers), but it is fair to say that the tests suggest a superiority of the BFS method over other available methods.

**Table 3.2    Termination Criteria**

| | |
|---|---|
| 1. $k > M$ | 3. $\|\nabla f(x^{(k)})\| \leqslant \varepsilon_1$ |
| 2. $\dfrac{\|\Delta x\|}{\|x^{(k)}\|} \leqslant \varepsilon_1$ | 4. $\nabla f(x^{(k)})s(x^{(k)}) > 0$ |
| | 5. $f(x^{(k+1)}) > f(x^{(k)})$ |

It is probably not appropriate at this point in our discussion to devote much attention to the theory of numerical experiments. On the other hand, we should mention that testing guidelines are now available [72], and that most of the results given in the literature prior to approximately 1977 should be used cautiously. Certainly ratings on the basis of number of function evaluations *only* can be misleading, as shown recently by Miele and Gonzalez [73].

We have performed some numerical experiments designed to demonstrate the relative utility of the Cauchy, Fletcher-Reeves (FR) DFP, and BFS methods on a CDC-6500 using single precision arithmetic. The FR method is restarted every $N + 1$ steps, and the algorithms terminate when a condition in Table 3.2 is met. Table 3.3 gives the results for Rosenbrock's function:

$$f(x) = 100\left(x_2 - x_1^2\right)^2 + \left(1 - x_1\right)^2 \tag{3.92}$$

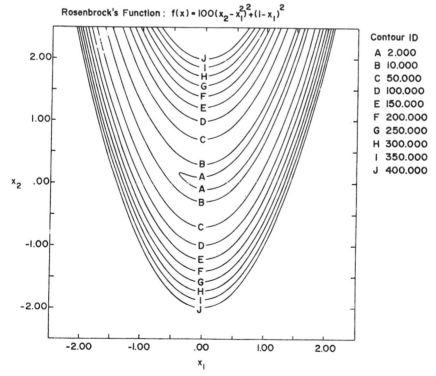

Figure 3.16.    Rosenbrock's function.

**Table 3.3 Rosenbrock's Function ($x^{(0)} = [-1.2, 1.0]^T$)**

| Method | Half | Gold | Coggin | Cubic |
|---|---|---|---|---|
| Cauchy | 1.10E-10/7.47/38,424[a] | 1.25E-10/1.382/4,066 | 8.30E-8/1.074/3,570 | 6.19E-9/2.631/10,685 |
| FR | 3.24E-6/.18/988 | 5.91E-6/.57/805 | 5.96E-5/.109/370 | 2.77E-7/.059/273 |
| DFP | 2.45E-8/.173/977 | 2.39E-8/.133/656 | 6.6E-8/.115/331 | 4.3E-8/.063/239 |
| BFS | 5.6E-8/.169/932 | 3.6E-8/.161/740 | 2.1E-8/.115/315 | 3.9E-9/.065/204 |

[a] $f(x^*)$/time/N.F., where N.F. = number of function evaluations required to terminate, time = seconds of CPU time to termination, and $f(x^*) = f(x)$ at termination.

which has been devised to be a specific challenge to gradient-based methods. The results with various line-searching strategies are given, and numerical approximations to the gradient were used throughout. Notice that the Cauchy/half combination supplied the best value of $f$ at termination, but at the most expense. The most efficient with regard to number of function evaluations is seen to be the BFS/cubic combination. "Half" is the interval-halving method, "gold" is golden section, "Coggin" is a particular version of quadratic interpolation [21], and the "cubic" method involves the fitting of a single cubic at each stage.

Table 3.4 contains the results for a more practical problem, the minimization of a gear train inertia [74]. The problem (see Figure 3.17) is one of a set used by Eason and Fenton to compare various algorithms. The objective function is

$$f(x) = \left\{ 12 + x_1^2 + \frac{1 + x_2^2}{x_1^2} + \frac{x_1^2 x_2^2 + 100}{(x_1 x_2)^4} \right\} \left( \frac{1}{10} \right) \qquad (3.93)$$

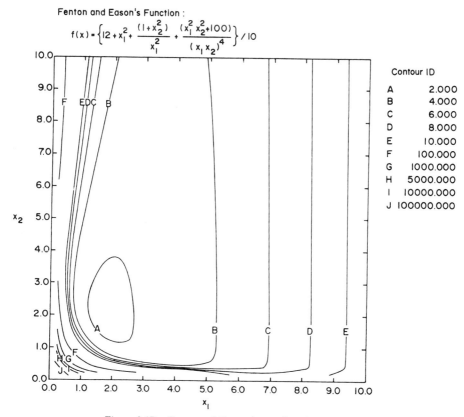

Figure 3.17.   Eason and Fenton's gear inertia problem.

**Table 3.4  Eason and Fenton Problem No. 10 ($x^{(0)} = [.5, .5]^T$)**

| Method | Half | Gold | Coggin | Cubic |
|---|---|---|---|---|
| CAUCHY | 1.744/.299/1026[a] | 1.744/.219/688 | 1.744/.181/312 | 1.744/.091/171 |
| FR | 1.744/.066/249 | 1.744/.053/199 | 1.744/12.21/28141 | 1.744/.025/92 |
| DFP | 1.744/.056/232 | 1.744/.051/184 | 1.744/.09/208 | 1.744/.079/262 |
| BFS | 1.744/.055/226 | 1.744/.051/176 | 1.744/.087/195 | 1.744/.043/133 |

[a]$f(x^*)$/time/N.F.

**Table 3.5  Wood's Function ($x^{(0)} = [-3, -1, -3, -1]^T$)**

| Method | Half | Gold | Coggin | Cubic |
|---|---|---|---|---|
| CAUCHY | 3.74E-9/13.466/39,503[a] | 1.77E-8/10.70/29/692 | 2.95E-9/5.844/12,392 | 8/36E-6/7.89/20,007 |
| FR | 1.3E-10/.401/1257 | 1.2E-8/.311/942 | 2.9E-8/.571/1468 | 2.0E-7/.083/244 |
| DFP | 3.9E-9/.810/2404 | 3.7E-9/.684/1895 | 2.9E-9/.403/747 | 9.5E-10/.298/727 |
| BFS | 2.2E-9/.759/2054 | 2.0E-9/.652/1650 | 2.0E-9/.444/723 | 2.3E-9/.240/410 |

[a]$f(x^*)$/time/N.F.

On this problem we note that the FR/cubic algorithm is most efficient in both time and number of function evaluations.

Table 3.5 gives the final group of results for Wood's function [75] (see Figure 3.18):

$$f(x) = 100(x_2 - x_1^2)^2 + (1 - x_1)^2 + 90(x_4 - x_3^2)^2 + (1 - x_3)^2$$

$$+ 10.1\left[(x_2 - 1)^2 + (x_4 - 1)^2\right] + 19.8(x_2 - 1)(x_4 - 1) \qquad (3.94)$$

and once again the FR/cubic seems to be most efficient. These tests have been carefully executed, but we see that the results do not agree with other published results. We have been much more successful with the FR method than previously reported. It seems that additional numerical results are called for.

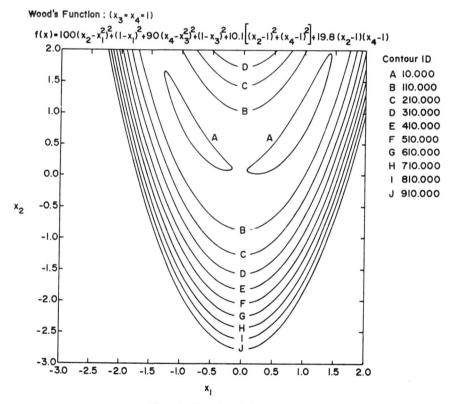

Figure 3.18.  Wood's function.

## 3.5   SUMMARY

In this chapter we have covered pertinent issues related to functions of several variables as they appear in an optimization setting. We have given the necessary and sufficient conditions for existence of a minimum of a function of several variables. On the other hand, most of the chapter is devoted to a (hopefully) orderly survey of appropriate methods. Some methods have been included because of their historical importance, while others represent the best that is currently available. Methods that require values of $f(x)$ only; $f(x)$ and $\nabla f(x)$; and $f(x)$, $\nabla f(x)$, and $\nabla^2 f(x)$ are included in the discussions. The important concept of conjugacy is considered in some depth, and the latest conjugate-gradient and quasi-Newton methods are cited. The discussion is not exhaustive, and many methods are omitted due to lack of space. The chapter closes with algorithmic considerations and brief numerical results.

## REFERENCES

1. Murray, W. (Ed.), *Numerical Methods for Unconstrained Optimization*, Academic Press, London, 1972.

2. Box, M. J., D. Davies, and W. H. Swann, *Non-Linear Optimization Techniques*, ICI Monograph 5, Oliver and Boyd, Edinburgh, 1969.

3. Krolak, P., and L. Cooper, "An Extension of Fibonaccian Search to Several Variables," *ACM*, **6**, 639–641 (1963).

4. Box, G. E. P., "Evolutionary Operation: A Method for Increasing Industrial Productivity," *Appl. Stat.*, **6**, 81–101 (1957).

5. Spendley, W., G. R. Hext, and F. R. Himsworth, "Sequential Application of Simplex Designs in Optimization and Evolutionary Operation," *Technometrics*, **4**, 441–461 (1962).

6. Nelder, J. A., and R. Mead, "A Simplex Method for Function Minimization," *Computer J.*, **7**, 308–313 (1965).

7. Box, G. E. P., and N. R. Draper, *Evolutionary Operation*, Wiley, New York, 1969.

8. Parkinson, J. M., and D. Hutchinson, "An Investigation into the Efficiency of Variants on the Simplex Method," in *Numerical Methods for Non-Linear Optimization* (F. A. Lootsma, Ed.), Academic Press, London, 1972, pp. 115–135.

9. Powell, M. J. D., "On Search Directions for Minimization Algorithms," *Math. Prog.*, **4** (2), 193–201 (1973).

10. Hooke, R., and T. A. Jeeves, "Direct Search of Numerical and Statistical Problems," *J. ACM*, **8**, 212–229 (1966).

11. Bandler, J. W., and P. A. McDonald, "Optimization of Microwave Networks by Razor Search," *IEEE Trans. Microwave Theory Tech.*, **17**, 552–562 (1969).

12. Emery, F. E., and M. O. O'Hagan, "Optimal Design of Matching Networks for Microwave Transistor Amplifiers," *IEEE Trans. Microwave Theory Tech.*, **14**, 696–698 (1966).

13. Rosenbrock, H. H., "An Automated Method for Finding the Greatest or Least Value of a Function," *Computer J.*, **3** (3), 175–184 (1960).

14. Swann, W. H., "Report on the Development of a New Direct Search Method of Optimization," ICI Ltd., *Central Instr. Res. Lab.*, *Res. Note* **64 / 3**, 1964.

15. Powell, M. J. D., "An Efficient Method for Finding the Minimum of a Function of Several Variables without Calculating Derivatives," *Computer J.*, 7, 155–162 (1964).

16. Zangwill, W. I., "Minimizing a Function without Calculating Derivatives," *Computer J.*, 10, 293–296 (1967).

17. Brent, R. P., *Algorithms for Minimization without Derivatives*, Prentice-Hall, Englewood Cliffs, NJ, 1973.

18. Gill, P. E., W. Murray, and M. H. Wright, *Practical Optimization*, Academic Press, New York, 1981, p. 57.

19. Fletcher, R., *Practical Methods of Optimization*, Vol 1: *Unconstrained Optimization*, Wiley, New York, 1980, p. 17.

20. Cauchy, A., "Method generale pour la resolution des systemes d'equations simultanees," *Compt. Rend. Acad. Sci.*, 25, 536–538 (1847).

21. Coggins, G. F., "Univariate Search Methods," Imperial Chemical Industries Ltd., *Central Instr. Res. Lab.*, *Res. Note* 64 / 11, 1964.

22. Gabriele, G. A., and K. M. Ragsdell, *OPTLIB: An Optimization Program Library*, *Purdue Research Foundation*, *January* 1977.

23. Ragsdell, K. M., R. R. Root, and G. A. Gabriele, "Newton's Method: A Classical Technique of Contemporary Value," *ASME Design Technol. Transfer Conf. Proc.*, New York, October 1974, p. 137.

24. Mangasarian, O. L., "Techniques of Optimization," ASME paper 71-vibr.-118, *ASME Design Automat. Conf.*, Sept. 8–10, 1971.

25. Marquardt, D. W., "An Algorithm for Least Squares Estimation of Non-Linear Parameters," *SIAM J.*, 11, 431–441 (1963).

26. Powell, M. J. D., "Problems Related to Unconstrained Optimization," in *Numerical Methods for Unconstrained Optimization*, Academic Press, New York, 1972, chap. 3, p. 29.

27. Bard, Y., *Nonlinear Parameter Estimation*, Academic Press, New York, 1974.

28. Hestenes, M. R., and E. Stiefel, "Methods of Conjugate Gradients for Solving Linear Systems," *NBS Res. J.*, 49, 409–436 (1952).

29. Fletcher, R., and C. M. Reeves, "Function Minimization by Conjugate Gradients," *Computer J.*, 7, 149–154 (1964).

30. Fried, I., and J. Metzler, "SOR vs Conjugate Gradients in a Finite Element Discretization," *Int. J. Numerical Methods Eng.*, 12, 1329–1342, (1978).

31. Miele, A., and J. W. Cantrell, "Study on a Memory Gradient Method for Minimization of Functions," *JOTA*, 3 (6), 459 (1969).

32. Polak, E., and G. Ribiere, "Note sur la convergence de methods de directions conjuguees," *Rev. Fr. Inform. Rech. Operat.*, 16, 35–43 (1969).

33. Davison, E. J., and P. Wong, "A Robust Conjugate-Gradient Algorithm which Minimizes L-Functions," *Auotmatica*, 11, 297–308 (1975).

34. Boland, W. R., E. R. Kamgnia, and J. S. Kowalik, "A Conjugate Gradient Optimization Method Invariant to Nonlinear Scaling," *JOTA*, 27 (2), 221–230 (1979).

35. Boland, W. R., and J. S. Kowalik, "Extended Conjugate-Gradient Methods with Restarts," *JOTA*, 28 (1), 1–9 (1979).

36. Crowder, H. P., and P. Wolfe, "Linear Convergence of the Conjugate Gradient Method," *IBM J. Res. Develop.*, 16, 431–433 (1972).

37. Powell, M. J. D., "Some Convergence Properties of the Conjugate Gradient Method," *Math. Prog.*, 11, 42–49 (1976).

38. Powell, M. J. D., "Restart Procedures for the Conjugate Gradient Method," *Math. Prog.*, 12, 241–254 (1977).

39. Beale, E. M. L., "A Derivation of Conjugate Gradients," in *Numerical Methods for Non-Linear Optimization* (F. A. Lootsma, Ed.), Academic Press, New York, 1972, pp. 39–43.

40. Shanno, D. F., "Conjugate Gradient Methods with Inexact Searches," *Math. Oper. Res.*, **3** (3), 244–256 (1978).

41. Gabriele, G. A., and K. M. Ragsdell, "Large Scale Nonlinear Programming Using the Generalized Reduced Gradient Method," ASME *J. Mech. Des.*, **102** (3), 566–573 (1980).

42. Avriel, M., *Nonlinear Programming: Analysis and Methods*, Prentice-Hall, Englewood Cliffs, NJ, 1976, p. 322.

43. Davidon, W. C., "Variable Metric Method for Minimization," *AEC Res. Develop. Rep.*, **ANL-599**, 1959.

44. Fletcher, R., and M. J. D. Powell, "A Rapidly Convergent Descent Method for Minimization," *Computer J.*, **6**, 163 (1963).

45. Broyden, G. G., "The Convergence of a Class of Double-Rank Minimization Algorithms," *J. Inst. Math. Appl.*, **6**, 76–90, 222–231 (1970).

46. Fletcher, R., "A New Approach to Variable Metric Algorithms," *Computer J.*, **13**, 317–322 (1970).

47. Shanno, D. F., "Conditioning of Quasi-Newton Methods for Function Minimization," *Math. Comp.*, **24**, 647–657 (1970).

48. Powell, M. J. D., "A View of Unconstrained Optimization," presented at IMA Conf. "Optimization in Action," January 1975.

49. Davidon, W. C., "Optimally Conditioned Optimization Algorithms without Line Searches," *Math. Prog.*, **9**, 1–30 (1975).

50. Powell, M. J. D., "Quadratic Termination Properties of Davidon's New Variable Metric Algorithm," *Math. Prog.*, **12**, 141–147 (1977).

51. Goldfarb, D., "Generating Conjugate Directions without Line Searches Using Factorized Variable Metric Updating Formulas," *Math Prog.*, **13**, 94–110 (1977).

52. McCormick, G. P., "Methods of Conjugate Directions versus Quasi-Newton Methods," *Math. Prog.*, **3** (1), 101–116 (1972).

53. Shanno, D. F., and K. H. Phua, "A Variable Method Subroutine for Unconstrained Nonlinear Minimization," Manage. Inform. Syst. Rep. **28**, The University of Arizona, 1979.

54. Dixon, L. C. W., "Quasi-Newton Algorithms Generate Identical Points," *Math. Prog.*, **2** (3), 383–387 (1972).

55. Dixon, L. C. W., "Quasi-Newton Algorithms Generate Identical Points II: The Proofs of Four New Theorems," *Math. Prog.*, **3** (3), 345–358 (1972).

56. Shanno, D. F., and K. H. Phua, "Matrix Conditioning and Nonlinear Optimization," *Math. Prog.*, **14**, 149–160 (1978).

57. Shanno, D. F., and K. H. Phua, "Numerical Comparison of Several Variable-Metric Algorithms," *JOTA*, **25** (4), 507–518 (1978).

58. Shanno, D. F., and K. H. Phua, "Algorithms 500: Minimization of Unconstrained Multivariate Functions [E4], "*ACM Trans. Math. Software*, **2** (1), 87–94 (1976).

59. LaFrance, L. J., J. F. Hamilton, and K. M. Ragsdell, "On the Performance of a Combined Generalized Reduced Gradient and Davidon-Fletcher-Powell Algorithm: GRGDFP," *Eng. Optimization*, **2**, 269–278 (1977).

60. Root, R. R., and K. M. Ragsdell, "Computational Enhancements to the Method of Multipliers," *ASME J. Mech. Des.*, **102** (3), 517–523 (1980).

61. Han, S. P., "Superlinearly Convergent Variable Metric Algorithms for General Nonlinear Programming Problems," *Math. Prog.*, **11**, 263–282 (1976).

62. Powell, M. J. D., "The Convergence of Variable Metric Method for Nonlinear Constrained Optimization Calculations," in *Nonlinear Programming 3*, (O. L. Mangasarian, R. R. Meyer, and S. M. Robinson, Eds.), Academic Press, New York, 1978, pp. 27–64.

63. Toint, Ph. L., "On Sparse and Symmetric Matrix Updating Subject to a Linear Equation," *Math. Comp.*, **31** (140), 954–961 (1977).

64. Powell, M. J. D., and Ph. L. Toint, "On the Estimation of Sparse Hessian Matrices," *SIAM J. Numer. Anal.*, **16**, 1060–1074 (1979).

65. Shanno, D. F., "On Variable-Metric Methods for Sparse Hessians", *Math. Comp.*, **34** (150), 499–514 (1980).

66. Shanno, D. F., "Computational Experience with Methods for Estimating Sparse Hessians for Nonlinear Optimization," *JOTA*, **35** (2), 183–193 (Oct. 1981).

67. Stewart, G. W., "A Modification of Davidon's Method to Accept Difference Approximations of Derivatives," *JACM*, **14**, 72–83 (1967).

68. Gill, P. E., W. Murray, M. A. Saunders, and M. H. Wright, "Computing Finite-Difference Approximations to Derivatives for Numerical Optimization," Stanford Systems Optimization Lab. Tech. Rep. **SOL80-6**, 1980.

69. Himmelblau, D. M., *Applied Nonlinear Programming*, McGraw-Hill, New York, 1972, p. 190.

70. Sargent, R. W. H., and D. J. Sebastian, "Numerical Experience with Algorithms for Unconstrained Minimization," in *Numerical Methods for Non-Linear Optimization* (F. A. Lootsma, Ed.), Academic Press, New York, 1971, Chap. 5, p. 45.

71. Carpenter, W. C., and E. A. Smith, "Computational Efficiency in Structural Optimization," *Eng. Optimization*, **1** (3), 169–188 (1975).

72. Crowder, H., R. S. Dembo, and J. M. Mulvey, "On Reporting Computational Experiments with Mathematical Software," ACM *Trans. Math. Software*, **5** (2), 198–203 (1979).

73. Miele, A., and S. Gonzalez, "On the Comparative Evaluation of Algorithms for Mathematical Programming Problems," in *Nonlinear Programming III* (Mangasarian, Meyer, and Robinson, Eds.) Academic Press, New York, 1978, pp. 337–359.

74. Eason, E. D., and R. G. Fenton, "A Comparison of Numerical Optimization Methods for Engineering Design," *J. Eng. Ind.*, *Trans. ASME*, **96** (1), 196–200 (1974).

75. Colville, A. R., "A Comparative Study of Nonlinear Programming Codes," *Tech. Rep.* **320-2949**, IBM New York Scientific Center, 1968.

76. Sherwood, T. K., *A Course in Process Design*, MIT Press, Cambridge, MA, 1963, p. 84.

## PROBLEMS

**3.1.** Explain why the search directions used in a direct search algorithm such as Hooke–Jeeves must be linearly independent. How many direction vectors must be used?

**3.2.** Cite two situations under which the Simplex search method would be preferable to Powell's conjugate direction method.

**3.3.** Why do we use quadratic functions as the basis for developing nonlinear optimization algorithms?

**3.4.** What is the utility of the parallel subspace property of quadratic functions?

**3.5.** What is the descent property? Why is it necessary? Cite one algorithm which does have and one which does not have this property.

**3.6.** Why is it necessary for the $A^k$ matrix of quasi-Newton methods to be positive definite in a minimization problem?

**3.7.** Explain the relationship of Marquardt's method to the Cauchy and Newton methods. Of the three, which is preferred?

**3.8.** Is it possible that DFP and Fletcher–Reeves methods generate identical points if both are started at the same point with a quadratic function? If so, under what conditions? If not, why not?

**3.9.** Explain quadratic convergence. Cite one algorithm which does have and one which does not have this property.

**3.10.** Show that the function

$$f(x) = 3x_1^2 + 2x_2^2 + x_3^2 - 2x_1x_2 - 2x_1x_3$$

$$+ 2x_2x_3 - 6x_1 - 4x_2 - 2x_3$$

is convex.

**3.11.** Find and classify the stationary points of

$$f(x) = 2x_1^3 + 4x_1x_2^2 - 10x_1x_2 + x_2^2$$

as shown in Figure 3.19.

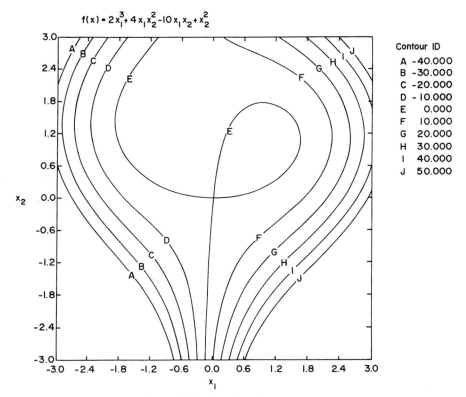

Contour ID

A  -40.000
B  -30.000
C  -20.000
D  -10.000
E    0.000
F   10.000
G   20.000
H   30.000
I   40.000
J   50.000

**Figure 3.19.** Function of Problem 3.

**3.12.** Investigate the definiteness of the following quadratic forms:

$$Q_1(x) = x_1^2 + 2x_2^2 - 3x_3^2 - 6x_1x_2 + 8x_1x_3 - 4x_2x_3$$

$$Q_2(x) = 2ax_1x_2 + 2bx_2x_3 + 2cx_3x_1$$

$$Q_3(x) = x_1^2 + 5x_2^2 + 3x_3^2 + 4x_1x_2 - 2x_2x_3 - 2x_1x_3$$

**3.13.** Use Gaussian elimination to "complete the square" and show that

$$Q(x) = x_1^2 + 2x_1x_2 + 4x_1x_3 + 3x_2^2 + 2x_2x_3 + 5x_3^2$$

is a positive-definite quadratic form.

**3.14.** Suppose at $x = \bar{x}$, $\nabla f(\bar{x}) = 0$. What can you say about $\bar{x}$ if
(a) $f(x)$ is convex?
(b) $f(x)$ is concave?
(c) $\nabla^2 f(\bar{x})$ is indefinite?
(d) $\nabla^2 f(\bar{x})$ is positive semidefinite?
(e) $\nabla^2 f(\bar{x})$ is negative semidefinite?

**3.15.** *Peano's Counterexample.* Consider the function:

$$f(x) = \left(x_1 - a_1^2 x_2^2\right)\left(x_1 - a_2^2 x_2^2\right)$$

with $a_1$ and $a_2$ constant.
(a) Classify the point $x = [0, 0]^T$.
(b) Show that $f(x)$ attains a maximum at the origin for all points on the curve

$$x_1 = \tfrac{1}{2}\left(a_1^2 + a_2^2\right)x_2^2$$

(c) Draw several contours of the function in the neighborhood of the origin.

**3.16.** While searching for the minimum of

$$f(x) = \left[x_1^2 + (x_2 + 1)^2\right]\left[x_1^2 + (x_2 - 1)^2\right]$$

we terminate at the following points:
(a) $x^{(1)} = [0, 0]^T$
(b) $x^{(2)} = [0, 1]^T$
(c) $x^{(3)} = [0, -1]^T$
(d) $x^{(4)} = [1, 1]^T$
Classify each point.

**3.17.** Suppose you are a fill-dirt contractor and you have been hired to transport 400 cu yd of material across a large river. In order to transport the material, a container must be constructed. The following cost data are available:

Each round trip costs $4.20.

Material costs for container:

| | |
|---|---|
| Bottom: | 20.00/sq yd |
| Sides: | 5.00/sq yd |
| Ends: | 20.00/sq yd |

Design a container to minimize cost that will perform the assigned task.

**3.18.** Consider Rosenbrock's function:

$$f(x) = 100(x_2 - x_1^2)^2 + (1 - x_1)^2$$

and the starting point

$$x^{(0)} = [-1.2, 0]^T$$

Find $x^*$ such that $f(x^*)$ is a minimum,

(a) Using the Nelder and Mead simplex search method, perform four iterations by hand, and then using $x^{(0)}$ again use SPX in OPTLIB [22] or other appropriate software of your choice.

(b) Using the Hooke-Jeeves method (PS in OPTLIB).

(c) Using Powell's conjugate directions method (PCD in OPTLIB).

**3.19.** Consider the grain storage bin given in Figure 3.20. Choose $h$, $d$, and $\phi$ such that the bin holds a prescribed volume ($v^* = 10$ m$^3$) and is of minimum cost. Assume the base is made of wood at $C_1 = 1$ $/m$^2$, and the remainder is constructed of sheet metal at $C_2 = 1.5$ $/m$^2$. Use the volume constraint to eliminate one variable, and solve the resulting two-variable problem using the pattern search method ($\theta = 30°$). *Hint*:

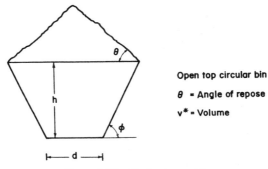

Open top circular bin

$\theta$ = Angle of repose

$v^*$ = Volume

**Figure 3.20.** Grain storage bin.

You may find it useful to describe the bin with another (but equivalent) set of design variables.

**3.20.** Consider a gas pipeline transmission system as shown in Figure 3.21 (Sherwood [76]), where compressor stations are placed $L$ miles apart.

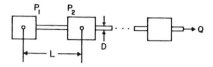

**Figure 3.21.** Gas pipeline.

Assume that the total annual cost of the transmission system and its operation is

$$C(D, P_1, L, r) = 7.84D^2P_1 + 450{,}000 + 36{,}900D + \frac{(6.57)(10^6)}{L}$$

$$+ \frac{(772)(10^6)}{L}(r^{0.219} - 1) \qquad (\$/\text{yr}) \qquad (1)$$

where  $D$ = pipe I.D., in.
  $P_1$ = compressor discharge pressure, psia
  $L$ = length between compressor stations, miles
  $r$ = compression ratio = $P_1/P_2$

Furthermore, assume that the flowrate is

$$Q = 3.39\left[\frac{(P_1^2 - P_2^2)D^5}{fL}\right]^{1/2} \qquad (\text{SCF}/\text{hr}) \qquad (2)$$

with $f$ = friction factor = $.008\ D^{-1/3}$. Let the flow rate be $100 \times 10^6$ SCF/day, and use expression (2) to eliminate $P_1$ from (1). Then find the minimum cost design using the simplex search method of Nelder and Mead and Powell's conjugate directions direct-search method.

**3.21.** Locate the minima of Himmelblau's function:

$$f(x) = \left(x_1^2 + x_2 - 11\right)^2 + \left(x_1 + x_2^2 - 7\right)^2$$

(see Figure 3.1) from

$$x^{(1)} = [5,5]^T \qquad x^{(2)} = [5,-5]^T \qquad x^{(3)} = [0,0]^T$$

$$x^{(4)} = [-5,-5]^T \qquad x^{(5)} = [5,0]^T$$

using the pattern search method of Hooke-Jeeves. Terminate when the elements of $x$ are accurate to three decimal places.

**3.22.** Given the current iterate $x^{(k)}$, search direction $S(x^{(k)})$, and an iteration of the form in (3.42), show that

$$\alpha^* = \frac{\nabla q(x^{(k)})^T S(x^{(k)})}{s(x^{(k)})^T C S(x^{(k)})}$$

assuming the objective function is a quadratic:

$$f(x) = q(x) = a + b^T x + \tfrac{1}{2} x^T C x$$

**3.23.** Find the dimensions of the minimum-cost open-top rectangular container. (Let $v^* = 10 \text{ m}^3$.)

**3.24.** Given $q(x) = 8x_1^2 + 4x_1 x_2 + 5x_2^2$, $x^{(0)} = [10, 10]^T$, and two linearly independent directions,

$$d^{(0)} = \nabla q(x^{(0)}) = [200, 140]^T \qquad d^{(1)} = [7, -10]^T$$

Form a new direction $(S^{(1)} = d^{(1)} + \beta_{11} \Delta g^{(1)})$ conjugate to $d^{(0)}$. Search them both in turn to find $x^*$. That is, search along $S^{(0)} = d^{(0)}$ first; then, from this minimum point, search along $S^{(1)}$. How does $S^{(1)}$ compare to the Fletcher-Reeves direction?

**3.25.** Find a direction orthogonal to the vector

$$s_1 = \left[ \frac{1}{\sqrt{3}}, -\frac{1}{\sqrt{3}}, -\frac{1}{\sqrt{3}} \right]^T \qquad x = [0,0,0]^T$$

Also find a new direction $s_2$ conjugate to $s_1$ with respect to the objective function $f(x) = x_1 + 2x_2^2 - x_1 x_2 + 3x_1^2 - 2x_1 x_3 + x_3^2$ starting from the point $x$.

**3.26.** Locate and classify the stationary points of

$$f(x) = x_1^3 - x_1 x_2 + x_2^2 - 2x_1 + 3x_2 - 4$$

**3.27.** Eason has shown that the allocation of gear ratios in a triple reduction spur-gear train for minimum inertia with an overall reduction of 10 reduces to the minimization of the function in (3.93). Use the Hooke-Jeeves, Cauchy, and Fletcher-Reeves methods to approximate $x^*$ with $x^{(0)} = [.5, .5]^T$.

**3.28.** Given the function

$$f(x) = 100(x_2 - x_1^2)^2 + (1 - x_1)^2$$

and

$$x^{(0)} = [-1.2, 1]^T \qquad x^{(1)} = [-1.3, 1.07]^T$$

the first two points in a search for $x^*$, the minimum of $f$. Calculate the search direction at $x^{(1)}$ using the following gradient-based methods:

(a) Cauchy

(b) Modified Newton

(c) Fletcher-Reeves

(d) Marquardt ($\lambda^{(1)} = 100$)

**3.29.** We would like to study the effect of $\varepsilon_2$, the line search convergence criterion, on the performance of the Fletcher-Reeves, Davidon-Fletcher-Powell, and the Broyden-Fletcher-Shanno methods. Apply each method to Wood's function (3.94) from $x^{(0)} = [-3, -1, -3, -1]^T$. Let $\varepsilon_2 = \alpha \varepsilon_1$; $\alpha = .01, .1, 1,$ and 10; and $\varepsilon_1$, the overall convergence criterion, chosen appropriately for the computing environment. *Note:* On a CDC-6600, $\varepsilon_1 = 10^{-4}$ is appropriate.

**3.30.** Consider once again the grain storage bin of problem 10. Investigate the effect of the relative cost $c_2/c_1$ on the shape of the optimal bin. In particular, find the optimal design when $c_2/c_1 = .5, 1.5,$ and 3. Draw the resulting designs. What general conclusions can be drawn?

**3.31.** Perform three iterations of the Cauchy, modified Newton, and DFP methods on Powell's function:

$$f(x) = (x_1 + 10x_2)^2 + 5(x_3 - x_4)^2 + (x_2 - 2x_3)^4$$

$$+ 10(x_1 - x_4)^4$$

from

$$x^{(0)} = [3, -1, 0, 1]^T$$

**3.32.** An engineer has performed some experiments and observed a certain quantity $Q$ as a function of a variable $t$. He has good reasons to believe that there is a physical law relating $t$ and $Q$ that takes the form

$$Q(t) = a \sin t + b \tan t + c$$

He wants to have the "best" possible idea of the value of the coefficients $a$, $b$, and $c$ using the results of his $n$ experiments $(t_1, Q_1; t_2, Q_2; t_3, Q_3; \ldots; t_n, Q_n)$. Taking into account that his experiments are not perfect and also perhaps that the relation is not rigorous, he does not expect to find a perfect fit. So he defines an error term $e_i$, where

$$e_i = Q_i - Q(t_i)$$

His first idea is to minimize the function

$$R = \sum_{i=1}^{n} e_i^2$$

Show with detailed justifications that he will have to solve a system of linear equations in order to find the best values of $a$, $b$, and $c$. *Hint:* First show that the engineers' problem reduces to minimizing an unconstrained convex function.

**3.33.** From theoretical considerations it is believed that the dependent variable $y$ is related to variable $x$ via a two-parameter function:

$$y(x) = \frac{k_1 x}{1 + k_2 x}$$

The parameters $k_1$ and $k_2$ are to be determined by a least squares fit of the following experimental data:

| | |
|---|---|
| 1.0 | 1.05 |
| 2.0 | 1.25 |
| 3.0 | 1.55 |
| 4.0 | 1.59 |

Find $k_1$ and $k_2$, using the BFS variable metric method.

**3.34.** Show that $\gamma^{(i)} = 0$; $i = 0, 1, 2, \ldots, k - 2$, when the conjugate gradient update form (3.62) is employed with $S^{(0)} = -g^{(0)}$, (3.42), and $f(x)$ is quadratic.

# Chapter 4

# Linear Programming

The term *linear programming* defines a particular class of optimization problems in which the constraints of the system can be expressed as linear equations or inequalities and the objective function is a linear function of the design variables. Linear programming (LP) techniques are widely used to solve a number of military, economic, industrial, and societal problems. The primary reasons for its wide use are the availability of commercial software to solve very large problems and the ease with which data variation (sensitivity analysis) can be handled through LP models.

In a 1976 survey of American companies [1], linear programming came out as the most often used technique (74%) among all the optimization methods. It was reported in a 1979 issue of *Science News* [2] that about one-fourth of the computer time spent on scientific computations in recent years had been devoted to solving LP problems and their many variations!

## 4.1 FORMULATION OF LINEAR PROGRAMMING MODELS

*Formulation* refers to the construction of LP models of real problems. Model building is not a science; it is primarily an art that is developed mainly by experience. The basic steps involved in formulating an LP model are to identify the design/decision variables, express the constraints of the problem as linear equations or inequalities, and write the objective function to be maximized or minimized as a linear function. We shall illustrate the basic steps in formulation with some examples.

### Example 4.1 An Inspection Problem

A company has two grades of inspectors, 1 and 2, who are to be assigned to a quality control inspection. It is required that at least 1800 pieces be inspected per 8-hr day. Grade 1 inspectors can check pieces at the rate of 25 per hour, with an accuracy of 98 percent. Grade 2 inspectors check at the rate of 15 pieces per hour, with an accuracy of 95 percent.

The wage rate of a grade 1 inspector is \$4.00/hr, while that of a grade 2 inspector is \$3.00/hr. Each time an error is made by an inspector, the cost to the company is \$2.00. The company has available for the inspection job eight grade 1 inspectors, and ten grade 2 inspectors. The company wants to determine the optimal assignment of inspectors that will minimize the total cost of the inspection.

**Formulation.** Let $x_1$ and $x_2$ denote the number of grade 1 and grade 2 inspectors assigned to inspection. Since the number of available inspectors in each grade is limited, we have the following constraints:

$$x_1 \leqslant 8 \qquad \text{(grade 1)}$$
$$x_2 \leqslant 10 \qquad \text{(grade 2)}$$

The company requires that at least 1800 pieces be inspected daily. Thus, we get

$$8(25)x_1 + 8(15)x_2 = 200x_1 + 120x_2 \geqslant 1800$$

or

$$5x_1 + 3x_2 \geqslant 45$$

In order to develop the objective function, we note that the company incurs two types of costs during inspection: (1) wages paid to the inspector, and (2) cost of his inspection errors. The cost of each grade 1 inspector is

$$\$4 + \$2(25)(.02) = \$5/\text{hr}$$

Similarly, for each grade 2 inspector,

$$\$3 + \$2(15)(.05) = \$4.50/\text{hr}$$

Thus, the objective function is to minimize the daily cost of inspection given by

$$Z = 8(5x_1 + 4.50x_2) = 40x_1 + 36x_2$$

The complete formulation of the LP problem thus becomes

$$\text{Minimize} \qquad Z = 40x_1 + 36x_2$$

$$\text{Subject to} \quad x_1 \leqslant 8 \qquad x_2 \leqslant 10$$

$$5x_1 + 3x_2 \geqslant 45$$

$$x_1 \geqslant 0 \qquad x_2 \geqslant 0$$

## Example 4.2  Hydroelectric Power Systems Planning

An agency controls the operation of a system consisting of two water reservoirs with one hydroelectric power generation plant attached to each as shown in Figure 4.1. The planning horizon for the system is broken into two periods. When the reservoir is at full capacity, additional inflowing water is spilled over a spillway. In addition, water can also be released through a spillway as desired for flood protection purposes. Spilled water does not produce any electricity.

Assume that on an average 1 kilo-acre-foot (KAF) of water is converted to 400 megawatt hours (MWh) of electricity by power plant A and 200 MWh by power plant B. The capacities of power plants A and B are 60,000 and 35,000 MWh per period. During each period, up to 50,000 MWh of electricity can be sold at \$20.00/MWh, and excess power above 50,000 MWh can only be sold for \$14.00/MWh. The following table gives additional data on the reservoir operation and inflow in kilo-acre-feet:

|  | Reservoir A | Reservoir B |
|---|---|---|
| Capacity | 2000 | 1500 |
| Predicted inflow |  |  |
|    Period 1 | 200 | 40 |
|    Period 2 | 130 | 15 |
| Minimum allowable level | 1200 | 800 |
| Level at the beginning |  |  |
|    of period 1 | 1900 | 850 |

Develop a linear programming model for determining the optimal operating policy that will maximize the total revenue from electricity sales.

**Formulation.**  This example illustrates how certain nonlinear objective functions can be handled by LP methods. The nonlinearity in the objective function is due to the differences in the unit megawatt-hour revenue from electricity sales depending on the amount sold. The nonlinearity is apparent if a graph is plotted between total revenue and electricity sales. This is illustrated by the graph of Figure 4.2, which is a piecewise linear function, since it is

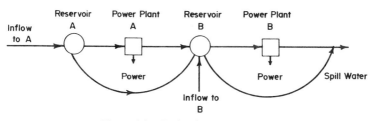

**Figure 4.1.**  Hydroelectric system.

linear in the regions $(0, 50000)$ and $(50000, \infty)$. Hence, by partitioning the quantity of electricity sold into two parts, the part that is sold for \$20.00/MWh and the part that is sold for \$14.00/MWh, the objective function could be represented as a linear function.

Let us first formulate the problem of reservoir operation for period 1 and then extend it to period 2.

| | |
|---|---|
| PH1 | Power sold at \$20/MWh, MWh |
| PL1 | Power sold at \$14/MWh, MWh |
| XA1 | Water supplied to power plant A, KAF |
| XB1 | Water supplied to power plant B, KAF |
| SA1 | Spill water drained from reservoir A, KAF |
| SB1 | Spill water drained from reservoir B, KAF |
| EA1 | Reservoir A level at the end of period 1, KAF |
| EB1 | Reservoir B level at the end of period 1, KAF |

The total power produced in period 1 = 400XA1 + 200XB1, while the total power sold = PH1 + PL1. Thus, the power generation constraint becomes

$$400XA1 + 200XB1 = PH1 + PL1$$

Since the maximum power that can be sold at the higher rate is 50,000 MWh, we get PH1 $\leqslant$ 50,000. The capacity of power plant A, expressed in water units (KAF) is

$$\frac{60,000}{400} = 150$$

Hence,

$$XA1 \leqslant 150$$

Similarly,

$$XB1 \leqslant 87.5$$

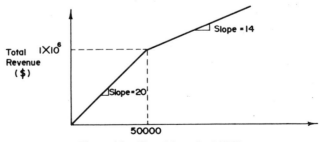

**Figure 4.2.**   Electricity sales MWH.

The conservation of water flow for reservoir A is given by

Water supplied to power plant A + spill water + ending reservoir level
= beginning reservoir level + predicted inflow

$$XA1 + SA1 + EA1 = 1900 + 200 = 2100$$

Since the capacity of reservoir A is 2000 KAF and the minimum allowable level is 1200, we get the following additional constraints:

$$EA1 \leqslant 2000$$

$$EA1 \geqslant 1200$$

Similarly, for reservoir B, we get

$$XB1 + SB1 + EB1 = 850 + 40 + XA1 + SA1$$

$$800 \leqslant EB1 \leqslant 1500$$

We can now develop the model for period 2, by defining the appropriate decision variables PH2, PL2, XA2, XB2, SA2, SB2, EA2, and EB2.

The power generation constraints for period 2 are given by

$$400XA2 + 200XB2 = PH2 + LH2$$

$$PH2 \leqslant 50,000$$

$$XA2 \leqslant 150$$

$$XB2 \leqslant 87.5$$

The water flow constraints are given by

$$XA2 + SA2 + EA2 = EA1 + 130$$

$$1200 \leqslant EA2 \leqslant 2000$$

$$XB2 + SB2 + EB2 = EB1 + 15 + XA2 + SA2$$

$$800 \leqslant EB2 \leqslant 1500$$

In addition to the above constraints, we have the nonnegativity restrictions on all the decision variables.

The objective function is to maximize the total revenue from sales, given by

$$\text{Maximize} \quad Z = 20(PH1 + PH2) + 14(PL1 + PL2)$$

Thus, the final model has 16 decision variables and 20 constraints, excluding the nonnegativity constraints.

## 4.2  GRAPHICAL SOLUTION OF LINEAR PROGRAMS IN TWO VARIABLES

In Section 4.1, some examples were presented to illustrate how practical problems can be formulated mathematically as LP problems. The next step after formulation is to solve the problem mathematically to obtain the best possible solution. In this section, a graphical procedure to solve LP problems involving only two variables is discussed. Although in practice such small problems are not usually encountered, the graphical procedure is presented to illustrate some of the basic concepts used in solving large LP problems.

*Example 4.3*

Recall the inspection problem given by Example 4.1:

$$\text{Minimize} \quad Z = 40x_1 + 36x_2$$

$$\text{Subject to} \quad x_1 \leqslant 8 \qquad x_2 \leqslant 10$$

$$5x_1 \quad + 3x_2 \geqslant 45$$

$$x_1 \geqslant 0, \qquad x_2 \geqslant 0$$

In this problem, we are interested in determining the values of the variables $x_1$ and $x_2$ that will satisfy all the restrictions and give the least value for the objective function. As a first step in solving this problem, we want to identify all possible values of $x_1$ and $x_2$ that are nonnegative and satisfy the constraints. For example, the point $x_1 = 8$, $x_2 = 10$ is positive and satisfies all the constraints. Such a point is called a *feasible solution*. The set of all feasible solutions is called the *feasible region*. Solution of a linear program is nothing but finding the best feasible solution in the feasible region. The best feasible solution is called an *optimal solution* to the LP problem. In our example, an optimal solution is a feasible solution that minimizes the objective function $40x_1 + 36x_2$. The value of the objective function corresponding to an optimal solution is called the *optimal value* of the linear program.

To represent the feasible region in a graph, every constraint is plotted, and all values of $x_1, x_2$ that will satisfy these constraints are identified. The nonnegativity constraints imply that all feasible values of the two variables will lie in the first quadrant. The constraint $5x_1 + 3x_2 \geqslant 45$ requires that any feasible solution $(x_1, x_2)$ to the problem should be on one side of the straight line $5x_1 + 3x_2 = 45$. The proper side is found by testing whether the origin satisfies the constraints. The line $5x_1 + 3x_2 = 45$ is first plotted by taking two convenient points (e.g., $x_1 = 0$, $x_2 = 15$, and $x_1 = 9$, $x_2 = 0$).

The proper side is indicated by an arrow directed above the line, since the origin does not satisfy the constraint. Similarly, the constraints $x_1 \leq 8$ and $x_2 \leq 10$ are plotted. The feasible region is given by the shaded region $ABC$ shown in Figure 4.3. Obviously, there are an infinite number of feasible points in this region. Our objective is to identify the feasible point with the lowest value of $Z$.

Observe that the objective function, given by $Z = 40x_1 + 36x_2$, represents a straight line if the value of $Z$ is fixed a priori. Changing the value of $Z$ essentially translates the entire line to another straight line parallel to itself. In order to determine an optimal solution, the objective function line is drawn for a convenient value of $Z$ such that it passes though one or more points in the feasible region. Initially $Z$ is chosen as 600. When this line is moved closer to the origin, the value of $Z$ is further decreased (Figure 4.3). The only limitation on this decrease is that the straight line $40x_1 + 36x_2 = Z$ must contain at least one point in the feasible region $ABC$. This clearly occurs at the corner point $A$

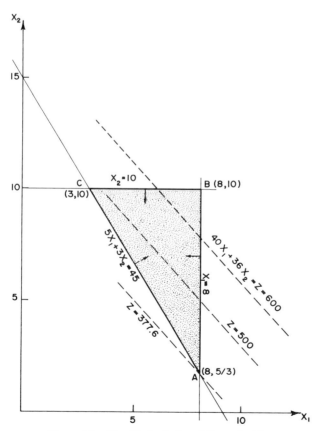

**Figure 4.3.** Graphical solution of Example 4.3.

given by $x_1 = 8$, $x_2 = 1.6$. This is the best feasible point, giving the lowest value of $Z$ as 377.60. Hence,

$$x_1 = 8 \qquad x_2 = 1.6$$

is an *optimal solution*, and $Z = 377.60$ is the *optimal value* for the linear program.

Thus for the inspection problem the optimal utilization is achieved by using eight grade 1 inspectors and 1.6 grade 2 inspectors. The fractional value $x_2 = 1.6$ suggests that one of the grade 2 inspectors is utilized for only 60 percent of the time. If this is not feasible, the normal practice is to round off the fractional values to get an approximate optimal integer solution as $x_1 = 8$, $x_2 = 2$.

**Unique Optimal Solution.**   In Example 4.3 the solution $x_1 = 8$, $x_2 = 1.6$ is the only feasible point with the lowest value of $Z$. In other words, the values of $Z$ corresponding to the other feasible solutions in Figure 4.3 exceed the optimal value of 377.60. Hence for this problem, the solution $x_1 = 8$, $x_2 = 1.6$ is called a *unique optimal solution*.

**Alternative Optimal Solutions.**   In some LP problems there may exist more than one feasible solution whose objective function values equal the optimal value of the linear program. In such cases, all these feasible solutions are optimal solutions, and the linear program is said to have *alternative* or *multiple optimal solutions*.

*Example 4.4*

$$\text{Maximize} \quad Z = x_1 + 2x_2$$

$$\text{Subject to} \quad x_1 + 2x_2 \leqslant 10$$

$$x_1 + x_2 \geqslant 1$$

$$x_2 \leqslant 4$$

$$x_1 \geqslant 0 \qquad x_2 \geqslant 0$$

The feasible region is shown in Figure 4.4. The objective function lines are drawn for $Z = 2$, 6, and 10. The optimal value for the linear program is 10, and the corresponding objective function line $x_1 + 2x_2 = 10$ coincides with side $BC$ of the feasible region. Thus, the corner-point feasible solutions $x_1 = 10$, $x_2 = 0$ ($B$), $x_1 = 2$, $x_2 = 4$ ($C$), and all other feasible points on the line $BC$ are optimal solutions.

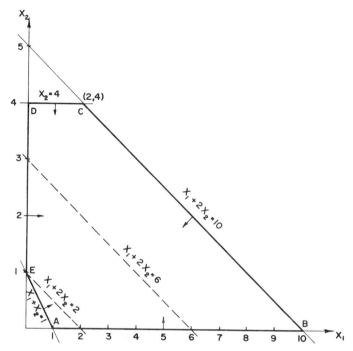

**Figure 4.4.**   Graphical solution of Example 4.4.

**Unbounded Optimum.**   It is possible for some LP problems not to have an optimal solution. In other words, it is possible to find better feasible solutions improving the objective function values continuously. This would have been the case if the constraint $x_1 + 2x_2 \leqslant 10$ were not given in Example 4.4. In this case, moving farther away from the origin increases the objective function $x_1 + 2x_2$, and maximum $Z$ would be $+\infty$. In cases where no finite maximum exists, the linear program is said to have an *unbounded optimum*.

It is inconceivable for a practical problem to have an unbounded solution, since this implies that one can make infinite profit from a finite amount of resources! If such a solution is obtained in a practical problem it usually means that one or more constraints have been omitted inadvertently during the initial formulation of the problem. These constraints would have prevented the objective function from assuming infinite values.

**Conclusion.**   In Example 4.3 the optimal solution was unique and occurred at the corner point $A$ in the feasible region. In Example 4.4 we had multiple optimal solutions to the problem, which included two corner points $B$ and $C$. In either case, one of the corner points of the feasible region was always an optimal solution. As a matter of fact, the following property is true for any LP problem.

### Property 1

If there exists an optimal solution to a linear programming problem, then at least one of the corner points of the feasible region will always qualify to be an optimal solution.

This is the fundamental property on which the simplex method for solving LP problems is based. Even though the feasible region of an LP problem contains an infinite number of points, an optimal solution can be determined by merely examining the finite number of corner points in the feasible region. For instance, in Example 4.4 the objective function was to maximize

$$Z = x_1 + 2x_2$$

The corner points of the feasible region were $A$ $(1,0)$, $B$ $(10,0)$, $C$ $(2,4)$, $D$ $(0,4)$, and $E$ $(0,1)$. Evaluating their $Z$ values, we get $Z(A) = 1$, $Z(B) = 10$, $Z(C) = 10$, $Z(D) = 8$, and $Z(E) = 2$. Since the maximum value of $Z$ occurs at corner points $B$ and $C$, both are optimal solutions by Property 1.

## 4.3   LINEAR PROGRAM IN STANDARD FORM

The standard form of an LP problem with $m$ constraints and $n$ variables can be represented as follows:

$$\text{Maximize or Minimize}\quad Z = c_1 x_1 + c_2 x_2 + \cdots + c_n x_n$$

$$\text{Subject to}\quad a_{11}x_1 + a_{12}x_2 + \cdots + a_{1n}x_n = b_1$$

$$a_{21}x_1 + a_{22}x_2 + \cdots + a_{2n}x_n = b_2$$

$$\vdots \qquad\qquad\qquad \vdots$$

$$a_{m1}x_1 + a_{m2}x_2 + \cdots + a_{mn}x_n = b_m$$

$$x_1 \geqslant 0 \qquad x_2 \geqslant 0 \qquad \cdots \qquad x_n \geqslant 0$$

$$b_1 \geqslant 0 \qquad b_2 \geqslant 0 \qquad \cdots \qquad b_m \geqslant 0$$

In matrix-vector notation, the standard LP problem can be expressed compactly as

$$\text{Maximize or minimize} \quad Z = cx$$

$$\text{Subject to} \quad \mathbf{A}x = b$$

$$x \geqslant 0$$

$$b \geqslant 0$$

where $\mathbf{A}$ is an $m \times n$ matrix, $x$ is an $n \times 1$ column vector, $b$ is an $m \times 1$ column vector, and $c$ is a $1 \times n$ row vector.

In practice, $\mathbf{A}$ is called the *coefficient matrix*, $x$ is the *decision vector*, $b$ is the *requirement vector*, and $c$ is the *profit (cost) vector* of the linear program.

The simplex method for solving LP problems requires that the problem be expressed in standard form. But not all linear programs come in standard form. Very often the constraints are expressed as inequalities rather than equations. In some problems, not all the design variables may be restricted to be nonnegative. Hence, the first step in solving a linear program is to convert it to a problem in standard form.

### 4.3.1   Handling Inequalities

Inequality constraints can be converted to equations by introducing *slack* or *surplus* variables. For example, an inequality of the type

$$x_1 + 2x_2 + 3x_3 + 4x_4 \leqslant 25 \tag{4.1}$$

can be converted to an equation by introducing a *slack* variable $S_1$ as follows:

$$x_1 + 2x_2 + 3x_3 + 4x_4 + S_1 = 25$$

The variable $S_1$ is nonnegative, and it represents the slack between the left-hand side and the right-hand side of inequality (4.1). Similarly, an inequality of the type

$$2x_1 + x_2 - 3x_3 \geqslant 12$$

can be converted to an equation by introducing a *surplus* variable $S_2$ as follows:

$$2x_1 + x_2 - 3x_3 - S_2 = 12$$

It should be emphasized here that slack and surplus variables are as much a part of the original problem as the design variables used in the formulation of the linear program. These variables can remain positive throughout, and their values in the optimal solution give useful information as to whether or not the inequality constraints are *binding* or *active* in the problem.

### 4.3.2  Handling Unrestricted Variables

In some situations, it may become necessary to introduce a variable that can assume both positive and negative values. Since the standard form requires all variables to be nonnegative, an unrestricted variable is generally replaced by the difference of two nonnegative variables. For example, if $s_1$ is an unrestricted variable, then the following transformation of variables is used for $s_1$:

$$s_1 = s_1^+ - s_1^-$$

$$s_1^+ \geqslant 0 \qquad s_1^- \geqslant 0$$

The value of $s_1$ is positive or negative depending on whether $s_1^+$ is larger or smaller than $s_1^-$.

As an illustration, consider the following nonstandard linear program.

***Example 4.5***

$$\text{Maximize} \quad Z = x_1 - 2x_2 + 3x_3$$

$$\text{Subject to} \qquad x_1 + x_2 + x_3 \leqslant 7 \qquad (4.2)$$

$$x_1 - x_2 + x_3 \geqslant 2 \qquad (4.3)$$

$$3x_1 - x_2 - 2x_3 = -5 \qquad (4.4)$$

$$x_1, x_2 \geqslant 0$$

where $x_3$ is unrestricted in sign.

To convert the above problem to standard form,

1.  Replace $x_3$ by $x_4 - x_5$, where $x_4, x_5 \geqslant 0$.
2.  Multiply both sides of Eq. (4.4) by $-1$.
3.  Introduce slack and surplus variables $x_6$ and $x_7$ to constraints (4.2) and (4.3), respectively.
4.  Assign zero profit for $x_6$ and $x_7$ so that the objective function is not altered.

Thus, the problem reduces to the following standard linear program:

$$\text{Maximize} \quad Z = x_1 - 2x_2 + 3x_4 - 3x_5$$

$$\text{Subject to} \quad x_1 + x_2 + x_4 - x_5 + x_6 \qquad = 7$$

$$x_1 - x_2 + x_4 - x_5 \qquad - x_7 = 2$$

$$-3x_1 + x_2 + 2x_4 - 2x_5 \qquad = 5$$

$$x_1, x_2, x_4, x_5, x_6, x_7 \geqslant 0$$

Let us review the basic definitions using the standard form of the linear programming problem given by

$$\text{Maximize} \quad Z = cx$$

$$\text{Subject to} \quad Ax = b \qquad x \geqslant 0$$

1. A *feasible solution* is a nonnegative vector $x$ satisfying the constraints $Ax = b$.
2. A *feasible region*, denoted by $S$, is the set of all feasible solutions. Mathematically,

$$S = \{x \mid Ax = b, x \geqslant 0\}$$

If the feasible set $S$ is empty, then the linear program is said to be *infeasible*.
3. An *optimal solution* is a vector $x^\circ$ such that it is feasible, and its value of the objective function $(cx^\circ)$ is larger than that of any other feasible solution. Mathematically, $x^\circ$ is optimal if and only if $x^\circ \in S$, and $cx^\circ \geqslant cx$ for all $x \in S$.
4. The *optimal value* of a linear program is the value of the objective function corresponding to the optimal solution. If $Z^\circ$ is the optimal value, then $Z^\circ = cx^\circ$.
5. *Alternate optimal solution.* When a linear program has more than one optimal solution, it is said to have *alternate optimal solutions*. In this case, there exists more than one feasible solution having the same optimal value $(Z^\circ)$ of the objective function.
6. *Unique optimum.* The optimal solution of a linear program is said to be *unique* when no other optimal solution exists.
7. *Unbounded optimum* When a linear program does not possess a finite optimum (i.e., max $Z \rightarrow +\infty$ or min $Z \rightarrow -\infty$), it is said to have an *unbounded optimum*.

## 4.4   PRINCIPLES OF THE SIMPLEX METHOD

Consider the general LP problem in standard form with $m$ equations and $n$ variables:

$$\text{Maximize} \quad Z = c_1 x_1 + c_2 x_2 + \cdots + c_n x_n$$

$$\text{Subject to} \quad a_{11} x_1 + a_{12} x_2 + \cdots + a_{1n} x_n = b_1$$

$$a_{21} x_1 + a_{22} x_2 + \cdots + a_{2n} x_n = b_2$$

$$\vdots$$

$$a_{m1} x_1 + a_{m2} x_2 + \cdots + a_{mn} x_n = b_m$$

$$x_1, \ldots, \quad x_n \geqslant 0$$

In general, there are more variables than equations (i.e., $m < n$), and consequently the problem has an infinite number of feasible solutions. Hence, the selection of the best feasible solution, which maximizes $Z$, is a nontrivial problem.

A classical method of generating the solutions to the constraint equations is the *Gauss-Jordan elimination scheme* (see Appendix C for a review). The basic principle of the Gauss-Jordan elimination scheme is to reduce the system of $m$ equations in $n$ unknowns to a *canonical* or *row-echelon form* by elementary row operations. One such canonical reduction using the first $m$ variables $(x_1, x_2, \ldots, x_m)$ is

$$x_1 \quad + \quad \bar{a}_{1, m+1} x_{m+1} + \cdots + \bar{a}_{1s} x_s + \cdots + \bar{a}_{1n} x_n = \bar{b}_1$$

$$x_r + \bar{a}_{r, m+1} x_{m+1} + \cdots + \bar{a}_{rs} x_s + \cdots + \bar{a}_{rn} x_n = \bar{b}_r \quad \text{(S1)}$$

$$x_m + \bar{a}_{m, m+1} x_{m+1} + \cdots + \bar{a}_{ms} x_s + \cdots + \bar{a}_{mn} x_n = \bar{b}_m$$

The variables $x_1, \ldots, x_m$ which appear with a unit coefficient in only one equation and with zeros in all the other equations are called *basic* or *dependent* variables. In the canonical form, there is one such basic variable in every equation. The other $n - m$ variables $(x_{m+1}, \ldots, x_n)$ are called *nonbasic* or *independent* variables.

The advantage of obtaining a canonical form is that multiple solutions to the constraint equations can be obtained by simply choosing different values for the independent variables and solving the canonical system for the dependent variables. There are two types of elementary row operations that can be used to obtain a canonical or row echelon form:

1.  Multiply any equation in the system by a positive or negative number.
2.  Add to any equation a constant multiple (positive or negative) of any other equation in the system.

## Definition

A *pivot operation* is a sequence of elementary row operations that reduces the coefficient of a specified variable to unity in one of the equations and zero elsewhere.

## Definition

The solution obtained from a canonical system by setting the nonbasic or independent variables to zero is called a *basic solution*.

For example, in system (S1) a basic solution is given by $x_1 = \bar{b}_1, \ldots, x_m = \bar{b}_m$, $x_{m+1} = \cdots = x_n = 0$. If the value of $\bar{b}_1, \ldots, \bar{b}_m$ are also nonnegative, then this basic solution will be called a basic feasible solution.

## Definition

A *basic feasible solution* is a basic solution in which the values of the basic or dependent variables are nonnegative.

In system (S1), a canonical form was obtained using the first $m$ variables as basic variables. The choice of $x_1, \ldots, x_m$ as basic variables was purely for convenience. As a matter of fact, any set of $m$ variables could have been selected as basic variables out of the possible $n$ variables to obtain a canonical system and a basic solution. This means that with $m$ constraints and $n$ variables, the maximum number of basic solutions to the standard linear program is finite and is given by

$$\binom{n}{m} = \frac{n!}{m!(n-m)!}$$

By definition, every basic feasible solution is also a basic solution. Hence, the maximum number of basic feasible solutions is also limited by the above expression.

At the end of Section 4.2 it was pointed out that whenever there is an optimal solution to a linear program, one of the corner points of the feasible region is always optimal. It can be shown that every corner point of the feasible region corresponds to a basic feasible solution of the constraint equations. This means that an optimal solution to a linear program can be obtained by merely examining its basic feasible solutions. This will be a finite process since the number of basic feasible solutions cannot exceed $\binom{n}{m}$.

A naive approach to solving a linear program (which has an optimal solution) would be to generate all possible basic feasible solutions through canonical reduction and determine which one gives the best objective function value. But the *simplex method* for solving linear programs does this in a more efficient manner by examining only a fraction of the total number of basic feasible solutions! We shall now develop the details of the simplex method.

The simplex method as developed by G. B. Dantzig is an iterative procedure for solving LP problems expressed in standard form. In addition to the standard form, the simplex method requires that the constraint equations be expressed in canonical form from which a basic feasible solution could be obtained readily. The general steps of the simplex method are as follows:

1. Start with an initial basic feasible solution in canonical form.
2. Improve the initial solution if possible by finding another basic feasible solution with a better objective function value. At this step the simplex method implicitly eliminates from consideration all those basic feasible solutions whose objective function values are worse than the present one. This makes the procedure more efficient than the naive approach mentioned earlier.
3. Continue to find better basic feasible solutions, improving the objective function values. When a particular basic feasible solution cannot be improved further it becomes an optimal solution, and the simplex method terminates.

We begin by considering the details of step 2 of the simplex method by assuming that the basic solution given by system (S1) is feasible. Thus, we have an initial basic feasible solution in canonical form as follows:

$$\text{Basic:} \quad x_i = \bar{b}_i \geqslant 0 \quad \text{for } i = 1, 2, \ldots, m$$

$$\text{Nonbasic:} \quad x_j = 0 \quad \text{for } j = m + 1, \ldots, n$$

The set of basic variables will be called a *basis* and will be denoted $x_B$. Let the objective function coefficients of the basic variables be denoted $c_B$. For the initial basis,

$$x_B = (x_1, \ldots, x_m) \quad \text{and} \quad c_B = (c_1, c_2, \ldots, c_m)$$

Since the nonbasic variables are zero, the value of the objective function $Z$

corresponding to the initial basic feasible solution is given by

$$Z = c_B x_B = c_1 \bar{b}_1 + \cdots + c_m \bar{b}_m$$

Given this, the simplex method checks whether it is possible to find a better basic feasible solution with a higher value of $Z$. This is done by first examining whether the present solution is optimal. In case it is not optimal, the simplex method obtains an *adjacent basic feasible solution* with a larger value of $Z$ (or at least as large).

### Definition

An *adjacent basic feasible solution* differs from the present basic feasible solution in exactly one basic variable.

To obtain an adjacent basic feasible solution, the simplex method makes one of the basic variables a nonbasic variable and brings in a nonbasic variable as a basic variable in its place. The problem is to select the appropriate basic and nonbasic variables such that an exchange between them will give the maximum improvement to the objective function.

In any basic feasible solution, the basic variables can assume positive values while the nonbasic variables are always held at zero. Hence, making a nonbasic variable a basic variable is equivalent to increasing its value from zero to a positive quantity. Of course, the choice is made based on which nonbasic variable can improve the value of $Z$. This is determined by increasing the nonbasic variable by one unit and examining the resulting change in the value of the objective function.

To illustrate, consider the nonbasic variable $x_s$. Let us increase its value from 0 to 1 and study its effect on the objective function. Since we are interested in examining adjacent basic feasible solutions only, the values of the other nonbasic variables will continue to remain zero and system (S1) can be rewritten as follows:

$$x_1 \qquad\qquad + \bar{a}_{1s} x_s \ = \bar{b}_1$$

$$\vdots$$

$$x_r \qquad + \bar{a}_{rs} x_s \ = \bar{b}_r \qquad\qquad \text{(S2)}$$

$$\vdots$$

$$x_m + \bar{a}_{ms} x_s = \bar{b}_m$$

From system (S2), we obtain the new solution when $x_s$ increases from 0 to 1:

$$x_i = \bar{b}_i - \bar{a}_{is} \qquad \text{for } i = 1, \dots, m$$

$$x_s = 1$$

$$x_j = 0 \qquad \text{for } j = m + 1, \dots, n \quad \text{and} \quad j \neq s$$

The new value of the objective function is

$$Z = \sum_{i=1}^{m} c_i\left(\bar{b}_i - \bar{a}_{is}\right) + c_s$$

Hence, the net change in the value of $Z$ per unit increase in $x_s$, denoted $\bar{c}_s$, is

$$\bar{c}_s = \text{new value of } Z - \text{old value of } Z$$

$$= \sum_{i=1}^{m} c_i\left(\bar{b}_i - \bar{a}_{is}\right) + c_s - \sum_{i=1}^{m} c_i \bar{b}_i$$

$$= c_s - \sum_{i=1}^{m} c_i \bar{a}_{is} \tag{4.5}$$

$\bar{c}_s$ is called the *relative profit* of the nonbasic variable $x_s$, as opposed to its actual profit $c_s$ in the objective function. If $\bar{c}_s > 0$, then the objective function $Z$ can be increased further by making $x_s$ a basic variable. Equation (4.5), which gives the formula for calculating the relative profits is known as the *inner product rule*.

### Inner Product Rule

The relative profit coefficient of a nonbasic variable $x_j$, denoted by $\bar{c}_j$, is given by

$$\bar{c}_j = c_j - c_B \bar{P}_j$$

where $c_B$ corresponds to the profit coefficients of the basic variables and $\bar{P}_j$ corresponds to the $j$th column in the canonical system of the basis under consideration.

### Condition of Optimality

In a maximization problem, a basic feasible solution is optimal if the relative profits of its nonbasic variables are all *negative or zero*.

It is clear that when all $\bar{c}_j \leqslant 0$ for nonbasic variables, then every adjacent basic feasible solution has an objective function value lower than the present solution. This implies that we have a local maximum at this point. Since the objective function $Z$ is linear, a local maximum automatically becomes the global maximum.

For the sake of illustration, let us assume that $\bar{c}_s = \max \bar{c}_j > 0$, and hence the initial basic feasible solution is not optimal. The positive relative profit of $x_s$ implies that $Z$ will be increased by $\bar{c}_s$ units for every unit increase on the nonbasic variable $x_s$. Naturally, we would like to increase $x_s$ as much as possible so as to get the largest increase in the objective function. But as $x_s$ increases, the values of the basic variables change, and their new values are obtained from system (S2) as

$$x_i = \bar{b}_i - \bar{a}_{is}x_s \qquad \text{for } i = 1,\ldots, m \qquad (4.6)$$

If $\bar{a}_{is} < 0$, then $x_i$ increases as $x_s$ is increased, and if $\bar{a}_{is} = 0$, the value of $x_i$ does not change. However, if $\bar{a}_{is} > 0$, $x_i$ decreases as $x_s$ is increased and may turn negative (making the solution infeasible) if $x_s$ is increased indefinitely. Hence, the maximum increase in $x_s$ is given by the following rule:

$$\max x_s = \min_{\bar{a}_{is}>0}\left[\frac{\bar{b}_i}{\bar{a}_{is}}\right] \qquad (4.7)$$

Let

$$\frac{\bar{b}_r}{\bar{a}_{rs}} = \min_{\bar{a}_{is}>0}\left[\frac{\bar{b}_i}{\bar{a}_{is}}\right]$$

Hence, when $x_s$ is increased to $\bar{b}_r/\bar{a}_{rs}$, the basic variable $x_r$ turns zero first and is replaced by $x_s$ in the basis. $x_s$ becomes the new basic variable in row $r$, and the new basic feasible solution is given by

$$x_i = \bar{b}_i - \bar{a}_{is}\left(\frac{\bar{b}_r}{\bar{a}_{rs}}\right) \qquad \text{for all } i$$

$$x_s = \frac{\bar{b}_r}{\bar{a}_{rs}} \qquad (4.8)$$

$$x_j = 0 \qquad \begin{array}{l}\text{for all other nonbasic variables}\\ j = m + 1,\ldots, n \quad \text{and} \quad j \neq s\end{array}$$

Since a unit increase in $x_s$ increases the objective function $Z$ by $\bar{c}_s$ units, the total increase in $Z$ is given by

$$(\bar{c}_s)\left(\frac{\bar{b}_r}{\bar{a}_{rs}}\right) > 0$$

Equation (4.7), which determines the basic variables to leave the basis, is known as the *minimum ratio rule* in the simplex algorithm. Once again, the simplex algorithm checks whether the basic feasible solution given by (4.8) is optimal by calculating the relative profits for all the nonbasic variables, and the cycle is repeated until the optimality conditions are reached.

Let us now illustrate the basic steps of the simplex algorithm with an example.

### Example 4.6

Let us use the simplex method to solve the following problem:

$$\text{Maximize} \quad Z = 3x_1 + 2x_2$$

$$\text{Subject to} \quad -x_1 + 2x_2 \leqslant 4$$

$$3x_1 + 2x_2 \leqslant 14$$

$$x_1 - x_2 \leqslant 3$$

$$x_1 \geqslant 0 \qquad x_2 \geqslant 0$$

Converting the problem to standard form by the addition of slack variables, we get

$$\text{Maximize} \quad Z = 3x_1 + 2x_2$$

$$\text{Subject to} \quad -x_1 + 2x_2 + x_3 \qquad\qquad = 4$$

$$3x_1 + 2x_2 \qquad + x_4 \qquad = 14$$

$$x_1 - x_2 \qquad\qquad + x_5 = 3$$

$$x_1 \geqslant 0 \qquad x_2 \geqslant 0 \qquad x_3 \geqslant 0$$

$$x_4 \geqslant 0 \qquad x_5 \geqslant 0$$

We have a basic feasible solution in canonical form, with $x_3$, $x_4$, and $x_5$ as basic variables. The various steps of the simplex method can be carried out in a compact manner by using a tableau form to represent the constraints and the objective function. In addition, the various calculations (e.g., inner product rule, minimum ratio rule, pivot operation) can be made mechanical. The use of the tableau form has made the simplex method more efficient and convenient for computer implementation.

The tableau presents the problem in a detached coefficient form. Tableau 1 gives the initial basic feasible solution for Example 4.6. Note that only the coefficients of the variables are written for each constraint. The objective function coefficients $c_j$ are written above their respective $x_j$'s. *Basis* denotes the basic variables of the initial tableau, namely $x_3$ for constraint 1, $x_4$ for constraint 2, and $x_5$ for constraint 3. $c_B$ denotes their respective $c_j$ values. From Tableau 1, we obtain initial basic feasible solution immediately as $x_3 = 4$, $x_4 = 14$, $x_5 = 3$, and $x_1 = x_2 = 0$. The value of the objective function $Z$ is given by the inner product of the vectors $c_B$ and constants:

$$Z = (0,0,0)\begin{pmatrix} 4 \\ 14 \\ 3 \end{pmatrix} = 0$$

In order to check whether the above basic feasible solution is optimal, we calculate the relative profits using the inner product rule. Note that the relative profits $\bar{c}_3, \bar{c}_4, \bar{c}_5$ will be zero, since they are the basic variables. Both the nonbasic variables $x_1$ and $x_2$ have positive relative profits. Hence the current basic feasible solution is not optimal. The nonbasic variable $x_1$ gives the largest per-unit increase in $Z$, and hence we choose it as the new basic variable to enter the basis.

In order to decide which basic variable is going to be replaced, we apply the minimum ratio rule [Eq. (4.7)] by calculating the ratios for each constraint that has a positive coefficient under the $x_1$ column as follows:

| Row Number | Basic Variable | Ratio |
|:---:|:---:|:---:|
| 1 | $x_3$ | $\infty$ |
| 2 | $x_4$ | $\frac{14}{3}$ |
| 3 | $x_5$ | $\frac{3}{1}$ |

**Tableau 1**

| $c_B$ | Basis | | | | | | Constants |
|---|---|---|---|---|---|---|---|

| $c_B$ | Basis | $x_1$ | $x_2$ | $x_3$ | $x_4$ | $x_5$ | Constants |
|:---:|:---:|:---:|:---:|:---:|:---:|:---:|:---:|
| | | \(c_j\) | | | | | |
| | | 3 | 2 | 0 | 0 | 0 | |
| 0 | $x_3$ | $-1$ | 2 | 1 | 0 | 0 | 4 |
| 0 | $x_4$ | 3 | 2 | 0 | 1 | 0 | 14 |
| 0 | $x_5$ | ① | $-1$ | 0 | 0 | 1 | 3 |
| | $\bar{c}$ row | 3 | 2 | 0 | 0 | 0 | $Z = 0$ |

Note that no ratio is formed for the first row or is set to $\infty$, since the coefficient for $x_1$ in row 1 is negative. This means that $x_1$ can be increased to any amount without making $x_3$ negative. On the other hand, the ratio $14/3$ for the second row implies that $x_4$ will become zero when $x_1$ increases to $14/3$. Similarly, as $x_1$ increases to 3, $x_5$ turns zero. The minimum ratio is 3, or as $x_1$ is increased from 0 to 3, the basic variable $x_5$ will turn zero first and will be replaced by $x_1$. Row 3 is called the *pivot row*, and the coefficient of $x_1$ in the pivot row is called the *pivot element* (circled in Tableau 1). We obtain the new basic variables are $x_3$, $x_4$, and $x_1$, and the new canonical system by performing a pivot operation as follows:

1. Add the pivot row (row 3) to the first row to eliminate $x_1$.
2. Multiply the pivot row by $-3$, and add it to the second row to eliminate $x_1$.

In order to check whether the basic feasible solution given by Tableau 2 is optimal, we must calculate the new relative profit coefficients ($\bar{c}$ row). We can do this by applying the inner product rule as before. On the other hand, it is also possible to calculate the new $\bar{c}$ row through the pivot operation. Since $x_1$ is the new basic variable, its relative profit coefficient in Tableau 2 should be zero. To achieve this, we multiply the third row of Tableau 1 (the pivot row) by $-3$ and add it to the $\bar{c}$ row. This will automatically give the new $\bar{c}$ row for Tableau 2!

The movement from Tableau 1 to Tableau 2 is illustrated pictorially in Figure 4.5. The feasible region of Example 4.6 is denoted $ABCDE$. The dotted lines correspond to the objective function lines at $Z = 3$ and $Z = 9$. Note that the basic feasible solution represented by Tableau 1 corresponds to the corner point $A$. The $\bar{c}$ row in Tableau 1 indicates that either $x_1$ or $x_2$ may be made a basic variable to increase $Z$. From Figure 4.5 it is clear that either $x_1$ or $x_2$ can

**Tableau 2**

| $c_B$ | Basis | $c_j$ | | | | | Constants |
|---|---|---|---|---|---|---|---|
| | | 3 | 2 | 0 | 0 | 0 | |
| | | $x_1$ | $x_2$ | $x_3$ | $x_4$ | $x_5$ | |
| 0 | $x_3$ | 0 | 1 | 1 | 0 | 1 | 7 |
| 0 | $x_4$ | 0 | ⑤ | 0 | 1 | $-3$ | 5 |
| 3 | $x_1$ | 1 | $-1$ | 0 | 0 | 1 | 3 |
| $\bar{c}$ row | | 0 | 5 | 0 | 0 | $-3$ | $Z = 9$ |

be increased to increase $Z$. Having decided to increase $x_1$, it is clear that we cannot increase $x_1$ beyond three units (point $B$) in order to remain in the feasible region. This was essentially the minimum ratio value that was obtained by the simplex method. Thus, Tableau 2 is simply corner point $B$ in Figure 4.5.

Still the $\bar{c}$ row has a positive element, indicating that the nonbasic variable $x_2$ can improve the objective function further. To apply the minimum ratio rule, we find the minimum of $(\frac{7}{1}, \frac{5}{3}, \infty)$. This implies that $x_2$ replaces the basic variable $x_4$. The next basic feasible solution after the pivot operation is given by Tableau 3. Since none of the coefficients in the $\bar{c}$ row is positive, this tableau is optimal. An optimal solution is given by $x_1 = 4$, $x_2 = 1$, $x_3 = 6$, $x_4 = 0$, $x_5 = 0$, and the optimal value of $Z$ is 14.

**Alternate Optima.**   In Tableau 3, the nonbasic variable $x_5$ has a relative profit of zero. This means that no increase in $x_5$ will produce any change in the objective function value. In other words $x_5$ can be made a basic variable and the resulting basic feasible solution will also have $Z = 14$. By definition, any feasible solution whose value of $Z$ equals the optimal value is also an optimal solution. Hence, we have an alternate optimal solution to this linear program.

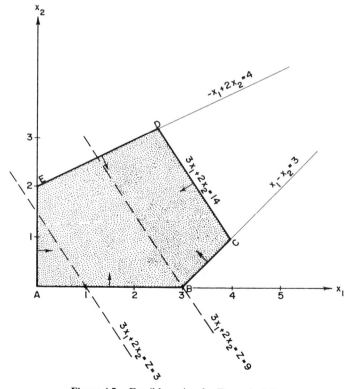

**Figure 4.5.**   Feasible region for Example 4.6.

LINEAR PROGRAMMING

## Tableau 3

| $c_B$ | Basis | $c_j$ | | | | | Constants |
|---|---|---|---|---|---|---|---|
| | | 3 | 2 | 0 | 0 | 0 | |
| | | $x_1$ | $x_2$ | $x_3$ | $x_4$ | $x_5$ | |
| 0 | $x_3$ | 0 | 0 | 1 | $-\frac{1}{5}$ | $\frac{8}{5}$ | 6 |
| 2 | $x_2$ | 0 | 1 | 0 | $\frac{1}{5}$ | $-\frac{3}{5}$ | 1 |
| 3 | $x_1$ | 1 | 0 | 0 | $\frac{1}{5}$ | $\frac{2}{5}$ | 4 |
| $\bar{c}$ row | | 0 | 0 | 0 | $-1$ | 0 | $Z = 14$ |

This can be obtained by making $x_5$ a basic variable. By the minimum ratio rule, $x_3$ leaves the basis, and we have an alternative optimal tableau as shown in Tableau 4. Thus the alternate optimal solution is given by $x_1 = \frac{5}{2}$, $x_2 = \frac{13}{4}$, $x_3 = 0$, $x_4 = 0$, and $x_5 = \frac{15}{4}$.

In general, an alternate optimal solution is indicated whenever there exists a nonbasic variable whose relative profit ($\bar{c}_j$ coefficient) is zero in the optimal tableau. Referring to Figure 4.5, the movement from Tableau 2 to Tableau 3 by the simplex method corresponds to moving along the edge $BC$ of the feasible region. Using the objective function $Z = 3x_1 + 2x_2$, it is clear that corner points $C$ and $D$, as well as the line $CD$, are optimal.

### Summary of Computational Steps

In summary, the computational steps of the simplex method in tableau form for a *maximization problem* are as follows:

**Step 1.** Express the problem in standard form.

## Tableau 4

| $c_B$ | Basis | $c_j$ | | | | | Constants |
|---|---|---|---|---|---|---|---|
| | | 3 | 2 | 0 | 0 | 0 | |
| | | $x_1$ | $x_2$ | $x_3$ | $x_4$ | $x_5$ | |
| 0 | $x_5$ | 0 | 0 | $\frac{5}{8}$ | $-\frac{1}{8}$ | 1 | $\frac{15}{4}$ |
| 2 | $x_2$ | 0 | 1 | $\frac{3}{8}$ | $\frac{1}{8}$ | 0 | $\frac{13}{4}$ |
| 3 | $x_1$ | 1 | 0 | $-\frac{1}{4}$ | $\frac{1}{4}$ | 0 | $\frac{5}{2}$ |
| $\bar{c}$ row | | 0 | 0 | 0 | $-1$ | 0 | $Z = 14$ |

**Step 2.** Start with an initial basic feasible solution in canonical form, and set up the initial tableau. (We shall discuss later in the section how to find an initial basic feasible solution in canonical form if none exists by inspection.)

**Step 3.** Use the inner product rule to find the relative profit coefficients ($\bar{c}$ row).

**Step 4.** If all the $\bar{c}_j$ coefficients are nonpositive, the current basic feasible solution is optimal. Otherwise, select the nonbasic variable with the most positive $\bar{c}_j$ value to enter the basis.

**Step 5.** Apply the minimum ratio rule to determine the basic variable to leave the basis.

**Step 6.** Perform the pivot operation to get the new tableau and the basic feasible solution.

**Step 7.** Compute the relative profit coefficients by using the pivot operation or the inner product rule. Return to step 4.

Each sequence of steps 4 through 7 is called an *iteration* of the simplex method. Thus each iteration gives a new tableau and an improved basic feasible solution. The efficiency of the simplex method depends on the number of basic feasible solutions it examines before reaching the optimal solution. Hence, the number of iterations is an important factor in simplex calculations.

### 4.4.1  Minimization Problems

There are two approaches to solving a minimization problem:

**Approach 1.** Convert the minimization problem to an equivalent maximization problem by multiplying the objective function by $-1$, and then use the simplex method as outlined for a maximization problem.

**Approach 2.** Recall that the coefficients in the $\bar{c}$ row give the net change in the value of $Z$ per unit increase in the nonbasic variable. A negative coefficient in the $\bar{c}$ row indicates that the corresponding nonbasic variable (when increased) will reduce the value of the objective function. Hence, in minimization problems, only those nonbasic variables with negative $\bar{c}_j$ values are eligible to enter the basis and improve the objective function. The optimal solution is obtained when all the coefficients in the $\bar{c}$ row are nonnegative. Thus all seven steps of the simplex method can be used for solving minimization problems with a minor modification in step 4.

**Modified Step 4.** If all the coefficients in the $\bar{c}$ row are positive or zero, the current basic feasible solution is optimal. Otherwise, select the nonbasic variable with the lowest (most negative) value in the $\bar{c}$ row to enter the basis.

### 4.4.2 Unbounded Optimum

A linear program is said to have an *unbounded optimum* when its optimal value is not finite and it tends to $+\infty$ or $-\infty$ depending on whether the objective is maximized or minimized, respectively.

Existence of an unbounded optimum is recognized by the simplex method when the minimum ratio rule [given by Eq. (4.7)] fails to identify the basic variable to leave the basis. This happens when none of the constraint coefficients of the nonbasic variable selected to enter the basis is positive. This means that no finite ratios can be formed in Eq. (4.7), and the nonbasic variable selected to improve the objective function can be increased indefinitely without violating the feasibility of the constraint set. This implies that the optimal value of the linear program is not finite.

### *Example 4.7*

Consider the simplex tableau of a *maximization* problem shown in Tableau 5. Since $\bar{c}_1 > 0$, the nonbasic variable $x_1$ can enter the basis to increase Z. But the minimum ratio rule fails, as there is no positive entry under the $x_1$ column. In other words, as $x_1$ increases, both the basic variables $x_3$ and $x_2$ will also increase and hence can never become negative to limit the increase of $x_1$. This means that $x_1$ can be increased indefinitely. Since each unit increase on $x_1$ increases Z by 11 units, Z can be increased indefinitely, and we have an unbounded optimum to the linear program. Note that an unbounded optimum for a practical problem implies an error in formulation of the linear program, in particular, omission of some important constraints.

### Tableau 5

| $c_B$ | Basis | $c_j$ | | | | Constants |
|---|---|---|---|---|---|---|
| | | 2 | 3 | 0 | 0 | |
| | | $x_1$ | $x_2$ | $x_3$ | $x_4$ | |
| 0 | $x_3$ | $-2$ | 0 | 1 | 0 | 6 |
| 3 | $x_2$ | $-3$ | 1 | 0 | 1 | 4 |
| | $\bar{c}$ row | 11 | 0 | 0 | $-3$ | $Z = 12$ |

### 4.4.3  Degeneracy and Cycling

A basic feasible solution in which one or more of the basic variables are zero is called a *degenerate basic feasible solution.* In contrast, a basic feasible solution in which all the basic variables are positive is said to be *nondegenerate.*

Degeneracy may be present in the initial formulation of the problem itself (when one or more right-hand side constants are zero), or it may be introduced during simplex calculations whenever two or more rows *tie* for the minimum ratio.

When the basic feasible solution is degenerate, it is quite possible for the minimum ratio to be zero. This implies that when a *basis change* is performed, there will not be any real improvement in the value of the objective function. As a matter of fact, it is conceivable for the simplex method to go through a series of iterations with no improvement in $Z$ value. This means that costly simplex calculations will have to be performed in the tableaus without any real return. This naturally reduces the computational efficiency of the simplex method. But a more important question is whether it is possible for the simplex method to go on indefinitely without improving the objective function. In fact, examples have been constructed to show that such a thing is theoretically possible. In such situations the simplex method may get into an infinite loop and will fail to reach the optimal solution. This phenomenon is called *classical cycling* or simply *cycling* in the simplex algorithm. Fortunately, such cycling does not happen in practical problems in spite of the fact that many practical problems have degenerate solutions. However, in a small number of cases, practical problems have been observed to fail to converge on a computer. This is mainly due to certain characteristics of the computer software and the way in which the simplex method is programmed. Gass [3] has called this *computer cycling.* Such problems invariably converge when solved on a different computer system.

### 4.4.4  Use of Artificial Variables

A major requirement of the simplex method is the availability of an initial basic feasible solution in canonical form. Without it the initial simplex tableau cannot be formed. There is a systematic way of getting a canonical system with a basic feasible solution when none is available by inspection. First the LP problem is converted to standard form such that all the variables are nonnegative, the constraints are equations, and all the right-hand side constants are nonnegative. Then each constraint is examined for the existence of a basic variable. If none is available, a new variable is added to act as the basic variable in that constraint. In the end, all the constraints will have a basic variable, and by definition we have a canonical system. Since the right-hand side elements are nonnegative, an initial simplex tableau can be formed readily. Of course, the additional variables have no relevance or meaning to the original problem. They are merely added so that we will have a ready canonical system to start the simplex method. Hence, these variables are termed *artificial variables* as opposed to the real decision variables in the problem. Eventually they will be forced to zero lest they unbalance the equations. To illustrate the use of artificial variables, consider the following LP problem.

### Example 4.8

$$\text{Minimize} \quad Z = -3x_1 + x_2 + x_3$$

$$\text{Subject to} \quad x_1 - 2x_2 + x_3 \leqslant 11$$

$$-4x_1 + x_2 + 2x_3 \geqslant 3$$

$$2x_1 \qquad\quad - x_3 = -1$$

$$x_1 \geqslant 0 \qquad x_2 \geqslant 0 \qquad x_3 \geqslant 0$$

First the problem is converted to the standard form as follows:

$$\text{Minimize} \quad Z = -3x_1 + x_2 + x_3$$

$$\text{Subject to} \quad x_1 - 2x_2 + x_3 + x_4 \qquad = 11 \qquad\qquad (4.9)$$

$$-4x_1 + x_2 + 2x_3 \qquad - x_5 = 3 \qquad\qquad (4.10)$$

$$-2x_1 \qquad + x_3 \qquad\qquad = 1 \qquad\qquad (4.11)$$

$$x_1 \geqslant 0 \qquad x_2 \geqslant 0 \qquad x_3 \geqslant 0 \qquad x_4 \geqslant 0 \qquad x_5 \geqslant 0$$

In Eq. (4.9) the slack variable $x_4$ is a basic variable. Since there are no basic variables in the other equations, we add artificial variables $x_6$ and $x_7$ to Eq. (4.10) and (4.11), respectively. To retain the standard form, $x_6$ and $x_7$ will be restricted to be nonnegative. We now have an "artificial system" given by

$$x_1 - 2x_2 + x_3 + x_4 \qquad\qquad = 11$$

$$-4x_1 + x_2 + 2x_3 \qquad - x_5 + x_6 \qquad = 3$$

$$-2x_1 \qquad + x_3 \qquad\qquad + x_7 = 1$$

$$x_1 \geqslant 0 \qquad \cdots \qquad x_7 \geqslant 0$$

The artificial system has a basic feasible solution in canonical form given by $x_1 = x_2 = x_3 = 0$, $x_4 = 11$, $x_5 = 0$, $x_6 = 3$, $x_7 = 1$. But this is not a feasible solution to the original problem, due to the presence of the artificial variables $x_6$ and $x_7$ at positive values. On the other hand, it is easy to see that any basic feasible solution to the artificial system, in which the artificial variables $x_6$ and $x_7$ are zero, is automatically a basic feasible solution to the original problem. Hence, the objective is to reduce the artificial variables to zero as soon as possible. This can be accomplished by the *two-phase* simplex method.

### 4.4.5 The Two-Phase Simplex Method

This is an approach to handle the artificial variables whenever they are added. Here the LP problem is solved in two phases.

**Phase 1.** This phase consists of finding an initial basic feasible solution to the original problem. In other words, the removal of the artificial variables is taken up first. For this an artificial objective function is created that is the sum of all the artificial variables. The artificial objective is then minimized using the simplex method.

In Example 4.8, the two-phase simplex method would create the following Phase 1 linear program with $W$ representing the artificial objective function.

*Phase 1 Problem*

$$\text{Minimize} \quad W = x_6 + x_7$$

$$\text{Subject to} \quad x_1 \qquad -2x_2 + x_3 + x_4 \qquad\qquad = 11$$

$$-4x_1 \qquad + x_2 + 2x_3 \qquad -x_5 + x_6 \qquad = 3$$

$$-2x_1 \qquad\qquad + x_3 \qquad\qquad +x_7 = 1$$

$$x_1 \geqslant 0 \quad \cdots \quad x_7 \geqslant 0$$

The original objective function, $Z = -3x_1 + x_2 + x_3$, is temporarily set aside during the Phase 1 solution.

If the minimum value of the artificial problem were zero, then all the artificial variables would have been reduced to zero, and we would have a basic feasible solution to the original problem. (*Note*: If the sum of nonnegative variables is zero, then each variable must be identically equal to zero.) We then go to Phase 2.

In case the minimum value of the artificial problem is positive, then at least one of the artificial variables is positive. This means that the original problem without the artificial variables is infeasible, and we terminate.

**Phase 2.** In this phase, the basic feasible solution found at the end of Phase 1 is optimized with respect to the original objective function. In other words, the final tableau of Phase 1 becomes the initial tableau for Phase 2 after changing the objective function. The simplex method is once again applied to determine the optimal solution.*

## 4.5 COMPUTER SOLUTION OF LINEAR PROGRAMS

Many practical problems formulated as linear programs run into hundreds of constraints and thousands of decision variables. These invariably have to be solved using a digital computer. In the tableau form, the simplex method is an iterative procedure and can be applied mechanically to any problem. Hence, it is ideally suited for computer implementation.

### 4.5.1 Computer Codes

Commercial LP computer codes are available from many computer manufacturers and private companies who specialize in marketing LP software for major computer systems. Depending on their capabilities, these codes vary in their complexity, ease of use, and cost. The need to solve large LP problems has led to the development of very complex and sophisticated LP computer codes, called mathematical programming systems (e.g., IBM MPSX-370, CDC APEX III, Management Science Systems MPS-III). These have sophisticated data-handling and analytical tools and can solve problems of the order of 8000–16,000 constraints and an unlimited number of variables. An LP problem with 50,000 constraints has been successfully solved by MPS-III.

Linear programming problems with 5000 or more constraints are considered large, and in their successful solution depends on problem structure and any special properties a problem may have. Typical LP problems in practice would have 500–1500 constraints and several thousand variables. These can be solved without much difficulty by any of the advanced LP computing systems, which are available on a fee basis in addition to the normal computer rental charges. If the user is planning to solve the LP problem frequently, then it may be cost effective to acquire the LP software. For infrequent use, it is better to employ a local qualified computer service bureau or rent a remote terminal connected to a major computer center that has the LP software available for its clients. An excellent survey of modern LP code characteristics, including data management, is given by Orchard-Hays [5]. White [6] presents a status report on computing algorithms for mathematical programming systems. All commercial computer codes use the simplex algorithm and many of its variants as the basic method for solving LP problems. For a more detailed discussion on LP software, including matrix generators and report writers, the reader is referred to Ravindran [7].

---

*The reader may refer to Phillips, Ravindran, and Solberg [4, pp. 51–52] for the complete solution of Example 4.8 by the two-phase simplex method.

### 4.5.2    Computer Implementation of the Simplex Method

The early computer codes for the simplex method performed calculations on the entire tableau as discussed in the earlier sections. This required fast access to all elements of the tableau to reduce computational effort and time. This was not possible unless the entire tableau was kept in the fast memory of the computer. But the limited core memory precluded this, and the solution of large linear programs became inefficient and expensive. Hence, further refinements were made in the simplex calculations so that the simplex method could be implemented more efficiently on a computer. This led to the development of the *revised simplex method*, which is currently implemented in all commercial computer codes.

The revised simplex method uses the same basic principles as the regular simplex method, but the entire simplex tableau need not be calculated at each iteration. The relevant information required to move from one basic feasible solution to another is generated directly from the original system of equations, so that less computational effort and time is needed in the simplex calculations. Currently all commercial computer codes use the revised simplex method for solving LP problems. It is possible to solve problems because the revised method uses less of central memory. For the complete details on the revised simplex method and its advantages, the reader is referred to the text by Phillips, Ravindran, and Solberg [4]. There are also a few minor modifications in computer implementation of the simplex method. One involves the rule for selecting the nonbasic variable to enter the basis. In general, codes for large computers do not compute the entire $\bar{c}$ row and then select the nonbasic variable with the most positive (in maximization problems) or the most negative (in minimization problems) $\bar{c}_j$ coefficient. In a large problem with thousands of nonbasic variables, this could take a considerable amount of computer time. Instead, the elements of the $\bar{c}$ row are calculated one at a time, and the first nonbasic variable that shows a possible improvement in $Z$ is selected as the variable to enter. This eliminates further calculations on the $\bar{c}$ row and an expensive search. Of course, it may result in an increase in the total number of iterations to solve the problem, but the reduction in the time needed for each iteration generally offsets this deficiency.

### 4.5.3    Computational Efficiency of the Simplex Method

It has been pointed out by many researchers in mathematical programming that the simplex method, viewed purely from a theoretical standpoint, is not an efficient method, since it examines adjacent basic feasible solutions only (i.e., changing one basic variable at a time). It is felt that the method can move faster to an optimal solution if it examines nonadjacent solutions as well (i.e., changing more than one basic variable at a time). But many of the suggested variants to the simplex method did not produce any appreciable change in the total computational time. Hence the basic simplex method is still considered to be the best procedure for solving LP problems. A survey of many of the variants of the simplex method is given by Barnes and Crisp [8].

The computational efficiency of the simplex method depends on (1) the number of iterations (basic feasible solutions) required to reach the optimal solution, and (2) the total computer time needed to solve the problem. Much effort has been spent on studying computational efficiency with regard to the number of constraints and the decision variables in the problem.

Empirical experience with thousands of practical problems shows that the number of iterations of a standard linear program with $m$ constraints and $n$ variables varies between $m$ and $3m$, the average being $2m$. A practical upper bound for the number of iterations is $2(m + n)$, although occasionally problems have violated this bound.

Computational time is found to vary approximately in relation to the cube of the number of constraints in the problem ($m^3$). For example, if Problem A has twice as many constraints as Problem B, then the computer time for Problem A will be about eight times that of Problem B.

The computational efficiency of the simplex method is more sensitive to the number of constraints than to the number of variables. Hence the general recommendation is to keep the number of constraints as small as possible by avoiding unnecessary or redundant constraints in the formulation of the LP problem.

Recently, a lot of interest has been generated in a new LP algorithm by the Russian mathematician Khachian [9]. Even though Khachian's algorithm is theoretically an important breakthrough, it does not look promising for computer implementation (see Dantzig [10]).

## 4.6 SENSITIVITY ANALYSIS IN LINEAR PROGRAMMING

In all LP models the coefficients of the objective function and the constraints are supplied as input data or as parameters to the model. The optimal solution obtained by the simplex method is based on the values of these coefficients. In practice the values of these coefficients are seldom known with absolute certainty, because many of them are functions of some uncontrollable parameters. For instance, future demands, the cost of raw materials, or the cost of energy resources cannot be predicted with complete accuracy before the problem is solved. Hence the solution of a practical problem is not complete with the mere determination of the optimal solution.

Each variation in the values of the data coefficients changes the LP problem, which may in turn affect the optimal solution found earlier. In order to develop an overall strategy to meet the various contingencies, one has to study how the optimal solution will change with changes in the input (data) coefficients. This is known as *sensitivity analysis* or *post-optimality analysis*.

Other reasons for performing a sensitivity analysis are:

1. Some data coefficients or parameters of the linear program may be controllable; for example, availability of capital, raw material, or machine capacities. Sensitivity analysis enables us to study the effect of changes in these parameters on the optimal solution. If it turns out that the optimal value (profit/cost) changes (in our favor) by a considerable amount for a small change in the given parameters, then it may be worthwhile to implement some of these changes. For example, if increasing the availability of labor by allowing overtime contributes to a greater increase in the maximum return, as compared to the increased cost of overtime labor, then we might want to allow overtime production.

2. In many cases, the values of the data coefficients are obtained by carrying out statistical estimation procedures on past figures, as in the case of sales forecasts, price estimates, and cost data. These estimates, in general, may not be very accurate. If we can identify which of the parameters affect the objective value most, then we can obtain better estimates of these parameters. This will increase the reliability of our model and the solution.

We shall now illustrate the practical uses of sensitivity analysis with the help of an example.

### Example 4.9

A factory manufactures three products, which require three resources—labor, materials, and administration. The unit profits on these products are \$10, \$6, and \$4, respectively. There are 100 hr of labor, 600 lb of material, and 300 hr of administration available per day. In order to determine the optimal product mix, the following LP model is formulated and solved:

$$\text{Maximize} \quad Z = 10x_1 + 6x_2 + 4x_3$$

$$\text{Subject to} \quad x_1 + x_2 + x_3 \leqslant 100 \quad \text{(labor)}$$

$$10x_1 + 4x_2 + 5x_3 \leqslant 600 \quad \text{(material)}$$

$$2x_1 + 2x_2 + 6x_3 \leqslant 300 \quad \text{(administration)}$$

$$x_1, x_2, x_3 \geqslant 0$$

where $x_1$, $x_2$, and $x_3$ are the daily production levels of products 1, 2, and 3 respectively.

A computer output of the solution of Example 4.9 is given in Table 4.1. We note from the optimal solution that the optimal product mix is to produce products 1 and 2 only at levels 33.33 and 66.67 units, respectively.

The *shadow prices* give the net impact in the maximum profit if additional units of certain resources can be obtained. Labor has the maximum impact, providing $3.33 increase in profit for each additional hour of labor. Of course, the shadow prices on the resources apply as long as their variations stay within the prescribed ranges on RHS constants given in Table 4.1. In other words, a $3.33/hr increase in profit is achievable as long as the labor hours are not increased beyond 150 hr. Suppose it is possible to increase the labor hours by 25% by scheduling overtime that incurs an additional labor cost of $50. To see whether it is profitable to schedule overtime, we first determine the net increase in maximum profit due to 25 hr of overtime as (25)($3.33) = $83.25. Since this is more than the total cost of overtime, it is economical to schedule overtime. It is important to note that when any of the RHS constants is changed, the optimal solution will change. However, the optimal product mix will be unaffected as long as the RHS constant varies within the specified range. In other words, we will still be making products 1 and 2 only, but their quantities may change.

The ranges on the objective function coefficients given in Table 4.1 exhibit the sensitivity of the *optimal solution* with respect to changes in the unit profits of the three products. It shows that the optimal solution will not be affected as long as the unit profit of product 1 stays between $6 and $15. Of course, the

**Table 4.1**

| Optimal Solution: | $x_1 = 33.33$, $x_2 = 66.67$, $x_3 = 0$ |
| Optimal Value: | maximum profit = $733.33 |
| Shadow Prices: | For row 1 = $3.33, for row 2 = 0.67, for row 3 = 0 |
| Opportunity Costs: | For $x_1 = 0$, for $x_2 = 0$, for $x_3 = 2.67$ |

Ranges on Objective Function Coefficients

| Variable | Lower Limit | Present Value | Upper Limit |
| --- | --- | --- | --- |
| $x_1$ | 6 | 10 | 15 |
| $x_2$ | 4 | 6 | 10 |
| $x_3$ | $-\infty$ | 4 | 6.67 |

Ranges on Right-Hand-Side (RHS) Constants

| Row | Lower Limit | Present Value | Upper Limit |
| --- | --- | --- | --- |
| 1 | 60 | 100 | 150 |
| 2 | 400 | 600 | 1000 |
| 3 | 200 | 300 | $\infty$ |

maximum profit will be affected by the change. For example, if the unit profit on product 1 increases from \$10 to \$12, the optimal solution will be the same but the maximum profit will increase to \$733.33 + (12 − 10)(\$33.33) = \$799.99.

Note that product 3 is not economical to produce. Its *opportunity cost* measures the negative impact of producing product 3 to the maximum profit. Hence, a further decrease in its profit contribution will not have any impact on the optimal solution or maximum profit. Also, the unit profit on product 3 must increase to \$6.67 (present value + opportunity cost) before it becomes economical to produce.

## Simultaneous Variations in Parameters [11]

The sensitivity analysis output on profit and RHS ranges are obtained by varying only one of the parameters and holding all others fixed at their current values. However, it is possible to use the sensitivity analysis output when several parameters are changed simultaneously. This is done with the help of the "100% *rule.*"

### 100% Rule (For Objective Function Coefficients)

$$\sum_j \frac{\delta c_j}{\Delta c_j} \leqslant 1 \tag{4.12}$$

where $\delta c_j$ is the actual increase (decrease) in the objective function coefficient of variable $x_j$, and $\Delta c_j$ is the maximum increase (decrease) allowed by sensitivity analysis. As long as inequality (4.12) is satisfied, the optimal solution to the LP problem will not change.

For example, suppose the unit profit on product 1 decreases by \$1 but increases by \$1 for both products 2 and 3. This simultaneous variation satisfies the *100% rule.* Since $\delta c_1 = -1$, $\Delta c_1 = -4$, $\delta c_2 = 1$, $\Delta c_2 = 4$, $\delta c_3 = 1$, $\Delta c_3 = 2.67$, and

$$\frac{-1}{-4} + \frac{1}{4} + \frac{1}{2.67} = 0.875 < 1$$

Hence, the optimal solution will not change, but the maximum profit will change by $(-1)(\$33.33) + 1(\$66.67) + 1(\$0) = \$33.34$.

### 100% Rule (For RHS Constants)

$$\sum_i \frac{\delta b_i}{\Delta b_i} \leqslant 1 \tag{4.13}$$

where $\delta b_i$ is the actual increase (decrease) in the RHS constant of the $i$th constraint, and $\Delta b_i$ is the maximum increase (decrease) allowed by sensitivity analysis. If inequality (4.13) is satisfied, then the optimal product mix remains the same and the shadow prices apply, but the optimal solution and maximum profit will change. Of course, the net change in the maximum profit can be obtained using the shadow prices.

For example, consider the simultaneous variation of 10 hr decrease in labor availability, 100 lb increase in material, and 50 hr decrease in administration. This implies that $\delta b_1 = -10$, $\delta b_2 = 100$, $\delta b_3 = -50$, while $\Delta b_1 = -40$, $\Delta b_2 = 400$, and $\Delta b_3 = -100$. The 100% rule is satisfied, since

$$\frac{-10}{-40} + \frac{100}{400} + \frac{-50}{-100} = 1$$

Hence, the optimal basis and the product mix will not be affected. Of course, the optimal solution will change. But the net change in the maximum profit can be obtained using the shadow prices:

$$\text{Change in optimal profit} = (\$3.33)(-10) + (\$0.67)(100) - (0)(50)$$

$$= \$33.70$$

*Warning*: Failure of the 100% rule does not necessarily imply that the LP solution will be affected. In other words, inequality (4.12) may be violated while the optimal solution remains unchanged.

## 4.7  APPLICATIONS

Linear programming models are used widely to solve a variety of military, economic, industrial, and social problems. In a 1976 survey of American companies [1], linear programming came out as the most often used technique (74%) among all the optimization methods. It has been reported in *Science News* [2] that about one-fourth of computer time spent on scientific computations in recent years has been devoted to solving LP problems and their many variations. The oil companies are among the foremost users of very large LP models, using them in petroleum refining, distribution, and transportation. The number of LP applications has grown so much in the last 20 years that it will be impossible to survey all the different applications. Instead, the reader is referred to two excellent textbooks, Gass [12] and Salkin and Saha [13], which are devoted solely to LP applications in diverse areas like defense, industry, retail business, agriculture, education, and environment. Many of the applications also contain a discussion of the experiences in using the LP model in practice.

Discussion of various computer solutions to large LP problems in industry and computational features of LP software are discussed in Driebeek [14],

Daellenbach and Bell [15], and Bradley, Hax, and Magnanti [11]. An excellent bibliography on LP applications is available in Gass [16], which contains a list of references arranged by area (e.g., agriculture, industry, military, production, transportation). In the area of industrial application the references have been further categorized by industry (e.g., chemical, coal, airline, iron and steel, paper, petroleum, railroad). For another bibliography on LP applications, the readers may refer to a survey by Gray and Cullinan-James [17]. For more recent applications of LP in practice, the reader should check the recent issues of *Interfaces, AIIE Transactions, Decision Sciences, European Journal of Operational Research, Management Science, Operations Research, Operational Research* (U.K.), *Naval Research Logistics Quarterly*, and *OpSearch* (India).

## 4.8 ADDITIONAL TOPICS IN LINEAR PROGRAMMING

In this section, we briefly discuss some of the advanced topics in linear programming. These include *duality theory, the dual simplex method*, and *integer programming*. For a full discussion, see references 4, 18, or 19.

### 4.8.1 Duality Theory

From both the theoretical and practical points of view, the theory of duality is one of the most important concepts in linear programming. The basic idea behind the duality theory is that every linear program has an associated linear program called its *dual* such that a solution to one gives a solution to the other. There are a number of important relationships between the solution to the original problem and its dual. These are useful in investigating the general properties of the optimal solution to a linear program and in testing whether a feasible solution is optimal. In addition, the optimal dual solution can be interpreted as the price one pays for the constraint resources, which is known as the *shadow price*. This plays an important role in postoptimality analysis (Section 4.6). Moreover, the concept of duality and the various duality theorems are extremely useful in the study of advanced mathematical programming topics. For a complete discussion of duality theory and its application, see reference 4.

### 4.8.2 The Dual Simplex Method

The dual simplex method is a modified version of the simplex method using duality theory. There are instances when the dual simplex method has an obvious advantage over the regular simplex method. It plays an important part in sensitivity analysis, parametric programming, solution of integer programming problems, many of the variants of the simplex method, and solution of some nonlinear programming problems. Details of the dual simplex method are discussed in reference 4.

### 4.8.3  Integer Programming

In practice, many LP problems do require integer solutions for some of the variables. For instance, it is not possible to employ a fractional number of workers or produce a fractional number of cars. The term (linear) integer programming refers to the class of LP problems wherein some or all of the decision variables are restricted to be integers. But solutions of integer programming problems are generally difficult, too time consuming, and expensive. Hence, a practical approach is to treat all integer variables as continuous and solve the associated linear program by the simplex method. We may be fortunate to get some of the values of the variables as integers automatically, but when the simplex method produces fractional solutions for some integer variables, they are generally rounded off to the nearest integer such that the constraints are not violated. This is very often used in practice, and it generally produces a good integer solution close to the optimal integer solution, especially when the values of the integer variables are large.

There are situations when a model formulation may require the use of binary integer variables, which can take only the values 0 or 1. For these integer variables, rounding off produces poor integer solutions, and one does need other techniques to determine the optimal integer solution directly. Some of these techniques and many practical applications of integer programming are discussed in Chapter 11.

## 4.9  SUMMARY

In this chapter, we discussed linear programming, a special class of optimization problems with linear constraints and linear objective functions. Linear programming (LP) models are easy to solve computationally and have a wide range of applications in diverse fields. We also developed the simplex algorithm for solving general LP problems. We then discussed how to perform post-optimality analysis on the LP solutions. The ability to perform an analysis of this type is the main reason for the success of linear programming in practice. Other practical uses of linear programming are in solving integer and nonlinear programming problems as a sequence of LP problems.

## REFERENCES

1. Fabozzi, E. J., and J. Valente, "Mathematical Programming in American Companies: A Sample Survey," *Interfaces*, **7**(1), 93–98 (Nov. 1976).
2. Steen, L. A., "Linear Programming: Solid New Algorithm," *Science News*, **116**, 234–236 (Oct. 6, 1979).
3. Gass, S. I., "Comments on the Possibility of Cycling with the Simplex Method," *Oper. Res.*, **27**(4), 848–852 (1979).

4. Phillips, D. T., A. Ravindran, and J. J. Solberg, *Operations Research: Principles and Practice*, Wiley, New York, 1976.

5. Orchard-Hays, W., "On the Proper Use of a Powerful MPS," in *Optimization Methods for Resource Allocation* (R. W. Cottle and J. Krarup, Eds.), English Universities Press, London, 1974, pp. 121–149.

6. White, W. W., "A Status Report on Computing Algorithms for Mathematical Programming," *Comput. Surv.* **5**(3), 135–166 (1973).

7. Ravindran, A., "Linear Programming," in *Handbook of Industrial Engineering* (G. Salvendy, Ed.), Wiley, New York, 1982, Chap. 14.2, pp. 14.2.1–14.2.11.

8. Barnes, J. W., and R. M. Crisp, Jr., "Linear Programming: A Survey of General Purpose Algorithms," *Amer. Inst. Ind. Eng. Trans.*, **7**(3), 212–221 (1975).

9. Khachian, L. G., "A Polynomial Algorithm in Linear Programming," *Soviet Math. Dokl.*, **20**(1), 191–194 (1979).

10. Dantzig, G. B., "Comments on Khachian's Algorithm for Linear Programming," Tech. Rep. SOL 79-22, Department of Operations Research, Stanford University, Stanford, CA, 1979.

11. Bradley, S. P., A. C. Hax, and T. L. Magnanti, *Applied Mathematical Programming*, Addison-Wesley, Reading, MA, 1977.

12. Gass, S. I., *An Illustrated Guide to Linear Programming*, McGraw-Hill, New York, 1970.

13. Salkin, H. M., and J. Saha, *Studies in Linear Programming*, North Holland/American Elsevier, New York, 1975.

14. Driebeek, N. J., *Applied Linear Programming*, Addison-Wesley, Reading, MA, 1969.

15. Daellenbach, H. G., and E. J. Bell, *User's Guide to Linear Programming*, Prentice-Hall, Englewood Cliffs, NJ, 1970.

16. Gass, S., *Linear Programming*, 4th ed., McGraw Hill, New York, 1975.

17. Gray, P., and C. Cullinan-James, "Applied Optimization—A Survey," *Interfaces*, **6**(3), 24–41 (May 1976).

18. Dantzig, G. B., *Linear Programming and Extensions*, Princeton University Press, Princeton, NJ, 1963.

19. Murty, K. G., *Linear and Combinatorial Programming*, Wiley, New York, 1976.

## PROBLEMS

**4.1.** What is the difference between the Simplex method and exhaustive enumeration of feasible corner points of the constrained region?

**4.2.** What is the function of the minimum ratio rule in the Simplex method?

**4.3.** How do you recognize that an LP problem is unbounded while using the Simplex method?

**4.4.** What is the function of the Inner Product Rule in the Simplex method?

**4.5.** What are artificial variables and why do we need them? How do they differ from slack–surplus variables?

**4.6.** Describe a systematic procedure for finding a feasible solution to a system of linear inequalities.

**4.7.** What is sensitivity analysis and why do we perform it?

**4.8.** What are shadow prices? What are their practical uses?

**4.9.** When do we use the "100% Rule" and why?

**4.10.** How does an addition of constant $K$ to the objective function of an LP affect (i) its optimal solution, (ii) its optimal value?

**4.11.** Two products, A and B, are made that each involve two chemical operations. Each unit of product A requires 2 hr on operation 1 and 4 hr on operation 2. Available time for operation 1 is 16 hr, and for operation 2 it is 24 hr. The production of product B also results in a by-product C at no extra cost. Although some of this by-product can be sold at a profit, the remainder has to be destroyed. Product A sells for $4 profit per unit, while product B sells for $10 profit per unit. By-product C can be sold at a unit profit of $3, but if it cannot be sold the destruction cost is $2 per unit. Forecasts show that up to five units of C can be sold. The company gets two units of C for each unit of B produced.

Determine the production quantity of A and B, keeping C in mind, so as to make the largest profit.

**4.12.** Consider the problem of scheduling the weekly production of a certain item for the next 4 weeks. The production cost of the item is $10 for the first 2 weeks and $15 for the last 2 weeks. The weekly demands are 300, 700, 900, and 800 units, which must be met. The plant can produce a maximum of 700 units each week. In addition, the company can employ overtime during the second and third weeks. This increases the weekly production by an additional 200 units, but the cost of production increases by $5 per item. Excess production can be stored at a cost of $3 an item. How should the production be scheduled so as to minimize the total costs? Formulate this as an LP problem.

**4.13.** A refinery has four different crudes, which are to be processed to yield four products: gasoline, heating oil, jet fuel, and lube oil. There are maximum limits both on product demand (what can be sold) and crude availability. A schematic of the processing operation is given in Figure 4.6. Given the tabulated profits, costs, and yields (see Table 4.3):

(a) Set up the model appropriate for scheduling the refinery for maximum profit.

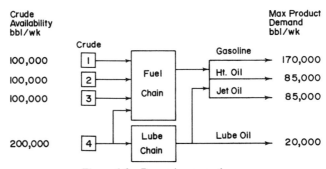

**Figure 4.6.** Processing operation.

**Table 4.3  Profits, Costs, Yields**

| | 1 | 2 | 3 | Fuel Process | Lube Process | Product Value $/bbl |
|---|---|---|---|---|---|---|
| | | | | **4** | | Product |
| Yields, bbl product per bbl crude | | | | | | |
| Gasoline | 0.6 | 0.5 | 0.3 | 0.4 | 0.4 | 45.00 |
| Heating oil | 0.2 | 0.2 | 0.3 | 0.3 | 0.1 | 30.00 |
| Jet fuel | 0.1 | 0.2 | 0.3 | 0.2 | 0.2 | 15.00 |
| Lube oil | 0 | 0 | 0 | 0 | 0.2 | 60.00 |
| Other* | 0.1 | 0.1 | 0.1 | 0.1 | 0.1 | |
| Crude cost, $/bbl | 15.00 | 15.00 | 15.00 | 25.00 | 25.00 | |
| Operating cost, $/bbl | 5.00 | 8.50 | 7.50 | 3.00 | 2.50 | |

*"Other" refers to losses in processing.

(b) Suppose the given maximum product demands are changed to be minimum requirements. Set up the model appropriate for scheduling the refinery for minimum cost operation.

**4.14.** A company manufactures three products: A, B, and C. Each unit of product A requires 1 hr of engineering service, 10 hr of direct labor, and 3 lb of material. To produce one unit of product B, it requires 2 hr of engineering, 4 hr of direct labor, and 2 lb of material. Each unit of product C requires 1 hr of engineering, 5 hr of direct labor, and 1 lb of material. There are 100 hr of engineering, 700 hr of labor, and 400 lb of material available. Since the company offers discounts for bulk purchases, the profit figures are as shown below:

| Product A | | Product B | | Product C | |
|---|---|---|---|---|---|
| Sales, units | Unit profit, $ | Sales, units | Unit Profit, $ | Sales, units | Unit Profit $ |
| 0–40 | 10 | 0–50 | 6 | 0–100 | 5 |
| 40–100 | 9 | 50–100 | 4 | Over 100 | 4 |
| 100–150 | 8 | Over 100 | 3 | | |
| Over 150 | 7 | | | | |

For example, if 120 units of A are sold, the first 40 units will earn a profit of $10/unit, the next 60 will earn $9/unit, and the remaining 20 units $8/unit. Formulate a linear program to determine the most profitable product mix.

**4.15.** A scientist has observed a certain quantity $Q$ as a function of a variable $t$. He is interested in determining a mathematical relationship relating $t$

and $Q$, which takes the form

$$Q(t) = at^3 + bt^2 + ct + d$$

from the results of his $n$ experiments $(t_1, Q_1), (t_2, Q_2), \ldots, (t_n, Q_n)$. He discovers that the values of the unknown coefficients $a$, $b$, $c$, and $d$ must be nonnegative and should add up to 1. To account for errors in his experiments, he defines an error term,

$$e_i = Q_i - Q(t_i)$$

He wants to determine the best values for the coefficients $a$, $b$, $c$, and $d$ using the following criterion functions:

$$\text{Criterion 1:} \quad \text{Minimize} \quad Z = \sum_{i=1}^{n} |e_i|$$

$$\text{Criterion 2:} \quad \text{Minimize} \quad \max_i |e_i|$$

where $e_i$ is the value of the error associated with the $i$th experiment. Show that the scientist's problem reduces to an LP problem under both criterion 1 and criterion 2.

**4.16.** A machine shop has one drill press and five milling machines, which are to be used to produce an assembly consisting of parts 1 and 2. The productivity of each machine for the two parts is given below:

Production Time, min/piece

| Part | Drill | Mill |
|------|-------|------|
| 1 | 3 | 20 |
| 2 | 5 | 15 |

It is desired to maintain a balanced loading on all machines such that no machine runs more than 30 min/day longer than any other machine (assume that the milling load is split evenly among all five milling machines). Divide the work time of each machine to obtain the maximum number of completed assemblies, assuming an 8-hr working day.

**4.17.** A company makes a specialty solvent at two levels of purity, which it sells in gallon containers. Product A is of higher purity than product B, and profits are $0.40/gal on A and $0.30/gal on B.

Product A requires *twice* the processing time of B. If the company produced only B it could make 1000 gal/day. However, process throughput limitations permit production of only 800 gal/day of both

A and B combined. Contract sales require that at least 200 gal/day of B be produced.

Assuming all of the product can be sold, what volumes of A and B should be produced? Solve, using the tableau form of the simplex method. Confirm your solution using graphical means.

**4.18.** Transform the following linear program to the standard form:

$$\text{Minimize} \quad Z = -3x_1 + 4x_2 - 2x_3 + 5x_4$$

$$\text{Subject to} \quad 4x_1 - x_2 + 2x_3 - x_4 = -2$$

$$x_1 + x_2 + 3x_3 - x_4 \leqslant 14$$

$$-2x_1 + 3x_2 - x_3 + 2x_4 \geqslant 2$$

$$x_1 \geqslant 0 \qquad x_2 \geqslant 0 \qquad x_3 \leqslant 0 \qquad x_4 \text{ unrestricted in sign}$$

**4.19.** Consider a system of two equations in five unknowns as follows:

$$x_1 + 2x_2 + 10x_3 + 4x_4 - 2x_5 = 5$$

$$x_1 + x_2 + 4x_3 + 3x_4 + x_5 = 8$$

$$x_1, \ldots, x_5 \geqslant 0$$

(a) Reduce the system to canonical form with respect to $(x_1, x_2)$ as basic variables. Write down the basic solution. Is it feasible? Why or why not?

(b) What is the maximum number of basic solutions possible?

(c) Find a canonical system that will give a basic feasible solution to the above system by trial and error.

**4.20.** Consider the following linear program:

$$\text{Maximize} \quad Z = 2x_1 - x_2 + x_3 + x_4$$

$$\text{Subject to} \quad -x_1 + x_2 + x_3 \qquad + x_5 \qquad = 1$$

$$x_1 + x_2 \qquad + x_4 \qquad = 2$$

$$2x_1 + x_2 + x_3 \qquad + x_6 = 6$$

$$x_1, \ldots, x_6 \geqslant 0$$

(a) Write down the initial basic feasible solution by inspection.

(b) Find a feasible solution by increasing the nonbasic variable $x_1$ by one unit, while holding $x_2$ and $x_3$ as zero. What will be the net change in the objective function?

(c) What is the maximum possible increase in $x_1$ subject to the constraints?

(d) Find the new basic feasible solution when $x_1$ is increased to its maximum value found in (c).

(e) Is the new basic feasible solution obtained in (d) optimal? Why or why not?

**4.21.**  Use the simplex method to solve

$$\text{Maximize} \quad Z = x_1 + 3x_2$$

$$\text{Subject to} \quad x_1 \qquad \leqslant 5$$

$$x_1 + 2x_2 \leqslant 10$$

$$x_2 \leqslant 4$$

$$x_1, x_2 \geqslant 0$$

Plot the feasible region using $x_1$ and $x_2$ as coordinates. Follow the solution steps of the simplex method graphically by interpreting the shift from one basic feasible solution to the next in the feasible region.

**4.22.**  Find an optimal solution to the following linear program by *inspection*:

$$\text{Minimize} \quad Z = x_1 - 3x_2 + 2x_3$$

$$\text{Subject to} \quad -2 \leqslant x_1 \leqslant 3$$

$$0 \leqslant x_2 \leqslant 4 \qquad 2 \leqslant x_3 \leqslant 5$$

**4.23.**  Use the simplex method to solve:

$$\text{Minimize} \quad Z = 3x_1 + x_2 + x_3 + x_4$$

$$\text{Subject to} \quad -2x_1 + 2x_2 + x_3 \qquad = 4$$
$$3x_1 + x_2 + x_4 = 6$$

$$x_1, x_2, x_3, x_4 \geqslant 0$$

Find an alternative optimal solution if one exists.

**4.24.** Use the simplex method to solve:

$$\text{Maximize} \quad Z = x_1 + 2x_2 + 3x_3 + 4x_4$$

$$\text{Subject to} \qquad x_1 + 2x_2 + 2x_3 + 3x_4 \leqslant 20$$

$$2x_1 + x_2 + 3x_3 + 2x_4 \leqslant 20$$

$$x_1, \ldots, x_4 \geqslant 0$$

Is the optimal solution unique? Why or why not?

**4.25.** Use the simplex method to verify that the following problem has no optimal solution:

$$\text{Maximize} \quad Z = x_1 + 2x_2$$

$$\text{Subject to} \quad -2x_1 + x_2 + x_3 \leqslant 2$$

$$-x_1 + x_2 - x_3 \leqslant 1$$

$$x_1, x_2, x_3 \geqslant 0$$

From the final simplex tableau, construct a feasible solution whose value of the objective function is greater than 2000.

**4.26.** Consider the standard LP problem

$$\text{Minimize} \quad Z = cx$$

$$\text{Subject to} \quad Ax = b \qquad x \geqslant 0$$

Let the vectors $x^{(1)}$ and $x^{(2)}$ be two optimal solutions to the above problem. Show that the vector $x(\lambda) = \lambda x^{(1)} + (1 - \lambda)x^{(2)}$ is also an optimal solution for any value of $\lambda$ between 0 and 1. (*Note*: The above result is very useful in linear programming. Once we have two optimal solutions to a linear program, then we can generate an infinite number of optimal solutions by varying $\lambda$ between 0 and 1.)

**4.27.** In the tableau for the maximization problem below, the values of the six constants $\alpha_1, \alpha_2, \alpha_3, \beta, \rho_1, \rho_2$ are unknown (assume there are no artificial variables):

| Basis | $x_1$ | $x_2$ | $x_3$ | $x_4$ | $x_5$ | $x_6$ | Constants |
|-------|-------|-------|-------|-------|-------|-------|-----------|
| $x_3$ | 4 | $\alpha_1$ | 1 | 0 | $\alpha_2$ | 0 | $\beta$ |
| $x_4$ | $-1$ | $-5$ | 0 | 1 | $-1$ | 0 | 2 |
| $x_6$ | $\alpha_3$ | $-3$ | 0 | 0 | $-4$ | 1 | 3 |
| $\bar{c}$ row | $\rho_1$ | $\rho_2$ | 0 | 0 | $-3$ | 0 | |

$$x_1, \ldots, x_6 \geqslant 0$$

State restrictions on the six unknowns $(\alpha_1, \alpha_2, \alpha_3, \beta, \rho_1, \rho_2)$ that would make the following statements true about the given tableau:

(a) The current solution is optimal, but an alternate optimum exists.

(b) The current solution is infeasible. (State which variable.)

(c) One of the constraints is inconsistent.

(d) The current solution is a degenerate basic feasible solution. (Which variable causes degeneracy?)

(e) The current solution is feasible, but the problem has no finite optimum.

(f) The current solution is the unique optimum solution.

(g) The current solution is feasible, but the objective can be improved by replacing $x_6$ by $x_1$. What will be the total change in the objective function value after the pivot?

**4.28.** A company that manufacturers three products A, B, and C using three machines M1, M2, and M3 wants to determine the optimal production schedule that maximizes the total profit. Product A has to be processed by machines M1, M2, and M3; product B requires M1 and M3; while product C requires M1 and M2. The unit profits on the three products are \$4, \$2, and \$5, respectively. The following linear program is formulated to determine the optimal product mix:

$$\text{Maximize} \quad Z = 4x_1 + 2x_2 + 5x_3$$

$$\text{Subject to} \quad x_1 + 2x_2 + x_3 \leqslant 430 \quad \text{(machine 1)}$$

$$3x_1 \quad\quad + 2x_3 \leqslant 460 \quad \text{(machine 2)}$$

$$x_1 + 4x_2 \quad\quad \leqslant 450 \quad \text{(machine 3)}$$

$$x_1, x_2, x_3 \geqslant 0$$

where $x_1$, $x_2$, and $x_3$ are the amounts of products A, B, and C, and the constraints reflect the available capacities of M1, M2, and M3. The computer prints out the following solution:

| | |
|---|---|
| Optimal solution | $x_1 = 0$, $x_2 = 100$, $x_3 = 230$ |
| Optimal value | max $Z = 1350$ |
| Shadow prices | 1.0, 2.0, and 0.0 for constraints 1, 2, and 3, respectively |
| Opportunity costs | 3.0, 0, and 0 for variables $x_1$, $x_2$, $x_3$, respectively |

### Ranges on Objective Function Coefficients

| Variable | Lower Limit | Present Value | Upper Limit |
|----------|-------------|---------------|-------------|
| $x_1$ | $-\infty$ | 4.0 | 7.0 |
| $x_2$ | 0 | 2.0 | 10.0 |
| $x_3$ | 2.333 | 5.0 | $\infty$ |

### Ranges on RHS Constants

| Row | Lower Limit | Present Value | Upper Limit |
|-----|-------------|---------------|-------------|
| 1 | 230 | 430 | 440 |
| 2 | 440 | 460 | 860 |
| 3 | 400 | 450 | $\infty$ |

(a) Because of an increase in the cost of raw material used for product C, its unit profit drops to $4. Determine the new optimal solution and the maximum profit.

(b) Suppose it is possible to increase the capacity of one of the machines. Which one would you recommend for expansion, and why?

(c) Due to an improvement in product design, the unit profit on product A can be increased to $6. Is it worthwhile to produce product A now? Explain.

(d) Suppose the capacity of machine 2 can be increased by another 200 min at a cost of $250. Is it economical to do so? Explain.

(e) Due to an increase in the cost of energy for operating the machines, the units profits of A, B, and C decrease by $2.0, $0.50, and $1.0, respectively. How will it affect the optimal solution and maximum profit?

# Chapter 5

# Constrained Optimality Criteria

In Chapters 2 and 3, we discussed the necessary and sufficient optimality criteria for unconstrained optimization problems. But most engineering problems involve optimization subject to several constraints on the design variables. The presence of constraints essentially reduces the region in which we search for the optimum. At the outset, it may appear that the reduction in the size of the feasible region should simplify the search for the optimum. On the contrary, the optimization process becomes more complicated, since many of the optimality criteria developed earlier need not hold in the presence of constraints. Even the basic condition that an optimum must be at a stationary point, where the gradient is zero, may be violated. For example, the *unconstrained minimum* of $f(x) = (x - 2)^2$ occurs at the stationary point $x = 2$. But if the problem has a constraint that $x \geqslant 4$, then the *constrained minimum* occurs at the point $x = 4$. This is *not* a stationary point of $f$, since $f'(4) = 4$. In this chapter we develop necessary and sufficient conditions of optimality for constrained problems. We begin with optimization problems involving only *equality constraints*.

## 5.1  EQUALITY-CONSTRAINED PROBLEMS

Consider an optimization problem involving several equality constraints:

$$\text{Minimize} \quad f(x_1, x_2, \ldots, x_N)$$

$$\text{Subject to} \quad h_k(x_1, \ldots, x_N) = 0 \qquad k = 1, \ldots, K$$

In principle, this problem can be solved as an unconstrained optimization problem by explicitly eliminating $K$ independent variables using the equality constraints. In effect, the presence of equality constraints reduces the *dimensionality* of the original problem from $N$ to $N - K$. Once the problem is reduced to an unconstrained optimization problem, the methods of Chapters 2 and 3 can be used to identify the optimum. To illustrate this, consider the following example.

***Example 5.1***

$$\text{Minimize} \quad f(x) = x_1 x_2 x_3$$

$$\text{Subject to} \quad h_1(x) = x_1 + x_2 + x_3 - 1 = 0$$

Eliminating the variable $x_3$ with the help of $h_1(x) = 0$, we get an unconstrained optimization problem involving two variables:

$$\min f(x_1, x_2) = x_1 x_2 (1 - x_1 - x_2)$$

The methods of Chapter 3 can now be applied to identify the optimum.

The *variable-elimination method* is applicable as long as the equality constraints can be solved explicitly for a given set of independent variables. In the presence of several equality constraints, the elimination process may become unwieldy. Moreover, in certain situations it may not be possible to solve the constraints explicitly to eliminate a variable. For instance, in Example 5.1 if the constraint $h_1(x) = 0$ were given as

$$h_1(x) = x_1^2 x_3 + x_2 x_3^2 + x_2^{-1} x_1 = 0$$

then no explicit solution of one variable in terms of the others would be possible. Hence, in problems involving several complex equality constraints, it is better to use the method of Lagrange multipliers, which is described in the next section, for handling the constraints.

## 5.2  LAGRANGE MULTIPLIERS

The method of Lagrange multipliers essentially gives a set of necessary conditions to identify candidate optimal points of equality-constrained optimization problems. This is done by converting the constrained problem to an equivalent unconstrained problem with the help of certain unspecified parameters known as *Lagrange multipliers*.

Consider the minimization of a function of $n$ variables subject to one equality constraint:

$$\text{Minimize} \quad f(x_1, x_2, \ldots, x_N) \tag{5.1}$$

$$\text{Subject to} \quad h_1(x_1, x_2, \ldots, x_N) = 0 \tag{5.2}$$

The method of Lagrange multipliers converts this problem to the following unconstrained optimization problem:

$$\text{Minimize} \quad L(x, v) = f(x) - vh_1(x) \tag{5.3}$$

The unconstrained function $L(x; v)$ is called the *Lagrangian function*, and $v$ is an unspecified constant called the *Lagrange multiplier*. There are no sign restrictions on the value of $v$.

Suppose for a given fixed value of $v = v^\circ$, the unconstrained minimum of $L(x; v)$ with respect to $x$ occurs at $x = x^\circ$ and $x^\circ$ satisfies $h_1(x^\circ) = 0$. Then, it is clear that $x^\circ$ minimizes Eq. (5.1) subject to (5.2) because for all values of $x$ that satisfy (5.2), $h_1(x) = 0$ and min $L(x; v) = $ min $f(x)$.

Of course, the challenge is to determine the appropriate value for $v = v^\circ$ so that the unconstrained minimum point $x^\circ$ satisfies (5.2). But this can be done by treating $v$ as a variable, finding the unconstrained minimum of (5.3) as a function of $v$, and adjusting $v$ such that (5.2) is satisfied. We illustrate this with the following example:

**Example 5.2**

$$\text{Minimize} \quad f(x) = x_1^2 + x_2^2$$

$$\text{Subject to} \quad h_1(x) = 2x_1 + x_2 - 2 = 0$$

The unconstrained minimization problem becomes

$$\text{Minimize} \quad L(x; v) = x_1^2 + x_2^2 - v(2x_1 + x_2 - 2)$$

**Solution.** Setting the gradient of $L$ with respect to $x$ equal to zero,

$$\frac{\partial L}{\partial x_1} = 2x_1 - 2v = 0 \Rightarrow x_1^\circ = v$$

$$\frac{\partial L}{\partial x_2} = 2x_2 - v = 0 \Rightarrow x_2^\circ = \frac{v}{2}$$

To test whether the stationary point $x^\circ$ corresponds to a minimum, we compute the Hessian matrix of $L(x; v)$ with respect to $x$ as

$$H_L(x; v) = \begin{bmatrix} 2 & 0 \\ 0 & 2 \end{bmatrix}$$

which is positive definite. This implies that $H_L(x; v)$ is a convex function for all $x$. Hence $x_1^\circ = v$, $x_2^\circ = v/2$ corresponds to the global minimum. To determine the optimal $v$, we substitute the values of $x_1^\circ$ and $x_2^\circ$ in the constraint

$2x_1 + x_2 = 2$ to get $2v + v/2 = 2$ or $v^\circ = \frac{4}{5}$. Hence the constrained minimum is attained at $x_1^\circ = \frac{4}{5}$, $x_2^\circ = \frac{2}{5}$, and min $f(x) = \frac{4}{5}$.

In the solution of Example 5.2, we treated $L(x; v)$ as a function of two variables $x_1$ and $x_2$ and considered $v$ as a parameter whose value is "adjusted" to satisfy the constraint. In problems where it is difficult to get an explicit solution to

$$\frac{\partial L}{\partial x_j} = 0 \qquad \text{for } j = 1, 2, \ldots, N$$

as a function of $v$, the values of $x$ and $v$ can be determined simultaneously by solving the following system of $N + 1$ equations in $N + 1$ unknowns:

$$\frac{\partial L}{\partial x_j} = 0 \qquad \text{for } j = 1, 2, \ldots, N$$

$$h_1(x) = 0$$

Any appropriate numerical search technique discussed in Chapter 3 (e.g., Newton's method) could be used to determine all possible solutions. For each of the solutions $(x^\circ; v^\circ)$, the Hessian matrix of $L$ with respect to $x$ has to be evaluated to determine whether it is positive definite for a local minimum (or negative definite for a local maximum).

### Example 5.3

$$\text{Maximize} \quad f(x) = x_1 + x_2$$

$$\text{Subject to} \quad x_1^2 + x_2^2 = 1$$

**Solution**

$$L(x; v) = x_1 + x_2 - v\left(x_1^2 + x_2^2 - 1\right)$$

$$\frac{\partial L}{\partial x_1} = 1 - 2vx_1 = 0$$

$$\frac{\partial L}{\partial x_2} = 1 - 2vx_2 = 0$$

$$h_1(x) = x_1^2 + x_2^2 - 1 = 0$$

There are two solutions to this system of three equations in three variables, given by

$$\left(x^{(1)}; v_1\right) = \left(-\frac{1}{\sqrt{2}}, -\frac{1}{\sqrt{2}}; -\sqrt{\frac{1}{2}}\right)$$

$$\left(x^{(2)}; v_2\right) = \left(\frac{1}{\sqrt{2}}, \frac{1}{\sqrt{2}}; \sqrt{\frac{1}{2}}\right)$$

The Hessian matrix of $L(x; v)$ with respect to $x$ is given by

$$H_L(x; v) = \begin{bmatrix} -2v & 0 \\ 0 & -2v \end{bmatrix}$$

Evaluating the matrix $H$ at the two solutions, we find that

$$H_L\left(x^{(1)}; v_1\right) = \begin{bmatrix} \sqrt{2} & 0 \\ 0 & \sqrt{2} \end{bmatrix} \quad \text{is positive definite}$$

and

$$H_L\left(x^{(2)}; v_2\right) = \begin{bmatrix} -\sqrt{2} & 0 \\ 0 & -\sqrt{2} \end{bmatrix} \quad \text{is negative definite}$$

Hence, $(x^{(2)}; v_2)$ corresponds to the maximum of $L$ with respect to $x$, and the optimal solution is $x_1^o = x_2^o = 1/\sqrt{2}$. [Note that $(x^{(1)}; v_1)$ corresponds to the minimum of $L$.]

It is to be emphasized here that if we consider $L$ as a function of three variables, namely, $x_1$, $x_2$ and $v$, then the points $(x^{(1)}; v_1)$ and $(x^{(2)}; v_2)$ do not correspond to a minimum or maximum of $L$ with respect to $x$ and $v$. As a matter of fact, they become *saddlepoints* of the function $L(x, v)$. We shall discuss saddlepoints and their significance later in this chapter.

The Lagrange multiplier method can be extended to several equality constraints. Consider the general problem

$$\text{Minimize} \quad f(x)$$

$$\text{Subject to} \quad h_k(x) = 0 \qquad k = 1, 2, \dots, K$$

The Lagrange function becomes

$$L(x; v) = f(x) - \sum_{k=1}^{K} v_k h_k$$

Here $v_1, v_2, \ldots, v_K$ are the Lagrange multipliers, which are unspecified parameters whose values will be determined later. By setting the partial derivatives of $L$ with respect to $x$ equal to zero, we get the following system of $N$ equations in $N$ unknowns:

$$\frac{\partial L(x; v)}{\partial x_1} = 0$$

$$\frac{\partial L(x; v)}{\partial x_2} = 0$$

$$\vdots$$

$$\frac{\partial L(x; v)}{\partial x_N} = 0$$

It is difficult to solve the above system as a function of the vector $v$ explicitly, we can augment the system with the constraint equations

$$h_1(x) = 0$$

$$h_2(x) = 0$$

$$\vdots$$

$$h_K(x) = 0$$

A solution to the augmented system of $N + K$ equations in $N + K$ variables gives a stationary point of $L$. This can then be tested for minimum or maximum by computing the Hessian matrix of $L$ with respect to $x$ as discussed in the single-constraint case.

There may exist some problems for which the augmented system of $N + K$ equations in $N + K$ unknowns may not have a solution. In such cases, the Lagrange multiplier method would fail. However, such cases are rare in practice.

## 5.3   ECONOMIC INTERPRETATION OF LAGRANGE MULTIPLIERS

So far in our discussion, we have treated Lagrange multipliers as adjustable parameters whose values are adjusted to satisfy the constraints. In fact, the Lagrange multipliers do have an important economic interpretation as *shadow prices* of the constraints, and their optimal values are very useful in sensitivity analysis. To exhibit this interpretation, let us consider the following optimization problem involving two variables and one equality constraint:

$$\text{Minimize} \quad f(x_1, x_2)$$

$$\text{Subject to} \quad h_1(x_1, x_2) = b_1$$

where $b_1$ corresponds to the availability of a certain scarce resource.

$$L(x; v_1) = f(x) - v_1[h_1(x) - b_1]$$

Let us assume that the stationary point of $L$ corresponds to the global minimum.

$$\frac{\partial L}{\partial x_1} = \frac{\partial f}{\partial x_1} - v_1\frac{\partial h_1}{\partial x_1} = 0 \tag{5.4}$$

$$\frac{\partial L}{\partial x_2} = \frac{\partial f}{\partial x_2} - v_1\frac{\partial h_1}{\partial x_2} = 0 \tag{5.5}$$

Let $v_1^\circ$ be the optimal Lagrange multiplier and $x^\circ$ be the optimal solution. Let the minimum of $L(x; v)$ for $v_1 = v_1^\circ$ occur at $x = x^\circ$ such that $h_1(x^\circ) = b_1$ and $f(x^\circ) = L(x^\circ; v_1^\circ) = f^\circ$. It is clear that the optimal values $(x^\circ; v_1^\circ)$ are a function of $b_1$, the limited availability of the scarce resource.

The change in $f^\circ$, the optimal value of $f$, due to a change in $b_1$ is given by the partial derivative $\partial f^\circ/\partial b_1$. By the chain rule,

$$\frac{\partial f^\circ}{\partial b_1} = \frac{\partial f^\circ}{\partial x_1^\circ} \cdot \frac{\partial x_1^\circ}{\partial b_1} + \frac{\partial f^\circ}{\partial x_2^\circ} \cdot \frac{\partial x_2^\circ}{\partial b_1} \tag{5.6}$$

The partial derivative of the constraint function $h_1(x) - b_1 = 0$ is given by

$$\frac{\partial h_1}{\partial x_1^\circ} \cdot \frac{\partial x_1^\circ}{\partial b_1} + \frac{\partial h_1}{\partial x_2^\circ} \cdot \frac{\partial x_2^\circ}{\partial b_1} - 1 = 0 \tag{5.7}$$

Multiply both sides of (5.7) by $v_1^\circ$ and subtract from (5.6) to get

$$\frac{\partial f^\circ}{\partial b_1} = v_1^\circ + \sum_{j=1}^{2}\left[\frac{\partial f^\circ}{\partial x_j^\circ} - v_1^\circ\frac{\partial h_1}{\partial x_j^\circ}\right]\frac{\partial x_j^\circ}{\partial b_1} \tag{5.8}$$

Since $x^\circ$ and $v_1^\circ$ satisfy (5.4) and (5.5), Eq. (5.8) reduces to

$$\frac{\partial f^\circ}{\partial b_1} = v_1^\circ \tag{5.9}$$

Thus, from (5.9) we note that the rate of change of the optimal value of $f$ with respect to $b_1$ is given by the optimal value of the Lagrange multiplier $v_1^o$. In other words, the change in the optimal value of the objective function per unit increase in the right-hand-side constant of a constraint, which we defined as *shadow price* in Chapter 4, is given by the Lagrange multiplier. Depending on the sign of $v_1^o$, $f^o$ may increase or decrease with a change in $b_1$.

For an optimization problem with $K$ constraints and $n$ variables given by

$$\text{Minimize} \quad f(x)$$

$$\text{Subject to} \quad h_k(x) = b_k \quad \text{for } k = 1, 2, \ldots, K$$

we can show, using a similar argument, that

$$\frac{\partial f^o}{\partial b_k} = v_k^o \quad \text{for } k = 1, 2, \ldots, K$$

## 5.4 KUHN-TUCKER CONDITIONS

In the previous section we found that Lagrange multipliers could be used in developing optimality criteria for the equality-constrained optimization problems. Kuhn and Tucker have extended this theory to include the general nonlinear programming (NLP) problem with both equality and inequality constraints. Consider the following general NLP problem:

$$\text{Minimize} \quad f(x) \quad \quad (5.10)$$

$$\text{Subject to} \quad g_j(x) \geqslant 0 \quad \text{for } j = 1, 2, \ldots, J \quad \quad (5.11)$$

$$h_k(x) = 0 \quad \text{for } k = 1, 2, \ldots, K \quad \quad (5.12)$$

$$x = (x_1, x_2, \ldots, x_N)$$

### Definition

The inequality constraint $g_j(x) \geqslant 0$ is said to be an *active* or *binding constraint* at the point $\bar{x}$ if $g_j(\bar{x}) = 0$; it is said to be *inactive or nonbinding* if $g_j(\bar{x}) > 0$.

If we can identify the inactive constraints at the optimum before solving the problem, then we can delete those constraints from the model and reduce the problem size. The main difficulty lies in identifying the inactive constraints before the problem is solved.

Kuhn and Tucker have developed the necessary and sufficient optimality conditions for the NLP problem assuming that the functions $f$, $g_j$, and $h_k$ are differentiable. These optimality conditions, commonly known as the *Kuhn-Tucker conditions* (KTC) may be stated in the form of finding a solution to a system of nonlinear equations. Hence, they are also referred to as the *Kuhn-Tucker problem* (KTP).

### 5.4.1  Kuhn-Tucker Conditions or Kuhn-Tucker Problem

Find vectors $x_{(N \times 1)}$, $u_{(1 \times J)}$, and $v_{(1 \times K)}$ that satisfy

$$\nabla f(x) - \sum_{j=1}^{J} u_j \nabla g_j(x) - \sum_{k=1}^{K} v_k \nabla h_k(x) = 0 \qquad (5.13)$$

$$g_j(x) \geqslant 0 \qquad \text{for } j = 1, 2, \ldots, J \qquad (5.14)$$

$$h_k(x) = 0 \qquad \text{for } k = 1, 2, \ldots, K \qquad (5.15)$$

$$u_j g_j(x) = 0 \qquad \text{for } j = 1, 2, \ldots, J \qquad (5.16)$$

$$u_j \geqslant 0 \qquad \text{for } j = 1, 2, \ldots, J \qquad (5.17)$$

Let us first illustrate the Kuhn-Tucker conditions with an example.

### Example 5.4

$$\text{Minimize} \quad f(x) = x_1^2 - x_2$$

$$\text{Subject to} \qquad x_1 + x_2 = 6$$

$$x_1 - 1 \geqslant 0 \qquad x_1^2 + x_2^2 \leqslant 26$$

**Solution.**   Expressing the above problem in the NLP problem format given by Eqs. (5.12) and (5.13), we get

$$f(x) = x_1^2 - x_2 \qquad\qquad \nabla f(x) = (2x_1, -1)$$

$$g_1(x) = x_1 - 1 \geqslant 0 \qquad\qquad \nabla g_1(x) = (1, 0)$$

$$g_2(x) = 26 - x_1^2 - x_2^2 \geqslant 0 \qquad \nabla g_2(x) = (-2x_1, -2x_2)$$

$$h_1(x) = x_1 + x_2 - 6 = 0 \qquad \nabla h_1(x) = (1, 1)$$

Equation (5.13) of the Kuhn-Tucker conditions reduces to

$$\frac{\partial f}{\partial x_j} - u_1 \frac{\partial g_1}{\partial x_j} - u_2 \frac{\partial g_2}{\partial x_j} - v_1 \frac{\partial h_1}{\partial x_j} = 0 \qquad \text{for } j = 1, 2$$

This corresponds to

$$2x_1 - u_1 + 2x_1u_2 - v_1 = 0$$

$$-1 \qquad + 2x_2u_2 - v_1 = 0$$

Equations (5.14) and (5.15) of KTP correspond to the given constraints of the NLP problem and are given by

$$x_1 - 1 \geqslant 0$$

$$26 - x_1^2 - x_2^2 \geqslant 0$$

$$x_1 + x_2 - 6 = 0$$

Equation (5.16) is known as the *complementary slackness condition* in KTP and is given by

$$u_1(x_1 - 1) = 0$$

$$u_2(26 - x_1^2 - x_2^2) = 0$$

$$u_1, u_2 \geqslant 0$$

Note that the variables $u_1$ and $u_2$ are restricted to be zero or positive, while $v_1$ is unrestricted in sign.

Thus, the Kuhn-Tucker conditions for Example 5.4 are given by

$$2x_1 - u_1 + 2x_1u_2 - v_1 = 0$$

$$-1 \qquad + 2x_2u_2 - v_1 = 0$$

$$x_1 - 1 \geqslant 0$$

$$26 - x_1^2 - x_2^2 \geqslant 0$$

$$x_1 + x_2 - 6 = 0$$

$$u_1(x_1 - 1) = 0$$

$$u_2(26 - x_1^2 - x_2^2) = 0$$

$$u_1 \geqslant 0, \qquad u_2 \geqslant 0, \qquad v_1 \text{ unrestricted}$$

## 5.4.2    Interpretation of Kuhn-Tucker Conditions

To interpret the Kuhn-Tucker conditions, consider the equality-constrained NLP problem

$$\text{Minimize} \quad f(x)$$

$$\text{Subject to} \quad h_k(x) = 0 \qquad k = 1, \ldots, K$$

The Kuhn-Tucker conditions are given by

$$\nabla f(x) - \sum_k v_k \nabla h_k(x) = 0 \tag{5.18}$$

$$h_k(x) = 0 \tag{5.19}$$

Now consider the Lagrangian function corresponding to the equality-constrained NLP problem:

$$L(x; v) = f(x) - \sum_k v_k h_k(x)$$

The first-order optimality conditions are given by

$$\nabla_L(x) = \nabla f(x) - \sum_k v_k \nabla h_k(x) = 0$$

$$\nabla_L(v) = h_k(x) = 0$$

We find that the Kuhn-Tucker conditions (5.18) and (5.19) are simply the first-order optimality conditions of the Lagrangian problem.

Let us consider the inequality-constrained NLP problem:

$$\text{Minimize} \quad f(x)$$

$$\text{Subject to} \quad g_j(x) \geqslant 0, \qquad j = 1, \ldots, J$$

The Kuhn-Tucker conditions are given by

$$\nabla f(x) - \sum_j u_j \nabla g_j(x) = 0$$

$$g_j(x) \geqslant 0 \qquad u_j g_j(x) = 0 \qquad u_j \geqslant 0$$

The Lagrangian function can be expressed as

$$L(x; u) = f(x) - \sum_j u_j g_j(x)$$

The first-order optimality conditions are given by

$$\nabla f(x) - \sum_j u_j \nabla g_j(x) = 0 \tag{5.20}$$

$$g_j(x) = 0 \qquad \text{for } j = 1, \ldots, J$$

Note that $u_j$ is the Lagrange multiplier corresponding to constraint $j$. In Section 5.3 we showed that $u_j$ represents the shadow price of constraint $j$; in other words, $u_j$ gives the change in the minimum value of the objective function $f(x)$ per unit increase in the right-hand-side constant.

If we assume that the $j$th constraint is inactive [i.e., $g_j(x) > 0$], then $u_j = 0$ and $u_j g_j(x) = 0$. On the other hand, if the $j$th constraint is active (i.e., $g_j(x) = 0$), then its shadow price $u_j$ need not necessarily be zero, but the value $u_j g_j(x) = 0$, since $g_j(x) = 0$. Hence,

$$u_j g_j(x) = 0 \qquad \text{for all } j = 1, \ldots, J$$

To determine the sign of $u_j$ or the sign of the shadow price of the constraint $g_j(x) \geqslant 0$, let us increase its right-hand-side value from 0 to 1. It is clear that this will constrain the problem further, because any solution that satisfied $g_j(x) \geqslant 1$ will automatically satisfy $g_j(x) \geqslant 0$. Hence, the feasible region will become smaller, and the minimum value of $f(x)$ cannot improve (i.e., will generally increase). In other words, the shadow price of the $j$th constraint $u_j$ is nonnegative, as given by the Kuhn-Tucker conditions.

## 5.5  KUHN-TUCKER THEOREMS

In the previous section we developed the Kuhn-Tucker conditions for constrained optimization problems. Using the theory of Lagrange multipliers, we saw intuitively that the Kuhn-Tucker conditions give the necessary conditions of optimality. In this section, we see the precise conditions under which the Kuhn-Tucker problem implies the necessary and sufficient conditions of optimality.

### Theorem 5.1  Kuhn-Tucker Necessity Theorem

Consider the NLP problem given by Eqs. (5.10)–(5.12). Let $f$, $g$, and $h$ be differentiable functions and $x^*$ be a feasible solution to NLP. Let $I = \{j \mid g_j(x^*) = 0\}$. Furthermore, $\nabla g_j(x^*)$ for $j \in I$ and $\nabla h_k(x^*)$ for $k = 1, \ldots, K$ are linearly independent. If $x^*$ is an optimal solution to NLP, then there exists a $(u^*, v^*)$ such that $(x^*, u^*, v^*)$ solves the Kuhn-Tucker problem given by Eqs. (5.13)–(5.17).

The proof of the theorem is beyond the scope of this text. Interested students may refer to Bazaraa and Shetty [1, Chapter 4].

The conditions that $\nabla g_j(x^*)$ for $j \in I$ and $\nabla h_k(x^*)$ for $k = 1, \ldots, K$ are linearly independent at the optimum is known as a constraint qualification. A constraint qualification essentially implies certain regularity conditions on the feasible region that are frequently satisfied in practical problems. However, in general, it is difficult to verify the constraint qualification, since it requires that the optimum solution be known beforehand. For certain special NLP problems, however, the constraint qualification is always satisfied:

1. When all the inequality and equality constraints are linear.
2. When all the inequality constraints are concave functions and the equality constraints are linear and there exists at least one feasible $x$ that is strictly inside the feasible region of the inequality constraints. In other words, there exists an $\bar{x}$ such that $g_j(\bar{x}) > 0$ for $j = 1, \ldots, J$ and $h_k(\bar{x}) = 0$ for $k = 1, \ldots, K$.

When the constraint qualification is not met at the optimum, there may not exist a solution to the Kuhn-Tucker problem.

***Example 5.5***

$$\text{Minimize} \quad f(x) = (x_1 - 3)^2 + x_2^2$$

$$\text{Subject to} \quad g_1(x) = (1 - x_1)^3 - x_2 \geq 0$$

$$g_2(x) = x_1 \geq 0$$

$$g_3(x) = x_2 \geq 0$$

**Solution.** Figure 5.1 illustrates the feasible region for this nonlinear program. It is clear that the optimal solution to this problem is $x_1^* = 1$, $x_2^* = 0$, and $f(x^*) = 4$. We shall now show that the constraint qualification is not satisfied at the optimum.

Since $g_1(x^*) = 0$, $g_2(x^*) > 0$, and $g_3(x^*) = 0$, $I = \{1, 3\}$. Now,

$$\nabla g_1(x^*) = \left[ -3(1 - x_1)^2, -1 \right]_{x=x^*} = (0, -1)$$

and

$$\nabla g_3(x^*) = (0, 1)$$

It is clear that $\nabla g_1(x^*)$ and $\nabla g_3(x^*)$ are not linearly independent. Hence, the constraint qualification is not satisfied at the point $x^* = (1, 0)$.

Let us now write the Kuhn-Tucker conditions to see whether they will be satisfied at $(1, 0)$. Equations (5.13), (5.16), and (5.17) of the Kuhn-Tucker

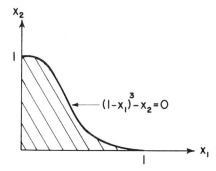

Figure 5.1.  Feasible region of Example 5.5.

conditions become

$$2(x_1 - 3) - u_1\left[-3(1 - x_1)^2\right] - u_2 = 0 \tag{5.21}$$

$$2x_2 - u_1(-1) - u_3 = 0 \tag{5.22}$$

$$u_1\left[(1 - x_1)^3 - x_2\right] = 0 \tag{5.23}$$

$$u_2 x_1 = 0 \tag{5.24}$$

$$u_3 x_2 = 0 \tag{5.25}$$

$$u_1, u_2, u_3 \geqslant 0 \tag{5.26}$$

At $x^* = (1, 0)$, Eq. (5.21) implies $u_2 = -4$, while to satisfy Eq. (5.24), $u_2 = 0$. Hence, there exists no Kuhn-Tucker point at the optimum.

Note that when the constraint qualification is violated, it does not necessarily imply that a Kuhn-Tucker point does not exist. To illustrate this, suppose the objective function of Example 5.5 is changed to $f(x) = (x_1 - 1)^2 + x_2^2$. The optimum still occurs at $x^* = (1, 0)$, and it does not satisfy the constraint qualifications. The Kuhn-Tucker conditions given by Eqs. (5.22)–(5.26) will remain the same, while Eq. (5.21) becomes

$$2(x_1 - 1) - u_1\left[-3(1 - x_1)^2\right] - u_2 = 0$$

The reader can easily verify that there exists a Kuhn-Tucker point given by $x^* = (1, 0)$ and $u^* = (0, 0, 0)$ that satisfies the Kuhn-Tucker conditions.

The Kuhn-Tucker necessity theorem helps to identify points that are not optimal. In other words, given a feasible point that satisfies the constraint qualification, we can use Theorem 5.1 to prove that it is not optimal if it does not satisfy the Kuhn-Tucker conditions. On the other hand, if it does satisfy the Kuhn-Tucker conditions, there is no assurance that it is optimal to the nonlinear program! For example, consider the following NLP problem.

***Example 5.6***

$$\text{Minimize} \quad f(x) = 1 - x^2$$

$$\text{Subject to} \quad -1 \leqslant x \leqslant 3$$

**Solution.**  Here,

$$g_1(x) = x + 1 \geqslant 0$$

$$g_2(x) = 3 - x \geqslant 0$$

The Kuhn-Tucker conditions are given by

$$-2x - u_1 + u_2 = 0 \tag{5.27}$$

$$-1 \leqslant x \leqslant 3 \tag{5.28}$$

$$u_1(x + 1) = 0 \tag{5.29}$$

$$u_2(3 - x) = 0 \tag{5.30}$$

$$u_1, u_2 \geqslant 0 \tag{5.31}$$

Since the constraints are linear, the constraint qualification is satisfied at all feasible points. It is clear that $x = 3$ is optimal. But consider the feasible solution $x = 2$. To prove that it is not optimal, let us try to construct a Kuhn-Tucker point at $x = 2$ that satisfies Eqs. (5.27)–(5.31). To satisfy Eqs. (5.29) and (5.30), $u_1 = u_2 = 0$; but $x = 2$, $u_1 = u_2 = 0$ violates Eq. (5.27). Hence, by Theorem 5.1, $x = 2$ cannot be optimal.

On the other hand, the solution $x = u_1 = u_2 = 0$ satisfies Eqs. (5.27)–(5.31) and hence is a Kuhn-Tucker point, but it is not optimal! By Theorem 5.1, we also know that the Kuhn-Tucker conditions must be satisfied at the optimal solution $x = 3$. It is easy to verify that the solution $x = 3$, $u_1 = 0$, $u_2 = 6$ satisfies the Kuhn-Tucker conditions.

The following theorem gives conditions under which a Kuhn-Tucker point automatically becomes an optimal solution to the NLP problem.

### Theorem 5.2   Kuhn-Tucker Sufficiency Theorem

Consider the NLP problem given by Eqs. (5.10)–(5.12). Let the objective function $f(x)$ be convex, the inequality constraints $g_j(x)$ be all concave functions for $j = 1, \ldots, J$, and the equality constraints $h_k(x)$ for $k = 1, \ldots, K$ be linear. If there exists a solution $(x^*, u^*, v^*)$ that satisfies the Kuhn-Tucker conditions given by Eqs. (5.13)–(5.17), then $x^*$ is an *optimal* solution to the NLP problem.

A rigorous proof of the Kuhn-Tucker sufficiency theorem can be found in Mangasarian [2].

When the sufficiency conditions of Theorem 5.2 hold, finding a Kuhn-Tucker point gives an optimal solution to an NLP problem. Theorem 5.2 can also be used to prove that a given solution to an NLP problem is optimal. To illustrate this, recall Example 5.4:

$$\text{Minimize} \quad f(x) = x_1^2 - x_2$$

$$\text{Subject to} \quad g_1(x) = x_1 - 1 \qquad \geqslant 0$$

$$g_2(x) = 26 - x_1^2 - x_2^2 \geqslant 0$$

$$h_1(x) = x_1 + x_2 - 6 = 0$$

We shall prove that $x_1^* = 1$, $x_2^* = 5$ is optimal by using Theorem 5.2. Now,

$$\nabla f(x) = (2x_1, -1) \quad \text{and} \quad \mathbf{H}_f(x) = \begin{bmatrix} 2 & 0 \\ 0 & 0 \end{bmatrix}$$

Since $\mathbf{H}_f(x)$ is positive semidefinite for all $x$, $f(x)$ is a convex function. The inequality constraint $g_1(x)$ is linear and hence both convex and concave. To show that $g_2(x)$ is concave, compute

$$\nabla g_2(x) = (-2x_1, -2x_2) \quad \text{and} \quad \mathbf{H}_{g_2}(x) = \begin{bmatrix} -2 & 0 \\ 0 & -2 \end{bmatrix}$$

Since $H_{g_2}(x)$ is negative definite, $g_2(x)$ is concave. The equality constraint $h_1(x)$ is linear. Hence all the sufficiency conditions of Theorem 5.2 are satisfied, and if we are able to construct a Kuhn-Tucker point using $x^* = (1, 5)$, the solution $x^*$ is indeed optimal.

The Kuhn-Tucker conditions of Example 5.4 are given below:

$$2x_1 - u_1 + 2x_1u_2 - v_1 = 0 \tag{5.32}$$

$$-1 \qquad + 2x_2u_2 - v_1 = 0 \tag{5.33}$$

$$x_1 - 1 \geqslant 0 \tag{5.34}$$

$$26 - x_1^2 - x_2^2 \geqslant 0 \tag{5.35}$$

$$x_1 + x_2 - 6 = 0 \tag{5.36}$$

$$u_1(x_1 - 1) = 0 \tag{5.37}$$

$$u_2(26 - x_1^2 - x_2^2) = 0 \tag{5.38}$$

$$u_1, u_2 \geqslant 0 \tag{5.39}$$

$x^* = (1, 5)$ satisfies Eqs. (5.34)–(5.36), and hence it is feasible. Equations (5.32) and (5.33) reduce to

$$2 - u_1 + 2u_2 - v_1 = 0$$

$$-1 + 10u_2 - v_1 = 0$$

By setting $v_1 = 0$, we can get a solution $u_2 = 0.1$ and $u_1 = 2.2$. Thus, the solution $x^* = (1, 5)$, $u^* = (2.2, 0.1)$, and $v_1^* = 0$ satisfies the Kuhn-Tucker conditions. Since the sufficiency conditions of Theorem 5.2 are satisfied, $x^* = (1, 5)$ is an optimal solution to Example 5.4. Note that there also exist other values of $u_1, u_2, v_1$ that satisfy Eqs. (5.32)–(5.39).

### Remarks

1. For practical problems, the constraint qualification will generally hold. If the functions are differentiable, a Kuhn-Tucker point is a possible candidate for the optimum. Hence, many of the NLP methods attempt to converge to a Kuhn-Tucker point. (Recall the analogy to the unconstrained optimization case wherein the algorithms attempt to determine a stationary point.)

2. When the sufficiency conditions of Theorem 5.2 hold, a Kuhn-Tucker point automatically becomes the global minimum. Unfortunately, the sufficiency conditions are difficult to verify, and often practical problems may not possess these nice properties. Note that the presence of one nonlinear equality constraint is enough to violate the assumptions of Theorem 5.2.

3. The sufficiency conditions of Theorem 5.2 have been generalized further to nonconvex inequality constraints, nonconvex objectives, and nonlinear equality constraints. These use generalizations of convex functions such as quasi-convex and pseudo-convex functions. Interested readers are referred to Mangasarian [2, chapters 9 and 10].

## 5.6 SADDLEPOINT CONDITIONS

The discussion of Kuhn-Tucker optimality conditions of Sections 5.4 and 5.5 assume that the objective function and the constraints are differentiable. We now discuss constrained optimality criteria for nondifferentiable functions.

### Definition

A function $f(x, y)$ is said to have a *saddlepoint* at $(x^*, y^*)$ if $f(x^*, y) \leqslant f(x^*, y^*) \leqslant f(x, y^*)$ for all $x$ and $y$.

The definition of a saddlepoint implies that $x^*$ minimizes the function $f(x, y^*)$ for all $x$ and $y^*$ maximizes the function $f(x^*, y)$ for all $y$. For example, consider a function $f(x, y) = x^2 - xy + 2y$ defined over all real values of $x$ and nonnegative values of $y$. It is easy to verify that the function possesses a saddlepoint at the point $x^* = 2, y^* = 4$. In other words,

$$f(2, y) \leq f(2,4) \leq f(x,4) \qquad \text{for all } y \geq 0 \text{ and all real } x$$

Recall the Lagrange multiplier method discussed in Section 5.2. It solves a constrained optimization problem of the form

$$\text{Minimize} \quad f(x)$$

$$\text{Subject to} \quad h_k(x) = 0 \qquad \text{for } k = 1,\ldots, K$$

The Lagrangian function is defined to be

$$L(x; v) = f(x) - \sum_k v_k h_k(x)$$

Suppose at $v = v^*$ the minimum of $L(x, v^*)$ occurs at $x = x^*$ such that $h_k(x^*) = 0$. We know then by the Lagrange multiplier method that $x^*$ is an optimal solution to the nonlinear program. It can be shown that $(x^*, v^*)$ is a saddlepoint of the Lagrangian function satisfying

$$L(x^*, v) \leq L(x^*, v^*) \leq L(x, v^*) \qquad \text{for all } x \text{ and } v$$

Consider the general NLP problem:

$$\text{Minimize} \quad f(x)$$

$$\text{Subject to} \quad g_j(x) \geq 0 \qquad \text{for } j = 1,\ldots, J$$

$$x \in S$$

The set $S$ may be used to impose additional restrictions on the design variables. For example, the design variables may all be integers or restricted to a certain discrete set.

The *Kuhn-Tucker saddlepoint problem* (KTSP) is as follows: Find $(x^*, u^*)$ such that

$$L(x^*, u) \leq L(x^*, u^*) \leq L(x, u^*)$$

$$\text{all } u \geq 0 \quad \text{and} \quad \text{all } x \in S$$

where

$$L(x, u) = f(x) - \sum_j u_j g_j(x)$$

### Theorem 5.3  Sufficient Optimality Theorem

If $(x^*, u^*)$ is a saddlepoint solution of a KTSP, then $x^*$ is an optimal solution to the NLP problem.

A proof of this theorem is available in Mangasarian [2, Chapter 3].

### Remarks

1. No convexity assumptions of the functions have been made in Theorem 5.3.
2. No constraint qualification is invoked.
3. Nonlinear equality constraints of the form $h_k(x) = 0$ for $k = 1, \ldots, K$ can be handled easily by redefining the Lagrangian function as

$$L(x, u, v) = f(x) - \sum_j u_j g_j(x) - \sum_k v_k h_k(x)$$

Here the variables $v_k$ for $k = 1, \ldots, K$ will be unrestricted in sign.

4. Theorem 5.3 provides only a sufficient condition. There may exist some NLP problems for which a saddlepoint does not exist even though the NLP problem has an optimal solution.

**Existence of Saddlepoints.**   There exist necessary optimality theorems that guarantee the existence of a saddlepoint solution without the assumption of differentiability. However, they assume that the constraint qualification is met and that the functions are convex.

### Theorem 5.4  Necessary Optimality Theorem

Let $x^*$ minimize $f(x)$ subject to $g_j(x) \geq 0, j = 1, \ldots, J$ and $x \in S$. Assume $S$ is a convex set, $f(x)$ is a convex function, and $g_j(x)$ are concave functions on $S$. Assume also that there exists a point $\bar{x} \in S$ such that $g_j(\bar{x}) > 0$ for all $j = 1, 2, \ldots, J$. Then there exists a vector of multipliers $u^* \geq 0$ such that $(x^*, u^*)$ is a saddlepoint of the Lagrangian function

$$L(x, u) = f(x) - \sum_j u_j g_j(x)$$

satisfying

$$L(x^*, u) \leq L(x^*, u^*) \leq L(x, u^*)$$

for all $x \in S$ and $u \geq 0$.

For a proof of this theorem, refer to the text by Lasdon [3, Chapter 1].

Even though Theorem 5.3 and the Kuhn-Tucker saddlepoint problem (KTSP) provide sufficient conditions for optimality without invoking differentiability and convexity, determination of a saddlepoint to a KTSP is generally difficult. However, the following theorem makes it computationally more attractive.

***Theorem 5.5***

A solution $(x^*, u^*)$ with $u^* \geq 0$ and $x^* \in S$ is a saddlepoint of a KTSP if and only if the following conditions are satisfied:

(i)  $x^*$ minimizes $L(x, u^*)$ over all $x \in S$

(ii)  $g_j(x^*) \geq 0$   for $j = 1, \ldots, J$

(iii)  $u_j g_j(x^*) = 0$   for $j = 1, \ldots, J$

For a proof, see Lasdon [3, Chapter 1].

Condition (i) of Theorem 5.5 amounts to finding an unconstrained minimum of a function, and any of the direct-search methods discussed in Chapter 3 could be used. Of course, this assumes prior knowledge of the value of $u^*$. However, a trial-and-error method can be used to determine $u^*$ and $x^*$ simultaneously and also satisfy conditions (ii) and (iii). One such method, due to Everett [4], is called the *generalized Lagrange multiplier method*. Theorems 5.3 and 5.5 also form the basis of many of the Lagrangian relaxation methods that have been developed for solving large-scale NLP problems [3]. It is important to note that saddlepoints may not exist for all NLP problems. The existence of saddlepoints is guaranteed only for NLP problems that satisfy the conditions of Theorem 5.4.

## 5.7  SECOND-ORDER OPTIMALITY CONDITIONS

In Sections 5.4–5.6, we discussed the first-order necessary and sufficient conditions, called the Kuhn-Tucker conditions, for constrained optimization problems using the gradients of the objective function and constraints. Second-order necessary and sufficient optimality conditions that apply to twice-differentiable functions have been developed by McCormick [5] whose main results are summarized in this section. Consider the following NLP problem.

**Problem P1**

$$\text{Minimize} \quad f(x)$$

$$\text{Subject to} \quad g_j(x) \geqslant 0 \quad j = 1, 2, \ldots, J$$

$$h_k(x) = 0 \quad k = 1, 2, \ldots, K$$

$$x \in R^N$$

The first-order Kuhn-Tucker conditions are given by

$$\nabla f(x) - \sum_j u_j \nabla g_j(x) - \sum v_k \nabla h_k(x) = 0 \tag{5.40}$$

$$g_j(x) \geqslant 0 \qquad j = 1, \ldots, J \tag{5.41}$$

$$h_k(x) = 0 \qquad k = 1, \ldots, K \tag{5.42}$$

$$u_j g_j(x) = 0 \qquad j = 1, \ldots, J \tag{5.43}$$

$$u_j \geqslant 0 \qquad j = 1, \ldots, J \tag{5.44}$$

### Definitions

$\bar{x}$ is a *feasible solution* to an NLP when $g_j(\bar{x}) \geqslant 0$ for all $j$ and $h_k(\bar{x}) = 0$ for all $k$.

$x^*$ is a *local minimum* to an NLP when $x^*$ is feasible and $f(x^*) \leqslant f(\bar{x})$ for all feasible $\bar{x}$ in some small neighborhood $\delta(x^*)$ of $x^*$.

$x^*$ is a *strict* (*unique* or *isolated*) *local minimum* when $x^*$ is feasible and $f(x^*) < f(\bar{x})$ for all feasible $\bar{x} \neq x^*$ in some small neighborhood $\delta(x^*)$ of $x^*$.

*A Kuhn-Tucker point* to an NLP is a vector $(x^*, u^*, v^*)$ satisfying Eqs. (5.40)–(5.44).

Let us first consider the basic motivation for the second-order optimality conditions. For simplicity, consider an equality-constrained NLP problem as follows:

$$\text{Minimize} \quad f(x)$$

$$\text{Subject to} \quad h_k(x) = 0, \qquad k = 1, 2, \ldots, K$$

The first-order Kuhn-Tucker conditions are given by $h_k(x) = 0$, $k = 1,\ldots,k$

$$\text{and} \quad \nabla f(x) - \sum_k v_k \nabla h_k(x) = 0 \qquad (5.45)$$

Consider a point $\bar{x}$ that satisfies the first-order conditions. To check further whether it is a local minimum, we can write down the Taylor series expansion at the point $\bar{x}$ using higher order terms for each function $f$ and $h_k$ as follows:

$$\Delta f(\bar{x}) = f(\bar{x} + \Delta x) - f(\bar{x})$$

$$= \nabla f(\bar{x})\,\Delta x + \tfrac{1}{2}\Delta x^T \mathbf{H}_f \Delta x + O(\Delta x) \qquad (5.46)$$

where $O(\Delta x)$ are very small higher order terms involving $\Delta x$.

$$\Delta h_k(\bar{x}) = h_k(\bar{x} + \Delta x) - h_k(\bar{x})$$

$$= \nabla h_k(\bar{x})\,\Delta x + \tfrac{1}{2}\Delta x^T \mathbf{H}_k \Delta x + O(\Delta x) \qquad (5.47)$$

where $\mathbf{H}_k$ is the Hessian matrix of $h_k(x)$ evaluated at $\bar{x}$. Multiply Eq. (5.47) by the Kuhn-Tucker multiplier $v_k$, and sum over all $k = 1,\ldots,K$. Subtracting this sum from Eq. (5.46), we obtain

$$\Delta f(\bar{x}) - \sum_k v_k\,\Delta h_k(\bar{x}) = \left[\nabla f(\bar{x}) - \sum_k v_k \nabla h_k(\bar{x})\right]\Delta x$$

$$+ \tfrac{1}{2}\Delta x^T \left[\mathbf{H}_f - \sum_k v_k \mathbf{H}_k\right]\Delta x + O(\Delta x) \qquad (5.48)$$

For $(\bar{x} + \Delta x)$ to be feasible,

$$\Delta h_k(\bar{x}) = 0 \qquad (5.49)$$

Assuming that the constraint qualification is satisfied at $\bar{x}$, the Kuhn-Tucker necessary theorem implies that

$$\nabla f(\bar{x}) - \sum_k v_k \nabla h_k(\bar{x}) = 0 \qquad (5.50)$$

Using Eqs. (5.49) and (5.50), Eq. (5.48) reduces to

$$\Delta f(\bar{x}) = \tfrac{1}{2}\Delta x^T \left[\mathbf{H}_f - \sum_k v_k \mathbf{H}_k\right]\Delta x + O(\Delta x) \qquad (5.51)$$

For $\bar{x}$ to be a local minimum, it is *necessary* that $\Delta f(\bar{x}) \geq 0$ for all feasible movement $\Delta x$ around $\bar{x}$. Using Eqs. (5.49) and (5.51), the above condition implies that

$$\Delta x^T \left[ \mathbf{H}_f - \sum_k v_k \mathbf{H}_k \right] \Delta x > 0 \qquad (5.52)$$

for all $\Delta x$ satisfying

$$\Delta h_k(\bar{x}) = 0 \qquad \text{for } k = 1, \ldots, K \qquad (5.53)$$

Using Eq. (5.47) and ignoring the second and higher order terms in $\Delta x$, Eq. (5.53) reduces to

$$\Delta h_k(\bar{x}) = \nabla h_k(\bar{x}) \Delta x = 0$$

Thus, assuming that the constraint qualification is satisfied at $\bar{x}$, the necessary conditions for $\bar{x}$ to be a local minimum are:

1. There exists $v_k$, $k = 1, \ldots, K$ such that $(\bar{x}, v)$ is a Kuhn-Tucker point.
2. $\Delta x^T [\mathbf{H}_f - \sum_k v_k \mathbf{H}_k] \Delta x \geqslant 0$ for all $\Delta x$ satisfying

$$\nabla h_k(\bar{x}) \Delta x = 0 \qquad \text{for } k = 1, \ldots, K$$

Similarly, the sufficient condition for $\bar{x}$ to be a strict local minimum is given by

$$\Delta f(\bar{x}) > 0 \qquad \text{for all feasible } \Delta x \text{ around } \bar{x}$$

This implies that

$$\Delta x^T \left[ \mathbf{H}_f - \sum_k v_k \mathbf{H}_k \right] \Delta x > 0$$

for all $\Delta x$ satisfying

$$\nabla h_k(\bar{x}) \Delta x = 0 \qquad \text{for all } k = 1, \ldots, K \qquad (5.54)$$

We shall now present the formal statements of second-order necessary and sufficient conditions for an NLP problem involving both equality and inequality constraints.

### Theorem 5.6   Second-Order Necessity Theorem

Consider the NLP problem given by Problem P1. Let $f$, $g$, and $h$ be twice-differentiable functions, and let $x^*$ be feasible for the nonlinear program. Let the active constraint set at $x^*$ be $I = \{j | g_j(x^*) = 0\}$. Furthermore, assume that $\nabla g_j(x^*)$ for $j \in I$ and $\nabla h_k(x^*)$ for $k = 1, 2, \ldots, K$ are linearly independent. Then the *necessary conditions* that $x^*$ be a *local minimum* to the NLP problem are that

1. There exists $(u^*, v^*)$ such that $(x^*, u^*, v^*)$ is a Kuhn-Tucker point.
2. For every vector $y_{(1 \times N)}$ satisfying

$$\nabla g_j(x^*) y = 0 \qquad \text{for } j \in I \tag{5.55}$$

$$\nabla h_k(x^*) y = 0 \qquad \text{for } k = 1, 2, \ldots, K \tag{5.56}$$

it follows that

$$y^T \mathbf{H}_L(x^*, u^*, v^*) y \geq 0 \tag{5.57}$$

where

$$L(x, u, v) = f(x) - \sum_{j=1}^{J} u_j g_j(x) - \sum_{k=1}^{K} v_k h_k(x)$$

- and $\mathbf{H}_L(x^*, u^*, v^*)$ is the Hessian matrix of the second partial derivatives of $L$ with respect to $x$ evaluated at $(x^*, u^*, v^*)$.

We shall illustrate Theorem 5.6 with an example in which the first-order necessary conditions are satisfied while the second-order conditions show that the point is not optimal.

*Example 5.7 [5]*

$$\text{Minimize} \quad f(x) = (x_1 - 1)^2 + x_2^2$$

$$\text{Subject to} \quad g_1(x) - -x_1 + x_2^2 \geq 0$$

Suppose we want to verify whether $x^* = (0, 0)$ is optimal.

**Solution**

$$\nabla f(x) = \left[ 2(x_1 - 1), 2x_2 \right]$$

$$\nabla g_1(x) = (-1, 2x_2) \qquad I = \{1\}$$

Since $\nabla g_1(x^*) = (-1, 0)$ is linearly independent, the constraint qualification is satisfied at $x^*$. The first-order Kuhn-Tucker conditions are given by

$$2(x_1 - 1) + u_1 = 0$$

$$2x_2 - 2x_2 u_1 = 0$$

$$u_1(-x_1 + x_2^2) = 0$$

$$u_1 \geqslant 0$$

$x^* = (0, 0)$ and $u_1^* = 2$ satisfy the above conditions. Hence, $(x^*, u^*) = (0, 0, 2)$ is a Kuhn-Tucker point and $x^*$ satisfies the first-order necessary conditions of optimality by Theorem 5.1. In other words, we do not know whether or not $(0, 0)$ is an optimal solution to the NLP problem!

Let us now apply the second-order necessary conditions to test whether $(0, 0)$ is a local minimum to the NLP problem. The first part of Theorem 5.6 is already satisfied, since $(x^*, u^*) = (0, 0, 2)$ is a Kuhn-Tucker point. To prove the second-order conditions, compute

$$\mathbf{H}_L(x, u) = \begin{bmatrix} 2 & 0 \\ 0 & 2 - 2u_1 \end{bmatrix}$$

At $(x^*, u^*)$,

$$\mathbf{H}_L(x^*, u^*) = \begin{bmatrix} 2 & 0 \\ 0 & -4 \end{bmatrix}$$

We have to verify whether

$$y^T \begin{bmatrix} 2 & 0 \\ 0 & -4 \end{bmatrix} y \geqslant 0$$

for all $y$ satisfying

$$\nabla g_1(x^*) y = 0 \quad \text{or} \quad (-1, 0)\begin{pmatrix} y_1 \\ y_2 \end{pmatrix} = 0$$

In other words, we need to consider only vectors $(y_1, y_2)$ of the form $(0, y_2)$ to satisfy Eq. (5.57). Now,

$$(0, y_2)\begin{bmatrix} 2 & 0 \\ 0 & -4 \end{bmatrix}\begin{bmatrix} 0 \\ y_2 \end{bmatrix} = -4y_2^2 < 0 \quad \text{for all } y_2 \neq 0$$

Thus, $x^* = (0, 0)$ does not satisfy the second-order necessary conditions, and hence it is *not* a local minimum for the NLP.

## Sufficient Conditions

When a point satisfies the second-order necessary conditions given by Theorem 5.6, it becomes a Kuhn-Tucker point and a candidate for a local minimum. In order to show that it is in fact a minimum point, we need the second-order sufficient conditions. Of course, when a nonlinear program satisfies the assumptions of Theorem 5.2 (Kuhn-Tucker sufficiency theorem), the Kuhn-Tucker point automatically becomes the global minimum. However, Theorem 5.2 requires that the objective function be convex, the inequality constraints concave, and the equality constraints linear. These assumptions are too rigid and may not be satisfied in practice very often. In such situations, the second-order sufficiency conditions may be helpful in showing that a Kuhn-Tucker point is a local minimum.

### Theorem 5.7   Second-Order Sufficiency Theorem

Sufficient conditions that a point $x^*$ is a *strict local minimum* of NLP problem P1, where $f$, $g_j$, and $h_k$ are twice-differentiable functions are that
(i)  There exists $(u^*, v^*)$ such that $(x^*, u^*, v^*)$ is a Kuhn-Tucker point.
(ii) For every nonzero vector $y_{(1 \times N)}$ satisfying

$$\nabla g_j(x^*)y = 0 \qquad j \in I_1 = \{j | g_j(x^*) = 0, u_j^* > 0\} \qquad (5.58)$$

$$\nabla g_j(x^*)y \geqslant 0 \qquad j \in I_2 = \{j | g_j(x^*) = 0, u_j^* = 0\} \qquad (5.59)$$

$$\nabla h_k(x^*)y = 0 \qquad k = 1, 2, \ldots, K \qquad (5.60)$$

$$y \neq 0$$

it follows that

$$y^T H_L(x^*, u^*, v^*)y > 0 \qquad (5.61)$$

*Note:*  $I_1 \cup I_2 = I$, the set of all active constraints at $x^*$.

Comparing Theorems 5.6 and 5.7, it is clear that the sufficient conditions add very few new restrictions to the necessary conditions, and no additional assumptions about the properties of the functions are needed. The minor changes are that Eq. (5.55) need not be satisfied for all active constraints and inequality (5.57) has to be satisfied as a strict inequality.

### Remark

The restrictions on vector $y$ given by Eqs. (5.58) and (5.59) use information on the multiplier $u^*$. Han and Mangasarian [6] have given an equivalent form without the use of the Kuhn-Tucker multiplier $u^*$. They prove that Eqs. (5.58) and (5.59) are equivalent to the following set of conditions:

$$\nabla f(x^*)y \leq 0$$

$$\nabla g_j(x^*)y \geq 0 \qquad \text{for } j \in I = \{j | g_j(x^*) = 0\}$$

We now illustrate the use of second-order sufficient conditions.

### Example 5.8

$$\text{Minimize} \quad f(x) = (x_1 - 1)^2 + x_2^2$$

$$\text{Subject to} \quad g_1(x) = -x_1 + \frac{x_2^2}{5} \geq 0$$

This problem is very similar to the one given in Example 5.7. Suppose we want to verify whether $x^* = (0,0)$ is a local minimum. Note that the region $S = \{x | g_1(x) \geq 0\}$ is not a convex set. For example, $\bar{x} = (0.2, 1)$ and $(0.2, -1)$ are feasible, but the midpoint $(0.2, 0)$ is not.

The Kuhn-Tucker conditions for this problem are

$$2(x_1 - 1) + u_1 = 0$$

$$2x_2 - \tfrac{2}{5}x_2 u_1 = 0$$

$$u_1\left(-x_1 + \frac{x_2^2}{5}\right) = 0$$

$$u_1 \geq 0$$

$x^* = (0,0)$, $u_1^* = 2$ satisfies the Kuhn-Tucker conditions. Using Theorem 5.1, we can conclude that $x^* = (0,0)$ satisfies the necessary conditions for the minimum. But we cannot conclude that $x^*$ is a local minimum, since the function $g_1(x)$ is convex and violates the assumptions of Theorem 5.2 (Kuhn-Tucker sufficiency theorem).

Using the second-order sufficient conditions, we find that

$$\mathbf{H}_L(x^*, u^*) = \begin{bmatrix} 2 & 0 \\ 0 & 1.2 \end{bmatrix}$$

The vector $y = (y_1, y_2)$ satisfying Eqs. (5.58) and (5.59), is of the form $(0, y_2)$

as in Example 5.8. Inequality (5.61) reduces to

$$(0, y_2) \begin{bmatrix} 2 & 0 \\ 0 & 1.2 \end{bmatrix} \begin{bmatrix} 0 \\ y_2 \end{bmatrix} = 1.2 y_2^2 > 0 \qquad \text{for all } y_2 \neq 0$$

Hence by Theorem 5.7, $x^* = (0, 0)$ is a strict local minimum.

Fiacco [7] has extended Theorem 5.7 to sufficient conditions for weak (not necessarily strict) minimum also.

## 5.8  SUMMARY

In this chapter we developed necessary and sufficient conditions of optimality for constrained optimization problems. We began with the discussion of Lagrangian optimality conditions for problems with equality constraints. These were then extended to inequality constraints in the form of Kuhn-Tucker optimality conditions, which are first-order conditions involving the gradients of the objective function and the constraints.

We learned that the Kuhn-Tucker conditions are necessary when the functions are differentiable and the constraints satisfy some regularity condition known as the constraint qualification. Kuhn-Tucker conditions become sufficient conditions for global minima when the objective function is convex, the "inequality" constraints are concave functions, and the equality constraints are linear. We also discussed saddlepoint optimality conditions that could be applicable if the functions were not differentiable.

Since there could be several points satisfying the Kuhn-Tucker necessary conditions, we developed second-order necessary conditions that must be satisfied for a point to be a local minimum. Similarly, the assumptions under which the Kuhn-Tucker sufficiency conditions hold are quite rigid. Hence, second-order sufficiency conditions were developed that do not require convexity of the functions and linearity of the equality constraints. Both the necessary and sufficient second-order conditions impose additional restrictions over and above those given by Kuhn-Tucker and hence can be useful in reducing the set of candidate optima.

## REFERENCES

1. Bazaraa, M. S., and C. M. Shetty, *Nonlinear Programming: Theory and Algorithms*, Wiley, New York, 1979.

2. Mangasarian, O. L., *Nonlinear Programming*, McGraw-Hill, New York, 1969.

3. Lasdon, L. S., *Optimization Theory for Large Systems*, Macmillan, New York, 1979.

4. Everett, H., "Generalized Lagrange Multiplier Method for Solving Problems of Optimum Allocation of Resources," *Oper. Res.* **11**, 399–471 (1963).

5.  McCormick, G. P., "Second Order Conditions for Constrained Optima," *SIAM J. Appl. Math.* **15**, 641–652 (1967).

6.  Han, S. P., and O. L. Mangasarian, "Exact Penalty Functions in Nonlinear Programming," *Math. Programming* **17**, 251–269 (1979).

7.  Fiacco, A. V., "Second Order Sufficient Conditions for Weak and Strict Constrained Minima," *SIAM J. Appl. Math.* **16**, 105–108 (1968).

## PROBLEMS

**5.1.** Explain the difficulties that will be encountered if Lagrange multipliers are used with problems involving nonnegative variables.

**5.2.** What is the significance of the constraint qualification?

**5.3.** What is a saddle point solution? What is its significance in constrained optimization?

**5.4.** Under what conditions do saddle point solutions exist for NLP problems?

**5.5.** What are the primary uses of second-order necessary and sufficient conditions?

**5.6.** Consider the minimization of $f(x)$ over a feasible region denoted by $S$. Suppose $\nabla f(x) \neq 0$ for every $x \in S$. What can you say about the nature of optimal solution to this problem and why?

**5.7.** Use the Lagrange multiplier method to solve

$$\text{Minimize} \quad z = x_1^2 + x_2^2 + x_3^2$$

$$\text{Subject to} \quad x_1 + 2x_2 + 3x_3 = 7$$

$$2x_1 + 2x_2 + x_3 = \tfrac{9}{2}$$

Explain (in two or three lines) why your procedure guarantees an optimum solution.

**5.8.** Use the Lagrange multiplier method to determine the global minimum and the global maximum of the problem

$$f(x) = x_2 - x_1^2$$

$$\text{Subject to} \quad x_1^2 + x_2^2 = 1 \quad \text{with } x_1, x_2 \text{ unrestricted in sign}$$

**5.9.** Find the shortest distance from the point $(1,0)$ to the parabola $y^2 = 4x$ by

(a) Eliminating the variable $y$

(b) The Lagrange multiplier technique

Explain why procedure (a) fails to solve the problem, while procedure (b) does not fail.

**5.10.** Given the problem

$$\text{Minimize} \quad f(x) = (x_1 - 3)^2 + (x_2 + 4)^2 + e^{5x_3}$$

$$\text{Subject to} \quad x_1 + x_2 + x_3 \leqslant 1$$

$$x_1, x_2, x_3 \geqslant 0$$

(a) Write the Kuhn-Tucker conditions for this problem.

(b) Show that Kuhn-Tucker conditions are sufficient for the existence of an optimum solution to the given problem.

(c) Prove, using (a) and (b), that $x = (1, 0, 0)$ is optimal.

**5.11.** Consider the nonlinear program

$$\text{Minimize} \quad Z = x_1^2 + 2x_2^2$$

$$\text{Subject to} \quad x_1^2 + x_2^2 \leqslant 5$$

$$2x_1 - 2x_2 = 1$$

(a) Write the equivalent Kuhn-Tucker problem.

(b) Using (a), what can you conclude about the following solutions to the nonlinear program?

(i) $x^{(1)} = (0, 0)$

(ii) $x^{(2)} = (1, \frac{1}{2})$

(iii) $x^{(3)} = (\frac{1}{3}, -\frac{1}{6})$

You must justify your conclusions. Quote any theorem that is applicable to your conclusions.

**5.12.** Consider the NLP problem given by

$$\text{Minimize} \quad f(x) = 1 - 2x_1 - 4x_1x_2$$

$$\text{Subject to} \quad x_1 + 4x_1x_2 \leqslant 4$$

$$x_1 \geqslant 0 \qquad x_2 \geqslant 0$$

(a) Show that the feasible region of the above NLP problem is *not* a convex set.

(b) Write down the Kuhn-Tucker conditions for this NLP problem.

(c) Now let $y_1 = x_1$ and $y_2 = x_1x_2$. Write the given problem in terms of $(y_1, y_2)$. What kind of mathematical program is this? Show that the feasible region of the transformed problem is a convex set. Solve the transformed problem for optimal values of $y$, say, $(\bar{y}_1, \bar{y}_2)$. Obtain the corresponding values $(\bar{x}_1, \bar{x}_2)$.

(d) Prove that the constraint qualification is satisfied at $\bar{x}$. (You must show this analytically.)

(e) Prove that $\bar{x}$ is a *candidate optimum* using the Kuhn-Tucker necessity theorem.

(f) Are the Kuhn-Tucker conditions sufficient to show $\bar{x}$ is an optimal solution to the NLP problem? Why or why not? Explain.

5.13. Consider the nonlinear programming problem

$$\text{Maximize} \quad z = a_1 x_1 + x_2$$

$$\text{Subject to} \quad x_1^2 + x_2^2 \leqslant 25$$

$$x_1 - x_2 \leqslant 1$$

$$x_1 \geqslant 0 \qquad x_2 \geqslant 0$$

(a) Write down the Kuhn-Tucker conditions.

(b) Show that the Kuhn-Tucker conditions are both necessary and sufficient.

(c) Using (b), determine the range of values of $a_1$ for which the solution $x_1 = 4$, $x_2 = 3$ is optimal.

5.14. (a) What are the practical uses of the Kuhn-Tucker (i) necessity theorem and (ii) sufficiency theorem? State under what assumptions they are applicable.

(b) Write down the Kuhn-Tucker conditions for the following problem. Simplify, and give the final form.

$$\text{Maximize} \quad z = 3x_1^2 - 2x_2$$

$$\text{Subject to} \quad 2x_1 + x_2 = 4$$

$$x_1^2 + x_2^2 \leqslant 40$$

$$x_1 \geqslant 0$$

$$x_2 \text{ unrestricted in sign}$$

**5.15.** Consider the NLP problem:

$$\text{Maximize} \quad z = \sum_{j=1}^{N} x_j$$

$$\text{Subject to} \quad \sum_{j=1}^{N} \frac{x_j^2}{K_j} \leqslant D$$

$$x_j \geqslant 0 \qquad \text{for all } j$$

where the $K_j$'s and $D$ are positive constants.
(a) Write down the Kuhn-Tucker conditions.
(b) Show that a solution to the Kuhn-Tucker conditions gives an optimal solution to the NLP problem.
(c) Using (b), find an optimal solution to the NLP problem.

**5.16.** Consider the problem

$$\text{Minimize} \quad f(x) = x_1^2 + x_2^2$$

$$\text{Subject to} \quad (x_1 - 1)^3 - x_2^2 = 0$$

(a) Show graphically that the optimal solution is at $x_1 = 1$, $x_2 = 0$.
(b) Apply the Lagrange multiplier method to solve this problem. Can you explain why it fails?

**5.17.** Consider the NLP problem given in problem 5.12. Verify whether or not the second-order necessary and sufficient conditions are satisfied at the candidate optimum $\bar{x}$).

**5.18.** Consider the following NLP problem:

$$\text{Minimize} \quad f(x) = 0.2x_1^{3/2} + 0.05x_2^2 - 3x_1 - 5x_2$$

$$\text{Subject to} \quad 0.9x_1 + 0.5x_2 = 100$$

Using the Kuhn-Tucker conditions and the second-order optimality conditions, what can you conclude about the following points:
(a) $x_1 = 101.94$, $x_2 = 16.508$
(b) $x_1 = 84.1$, $x_2 = 48.62$
(c) $x_1 = 111.111$, $x_2 = 0$
(d) $x_1 = 0$, $x_2 = 200$
Note: $f$ is not a convex function.

# Chapter 6

# Transformation Methods

In this chapter we begin our discussion of practical methods of solution for the nonlinear programming problem:

$$\text{Minimize} \qquad f(x); \, x \in R^N \qquad (6.1)$$

$$\text{Subject to} \quad g_j(x) \geqslant 0 \quad j = 1, 2, 3, \ldots, J \qquad (6.2)$$

$$h_k(x) = 0 \quad k = 1, 2, 3, \ldots, K \qquad (6.3)$$

and

$$x_i^{(l)} \leqslant x_i \leqslant x_i^{(u)} \qquad i = 1, 2, 3, \ldots, N \qquad (6.4)$$

We shall assume throughout that an initial estimate $x^{(0)}$ of the solution $x^*$ is available. This initial estimate may or may not be feasible, that is, satisfy Eqs. (6.2)–(6.4). We discuss algorithms that generate a sequence of points in $R^N$ from $x^{(0)}$ to $x^{(T)}$, where $x^{(t)}$ is the generic point and $x^{(T)}$, the limit point, is the best estimate of $x^*$ produced by the algorithm. The points $x^{(t)}, t = 1, 2, 3, \ldots, T$, are stationary points of an associated unconstrained function called a penalty function. The original constrained problem is *transformed* into a sequence of unconstrained problems via the penalty function. The structure of the penalty function along with the rules for updating the *penalty parameters* at the close of each unconstrained minimization stage define the particular method. The penalty function is *exact* if only one unconstrained minimization is required.

The Lagrange and Kuhn-Tucker theory of the previous chapter provides a powerful motivation and theoretical basis for the methods considered here. We have seen in the previous Lagrange discussion that it is useful to form an unconstrained (Lagrangian) function in order to develop conditions of optimality. In addition, Everett [1] has given a method involving alternating Lagrange multiplier estimation and unconstrained minimization stages. The primary difficulty with the Everett approach is the numerical difficulty associated with the Lagrange multiplier estimation. On the other hand, we are obviously motivated to use these same (or similar) unconstrained functions to develop recursions, which are used iteratively to locate the constrained minimum. The idea of converting the constrained problem into a sequence of appropriately formed unconstrained problems is very appealing, since unconstrained problems can be solved both efficiently and reliably. We hope, of course, that only a few unconstrained subproblems of moderate difficulty will be required to approximate the constrained solution with acceptable accuracy.

Penalty function methods are classified according to the procedure employed for handling inequality constraints, since essentially all transformation methods treat equality constraints in the same way. We will refer to these methods as *interior* or *exterior* point methods depending upon whether the sequence $x^{(t)}$ contains feasible or infeasible points, respectively. When the sequence of stationary points contains both feasible and infeasible points, we say the method is *mixed*.

## 6.1  THE PENALTY CONCEPT

Consider the *penalty function*

$$P(x, R) = f(x) + \Omega(R, g(x), h(x))  \tag{6.5}$$

where $R$ is a set of penalty parameters and $\Omega$, the *penalty term*, is a function of $R$ and the constraint functions. The exact way in which the penalty parameters and the constraint functions are combined and the rules for updating the penalty parameters specify the particular method. As we have said, these methods are classified according to the manner in which they handle inequality constraints. In every case $\Omega$ is constructed in a manner that tends to favor the selection of feasible points over infeasible points by the unconstrained search method. An interior point method is the result of selecting a form for $\Omega$ that will force stationary points of $P(x, R)$ to be feasible. Such methods are also called *barrier methods*, since the penalty term forms a barrier of infinite $P$ function values along the boundary of the feasible region. Any useful transformation technique should have the following desirable characteristics:

1.  The subproblem solutions should approach a solution of the NLP, that is, $\lim_{t \to T < \infty} x^{(t)} = x^*$.

2.  The problem of minimizing $P(x, R)$ should be similar in difficulty to minimizing $f(x)$. That is, the method will be less than useful if the unconstrained subproblems are excessively difficult to solve, no matter how strong the theoretical basis of convergence.

3.  $R^{(t+1)} = F(R^{(t)})$ should be simple. It seems reasonable to hope that the calculation overhead associated with updating the penalty parameters should be small compared to the effort associated with solving the

unconstrained subproblems. (*Note*: This may in fact not be desirable for problems with very complex objective and constraint functions. In this case, considerable effort in updating the penalty parameters may be justified.)

### 6.1.1   Various Penalty Terms

In this section we consider a number of penalty forms that have been used widely and represent different procedures for handling constraints in an unconstrained setting. First consider the *parabolic penalty* used for equality constraints shown in Figure 6.1:

$$\Omega = R\{h(x)\}^2 \qquad (6.6)$$

Notice that the parabolic penalty term equally discourages positive or negative violations of $h(x)$. In addition, it is clear that with increasing values of $R$ the stationary values of $P(x, R)$ will approach $x^*$, since in the limit as $R$ grows large, $h(x^{(T)}) = 0$. Note also that $\Omega$ is continuous and has continuous derivatives.

Next consider several penalty forms for inequality constraints. Possibly the simplest form is the *infinite barrier* shown in Figure 6.2. This term assigns an infinite penalty to all infeasible points, and no penalty to feasible points. The use of this term causes $P(x, R)$ to be discontinuous along the boundary of the feasible region, and, rather obviously, $\nabla P$ does not exist along the boundary. Infinite penalties are not machine realizable, but one can use a large positive constant consistent with the computer word length. Siddall [2, 3] and other investigators [4] have used this penalty form with some degree of success; for

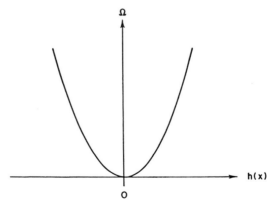

**Figure 6.1.**   Parabolic penalty.

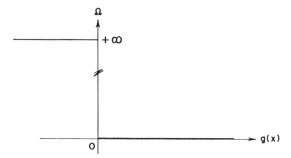

**Figure 6.2.**  Infinite barrier penalty.

instance, Siddall suggests

$$\Omega = 10^{20} \sum_{j \in \bar{J}} |g_j(x)| \tag{6.7}$$

where in this case $\bar{J}$ identifies the set of violated constraints, that is,

$$g_j(x) < 0 \qquad \text{for all} \quad j \in \bar{J} \tag{6.8}$$

Another useful form is the *log penalty*, as shown in Figure 6.3:

$$\Omega = -R \ln[g(x)] \tag{6.9}$$

Note that the term produces a positive penalty for all $x$ such that $0 < g(x) < 1$ and a negative penalty for all $x$ such that $g(x) > 1$. In this way, interior points are artificially favored over boundary points. The negative penalties could be avoided by setting $\Omega = 0$ for all $x$ such that $g(x) > 1$, but in so doing we introduce a discontinuity in $\nabla P$ in the interior region near the boundary. The log penalty is a barrier function and in fact is not defined for $x$ such that $g(x) < 0$. A special recovery procedure is needed to handle infeasible points

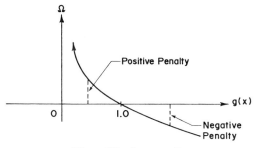

**Figure 6.3.**  Log penalty.

such as, say, $x^{(0)}$. Contrary to various reports in the literature, it is possible to generate an infeasible point using such a barrier function, for instance with an overly ambitious first step in a line search, and therefore special logic must be included to detect, prevent, and/or recover from such a possibility. Beginning with an initial feasible point and a positive value for $R$ (say, 10 or 100), $R$ is decreased after each unconstrained minimization and approaches zero in the limit.

The negative penalty is not present with the *inverse penalty* term shown in Figure 6.4:

$$\Omega = R\left[\frac{1}{g(x)}\right] \tag{6.10}$$

The inverse penalty term is also a barrier form, with similar difficulties associated with the evaluation of infeasible points. Rather obviously, a negative penalty is assigned to infeasible points, and special safeguards are in order. On the other hand, feasible points near the boundary are assigned quickly decreasing penalties as the interior region is penetrated. $P(x, R)$ and $\nabla P$ do not exist along the boundary of the feasible region. Beginning with an initial feasible point and $R$ positive, $R$ is decreased toward zero in the limit.

Finally consider the *bracket operator* term (Figure 6.5),

$$\Omega = R\langle g(x)\rangle^2 \tag{6.11}$$

where

$$\langle \alpha \rangle = \begin{cases} \alpha & \text{if} \quad \alpha \leq 0 \\ 0 & \text{if} \quad \alpha > 0 \end{cases} \tag{6.12}$$

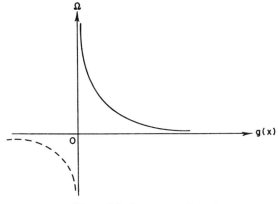

**Figure 6.4.** Inverse penalty.

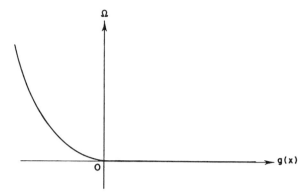

**Figure 6.5.**  Bracket operator.

Note first that the bracket operator term produces an exterior penalty function; stationary points of $P(x, R)$ may indeed be infeasible. On the other hand, note that feasible and infeasible points are handled equally well by the term, and in fact no penalty is assigned to boundary or feasible points. The bracket operator produces an appealing penalty form. $P(x, R)$ exists everywhere (for finite $x$) and is continuous. $R$ is chosen positive and is increased after each unconstrained stage. Let us demonstrate the use of these penalty terms with the following simple examples.

*Example 6.1   Parabolic Penalty Form*

Minimize $\qquad\qquad f(x) = (x_1 - 4)^2 + (x_2 - 4)^2$

Subject to $\qquad\quad h(x) = x_1 + x_2 - 5 = 0$

First form the penalty function using the quadratic penalty term:

$$P(x, R) = (x_1 - 4)^2 + (x_2 - 4)^2 + \frac{1}{R}(x_1 + x_2 - 5)^2 \qquad (6.13)$$

Now investigate the stationary points of $P(x, R)$ as a function of $R$.

$$\frac{\partial P}{\partial x_1} = 2(x_1 - 4) + \left(\frac{2}{R}\right)(x_1 + x_2 - 5) = 0$$

$$\frac{\partial P}{\partial x_2} = 2(x_2 - 4) + \left(\frac{2}{R}\right)(x_1 + x_2 - 5) = 0$$

$$x_1 = x_2 = \frac{10 + 8R}{4 + 2R} \qquad\qquad (6.14)$$

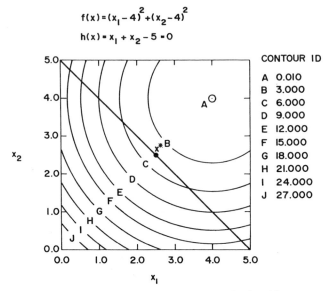

Figure 6.6. Two-variable equality-constrained problem.

Is the method convergent?

$$\lim_{R \to 0} \left( \frac{10 + 8R}{4 + 2R} \right) = \frac{10}{4} = 2.5 \tag{6.15}$$

The answer is yes, since by inspection of Figure 6.6 we see that $x^* = [2.5, 2.5]^T$, with $f(x) = 4.5$.

The stationary values of $P(x, R)$ for various values of $R$ are given in Table 6.1. Figures 6.7–6.9 give contours of $P(x, R)$ for various values of $R$. Notice that the contours of $P(x, R)$ do indeed change to align with $h(x)$ as $R$ approaches zero. This distortion of the shape of the contours of $P(x, R)$ has a positive effect, in that it is directly responsible for the convergence of the method.

Table 6.1   Stationary Values of Parabolic Penalty

| $R$ | $1/R$ | $x_1^{(t)} = x_2^{(t)}$ | $P(x^{(t)}, R)$ |
|---|---|---|---|
| $\infty$ | 0 | 4.0000 | 0.0000 |
| 10 | 0.1 | 3.7500 | 0.7500 |
| 1 | 1 | 3.0000 | 3.0000 |
| 0.1 | 10 | 2.5714 | 4.2857 |
| 0.01 | 100 | 2.5075 | 4.4776 |
| 0 | $\infty$ | 2.5000 | 4.5000 |

$$P(x, 1) = (x_1 - 4)^2 + (x_2 - 4)^2 + (x_1 + x_2 - 5)^2$$

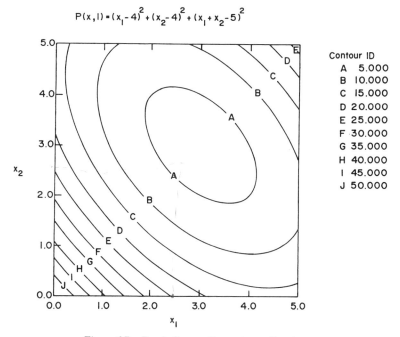

Contour ID

| | |
|---|---|
| A | 5.000 |
| B | 10.000 |
| C | 15.000 |
| D | 20.000 |
| E | 25.000 |
| F | 30.000 |
| G | 35.000 |
| H | 40.000 |
| I | 45.000 |
| J | 50.000 |

**Figure 6.7.** Parabolic penalty contours, $R = 1$.

Let us pause to examine the stationary point of $P(x, 1)$ in Example 6.1. Obviously it is not $x^*$ but is reasonably close. Note that $x^{(t)} = [3, 3]^T$; $f(x^{(t)}) = 2.0$ and $h(x^{(t)}) = 1.0$. In some engineering applications this answer might be close enough, but in general we would be concerned about the magnitude of the violation of $h(x)$. The question then arises, how shall we change the penalty function such that its stationary point is closer to $x^*$? Note that in finding the stationary point of $P(x, 1)$ exactly we have achieved a point that has too low a value for $f(x)$ and a corresponding violation of $h(x)$. But we built this difficulty into $P(x, 1)$ when we chose the penalty parameter $R = 1$, since $R$ tends to weight the equality constraint against the objective. That is, as $R$ grows small the stationary point of $P(x, R)$ will tend to satisfy $h(x)$ more closely and approach $x^*$, as can be seen in Table 6.1. Obviously, if we choose $R = 0$ we produce the exact solution in one unconstrained minimization. Hold it! How will we choose $R = 0$? Of course we cannot, but we could choose $R$ to be very small. The next problem then is to decide how small.

In Table 6.1 we see that $R = .01$ produces the correct solution to two places, when exact minimization is employed. But exact minimization will seldom be possible, and one of the unconstrained search methods previously considered will be employed. We are tempted to choose an even smaller value for $R$ so that the error in the result due to $R$ and the error due to inexact minimization will tend to add to an acceptable value. But there is a very real practical

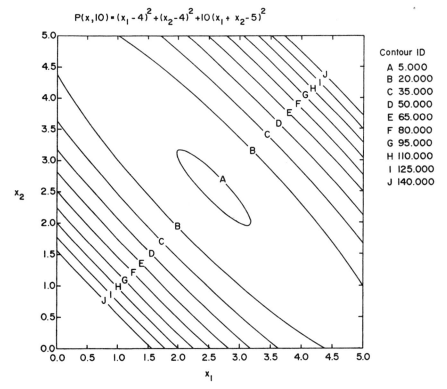

$$P(x,10) = (x_1 - 4)^2 + (x_2 - 4)^2 + 10(x_1 + x_2 - 5)^2$$

Contour ID

| | |
|---|---|
| A | 5.000 |
| B | 20.000 |
| C | 35.000 |
| D | 50.000 |
| E | 65.000 |
| F | 80.000 |
| G | 95.000 |
| H | 110.000 |
| I | 125.000 |
| J | 140.000 |

**Figure 6.8.** Parabolic penalty contours, $R = 10$.

difficulty with this approach, associated with the relationship $R$ has with the shape of the contours of $P(x, R)$. We see that as $R$ grows small, the contours of $P$ become more and more distorted. Therefore the unconstrained minimizations may become increasingly difficult. This thinking suggests a multistage strategy for approximating the solution. First choose $R$ to be a relatively large value, say 100, and minimize $P(x, 100)$; then decrease $R$ to, say, 10, and once again minimize $P(x, 10)$, using the result from $P(x, 100)$ to begin. We continue this decrease in $R$ until we notice acceptably small changes in the sequences $x^{(t)}$, $f(x^{(t)})$, and/or $P(x^{(t)}, R)$. We hope that the estimate of the minimizing point from the previous stage will be close enough to the minimizing point for the new stage to offset the increased minimization difficulty due to the decrease in $R$. This will often be the case if we do not decrease $R$ too rapidly, but when $R$ is decreased slowly many stages are required and the overall computing time to the solution can be great. This is the major difficulty with penalty functions, and much effort has been expended to develop rules for updating $R$ to increase the overall efficiency of the approach. We shall see that very similar difficulties occur when inequalities are present.

$$P(x,100) = (x_1 - 4)^2 + (x_2 - 4)^2 + 100(x_1 + x_2 - 5)^2$$

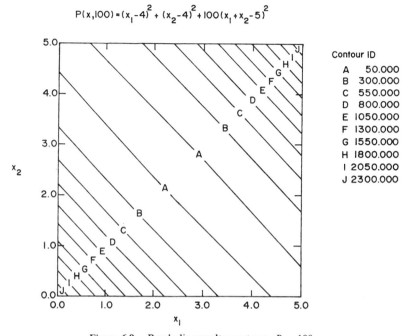

Figure 6.9. Parabolic penalty contours, $R = 100$.

It is of interest to examine the performance of several of the inequality constraint penalty forms on a similar problem.

### Example 6.2   Inequality Penalty Forms

Let us change the constraint in Example 6.1 to an inequality form such that $x^*$ remains the same:

$$g(x) = 5 - x_1 - x_2 \geqslant 0 \tag{6.16}$$

We handle this problem in a way that is very similar to the equality constraint case by employing the *bracket operator*:

$$P(x, R) = f(x) + R\langle g(x)\rangle^2 \tag{6.17}$$

or

$$P(x, R) = (x_1 - 4)^2 + (x_2 - 4)^2 + R\langle 5 - x_1 - x_2\rangle^2 \tag{6.18}$$

Next we investigate the stationary points of $P(x, R)$:

$$\frac{\partial P}{\partial x_1} = 2(x_1 - 4) + 2R\langle 5 - x_1 - x_2\rangle(-1) = 0$$

$$\frac{\partial P}{\partial x_2} = 2(x_2 - 4) + 2R\langle 5 - x_1 - x_2\rangle(-1) = 0$$

which leads to

$$x_1 = x_2 \tag{6.19}$$

as before. Therefore,

$$(x_1 - 4) - R\langle 5 - 2x_1\rangle = 0 \tag{6.20}$$

Assume that the argument of the bracket operator is $=$, $>$, and $<$ zero, respectively, and analyze the results. Therefore assume $2x_1 \geqslant 5$, which implies

$$(x_1 - 4) - R(5 - 2x_1) = 0$$

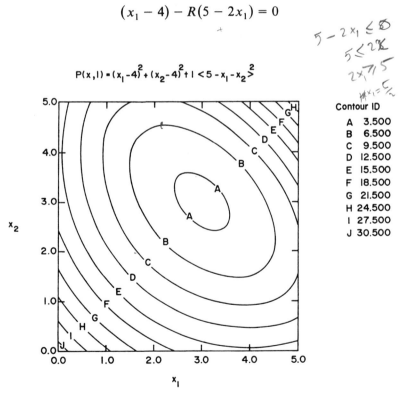

$$P(x, 1) = (x_1 - 4)^2 + (x_2 - 4)^2 + |<5 - x_1 - x_2>^2$$

Contour ID

| A | 3.500 |
| B | 6.500 |
| C | 9.500 |
| D | 12.500 |
| E | 15.500 |
| F | 18.500 |
| G | 21.500 |
| H | 24.500 |
| I | 27.500 |
| J | 30.500 |

**Figure 6.10.** Bracket operator penalty contours, $R = 1$.

or

$$x_1 = \frac{5R + 4}{2R + 1} \tag{6.21}$$

Therefore, $\lim_{R \to 0} x_1 = 2.5$, which agrees with the assumption. Therefore for $P(x, 1)$ the stationary point is $x^{(t)} = [3, 3]^T$, with $f(x^{(t)}) = 2$, $g(x^{(t)}) = -1$, and $P(x, 1) = 3$. As $R$ goes from 0 to $\infty$, the stationary points of $P(x, R)$ travel along the straight line from $[4, 4]^T$, the unconstrained solution, to the constrained solution at $[2.5, 2.5]^T$. For all values of $R$ the stationary points are infeasible, and hence this is an exterior form.

Comparing Eqs. (6.21) and (6.15) we see that the path to the solution with exact minimization is exactly the same for the parabolic and bracket operator forms. Plots of the contours of $P(x, R)$ for $R = 1$, 10, and 100 are given in Figures 6.10–6.12.

Consider now the application of the *log penalty* to this same inequality problem.

$$P(x, R) = f(x) - R \ln[g(x)] \tag{6.22}$$

or

$$P(x, R) = (x_1 - 4)^2 - R \ln(5 - x_1 - x_2) \tag{6.23}$$

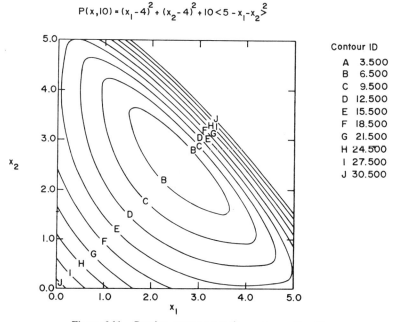

$$P(x, 10) = (x_1 - 4)^2 + (x_2 - 4)^2 + 10 < 5 - x_1 - x_2 >$$

| Contour ID | |
| --- | --- |
| A | 3.500 |
| B | 6.500 |
| C | 9.500 |
| D | 12.500 |
| E | 15.500 |
| F | 18.500 |
| G | 21.500 |
| H | 24.500 |
| I | 27.500 |
| J | 30.500 |

**Figure 6.11.**   Bracket operator penalty contours, $R = 10$.

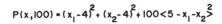

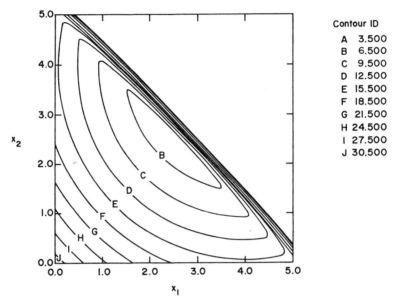

Contour ID

| | |
|---|---|
| A | 3.500 |
| B | 6.500 |
| C | 9.500 |
| D | 12.500 |
| E | 15.500 |
| F | 18.500 |
| G | 21.500 |
| H | 24.500 |
| I | 27.500 |
| J | 30.500 |

**Figure 6.12.** Bracket operator penalty contours, $R = 100$.

The stationary conditions are

$$\frac{\partial P}{\partial x_1} = 2(x_1 - 4) + R\left[\frac{1}{5 - x_1 - x_2}\right] = 0$$

$$\frac{\partial P}{\partial x_2} = 2(x_2 - 4) + R\left[\frac{1}{5 - x_1 - x_2}\right] = 0$$

which leads to

$$x_1 = x_2$$

Therefore,

$$2(x_1 - 4) + \left[\frac{R}{5 - 2x_1}\right] = 0 \qquad (6.24)$$

This leads to

$$2x_1^2 - 13x_1 + 20 - \frac{R}{2} = 0 \qquad (6.25)$$

**Table 6.2   Stationary Values of Log Penalty**

| $R$ | $x_1 = x_2$ | $f(x)$ | $g(x)$ | $-R\{g(x)\}$ | $P(x, R)$ |
|---|---|---|---|---|---|
| 100 | −1.8059 | 67.4170 | 8.6118 | −215.3133 | −147.8963 |
| 10 | 1.5000 | 12.5000 | 2.0000 | −6.9315 | 5.5685 |
| 1 | 2.3486 | 5.4542 | 0.3028 | +1.1947 | 6.6489 |
| 0.1 | 2.4835 | 4.5995 | 0.0034 | +0.3411 | 4.9406 |
| 0.01 | 2.4983 | 4.5100 | 0.0034 | +0.0568 | 4.5668 |
| 0 | 2.5000 | 4.5000 | 0.0000 | 0.0000 | 4.5000 |

and the roots produce the desired stationary values:

$$x_1 = \frac{13}{4} - \frac{1}{4}\sqrt{9 + 4R} \qquad (6.26)$$

where we reject the other root since it is infeasible. Now assume that $R$ begins

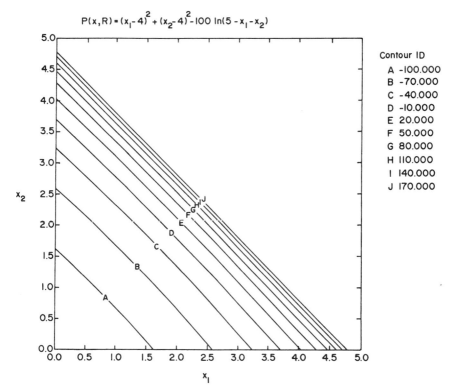

**Figure 6.13.**   Log penalty contours, $R = 100$.

large and is incremented toward zero:

$$\lim_{R \to 0} x_1 = 2.5 \qquad f(x^{(T)}) = 4.5$$

Once again we see that the method converges with exact minimization. The stationary values of $P(x, R)$ for various values of $R$ are given in Table 6.2, and plots of the contours of $P(x, R)$ are given in Figures 6.13–6.15. Notice from Table 6.2 that $P(x^{(t)}, R)$ does not increase monotonically to $f(x^*)$, but increases and then decreases. This is due to the negative penalty effect included in the log term.

Finally we consider the *inverse penalty*:

$$P(x, R) = f(x) + R\left[\frac{1}{g(x)}\right] \tag{6.27}$$

$$P(x, R) = (x_1 - 4)^2 + (x_2 - 4)^2 + R\left[\frac{1}{5 - x_1 - x_2}\right] \tag{6.28}$$

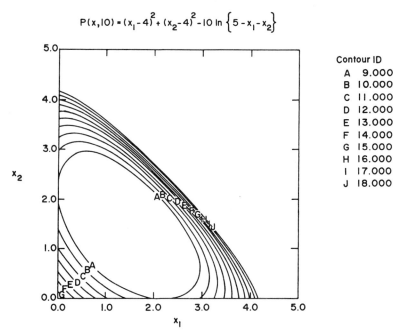

Figure 6.14.   Log penalty contours, $R = 10$.

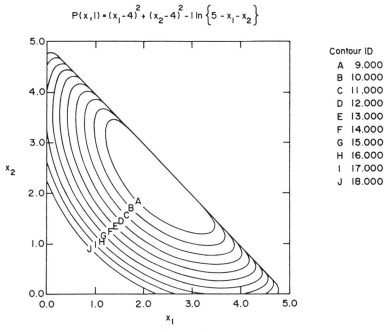

$$P(x, l) = (x_1 - 4)^2 + (x_2 - 4)^2 - l \ln \left\{ 5 - x_1 - x_2 \right\}$$

Contour ID

| | |
|---|---|
| A | 9.000 |
| B | 10.000 |
| C | 11.000 |
| D | 12.000 |
| E | 13.000 |
| F | 14.000 |
| G | 15.000 |
| H | 16.000 |
| I | 17.000 |
| J | 18.000 |

**Figure 6.15.** Log penalty contours, $R = 1$.

As before, examine the stationary values of $P(x, R)$ for various values of $R$:

$$\frac{\partial P}{\partial x_1} = 2(x_1 - 4) + \left[ \frac{R}{(5 - x_1 - x_2)^2} \right] = 0$$

$$\frac{\partial P}{\partial x_2} = 2(x_2 - 4) + \left[ \frac{R}{(5 - x_1 - x_2)^2} \right] = 0$$

**Table 6.3   Stationary Values of Inverse Penalty**

| $R$ | $x_1 = x_2$ | $f(x)$ | $g(x)$ | $R/g(x)$ | $P(x, R)$ |
|---|---|---|---|---|---|
| 100 | 0.5864 | 23.3053 | 3.8272 | 26.1288 | 49.4341 |
| 10 | 1.7540 | 10.0890 | 1.4920 | 6.7024 | 16.7914 |
| 1 | 2.2340 | 6.2375 | 0.5320 | 1.8797 | 8.1172 |
| 0.1 | 2.4113 | 5.0479 | 0.1774 | 0.5637 | 5.6116 |
| 0.01 | 2.4714 | 4.6732 | 0.0572 | 0.1748 | 4.8480 |
| 0.001 | 2.4909 | 4.5548 | 0.0182 | 0.0549 | 4.6097 |
| 0 | 2.5000 | 4.5000 | 0.0000 | 0.0000 | 4.5000 |

which leads to $x_1 = x_2$ and

$$4x_1^3 - 36x_1^2 + 105x_1 - 100 + \frac{R}{2} = 0 \qquad (6.29)$$

The stationary points are defined by the zeros of this nonlinear function. We use a zero-finding algorithm [5] to find the values given in Table 6.3, and plots of the contours of $P(x, R)$ for various $R$ values are given in Figures 6.16–6.18. Once again we see that the stationary values of the penalty function approach $x^*$ as $R \to 0$ with exact minimization.

The examples demonstrate both the utility of the various standard penalty approaches and their common weakness. That is, we have seen that convergence is associated with ever-increasing distortion of the penalty contours, which increases significantly the possibility of failure of the unconstrained search method. Recall that we have assumed in our discussion of convergence

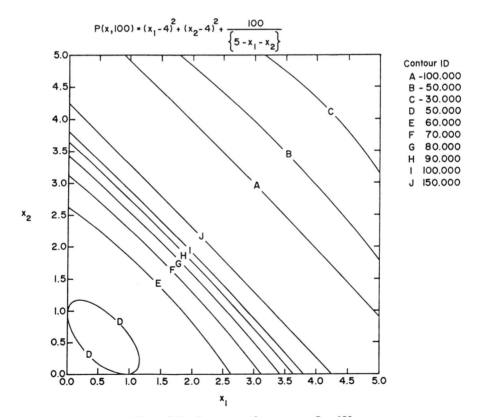

**Figure 6.16.** Inverse penalty contours, $R = 100$.

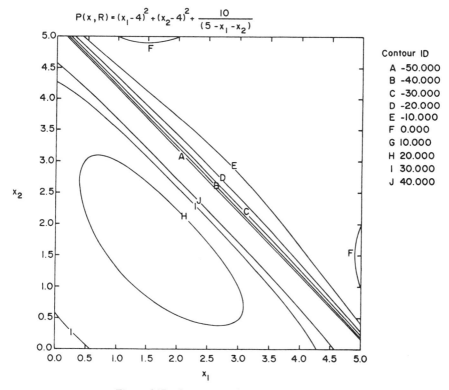

**Figure 6.17.** Inverse penalty contours, $R = 10$.

that the unconstrained searches could be completed successfully. Any inherent tendency that opposes this assumption seriously undermines the utility of the penalty function approach.

### 6.1.2 Choice of the Penalty Parameter, $R$

With each of the parametric penalty forms considered, it is necessary to choose an initial value for $R$ and to adjust this value after each unconstrained search to force convergence of the sequence of stationary points $x^{(t)}$. Hopefully it is clear that $R$ must be chosen so as to increase the weight of constraint violations for exterior forms and to decrease the weight for interior or barrier forms from stage to stage. In addition, we must devise a strategy for incrementing $R$ from stage to stage. For instance, with the simple parabolic penalty for equality constraints it would seem reasonable to begin with $R = 0$; that is, conduct an unconstrained search of $f(x)$ and increase $R$ by $\Delta R$ (say, powers of 10) at each stage. This would force the stationary points to approach satisfaction of the equality constraint. Experimentally we observe that the stationary points $x^{(t)}(R)$ converge smoothly to $x^*$ (with exact searches). Fiacco and McCormick

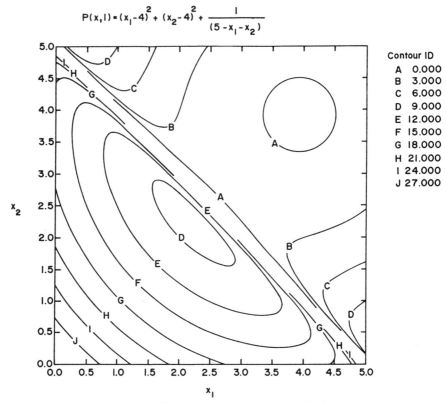

$$P(x,1) = (x_1-4)^2 + (x_2-4)^2 + \frac{1}{(5-x_1-x_2)}$$

Contour ID

| | |
|---|---|
| A | 0.000 |
| B | 3.000 |
| C | 6.000 |
| D | 9.000 |
| E | 12.000 |
| F | 15.000 |
| G | 18.000 |
| H | 21.000 |
| I | 24.000 |
| J | 27.000 |

**Figure 6.18.** Inverse penalty contours, $R = 1$.

[6, 7] and Lootsma [8, 9] have suggested that an approximating function be fitted to the sequence $x^{(t)}(R)$ and that this approximation be evaluated at $R = 0$ to provide a better estimate of $x^{(t+1)}$ to begin the next unconstrained search. It is hoped that in this way $R$ can be larger than would otherwise be possible. Fletcher and McCann [10] have conducted tests that suggest that a quadratic approximating function is best.

Even though this extrapolation technique may improve the performance of the basic penalty approach, it cannot remove the distortion in the shape of the contours of $P(x, R)$ as $R$ is adjusted. In fact, it is this very distortion that produces convergence of the method (with exact solution to the subproblems). On the other hand, the change in the penalty parameters forces the subproblems to be progressively less well-conditioned. That is, for every penalty form considered to this point the unconstrained subproblems will *automatically* become ill-conditioned as a natural consequence of the changes in $R$ needed to force convergence. This fact is easily observed from the shape of the contours in the previous figures and has been noted by several investigators, including

Murray [11] and Lootsma [12]. In fact, Lootsma shows that $M$ (the number of active constraints at $x^*$) eigenvalues of $\nabla^2 P(x, R)$ vary with $1/R$, which implies that $\nabla^2 P(x, R)$ becomes increasingly ill-conditioned as $R$ approaches zero. This means that the unconstrained subproblems become increasingly difficult to solve, and termination may be as a result of failure of the unconstrained method rather than convergence of the penalty method. Once again this is purely a numerical problem, since very strong convergence results can be proved in the presence of exact calculations.

## 6.2  ALGORITHMS, CODES, AND OTHER CONTRIBUTIONS

It is relatively straightforward to build a useful algorithm with the concepts previously discussed. A simple penalty function algorithm would take the following form.

### Penalty Function Algorithm

**Step 1.**  Define $N$, $J$, $K$, $\varepsilon_1$, $\varepsilon_2$, $\varepsilon_3$, $x^{(0)}$, and $R^{(0)}$,

where $\varepsilon_1$ = line-search termination criterion

$\varepsilon_2$ = unconstrained method termination criterion

$\varepsilon_3$ = penalty termination criterion

$x^{(0)}$ = initial estimate of $x^*$

$R^{(0)}$ = initial set of penalty parameters

**Step 2.**  Form $P(x, R) = f(x) + \Omega(R, g(x), h(x))$.

**Step 3.**  Find $x^{(t+1)}$ such that $P(x^{(t+1)}, R^{(t)}) \to$ min, with $R^{(t)}$ fixed. Terminate, using $\varepsilon_2$, and use $x^{(t)}$ to begin next search.

**Step 4.**  Is $|P(x^{(t+1)}, R^{(t)}) - P(x^{(t)}, R^{(t-1)})| \leqslant \varepsilon_3$?

Yes:  Set $x^{(t+1)} = x^{(T)}$ and terminate.

No:  Continue.

**Step 5.**  Choose $R^{(t+1)} = R^{(t)} + \Delta R^{(t)}$ according to a prescribed update rule, and go to 2.

Certainly a useful code would include additional steps and logic. For instance, it would seem wise to monitor several sequences $x^{(t)}$, $R^{(t)}$, $g(x^{(t)})$, $h(x^{(t)})$, $f(x^{(t)})$ in addition to the one given in step 4 in order to make a termination decision.

Possibly the first record of a penalty-type method in the literature is the work of Courant [13], but it is the "created response surface" idea of Carroll [14] that marks the beginning of the modern era of transformation techniques. The work of Fiacco and McCormick [15–17] is also important because it provided the first widely available NLP code, SUMT (which we discuss further in the next section). Somewhat later (1967), Zangwill [18] considered the general utility of the penalty approach and provided conditions of convergence for various penalty forms. The most complete examination of penalty methods has most probably been provided by Lootsma [19–21], who in particular has demonstrated the advantages and disadvantages of the various forms by examination of their properties along the boundary of the feasible region. In addition, McCormick [22] gives an interesting discussion of the relative merits of penalty and nonpenalty methods for constrained problems. For reasons previously discussed at length, we might expect to encounter difficulties in using some or all of the line-search techniques from Chapter 2 to minimize $P(x, R)$ along a prescribed direction, since $\nabla P(x, R)$ will take on extremely high values near the boundary. Many have recognized this difficulty, and Lasdon [23] has proposed an efficient technique especially designed to minimize penalty-type functions. It is not possible in this brief discussion to give a proper survey of the rich history of penalty methods; instead we have mentioned some of the more notable contributions. Much more complete surveys are given by Lootsma [4], Moe [24], Ryan [25], Fox [26], Himmelblau [27], and Avriel [28].

A number of penalty-type codes have been widely used; these include SUMT and OPTISEP, which are discussed shortly; COMPUTE II [29], a Gulf Oil Corporation software package; a package developed by Mayne and Afimiwala [30] at the State University of New York at Buffalo; and a package developed by Lill and Lootsma [31]. In every case a variety of penalty forms and unconstrained searching methods are available to the user. The utility of a number of the standard penalty forms is described in a comparative study by DeSilva and Grant [32], and the value of the approach is demonstrated by various applications given by Bracken and McCormick [33]. More recently, Rao and Gupta [34, 35] have employed penalty methods in the design of thermal systems.

Possibly the two most widely used penalty function packages are SUMT [36] by Fiacco, McCormick, and co-workers, and OPTISEP [3] by Siddall and his students at McMaster University. Sandgren [37] gives availability information and test results for these codes. The SUMT package has been released in at least two forms:

$$1967: \qquad P(x, R) = f(x) + R \sum_{j=1}^{J} \left[ \frac{1}{g_j(x)} \right] + \frac{1}{R} \sum_{k=1}^{K} \left[ h_k(x) \right]^2 \qquad (6.30)$$

$$1970: \qquad P(x, R) = f(x) - R \sum_{j=1}^{J} \ln\left[ g_j(x) \right] + \frac{1}{R} \sum_{k=1}^{K} \left[ h_k(x) \right]^2 \qquad (6.31)$$

where $R^{(0)} > 0$ and $R^{(0)} > R^{(1)} > R^{(2)} \to 0$. The 1970 version (SUMT-IV) is the one Sandgren tested (see Chapter 12). This package contains four separate penalty function algorithms, all employing the penalty form in (6.31) but with different rules for updating $R$ and/or performing the unconstrained search. In particular, the user can choose either the Cauchy, DFP, or one of two forms of the modified Newton method for the unconstrained searches. The penalty parameter $R$ is decreased after each unconstrained subproblem, and the extrapolation technique previously discussed is employed as well. The reader is referred to the SUMT-IV user's manual for more specific details.

The OPTISEP package contains a number of penalty methods, which differ primarily in the unconstrained method used. Two penalty forms are used:

$$P(x, R) = f(x) + 10^{20} \sum_{j=1}^{J} |\langle g_j(x)\rangle| + 10^{20} \sum_{k=1}^{K} |h_k(x)| \qquad (6.32)$$

and the form given in (6.30). SEEK1 employs (6.32) and a Hooke-Jeeves direct-search method. After termination, a number of trial points are generated randomly about $x^{(T)}$, and the search begins anew if a lower value of $P(x, R)$ is found. SEEK3 uses the 1967 SUMT form in (6.30) and uses (6.31) to locate an initial feasible point. Hooke-Jeeves is again used, and $R$ is reduced by a constant factor at each stage. SIMPLX employs (6.30) and the simplex direct search method. DAVID uses (6.30) and the Davidon-Fletcher-Powell gradient-based searching method. Finally, MEMGRD employs (6.30) and Miele's memory gradient method, which is a generalization of the Fletcher-Reeves conjugate gradient method. All programs use a similar format for problem entry.

## 6.3 THE METHOD OF MULTIPLIERS

It has been recognized for some time that the progressive ill-condition of the subproblems generated in the standard (SUMT-like) penalty approach limits the utility of the method for practical applications, particularly in competition with the linearization-based methods (to be discussed in Chapters 8 and 9). With this in mind, several investigators have suggested *fixed-parameter* SUMT-like penalty methods (most notably Kowalik [38], Fiacco and Mc-Cormick [39], Fiacco [40], Rosenbrock [41], Huard [42], and Staha [43]) and other extensions in order to mitigate this malady to some extent.

Of particular interest is Huard's *method of centers*, where we maximize

$$P(x, t) = [t - f(x)] \prod_{j=1}^{J} g_j(x) \qquad (6.33)$$

where $t$ is a moving truncation at each maximization stage, say, $t^{(l)} = f(x^{(l-1)})$, where $x^{(l-1)}$ is the maximum point of the previous stage. Note that $t$ is adjusted in the normal course of the calculations. It has been shown [28] that

Huard's product form is equivalent to the parameterless penalty form:

$$P(x) = -\ln\left[f(x^{(t-1)}) - f(x)\right] - \sum_{j=1}^{J} \ln g_j(x) \qquad (6.34)$$

where minimization is implied. Avriel comments that these parameter-free methods, although attractive on the surface, are exactly equivalent to SUMT with a particular choice of updating rule for $R$. In fact, Lootsma [44] shows that the convergence rates for SUMT and the method of centers are quite comparable.

Recall the role of the Lagrange function:

$$L(x, u, v) = f(x) - \sum_j u_j g_j(x) - \sum_k v_k h_k(x) \qquad (6.35)$$

in the discussion of the Kuhn-Tucker optimality conditions of Chapter 5. We are tempted to seek the values of $x$, $u$, and $v$ that cause the Lagrangian to take on a minimum value, but we *do not*, since we know that the solution $(x^*, u^*, v^*)$ is a stationary point of $L$ but not a minimum (in fact, it is a saddlepoint of $L$). On the other hand, it is productive to augment the Lagrangian (by, for instance, a quadratic loss penalty term) to form an unconstrained function whose *minimum* is a Kuhn-Tucker point of the original problem. In this section we discuss such an *augmented Lagrangian* method, called the method of multipliers, which can also be viewed as a remedy to the contour distortion malady previously discussed.

### 6.3.1   The Penalty Function

Consider the function

$$P(x, \sigma, \tau) = f(x)$$

$$+ R \sum_{j=1}^{J} \left\{ \langle g_j(x) + \sigma_j \rangle^2 - \sigma_j^2 \right\} + R \sum_{k=1}^{K} \left\{ \left[ h_k(x) + \tau_k \right]^2 - \tau_k^2 \right\}$$

$$(6.36)$$

where $R$ is a *constant* scale factor ($R$ may vary from constraint to constraint but remains constant from stage to stage), and the bracket operator is defined as before. The $\sigma_j$ and $\tau_k$ parameters are fundamentally different from $R$ in that they *bias* the penalty terms in a way that forces convergence of the iterates under rather mild conditions on the function. The method to be considered here is due to Schuldt [45, 46] and is similar to Rockafellar's modification [47] of the Powell-Hestenes [48, 49] multiplier method for equalities only. We shall see that the method of multipliers (MOM) is quite similar in philosophy to Everett's method [1], with the exception that MOM adjusts the Lagrange multiplier estimates ($\sigma$ and $\tau$) quite conveniently and automatically. We drop the $R$ in the penalty notation to underline the fact that $R$ is constant from stage to stage; only the $\sigma$ and $\tau$ elements change. Furthermore, $\sigma$ and $\tau$ are constant vectors during each unconstrained minimization. The updating rules (to be explained next) are applied to the elements of $\sigma$ and $\tau$ to prepare a new penalty function for the next unconstrained stage. It is not necessary for the starting vector $x^{(0)}$ to be feasible, and a convenient but completely arbitrary choice of parameters for the first stage is $\sigma = \tau = 0$. Thus the first minimization stage is identical to the first unconstrained minimization using standard exterior penalty terms previously discussed.

### 6.3.2   Multiplier Update Rule

Suppose that the vector $x^{(t)}$ minimizes the $t$th-stage penalty function:

$$P(x, \sigma^{(t)}, \tau^{(t)}) = f(x) + R \sum_{j=1}^{J} \left\{ \langle g_j(x) + \sigma_j^{(t)} \rangle^2 - \left[\sigma_j^{(t)}\right]^2 \right\}$$

$$+ R \sum_{k=1}^{K} \left\{ \left[h_k(x) + \tau_k^{(t)}\right]^2 - \left[\tau_k^{(t)}\right]^2 \right\} \qquad (6.37)$$

Multiplier estimates for the $(t + 1)$st stage are formed according to the following rules:

$$\sigma_j^{(t+1)} = \langle g_j(x^{(t)}) + \sigma_j^{(t)} \rangle \qquad j = 1, 2, 3, \ldots, J \qquad (6.38)$$

$$\tau_k^{(t+1)} = h_k(x^{(t)}) + \tau_k^{(t)} \qquad k = 1, 2, 3, \ldots, K \qquad (6.39)$$

Because of the bracket operator, $\sigma$ has no positive elements, whereas the elements of $\tau$ can take either sign. These parameters serve as a bias in the arguments of the penalty terms, and the updating rules tend to change the bias in a way that increases the penalty on violated constraints in successive stages, thus forcing the stationary points $x^{(t)}$ toward feasibility.

### 6.3.3   Penalty Function Topology

The effect of changes in the multipliers $\sigma$ and $\tau$ upon the shape of the contours of $P(x, \sigma, \tau)$ is not obvious from (6.37) but can be seen in an expression for $\nabla^2 P(x)$, the Hessian matrix of the penalty function. Using the definition of the bracket operator, we see the gradient to be

$$\nabla P(x) = \nabla f(x) + 2R \sum_{j=1}^{J} \{\langle g_j(x) + \sigma_j \rangle \nabla g_j(x)\}$$

$$+ 2R \sum_{k=1}^{K} \{[h_k(x) + \tau_k] \nabla h_k(x)\} \tag{6.40}$$

An additional differentiation produces the matrix of second derivatives:

$$\nabla^2 P(x) = \nabla^2 f(x) + 2R \sum_{j=1}^{J} \{\langle g_j(x) + \sigma_j \rangle \nabla^2 g_j(x) + [\nabla g_j(x)]^2\}$$

$$+ 2R \sum_{k=1}^{K} \{[h_k(x) + \tau_k] \nabla^2 h_k(x) + [\nabla h_k(x)]^2\} \tag{6.41}$$

It can be shown that the radius of curvature of a particular contour of $P(x)$ varies as $\nabla^2 P(x)$. Let the constraints $g(x)$ and $h(x)$ be linear; then,

$$\nabla^2 P(x) = \nabla^2 f(x) + 2R \sum_{j=1}^{J} \{\nabla g_j(x)\}^2 + 2R \sum_{k=1}^{K} \{\nabla h_k(x)\}^2 \tag{6.42}$$

So we see that $\nabla^2 P(x)$ is independent of $\sigma$ and $\tau$ when the constraints are linear, and therefore the contours do not change in *shape* from stage to stage in this case. The contours of $P(x)$ are simply shifted or biased relative to the contours of $f(x)$. This is an obviously desirable property, since convergence of the method does not depend on distortion of the contours of $P(x)$ and the associated infinite limit point of penalty parameters. Furthermore, we see that the contours of $P(x)$ do change in shape when the constraints are nonlinear, but the effect is second order rather than primary as in the previously considered methods.

### 6.3.4  Termination of the Method

Progress of the method is checked by monitoring the sequences $x^{(t)}$, $\sigma^{(t)}$, $\tau^{(t)}$, $f(x^{(t)})$, $g(x^{(t)})$, and $h(x^{(t)})$. It is assumed that the unconstrained algorithm can find the stationary points and has its own suitable criteria for termination (ending the stage). Termination occurs when the multiplier update and associated unconstrained minimization fail to produce change in one or more of the monitored sequences. Termination then produces the limit point $x^{(T)}$, which we now show to be a Kuhn-Tucker point.

Note that the gradient of the penalty function must vanish at the end of a stage. Also, according to expressions (6.38) and (6.39), the quantities $\langle g_j(x^{(t)})$ $+ \sigma_j^{(t)} \rangle$ and $[h_k(x^{(t)}) + \tau_k^{(t)}]$ become, respectively, $\sigma^{(t+1)}$ and $\tau^{(t+1)}$. Consider now the supposed limits, $\sigma^{(T)}$, $g_j(x^{(T)})$, $\tau^{(T)}$, and $h_k(x^{(T)})$. From (6.39) it is clear that $\tau^{(T)}$ can exist only if $h_k(x^{(T)}) = 0$. Similarly, (6.38) implies that the limits $\sigma^{(T)}$ and $g_j(x^{(T)})$ exist only if

$$g_j(x^{(T)}) > 0 \quad \text{and} \quad \sigma_j^{(T)} = 0 \tag{6.43}$$

or

$$g_j(x^{(T)}) = 0 \quad \text{and} \quad \sigma_j^{(T)} \leqslant 0 \tag{6.44}$$

Accordingly, the gradient given in (6.40) but evaluated at the limit point $x^{(T)}$ can be written equivalently:

$$\nabla P(x^{(T)}) = \nabla f(x^{(T)}) + 2R \sum_{j=1}^{J} \sigma_j^{(T)} \nabla g_j(x^{(T)}) + 2R \sum_{k=1}^{K} \tau_k^{(T)} \nabla h_k(x^{(T)}) \equiv 0$$

$$\tag{6.45}$$

with

$$g_j(x^{(T)}) \geqslant 0 \quad j = 1, 2, 3, \ldots, J \tag{6.46}$$

$$\sigma_j^{(T)} g_j(x^{(T)}) = 0 \quad j = 1, 2, 3, \ldots, J \tag{6.47}$$

$$\sigma_j^{(T)} \leqslant 0 \quad j = 1, 2, 3, \ldots, J \tag{6.48}$$

$$h_k(x^{(T)}) = 0 \quad k = 1, 2, 3, \ldots, K \tag{6.49}$$

Notice that expressions (6.45)–(6.49) are a restatement of the Kuhn-Tucker conditions gives in Chapter 5, and therefore the limit point $x^{(T)}$ is a Kuhn-Tucker point. Furthermore, we see from (6.45) that

$$u_j = -2R\sigma_j^{(t)} \tag{6.50}$$

$$v_k = -2R\tau_k^{(t)} \tag{6.51}$$

are Lagrange multiplier estimates and are obtained with essentially no additional calculation. The *dual variable* nature of these Lagrange multiplier estimates has been studied by a number of investigators, including Gould [50] and Gill [51].

### 6.3.5  MOM Characteristics

In summary, we note the following general characteristics of the method of multipliers:

**1.**    $\lim\limits_{t \to T < \infty} x^{(t)} = x^*$, a Kuhn-Tucker point

We have shown that this result follows assuming only that the unconstrained minimizations terminate. We expect the unconstrained subproblems to be less ill-conditioned than with the previously considered methods, because of the structure of the penalty form and the nature of the $\sigma$ and $\tau$ parameters. This leads one to observe that

**2.**    Minimize $P(x)$ is similar in difficulty to Minimize $f(x)$ for reasonable values of $R$.

**3.**    Finally we note that the $\sigma$ and $\tau$ update rules are simple, requiring essentially no additional calculation.

Root [52] and others [53, 54] have investigated the utility of more complex update rules for the multiplier estimates but have found in general that the additional computational overhead outweighs the decrease in the number of stages to termination. As previously noted, the value of this result is a strong function of the relative computational effort required to compute the various problem functions. In addition, Root [52] has given an "interiorlike" version of the method of multipliers and shown it to possess many of the positive characteristics of the exterior form considered here, with the additional pleasant feature that the iterates $x^{(t)}$ tend to remain in the feasible region.

On the other hand, the formulation considered here does have its weaknesses. As mentioned, the sequence of iterates $x^{(t)}$ will almost certainly approach $x^*$ from the infeasible region. In addition, we have the problem of choosing $R$ (which we discuss in Section 6.3.6), and recall our assumption of convergence of the unconstrained subproblems. Kort [54] shows the method to be globally convergent when applied to convex programs and gives a termination rule for the subproblems that preserves global convergence and rate of convergence with inexact subproblem solutions.

### *Example 6.3   Example 6.1 Revisited*

Consider once again the equality-constrained problem given in Example 6.1:

Minimize    $f(x) = (x_1 - 4)^2 + (x_2 - 4)^2$

Subject to $\quad h(x) = x_1 + x_2 - 5 = 0$

$$P(x, \tau) = (x_1 - 4)^2 + (x_2 - 4)^2 + \frac{1}{R}(x_1 + x_2 - 5 + \tau)^2 - \frac{1}{R}\tau^2$$

$$\frac{\partial P}{\partial x_1} = 2(x_1 - 4) + \frac{2}{R}(x_1 + x_2 - 5 + \tau) = 0$$

$$\frac{\partial P}{\partial x_2} = 2(x_2 - 4) + \frac{2}{R}(x_1 + x_2 - 5 + \tau) = 0$$

These expressions produce

$$x_1 = x_2 = \frac{5 + 4R - \tau}{2 + R} \tag{6.52}$$

For convenience, choose $R = 1$, to get

$$x_1 = x_2 = 3 - \frac{\tau}{3} \tag{6.53}$$

Now we use (6.39) to update the multiplier $\tau$ at the end of each constrained minimization of $P(x, \tau)$, which is here done exactly using (6.52). The results of these calculations are given in Table 6.4. Note that $P(x, \tau)$ increases as the iterations proceed. We are, in fact, locating a saddlepoint of $P(x, \tau)$, since we are *minimizing* with respect to $x$ but *maximizing* with respect to $\tau$. Contour maps of $P(x, \tau)$ for various values of $\tau$ are given in Figures 6.19–6.21. Once again, since the constraint is linear, we have no change in the shape of the contours of $P(x, \tau)$ as $\tau$ converges to its finite terminal value.

### 6.3.6 The Choice of $R$-Problem Scale

Poor problem scale is possibly the most frequently encountered difficulty in the application of modern transformation methods in engineering practice. A problem is poorly scaled (or ill-scaled) if the moduli of the variables and the

Table 6.4 MOM Results

| $t$ | $\tau$ | $x_1 = x_2$ | $f(x)$ | $P(x, \tau)$ |
|---|---|---|---|---|
| 0 | 0 | 3.0000 | 2.0000 | 3.0000 |
| 1 | 1.0000 | 2.6667 | 3.5556 | 4.3333 |
| 2 | 1.3333 | 2.5556 | 4.1728 | 4.4814 |
| 3 | 1.4444 | 2.5185 | 4.3896 | 4.4979 |
| 4 | 1.4818 | 2.5062 | 4.4629 | 4.4997 |
| — | 1.5000 | 2.5000 | 4.5000 | 4.5000 |

$$P(x, \tau) = (x_1 - 4)^2 + (x_2 - 4)^2 + (x_1 + x_2 - 5 + 1)^2 - (1)^2$$

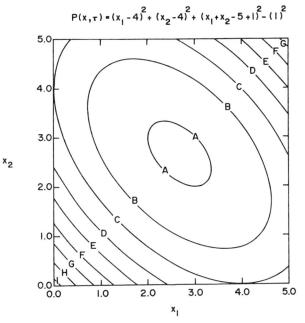

Contour ID

A  5.000
B  10.000
C  15.000
D  20.000
E  25.000
F  30.000
G  35.000
H  40.000
I  45.000
J  50.000

**Figure 6.19.**  MOM contours, $\tau = 1.0$.

$$P(x, \tau) = (x_1 - 4)^2 + (x_2 - 4)^2 + (x_1 + x_2 - 5 + \frac{13}{9})^2 - (\frac{13}{9})^2$$

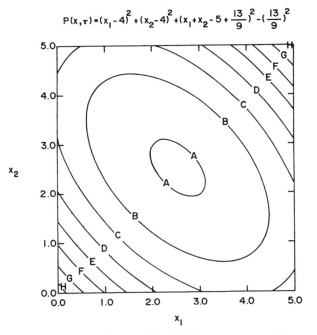

Contour ID

A   5.000
B   10.000
C   15.000
D   20.000
E   25.000
F   30.000
G   35.000
H   40.000
I   45.000
J   50.000

**Figure 6.20.**  MOM contours, $\tau = 13/9$.

$$P(x,\tau) = (x_1-4)^2 + (x_2-4)^2 + (x_1+x_2-5+1.5)^2 - (1.5)^2$$

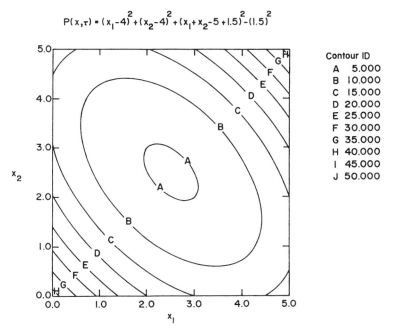

Figure 6.21. MOM contours, $\tau = 1.5$.

problem functions are not comparable. When this occurs, one or more of the design variables and/or constraints can artificially dominate the search and cause premature termination of *any* algorithm. In Chapter 13 we discuss a method by Root [55] that senses ill-scaling and automatically conditions the problem so that the minimization algorithm sees a better-conditioned problem. Keefer and Gottfried [56, 57] proposed a similar method in 1970 for use with the standard penalty approach, whereas Root was specifically interested in enhancing the performance of the MOM approach. Even though both methods were developed for use with transformation methods, they could be used to condition problems to be solved by any algorithm. It is worthy of note that the very nature of transformation methods makes problem scale an important issue. Accordingly, some effort to scale the problem variables and constraints seems advisable.

### 6.3.7 Variable Bounds

Most transformation approaches ignore the special nature of the variable bounds given in (6.4) and treat them as inequality constraints. Accordingly, for each design variable we must add two inequality constraints. Admittedly the additional constraints are simple, but a more direct approach seems desirable.

We seek a procedure that handles the bounds directly and does not increase the dimension (number of constraints) of the problem. The technique must be able to recognize that one or more variables is at or exceeds its bounds, and hold these variables on bounds if violation would otherwise occur. In addition, the technique must conveniently sense and allow feasible changes in bound variables.

One can imagine many approaches that would satisfy these general requirements. A particularly interesting question is the action to be taken when one or more bounds are violated. since the unconstrained searches are typically executed using a direction-generating method, we might simply back up along the current search direction until *all* variables were within bounds. This would preserve the line of search in the unconstrained algorithm but would require a zero-finding operation and would therefore be computationally expensive. A more direct, less elegant, and less expensive approach would be to simply set the out-of-bounds variables to their violated bounds simultaneously. Obviously this destroys the search direction, but our experience suggests that this is the best approach.

A related issue is the possibility of exploiting the presence of linear constraints. When some of the constraints are linear, certain economies of calculation are forthcoming. First, the rate of convergence will be enhanced due to the favorable impact on the penalty function topology. Second, the line-search algorithm can be made more efficient, since the linear boundaries are trivial to locate. In our own experiments with the method of multipliers [55]

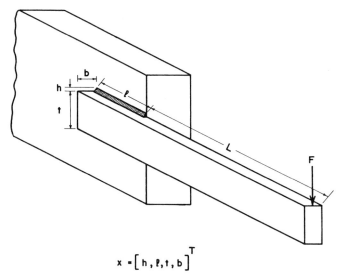

$$x = \left[ h, \ell, t, b \right]^T$$

**Figure 6.22.** The welded beam.

we have included scaling and variable-bound algorithms but we have not explored the efficiencies associated with linear constraint exploitation.

### Example 6.4   The Welded Beam

Consider once again the welded beam problem [58] described in Chapter 1 and shown in Figure 6.22.

Here we examine the solution using a MOM implementation called BIAS [59]. BIAS uses the simple multiplier update rule, scaling algorithm, and variable-bound procedure previously discussed. The objective remains the same, but three of the constraints ($g_4$ $g_5$, and $g_7$) are simply bounds on variables and are entered as such. The problem then becomes

$$\text{Minimize} \quad f(x) = (1 + c_3)x_1^2 x_2 + c_4 x_3 x_4 (L + x_2) \quad (6.54)$$

$$\text{Subject to} \quad g_1(x) = \tau_d - \tau(x) \geqslant 0$$

$$g_2(x) = \sigma_d - \sigma(x) \geqslant 0$$

$$g_3(x) = x_4 - x_1 \geqslant 0 \quad (6.55)$$

$$g_4(x) = P_c(x) - F \geqslant 0$$

$$g_5(x) = .25 - \text{DEL}(x) \geqslant 0$$

$$0.125 \leqslant x_1 \leqslant 10$$

$$0.100 \leqslant x_2 \leqslant 10$$

$$0.100 \leqslant x_3 \leqslant 10 \quad (6.56)$$

$$0.100 \leqslant x_4 \leqslant 10$$

where $x = [h, l, t, b]^T$
$c_3 = 0.10471$ ($/in.$^3$)
$c_4 = 0.04811$ ($/in.$^3$)

The lower bound on $x_1$ is from the problem statement, whereas the lower bounds on $x_2$, $x_3$ and $x_4$ are necessary to prevent selection of zero design variables, where functions in the constraint set are not defined. We prescale the constraints by dividing by suitable powers of 10 such that all constraint values

Table 6.5   Welded Beam Scale Factors

| $i, j$ | 1 | 2 | 3 | 4 | 5 |
|--------|-------|-------|-------|--------|--------|
| $\alpha_j$ | .8724 | .5324 | .9649 | 1.6730 | 7.2657 |
| $\eta_i$ | 1.0000 | .2704 | .1441 | 3.90 | |

are in the range $[0, 10]$ at $x^{(0)} = [1, 7, 4, 2]^T$, to get the following function values:

$$f(x^{(0)}) = 15.81545 \ (\$)$$

$$g_1(x^{(0)}) = 8.6785\text{E} - 01$$

$$g_2(x^{(0)}) = 1.4250\text{E} + 00$$

$$g_3(x^{(0)}) = 1.0000\text{E} + 00$$

$$g_4(x^{(0)}) = 1.8319\text{E} + 00$$

$$g_5(x^{(0)}) = 2.3285\text{E} - 01$$

Since $x^{(0)}$ is feasible and $K = 0$, one unconstrained DFP search is completed before the scale factors are calculated. The new scale factors are given in Table 6.5, where the $\alpha_j$ values scale the constraints,and the $\eta_i$ values scale the variables.

Table 6.6 gives the progress of the method at the end of each unconstrained DFP stage. At termination,

$$x^{(T)} = [.2444, 6.2187, 8.2915, .2444]^T$$

Table 6.6   BIAS Welded Beam Results ($R = 20$)

| Stage ($t$) | $x_1$ | $x_2$ | $x_3$ | $x_4$ | $f(x)$ | $P(x, \sigma)$ | Cumulative DFP Iterations |
|------|--------|--------|--------|-------|--------|--------|------|
| 1 | 0.2544 | 5.7098 | 8.3347 | .2417 | 2.3184 | 2.3482 | 27 |
| 2 | 0.2567 | 5.6319 | 8.2979 | .2435 | 2.3182 | 2.3476 | 36 |
| 3 | 0.2470 | 6.1130 | 8.2909 | .2444 | 2.3727 | 2.3804 | 44 |
| 4 | 0.2450 | 6.1937 | 8.2913 | .2444 | 2.3792 | 2.3810 | 64 |
| 5 | 0.2444 | 6.2177 | 8.2915 | .2444 | 2.3811 | 2.3811 | 71 |
| 6 | 0.2444 | 6.2187 | 8.2915 | .2444 | 2.3811 | 2.3811 | 76 |

and

$$f(x^{(T)}) = 2.38116 \ (\$)$$

$$g_1(x^{(T)}) = -6.5386\text{E} - 07$$

$$g_2(x^{(T)}) = -1.5824\text{E} - 07$$

$$g_3(x^{(T)}) = -2.6740\text{E} - 06$$

$$g_4(x^{(T)}) = 8.7002\text{E} - 08$$

$$g_5(x^{(T)}) = 2.324\text{E} - 01$$

The results indicate that the method found the vicinity of the solution quickly but did require additional iterations to focus the estimate of the solution before termination. Note that constraints 1–4 are tight at the solution, which seems logical. Even though only six iterations (stages) of the MOM algorithm are required, it should be noted in fairness that 1173 objective function and 1177 constraint set evaluations were required. The results given here are for a value of $R = 20$, and similar performance is obtained with values of $R$ in the range [10, 100]. Typically, higher $R$ values will produce fewer MOM stages with more DFP iterations per stage. The method will fail when values of $R$ are too small. For instance, when $R = 1$, BIAS terminated after 14 stages and 57 DFP iterations. At termination, only two of the constraints were tight and $x_3$ had not changed from $x_3^{(0)}$. This demonstrated the relationship between the scaling algorithm and the choice of $R$. Obviously, much work remains to be done here.

### 6.3.8  Other MOM-Type Codes

There are, of course, other good implementations of the method of multipliers. In fact, Schittkowski [60], in his recent comparative experiments, tested no less than 12 MOM codes. We mention a few of these codes here. ACDPAC, by M. J. Best and A. T. Bowler [61], uses the Rockafellar [47] augmented Lagrangian form with special provision for variable bounds and linear constraints. The Best and Ritter [62] conjugate direction method is employed for the subproblems. GAPFPR, by D. M. Himmelblau [63], uses the Hestenes [49] form, with inequalities transformed to equalities via slack variables. A quasi-Newton method with special provision for linear constraints is used for the subproblems. VF01A, by R. Fletcher, uses the Fletcher multiplier form [64] and a quasi-Newton method for the subproblems. Finally, P. E. Gill and W. Murray [65] have developed three related codes, SALQDR, SALQDF, and SALMNF, which employ an augmented Lagrangian form with inequalities transformed via slack variables. The various versions employ different unconstrained subproblem solution techniques and require either analytical or numerical derivatives. Additional data and discussion of the relative merits of these and other codes are given in Chapter 12.

## 6.4 SUMMARY

In this chapter we have examined transformation strategies with particular emphasis on augmented Lagrangian or multiplier methods. We have observed that the standard penalty forms—the log, inverse, and bracket operator—are simple in concept and implementation. In addition we noted the inevitable ill-condition of the unconstrained subproblems associated with the infinite limit value of the penalty parameter $R$ and the resulting distortion of the penalty function contours. The method of multipliers was examined as a remedy, and favorable computational characteristics were noted. Examples were given to support (and in some cases significantly contribute to) the discussion. Easy-to-use transformation-based codes are now available and are discussed further in Chapter 12.

## REFERENCES

1. Everett, H., "Generalized Lagrange Multiplier Method for Solving Problems of Optimum Allocation of Resources," *Oper. Res.*, **11**, 399–471 (1963).

2. Siddall, J. N., *Analytical Decision-Making in Engineering Design*, Prentice-Hall, Englewood Cliffs, NJ, 1972.

3. Siddall, J. N., "Opti-Sep: Designer's Optimization Subroutines," McMaster University, Canada, 1971.

4. Lootsma, F. A., "A Survey of Methods for Solving Constrained Minimization Problems via Unconstrained Minimization," in *Numerical Methods for Nonlinear Optimization* (F. A. Lootsma, Ed.), Academic Press, New York, 1972, pp. 313–348.

5. Ragsdell, K. M., R. R. Root, and G. A. Gabriele, Subroutine ZERO in *MELIB: A Subroutine Library for Mechanical Engineers*, 1975.

6. Fiacco, A. V., and G. P. McCormick, "Computational Algorithm for the Sequential Unconstrained Minimization Technique for Nonlinear Programming," *Manage. Sci.*, **10**, 601–617 (1964).

7. Fiacco, A. V., and G. P. McCormick, *Nonlinear Programming: Sequential Unconstrained Minimization Techniques*, Wiley, New York, 1968.

8. Lootsma, F. A., "Extrapolation in Logarithmic Programming," *Philips Res. Rep.*, **23**, 108–116 (1968).

9. Lootsma, F. A., "Boundary Properties of Penalty Functions for Constrained Minimization," *Philips Res. Rep. Suppl.*, **3**, 1970.

10. Fletcher, R., and A. P. McCann, "Acceleration Techniques for Nonlinear Programming," in *Optimization* (R. Fletcher, Ed.), Academic Press, New York, 1969, pp. 203–214.

11. Murray, W., "Ill-Conditioning in Barrier and Penalty Functions Arising in Constrained Nonlinear Programming," in *Proceedings of the Sixth Symposium on Mathematical Programming, Princeton, NJ*, 1967.

12. Lootsma, F. A., "Hessian Matrices of Penalty Functions for Solving Constrained Optin. tion Problems," *Philips Res. Rep.*, **24**, 322–331 (1969).

13. Courant, R., "Variational methods for the Solution of Problems of Equilibrium and Vibr. tions," *Bull. Am. Math. Soc.*, **49**, 1–23 (1943).

14. Carroll, C. W., "The Created Response Surface Technique for Optimizing Nonlinear Re-strained Systems," *Oper. Res.*, **9**, 169–184 (1961).

15. Fiacco, A. V., and G. P. McCormick, "The Sequential Unconstrained Minimization Tech-nique for Nonlinear Programming, a Primal-Dual Method," *Manage. Sci.*, **10**, 360–366 (1964).

16. Fiacco, A. V., and G. P. McCormick, "Computational Algorithm for the Sequential Uncon-strained Minimization Technique for Nonlinear Programming," *Manage. Sci.*, **10**, 601–617 (1964).

17. Fiacco, A. V., and G. P. McCormick, *Nonlinear Programming: Sequential Unconstrained Minimization Techniques*, Wiley, New York, 1968.

18. Zangwill, W. I., "Nonlinear Programming via Penalty Functions," *Manage. Sci.*, **13** (5), 344–358 (1967).

19. Lootsma, F. A., "Logarithmic Programming: A Method of Solving Nonlinear Programming Problems," *Philips Res. Rep.*, **22**, 329–344 (1967).

20. Lootsma, F. A., "Constrained Optimization via Penalty Functions," *Philips Res. Rep.*, **23**, 408–423 (1968).

21. Lootsma, F. A., "A Survey of Methods for Solving Constrained-Minimization Problems via Unconstrained Minimization," in *Optimization and Design* (M. Avriel, H. Rijckaert, and D. J. Wilde, Eds.), Prentice-Hall, Englewood Cliffs, NJ, 1973 pp. 88–118.

22. McCormick, G. P., "Penalty Function versus Non-Penalty Function Methods for Con-strained Nonlinear Programming Problems," *Math. Prog.*, **1** (2), 217–238 (1971).

23. Lasdon, L. S., "An Efficient Algorithm for Minimizing Barrier and Penalty Functions," *Math. Prog.*, **2** (1), 65–106 (1972).

24. Moe, J., "Penalty-Function Methods," in *Optimum Structural Design: Theory and Applica-tions* (R. H. Gallagher and O. C. Zienkiewicz, Eds.), Wiley, New York, 1973, chap. 9.

25. Ryan, D. M., "Penalty and Barrier Functions," in *Numerical Methods for Constrained Optimization* (P. E. Gill and W. Murray, Eds.), Academic Press, New York, 1974, pp. 175–190.

26. Fox, R. L., *Optimization Methods for Engineering Design*, Addison-Wesley, Reading, MA, 1971.

27. Himmelblau, D. M., *Applied Nonlinear Programming*, McGraw-Hill, New York, 1972, chap. 7, p. 299.

28. Avriel, M., *Nonlinear Programming: Analysis and Methods*, Prentice-Hall, Englewood Cliffs, NJ, 1976.

29. Gulf Oil Corporation, "Compute II, a General Purpose Optimizer for Continuous Nonlinear Models—Description and User's Manual," Gulf Computer Sciences Incorporated, Houston, TX, 1972.

30. Afimiwala, K. A., "Program Package for Design Optimization," Department of Mechanical Engineering, State University of New York at Buffalo, 1974.

31. Lill, S., The University of Liverpool Computer Laboratory: E04HAF, Numerical Algorithms Group, document no. 737, 1974.

32. DeSilva, B. M. E., and G. N. C. Grant, "Comparison of Some Penalty Function Based Optimization Procedures for the Synthesis of a Planar Truss," *Intern. J. Numerical Methods Eng.*, **7** (2), 155–173 (1973).

33. Bracken, J., and G. P. McCormick, *Selected Applications of Nonlinear Programming*, Wiley, New York, 1968.

34. Gupta, B. D., and S. S. Rao, "Optimum Design of Air-Conditioned Buildings," *ASME J. Mech. Des.*, **102** (3), 476–480 (1980).

35. Rao, S. S., and B. D. Gupta, "Application of External Distributions in the Design of Thermal Systems," *ASME J. Mech. Des.*, **102**, (3), 481–491 (1980).

36. Mylander, W. C., R. L. Holmes, and G. P. McCormick, *A Guide to SUMT—Version 4*, Research Analysis Corporation, Feb. 1974.

37. Sandgren, E., "The Utility of Nonlinear Programming Algorithms," Ph.D. dissertation, Purdue University, 1977; available from University Microfilm, 300 N. Zeeb Road, Ann Arbor, MI, document no. 7813115.

38. Kowalik, J., "Nonlinear Programming Procedures and Design Optimization," *Acta Polytech. Scand.*, **13**, 1966.

39. Fiacco, A. V., and G. P. McCormick, "The Sequential Unconstrained Minimization Technique Without Parameters," *Oper. Res.*, **15**, 820–827 (1967).

40. Fiacco, A. V., "Sequential Unconstrained Minimization Methods for Nonlinear Programming," thesis, Northwestern University, 1967.

41. Rosenbrock, H. H., "An Automatic Method for Finding the Greatest or Least Value of a Function," *Computer J.*, **3**, 175–184 (1960).

42. Huard, P., "Resolution of Mathematical Programming with Nonlinear Constraints by the Method of Centres," in *Nonlinear Programming* (J. Abodie, Ed.), North-Holland, Amsterdam, 1967, pp. 207–219.

43. Staha, R. L., "Documentation for Program Comet, A Constrained Optimization Code," The University of Texas, Austin, TX, 1973.

44. Lootsma, F. A., "Convergence Rates of Quadratic Exterior Penalty-Function Methods for Solving Constrained Minimization Problems," *Philips Res. Rep.*, **29** (1) (Feb. 1974).

45. Schuldt, S. B., "A Method of Multipliers for Mathematical Programming Problems with Equality and Inequality Constraints," *JOTA*, **17**, (1/2), 155–162 (1975).

46. Schuldt, S. B., G. A. Gabriele, R. R. Root, E. Sandgren, and K. M. Ragsdell, "Application of a New Penalty Function Method to Design Optimization," *ASME J. Eng. Ind.*, **99** (1), 31–36 (Feb. 1977).

47. Rockafellar, R. T., "A Dual Approach to Solving Nonlinear Programming Problems by Unconstrained Optimization," *Math. Programming*, **5** (3), 354–373 (1973).

48. Powell, M. J. D., "A Method for Nonlinear Constraints in Minimization Problems," in *Optimization* (R. Fletcher, Ed.), Academic Press, London, 1969 pp. 283–298.

49. Hestenes, M. R., "Multiplier and Gradient Methods," *JOTA*, **4** (5), 303–320 (1969).

50. Gould, F. J., and S. Howe, "A New Result on Interpreting Lagrange Multipliers as Dual Variables," Dept. Statistics, University of North Carolina, Chapel Hill, NC, 1971.

51. Gill, P. E., and W. Murray, "The Computation of Lagrange-Multiplier Estimates for Constrained Minimization," *Math. Programming*, **17** (1), 32–60 (1979).

52. Root, R. R., "An Investigation of the Method of Multipliers," Ph.D. dissertation, Purdue University, Dec. 1977.

53. Bertsekas, D. P., "On Penalty and Multiplier Methods for Constrained Minimization," in *Nonlinear Programming*, Vol. 2 (O. L. Mangasarian, R. R. Meyer, and S. M. Robinson, Eds.), Academic Press, New York, 1975, pp. 165–192.

54. Kort, B. W., "Rate of Convergence of the Method of Multipliers with Inexact Minimization," in *Nonlinear Programming*, Vol. 2 (Mangasarian, Meyer, and Robinson, Eds.), Academic Press, New York, 1975, pp. 193–214.

55. Root, R. R., and K. M. Ragsdell, "Computational Enhancements of the Method of Multipliers," *ASME J. Mech. Des.*, **102** (3), 517–523 (1980).

56. Keefer, D. L., and X. X. Gottfried, "Differential Constraint Scaling in Penalty Function Optimization," *AIIE Trans.*, **2** (4), 281 (1970).

57. Keefer, D. L., "Simpat: A Self-Bounding Direct Search Method for Optimization," *I & EC Proc. Des. Develop.*, **12**, 92 (1973).

58. Ragsdell, K. M., and D. T. Phillips, "Optimal Design of a Class of Welded Structures Using Geometric Programming," *ASME J. Eng. Ind.*, *Ser. B*, **98** (3), 1021–1025 (1976).

59. Root, R. R., and K. M. Ragsdell, "*BIAS*: A Nonlinear Programming Code in Fortran IV–Users Manual," Purdue Research Foundation, June 1976.

60. Schittkowski, K., "Nonlinear Programming Codes: Information, Tests, Performance," *Lecture Notes in Economics and Mathematical Systems*, Vol. 183, Springer-Verlag, Berlin, 1980.

61. Best, M. J., and A. T. Bowler, "ACDPAC: A Fortran IV Subroutine to Solve Differentiable Mathematical Programmes—Users Guide—Level 2.0," *Res. Rep.* **CORR75-26**, Dept. of Combinatorics and Optimization, University of Waterloo, Ontario, Canada, 1978.

62. Best, M. J., and Ritter, K., "An Accelerated Conjugate Direction Method to Solve Linearly Constrained Minimization Problems," *J. Computer Syst. Sci.*, **11** (3), 295–322 (1975).

63. Newell, J. S., and D. M. Himmelblau, "A New Method for Nonlinearly Constrained Optimization," *AIChE J.*, **21** (3), 479–486 (1975).

64. Fletcher, R., "An Ideal Penalty Function for Constrained Optimization," in *Nonlinear Programming*, Vol. 2 (O. L. Mangasarian, R. R. Meyer, and S. M. Robinson, Eds.), Academic Press, New York, 1975.

65. Gill, P. E., and W. Murray, (Eds.), *Numerical Methods for Constrained Optimization*, Academic Press, New York, 1974.

66. Freudenstein, F., "Approximate Synthesis of Four-Bar Linkages," *Trans. ASME*, 853–861 (Aug. 1955).

# PROBLEMS

**6.1.** Describe briefly the major difference between the method of multipliers and other transformation methods such as SUMT.

**6.2.** Under what conditions does the method of multipliers produce no distortion in the curvature of the penalty function contours from stage to stage?

**6.3.** Would it be possible and/or desirable to devise an *interior* version of the Method of Multiplier? Why or why not?

**6.4.** For the problem

$$\text{Minimize} \quad f(x)$$

$$\text{Subject to} \quad g_j(x) \geq 0 \quad j = 1, 2, 3, \ldots, J$$

we propose to use the two penalty functions

$$P_1(x, R) = f(x) + R \sum_{j=1}^{J} \langle g_j(x) \rangle^2$$

where $R$ is positive and increases in magnitude at the end of each unconstrained subproblem, and

$$P_2(x, R, \sigma) = f(x) + R \sum_{j=1}^{J} \langle g_j(x) + \sigma_j \rangle^2 - R \sum_{j=1}^{J} \sigma_j^2$$

with $R$ fixed and $\sigma_j$ adjusted after each unconstrained subproblem to force convergence. Discuss the relative merits of the two penalty forms, including any computational differences.

**6.5.** Consider the problem

Minimizes      $f(x) = 100(x_2 - x_1^2)^2 + (1 - x_1)^2$

Subject to      $g_1(x) = x_1 + 1 \geq 0$

$g_2(x) = 1 - x_2 \geq 0$

$g_3(x) = 4x_2 - x_1 - 1 \geq 0$

$g_2(x) = 1 - 0.5x_1 - x_2 \geq 0$

Construct $P$, using the log penalty, inverse penalty, and MOM penalty forms. Let $x^{(0)} = [-1, 1]^T$ and $[-0.5, 0.5]^T$ in each case. Discuss any difficulties encountered with these penalty forms at $x^{(0)}$.

**6.6.** For the inequality-constrained problem $(g_j(x) \geq 0, \ j = 1, 2, 3, \dots, J)$, suppose we start with an interior point. Explain whether the following functions act as interior or exterior penalty methods, and why.

(a) $P(x, R) = f(x) + R \sum_{j=1}^{J} \{g_j(x)\}^{-2}$

(b) $P(x, R) = f(x) + R[\min\langle 0, \min_j [g_j(x)]\rangle^2]$

**6.7.** Consider the problem

Minimize      $f(x) = (x_1 - 1)^2 + (x_2 - 1)^2$

Subject to      $g_1(x) = x_1 + x_2 - 1 \geq 0$

$g_2(x) = 1 - x_1 \geq 0$

$g_3(x) = 1 - x_2 \geq 0$

and the penalty function

$$P(x, R) = f(x) + R \sum_{j=1}^{J} \left[ \frac{1}{g_j(x)} \right]$$

(a) Find the stationary points of $P(x, R)$ as a function of $R$.

(b) Demonstrate the influence of $g_2$ and $g_3$ on the convergence rate of the method.

(c) Will the method converge more or less quickly when $g_2$ and $g_3$ are ignored?

**6.8.** Find the solution to the following problem:

$$\text{Minimize} \qquad f(x) = x_1^2 + x_2^2 + x_3^2$$

$$\text{Subject to} \qquad g_1(x) = 1 - x_2^{-1}x_3 \geqslant 0$$

$$g_2(x) = x_1 - x_3 \geqslant 0$$

$$h_1(x) = x_1 - x_2^2 + x_2x_3 - 4 = 0$$

$$0 \leqslant x_1 \leqslant 5$$

$$0 \leqslant x_2 \leqslant 3$$

$$0 \leqslant x_3 \leqslant 3$$

using an appropriate transformation method of your choice. Find $x$ to three decimal places.

**6.9.** We know that it is possible to define "curvature" $K$ and "radius of curvature" $R$ at a point of a function. Let the function be a particular contour value $\bar{f}$ of $f(x)$ in two dimensions. Develop a relationship between $\nabla^2 f(x)$ and $K$.

**6.10.** Discuss how one might use the relation in problem 6.9 to rate the quality of the various penalty forms.

**6.11.** *Minimum-Perimeter Enclosure Problem.* Consider the three circular plane objects in Figure 6.23. We wish to arrange the objects in the positive quadrant such that the circumscribed rectangle has a perimeter of minimum length. We could obviously extend the formulations to include objects of various shapes and numbers. The problem has a pleasing simplicity but is of practical value. One application is the planar packaging of electronic components. Another is the packing of a template for mass producing stamped metal parts.

Let the design variables be the $(x, y)$ locations of the object centers. The objective is the enclosure perimeter, $f(x) = 2(A + B)$, where $A$ and $B$ are the lengths of the rectangle sides. Force the objects to stay in the positive quadrant by writing constraints of the form $g_j(x) = x_a - R_a \geqslant 0$. To avoid object interference, insist that the distance between

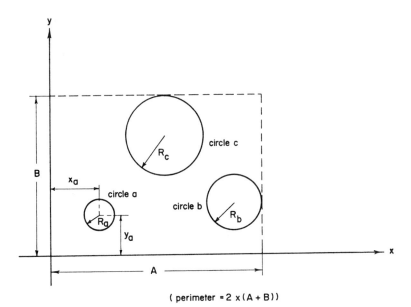

( perimeter = 2 × (A + B))

**Figure 6.23.** Minimum-perimeter enclosure problem.

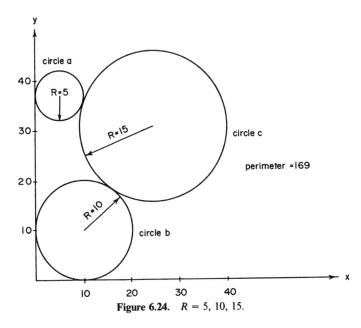

**Figure 6.24.** $R = 5, 10, 15$.

two objects is greater than or equal to the sum of their radii:

$$g_j(x) = \{(x_a - x_b)^2 + (y_a - y_b)^2\}^{1/2} - (R_a + R_b) \geqslant 0$$

Assume that the circle radii are given as shown in Figure 6.24.

(a) Formulate the complete NLP.

(b) Find the minimum perimeter enclosure, using a transformation method of your choice.

(c) Is the feasible region convex?

(d) What is the likelihood of encountering local solutions?

(e) Check your solution to (b) using a piece of paper, circle cutouts, a ruler, and experimentation by hand.

6.12. *Mechanism Precision Point Placement.* Consider the four-bar mechanism shown in Figure 6.25. Using this planar device, we wish to coordinate the angles $\theta$ and $\phi$ such that the desired function $y_d(x)$ is approximated. In particular, we wish to locate the three precision points $x_1, x_2, x_3$ such that the square of the error in the approximation is least.

Consider the generated and desired functions in Figure 6.26. and the resulting error function in Figure 6.27, where $E(x) = y_d(x) - y_g(x)$. The input-output relationship is constructed given $x_L, x_H, y_L, y_H$ and Freudenstein's equation [66]:

$$F(\phi) = R_1\cos\theta - R_2\cos\phi + R_3 - \cos(\theta - \phi) = 0$$

where $R_1 = d/c$, $R_2 = d/a$, $R_3 = (a^2 - b^2 + c^2 + d^2)/2ca$, by setting

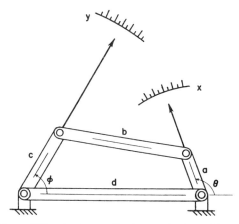

**Figure 6.25.** Angle coordination synthesis problem.

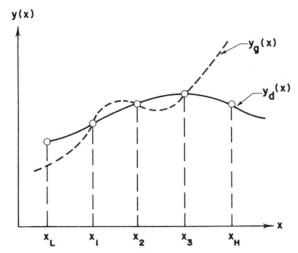

**Figure 6.26.** Precision point placement.

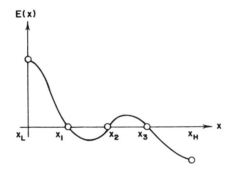

**Figure 6.27.** Output error function.

$E(x) = 0$ at $x_1$, $x_2$, and $x_3$. In addition, it is assumed that the linear relations $\theta = c_1 x + c_2$ and $\phi = c_3 y + c_4$ are known.

(a) Formulate an appropriate NLP for precision point placement.

(b) Let $y_d(x) = 1/x$, $x_L = 1$, $x_H = 2$, $y_L = 1$, $y_H = .5$, so that $\theta = 90x + 60$ and $\phi = -180y + 240$; and find the precision points $x_1$, $x_2$, $x_3$ given $x^{(0)} = [1.15, 1.50, 1.85]^T$.

**6.13.** Consider the problem

$$\text{Minimize} \quad f(x) = (x - 2)^2; \qquad x \in R^1$$

$$\text{Subject to} \qquad h(x) = x - 3 = 0$$

shown in Figure 6.28. Obviously the solution is $x^* = 3$ with $f(x^*) = 1.0$.

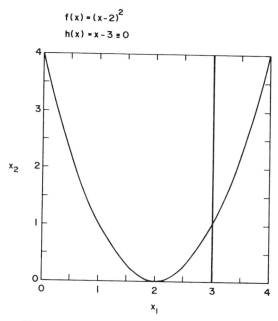

$$f(x) = (x-2)^2$$
$$h(x) = x - 3 \equiv 0$$

**Figure 6.28.** Single-variable constrained problem.

$$R = \infty, 10, 2, 1, 0.1$$

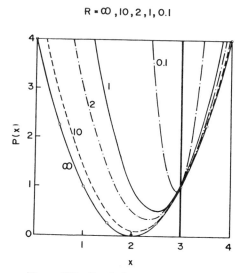

**Figure 6.29.** Parabolic penalty contours.

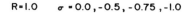

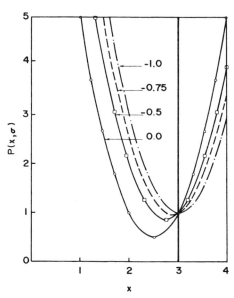

**Figure 6.30.**  MOM contours.

Develop relations for (a) the parabolic penalty and (b) the MOM penalty forms that support the contour plots given in Figures 6.29 and 6.30.

**6.14.** Solve problem 3.19 (without using the volume constraint to eliminate one of the design variables), using a penalty method of your choice.

**6.15.** (a) Solve problem 3.20 (without using the flow-rate constraint to eliminate one of the design variables), using the method of multipliers.

(b) Discuss the difficulties associated with using a gradient-based method for the subproblems.

# Chapter 7

# Constrained Direct Search

The transformation and Lagrangian based method discussed in Chapter 6 offer implicit ways of handling constraints that avoid consideration of the detailed structure of the constraint functions. The sole structural distinction is between equality and inequality constraints. By contrast, the methods to be considered in this and the next several chapters operate on the constraints explicitly, that is, they directly take into account the constraints in the course of the optimization iterations. In this chapter we focus on direct-search methods, that is, techniques for locating constrained optima that rely upon only objective and constraint function values to guide the search. As in the unconstrained case, the motivation for considering such methods is that in engineering applications it is frequently necessary to solve problems whose functions are discontinuous or nondifferentiable, and in such cases only function-value-based approaches are applicable.

We consider two families of methods: those that are essentially adaptations of the unconstrained direct-search methods discussed in Chapter 3, and those that rely upon the random selection of trial points. Both families of methods are heuristic in nature, that is, they are based on intuitive constructions unsupported by rigorous theory and thus offer no guarantees of convergence or estimates of rates of convergence. In spite of these limitations, we find these techniques to be quite reliable in yielding improved if not optimal solutions and easy to implement because of their simple search logic. We begin with a brief discussion of the steps required to prepare general constrained problems for solution via direct-search methods.

## 7.1  PROBLEM PREPARATION

The general constrained optimization problem involves inequality constraints, equality constraints, and variable upper and lower bounds. Intuitively, given a feasible point $x^0$, it should be possible to modify the function-value-based search logic of any unconstrained method to check that the trial points generated with $x^0$ as base also satisfy the problem constraints. For instance, if $v$ is a trial point that violates some inequality constraint or variable bound, then modify $v$ by reducing the step taken from $x^0$, so that these constraints are again satisfied. Equality constraints, however, are more troublesome, because even if $x^0$ and $v$ satisfy the equalities, the points on the line

$$x = x^0 + \alpha(v - x^0) \qquad \text{for } 0 \leqslant \alpha \leqslant 1$$

need not satisfy them, except in the case of linear equalities. If in addition $v$ is infeasible, then it is difficult to readjust the components of $v$ to ensure satisfaction of nonlinear equality constraints using only the constraint function values. It is, in fact, equivalent to solving the simultaneous equation set

$$h_k(x) = 0 \qquad k = 1, \ldots, K$$

given the initial estimate $v$, using only values of the $h_k(x)$ functions and adjusting the components of $v$ so that the resulting trial point is an improvement over the base point $x^0$. As a result of the difficulty of ensuring the satisfaction of equality constraints, virtually all direct-search methods require that problems be posed only in terms of inequality constraints. Equality constraints must be eliminated from the problem explicitly or implicity prior to solution.

A second, even more basic, problem is that direct-search methods must typically be initiated with a feasible point. To be sure, in many engineering applications a feasible starting point can be defined on the basis of prior knowledge about the system or based on the operating conditions of an existing system. However, this is not always the case, and thus systematic procedures to locate a feasible point are a matter of practical concern.

### 7.1.1 Treatment of Equality Constraints

The simplest approach to eliminating equality constraints is to solve each for one variable and use the resulting expression to elinimate that variable from the problem formulation by substitution. In doing so, however, care must be exercised to ensure that any specified bounds on the substituted variable are retained as inequality constraints.

*Example 7.1*

Consider the problem

$$\text{Minimize} \quad f(x) = x_1^2 + 4x_2^2 + x_3^2$$

$$\text{Subject to} \quad h_1(x) = x_1 + x_2^2 - 2 = 0$$

$$-1 \leqslant x_1 \leqslant 1$$

$$0 \leqslant (x_2, x_3) \leqslant 2$$

Clearly the constraint $h_1$ can be solved for $x_1$ to yield $x_1 = 2 - x_2^2$, and thus the objective function can be reduced by substitution to

$$f(x_2, x_3) = (2 - x_2^2)^2 + 4x_2^2 + x_3^2$$

To ensure satisfaction of the bounds on $x_1$, two inequality constraints must be imposed:

$$-1 \leqslant 2 - x_2^2 \leqslant 1$$

The resulting reduced-dimensionality problem becomes

$$\text{Minimize} \quad (2 - x_2^2)^2 + 4x_2^2 + x_3^2$$

$$\text{Subject to} \quad 1 \leqslant x_2 \leqslant \sqrt{3} \qquad 0 \leqslant x_3 \leqslant 2$$

Note that if the bounds on $x_1$ had been ignored, the minimum solution $x_2 = x_3 = 0$ of the reduced dimensionality problem would lead to an infeasible $x_1$, namely, $x_1 = 2$.

In implementing the equality-constraint elimination strategy, it is normally neither necessary nor desirable to actually carry out algebraic substitution. Instead, it is more convenient and much more reliable to simply include the expressions obtained by solving a constraint for a variable as statements defining the value of that variable in the routine for evaluating the problem's objective function and inequality constraints. In this way, values of the dependent variable can be saved for subsequent printing, and the possibility of algebraic substitution errors is minimized.

While substitution is an obvious way to eliminate equality constraints, it is not always possible to carry it out explicitly. In many cases it is either algebraically infeasible or inconvenient to actually solve an equation for one variable in terms of the others. In those cases, implicit elimination is possible by numerically solving the constraint for the value of the dependent variable corresponding to each selected trial value of the independent optimization variables. Such numerical treatment of equality constraints imposes a large computational overhead, particularly if the number of dependent variables is large, but must be accepted if direct-search methods are to be employed.

*Example 7.2*

The problem

$$\text{Minimize} \quad f(x) = x_1 x_2 x_3$$

$$\text{Subject to} \quad h_1(x) = x_1 + x_2 + x_3 - 1 = 0$$

$$h_2(x) = x_1^2 x_3 + x_2 x_3^2 + x_2^{-1} x_1 - 2 = 0$$

$$0 \leqslant (x_1, x_3) \leqslant \tfrac{1}{2}$$

involves two equality constraints and can therefore be viewed as consisting of two dependent variables and one independent variable. Clearly, $h_1$ can be solved for $x_1$ to yield

$$x_1 = 1 - x_2 - x_3$$

Thus, upon substitution the problem reduces to

Minimize    $(1 - x_2 - x_3)x_2 x_3$

Subject to    $(1 - x_2 - x_3)^2 x_3 + x_2 x_3^2 + x_2^{-1}(1 - x_2 - x_3) - 2 = 0$

$$0 \leqslant 1 - x_2 - x_3 \leqslant \tfrac{1}{2}$$

$$0 \leqslant x_3 \leqslant \tfrac{1}{2}$$

It is very difficult to solve the remaining equality constraint for one variable, say $x_3$, in terms of the other. Instead, for each value of the independent variable $x_2$, the corresponding value of $x_3$ must be calculated numerically using some root-finding method. Note that in this case, the bounds and inequalities involving $x_3$ become implicit functions of the independent variable $x_2$. That is, since $x_3$ and $x_2$ are related through some function $q(x_2) = x_3$, the constraints become

$$0 \leqslant 1 - x_2 - q(x_2) \leqslant \tfrac{1}{2}$$

$$0 \leqslant q(x_2) \leqslant \tfrac{1}{2}$$

which can be evaluated only when the equality constraint is solved.

In general, if the number of equality constraints is large, then it becomes appropriate to use equation-ordering and decomposition procedures to automate the selection of independent variables and the sequencing of the equality constraints for efficient solution. The discussion of these techniques is beyond the scope of this book. The reader is directed to texts dealing with the solution of equation sets [1, 2] or the excellent review given by Sargent [3].

### 7.1.2 Generation of Feasible Starting Points

Borrowing from the terminology introduced in the linear programming context, the process of generating a feasible starting point is often called a *Phase I* procedure. Three such procedures are in common use: random generation, minimization of an unconstrained penalty-type function, and sequential constrained minimization. In this section we review briefly only the first type of strategy. The other two are considered in Chapter 13.

Conceptually the simplest way of generating feasible starting points is to use pseudo-random number generation routines (available on most computer systems) to generate trial points, which are then tested for feasibility by evaluating the constraint functions. Specifically, the selected variable upper and lower bounds are used to calculate the components

$$x_i = x_i^{(L)} + r_i\left(x_i^{(U)} - x_i^{(L)}\right) \qquad i = 1,\ldots, N \qquad (7.1)$$

where $r_i$ is a random number distributed uniformly on the interval $(0, 1)$. If $x$ is an $N$-component vector, then $r_i$ can be sampled $N$ times and the $x_i$ calculated from (7.1). This point is then used to evaluate the problem constraint functions, and the process is continued until a point satisfying all constraints is found. Clearly, since the probability of finding a point simultaneously satisfying a set of equality constraints in this fashion is quite low, it is appropriate to use the form of the problem in which the equality constraints have been eliminated. The sampling is thus done only on the independent variables. Even in this case, random generation can require extensive computation time. Thus, for higher dimensionality and tightly constrained problems, the alternate strategies discussed in Chapter 13 may be necessary.

## 7.2 ADAPTATIONS OF UNCONSTRAINED SEARCH METHODS

In this section we briefly review the difficulties that can arise in attempting to adapt the direct-search methods of Chapter 3 to constrained problems. Throughout this discussion it is assumed that any equality constraints have been eliminated using the devices of Section 7.1.1. We consider in detail only one of the direct-search adaptations, the *complex method*, since this algorithm has enjoyed wide use in engineering applications.

### 7.2.1  Difficulties in Accommodating Constraints

In principle, all three of the basic direct-search methods considered in Chapter 3 could be modified to accommodate inequality constraints. In essence, before accepting any candidate point, the constraints could be checked, and if they are violated the current point could be rejected and another generated that would satisfy the constraints. In practice, however, considerable care must be exercised in defining the rules according to which infeasible points are modified to correct for infeasiblity. Otherwise, the resulting algorithm may not only suffer from a substantially reduced rate of convergence but may also terminate prematurely.

Let us first consider the adaptation of the *conjugate directions* method. Given an initial feasible point $x^0$ and a set of search directions $d_i, i = 1,\ldots, N$, the search would proceed with line searches along each of these direction vectors so as to both generate conjugate directions and improve the objective function value. In the presence of inequality constraints, any given line search may well lead to a one-parameter optimum that is infeasible. Clearly, to retain feasibility the search along the direction vector must be terminated short of the one-parameter optimum at the point at which the boundary of the feasible region is intersected. In this case, however, the generation of conjugate directions will no longer be guaranteed, and thus the search can degenerate to a series of line searches along a fixed set of direction vectors. As shown in Figure 7.1, this in turn can result in premature termination. In the figure, the line searches in the $d_1$ and $d_2$ directions from the point $x^{(t)}$ either lead outside the feasible region or to points with worse objective function values. Consequently, the search will terminate at $x^{(t)}$. Evidently, one way to escape the dilemma is to redefine the set of search directions whenever further progress is stymied. For instance, a new direction vector parallel to constraint $g_2$ would lead to improvement. Constructions of this nature have been proposed by Davies (cited in Chapter 5 of ref. 4,) and others, but efficient implementations typically require the use of constraint gradient information.

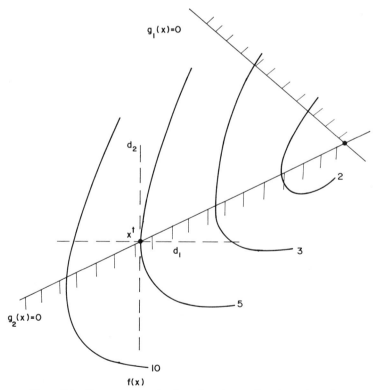

**Figure 7.1.**  Premature termination of conjugate directions adaptation.

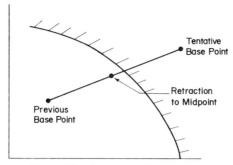

**Figure 7.2.** Retraction of pattern step.

Similar modifications have been proposed for use with the *pattern search* method. Recall that in this algorithm progress is made either through an exploratory move or through a pattern move. In the presence of constraints, either or both of these moves may lead to infeasible points, and thus some corrective rules must be added to accommodate these situations. As shown in Figure 7.2, if a pattern move results in an infeasible tentative base point, then the step can simply be reduced by a fixed fraction, say $1/2$, until a feasible point is found. Similarly, during exploration, infeasible points can simply be treated as points with a very high objective function value and thus rejected as shown in Figure 7.3. Eventually, a point will be reached at which all exploratory steps lead to either infeasible or feasible but worse points. In this case, the step sizes used in exploration can be reduced uniformly and the exploratory move restarted. However, as in the conjugate directions case (see Figure 7.1), it is nonetheless possible to reach base points for which step-size reduction does not lead to improvement. Again, this situation can be rectified only by developing search directions that lie parallel to the constraint surface. Proposals that accomplish this have been made by Klingman and Himmelblau [5] and others (see the discussion in ref. 6, chap. VII, as well as ref. 4). However, in all cases constraint and objective function gradient information is required to carry out the proposed constructions. If gradient values are available, then, of

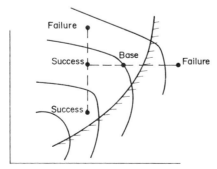

**Figure 7.3.** Rejection of infeasible exploratory step.

course, experience has shown that the methods of Chapters 8–10 are much to be preferred.

The conclusions we can draw from this brief consideration of direct search in the presence of constraints is that the simple device of retracting a step if it leads to infeasibility is an adequate way of ensuring trial-point feasibility. This device, of course, requires that the search be begun with a feasible point and that successive base points remain feasible so that retraction is always possible. On the other hand, the use of any set of search directions that does not adjust itself to allow movement along constraint surfaces is quite clearly unsatisfactory. This problem can, however, be mitigated if the search directions are widely scattered or in some sense randomized. This is effectively what is achieved in the adapation of the simplex direct-search method to accommodate inequality constraints that is discussed in the next section.

### 7.2.2 The Complex Method

The simplex direct-search method is based on the generation and maintenance of a pattern of search points and the use of projections of undesirable points through the centroid of the remaining points as the means of finding new trial points. In the presence of inequality constraints, it is evident that if the new point is infeasible, then it is a simple matter to retract it toward the centroid until it becomes feasible. Clearly, this will immediately destroy the regularity of the simplex, but, as was shown by Nelder and Meade, this is not necessarily a bad result. However, the generation of the initial simplex of points is problematic, because for any given simplex size parameter $\alpha$ it is highly likely that many of the points of the regular simplex will be infeasible. Evidently, as each point is calculated it must be tested for feasibility and, if infeasible, suitably adjusted. As a consequence of this adjustment, the regularity of the simplex construction is again destroyed. Moreover, the points must be generated sequentially rather than being essentially simultaneously defined using the formula for a regular simplex. These considerations have led to the simplex modification proposed by Box [7] and called the *complex method*.

Box proposed that the set of $P$ trial points be generated randomly and sequentially. The mechanics of random point generation proceeds as in Section 7.1.2. Given variable upper and lower bounds $x^{(U)}$ and $x^{(L)}$, the pseudo-random variable uniformly distributed on the interval $(0, 1)$ is sampled, and the point coordinates calculated via Eq. (7.1). $N$ samples are required to define a point in $N$ dimensions. Each newly generated point is tested for feasibility, and if infeasible it is retracted toward the centroid of the previously generated points until it becomes feasible. The total number of points to be used, $P$, should, of course, be no less than $N + 1$ but can be larger.

Given this set of points, the objective function is evaluated at each point, and the point corresponding to the highest value is rejected. A new point is generated by reflecting the rejected point a certain distance through the centroid of the remaining points. Thus, if $x^R$ is the rejected point and $\bar{x}$ is the centroid of the remaining points, then the new point is calculated via

$$x^m = \bar{x} + \alpha(\bar{x} - x^R)$$

The parameter $\alpha$ determines the distance of the reflection: $\alpha = 1$ corresponds to setting the distance $\|x^m - \bar{x}\|$ equal to $\|\bar{x} - x^R\|$; $\alpha > 1$ corresponds to an expansion; $\alpha < 1$ to a contraction.

Now, at the new point, the performance function and the constraints are evaluated. There are several alternatives:

1.  The new point is feasible and its function value is not the highest of the set of points. In this case, select the point that does correspond to the highest, and continue with a reflection.
2.  The new point is feasible and its function value is the highest of the current set of $P$ points. Rather than reflecting back again (which could cause cycling), retract the point by half the distance to the previously calculated centroid.
3.  The new point is infeasible. Retract the point by half the distance to the previously calculated centroid.

The search is terminated when the pattern of points has shrunk so that the points are sufficiently close together and/or when the differences between the function values at the points become small enough.

Box [7] has performed numerical experiments with this algorithm, and on this empirical basis recommends using $\alpha = 1.3$ and $P \approx 2N$. Biles [8], on the other hand, reports good results with $P = N + 2$. Most implementations follow Box's recommendation that if a point exceeds one of the bounds its corresponding coordinate be set equal to the bound. The $\alpha > 1$ compensates for the shrinking of the complex caused by halving the distances, while the large number of vertices is intended to prevent the complex from collapsing and flattening out along a constraint. The setting at the value of the bound is intended to avoid unnecessary contraction of the complex.

The complete algorithm can be summarized as follows.

### The Complex Method

Given an initial strictly feasible point $x^0$, reflection parameter $\alpha$, and termination parameters $\varepsilon$ and $\delta$.

**Step 1.** Generate the initial set of $P$ feasible points. For each point $p = 1,\ldots, P - 1$,

(a) Sample $N$ times to determine the point $x_i^p$.

(b) If $x^p$ is infeasible, calculate the centroid $\bar{x}$ of the current set of points and reset

$$x^p = x^p + \tfrac{1}{2}(\bar{x} - x^p)$$

Repeat until $x^p$ becomes feasible.

(c) If $x^p$ is feasible, continue with (a) until $P$ points are available.

(d) Evaluate $f(x^p)$, for $p = 0, \ldots, P - 1$.

**Step 2.** Carry out the reflection step.

(a) Select the point $x^R$ such that

$$f(x^R) = \max f(x^p) \equiv F_{max}$$

(b) Calculate the centroid $\bar{x}$ and the new point

$$x^m = \bar{x} + \alpha(\bar{x} - x^R)$$

(c) If $x^m$ is feasible and $f(x^m) \geqslant F_{max}$, retract half the distance to the centroid $\bar{x}$. Continue until $f(x^m) < F_{max}$.

(d) If $x^m$ is feasible and $f(x^m) < F_{max}$, go to step 4.

(e) If $x^m$ is infeasible, go to step 3.

**Step 3.** Adjust for feasibility.

(a) Reset violated variable bounds:
If $x_i^m < x_i^{(L)}$, set $x_i^m = x_i^{(L)}$.
If $x_i^m > x_i^{(U)}$, set $x_i^m = x_i^{(U)}$.

(b) If the resulting $x^m$ is infeasible, retract half the distance to the centroid. Continue until $x^m$ is infeasible, then go to step 2(c).

**Step 4.** Check for termination.

(a) Calculate

$$\bar{f} = \frac{1}{P}\sum f(x^p) \qquad \bar{x} = \frac{1}{P}\sum x^p$$

(b) If

$$\sum_{p}(f(x^p) - \bar{f})^2 \leqslant \varepsilon \qquad \text{and} \qquad \sum_{p}\|(x^p - \bar{x})^2\| \leqslant \delta$$

terminate. Otherwise, go to step 2(a).

Note that in order to carry out step 1, upper and lower bounds must be set on all variables. If such bounds do not naturally arise as part of the problem formulation, then it is necessary to provide some estimated values. These estimates must include the problem feasible region, and for computational efficiency the bounds should be as tight as possible.

The detailed calculations of the complex method are illustrated in the next example.

### Example 7.5

A rectangular structure with an open frontal face is to be designed to enclose a plant unit. The structure is to have a volume of 16,000 ft³, but the perimeter of the base should be no more than 220 ft. The depth is to be no more than 60 ft, and the width no more than 80 ft. Moreover, the width should not exceed three times the depth, and the height is to be no more than two-thirds of the width. The cost of the corrugated material of which the three sides and roof are to be constructed is $30/ft². Design the structure so as to minimize the materials cost.

### Formulation

As shown in Figure 7.4, we introduce the variables $x_1$ = depth (ft), $x_2$ = width (ft), $x_3$ = height (ft). The material cost will be given by

| | |
|---|---|
| Cost of roof: | $30x_1x_2$ |
| Cost of backwall: | $30x_2x_3$ |
| Cost of sides: | $2(30x_1x_3)$ |

Thus, our objective function becomes

$$f(x) = 30x_1x_2 + 30x_2x_3 + 60x_1x_3$$

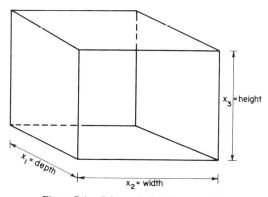

**Figure 7.4.**   Schematic for Example 7.5.

The equality constraints consist of the volume requirement,

$$x_1 x_2 x_3 = 16{,}000 \text{ ft}^3$$

The inequality constraints will include the perimeter limitation,

$$2(x_1 + x_2) \leqslant 220 \text{ ft}$$

the width limitation,

$$x_2 \leqslant 3x_1$$

the height limitation,

$$x_3 \leqslant \tfrac{2}{3}x_2$$

and the bounds,

$$0 \leqslant x_1 \leqslant 60 \text{ ft}$$

$$0 \leqslant x_2 \leqslant 80 \text{ ft}$$

In standard NLP form, the problem becomes

$$\text{Minimize} \quad f(x) = 30x_1 x_2 + 30x_2 x_3 + 60x_1 x_3$$

$$\text{Subject to} \quad h_1(x) = x_1 x_2 x_3 - 16{,}000 = 0$$

$$g_1(x) = 110 - x_1 - x_2 \geqslant 0$$

$$g_2(x) = 3x_1 - x_2 \geqslant 0$$

$$g_3(x) = \tfrac{2}{3}x_2 - x_3 \geqslant 0$$

$$0 \leqslant x_1 \leqslant 60$$

$$0 \leqslant x_2 \leqslant 80$$

$$0 \leqslant x_3$$

In order to solve this problem using the complex method, the equality constraint must be eliminated. In this case, explicit elimination is possible by just solving $h_1(x)$ for one of the variables, say $x_3$:

$$x_3 = \frac{16 \times 10^3}{x_1 x_2}$$

Using this equation to eliminate $x_3$, we obtain the reduced problem,

$$\text{Minimize} \quad f(x) = 30x_1x_2 + \frac{48 \times 10^4}{x_1} + \frac{96 \times 10^4}{x_2}$$

$$\text{Subject to} \quad g_1(x) = 110 - x_1 - x_2 \geqslant 0$$

$$g_2(x) = 3x_1 - x_2 \geqslant 0$$

$$g_3(x) = \tfrac{2}{3}x_2 - \frac{16 \times 10^3}{x_1 x_2} \geqslant 0$$

$$0 \leqslant x_1 \leqslant 60$$

$$0 \leqslant x_2 \leqslant 70$$

Observe that $g_3$ becomes nonlinear in $x_1$ and $x_2$. Note that since bounds are already imposed on $x_1$ and $x_2$, it is not necessary to impose estimated bounds for use with the complex strategy.

**Solution.** For purposes of initiating the complex method, suppose we use the point $x^{(1)} = (50, 50)$ as the initial feasible solution estimate. Since this is a two-dimensional problem, it will be necessary to generate three additional points. Each such point will be determined from the relations

$$x_1^{(k)} = 0 + r_1(60)$$

$$x_2^{(k)} = 0 + r_2(70)$$

where $r_1$ and $r_2$ are random numbers uniformly distributed on the interval 0 to 1.

Suppose the system pseudo-random number routine yields $r_1 = 0.5$ and $r_2 = 0.9$. Thus,

$$x^{(2)} = (30, 63)$$

Since this point is feasible, it will be retained as the second vertex.
Suppose the next pair of samplings of the random number generator yield

$$r_1 = 0.1 \quad \text{and} \quad r_2 = 0.15$$

Then,

$$x^{(3)} = (6, 10.5)$$

As shown in Figure 7.5, this point is infeasible. To render it feasible, retract it, toward the centroid of the remaining two feasible vertices $x^{(1)}$ and $x^{(2)}$. Since

$$\bar{x} = \tfrac{1}{2}(x^{(1)} + x^{(2)}) = (40, 56.5)$$

we have

$$x^{(3)} = \tfrac{1}{2}(x^{(3)} + \bar{x}) = (23, 33.5)$$

This point is feasible and is thus retained.

Finally, suppose we obtain $r_1 = 0.9$ and $r_2 = 0.33$, then $x^{(4)} = (54, 23.1)$, which is also a feasible point. Having identified four initial feasible vertex points, we next evaluate the objective function at these points and proceed with the first normal complex iteration.

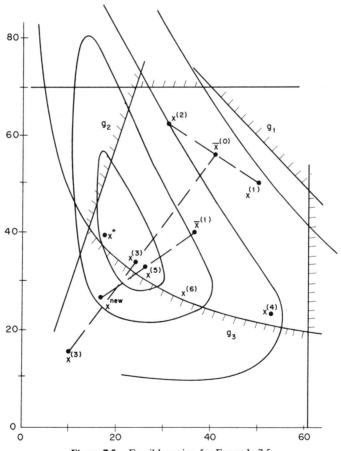

**Figure 7.5.** Feasible region for Example 7.5.

| Index | $x_1$ | $x_2$ | $f(x)$ |
|-------|-------|-------|--------|
| 1 | 50 | 50 | $103.8 \times 10^3$ |
| 2 | 30 | 63 | $87.94 \times 10^3$ |
| 3 | 23 | 33.5 | $72.64 \times 10^3$ |
| 4 | 54 | 23.1 | $87.87 \times 10^3$ |

From the tabulation, it is evident that $x^{(1)}$ is the worst point and should be reflected out.

The new candidate point obtained with step size $\alpha = 1.3$ will be

$$x^{(5)} = \bar{x} + 1.3(\bar{x} - x^{(1)}) = \begin{pmatrix} 35.67 \\ 39.87 \end{pmatrix} + 1.3\left\{ \begin{pmatrix} 35.67 \\ 39.87 \end{pmatrix} - \begin{pmatrix} 50 \\ 50 \end{pmatrix} \right\}$$

$$= (17.03, 26.69)$$

However, at $x^{(5)}$, $g_3$ is violated. The point is retracted toward the centroid, to obtain

$$x^{(5)} = \tfrac{1}{2}(x^{(5)} + \bar{x}) = (26.35, 33.28)$$

a feasible point with objective function value

$$f(x^{(5)}) = 73.37 \times 10^3$$

which is lower than the value at $x^{(1)}$.

The current set of vertices consists of $x^{(2)}$, $x^{(3)}$, $x^{(4)}$, and $x^{(5)}$. The worst of these is $x^{(2)}$, with the function value $87.94 \times 10^3$. The centroid of the remaining points is $\bar{x} = (34.45, 29.96)$, and the new point obtained after two successive retractions toward the centroid is $x^{(6)} = (34.81, 27.28)$, with $f(x^{(6)}) = 77.47 \times 10^3$. Observe, from Figure 7.5, that the current vertices $x^{(3)}$, $x^{(4)}$, $x^{(5)}$, and $x^{(6)}$ are all more or less flattened along constraint $g_3$. Subsequent iterations tend to move the search closer to the optimum solution $x^* = (20, 40)$. However, as evident from the iteration summary given in Table 7.1, the vertices tend to remain flattened against the constraint, and the search slows down considerably.

### 7.2.3  Discussion

While the complex method does not require continuity of the problem functions, since it makes no use of function value differences, strictly speaking it does require that the feasible region be a convex set (see Appendix A for definition). This requirement arises in two places: in the calculation of the centroid [steps 1(b) and 2(b)] and in the retraction of a feasible but unsatisfactory point. In the former case we assume that the centroid, that is, the convex combination of a set of feasible points, will itself be a feasible point. As shown

## Table 7.1  Iteration Sequence for Example 7.5

| Index | $x_1$ | $x_2$ | $f(x)$, $\times 10^{-3}$ | Vertex Replaced |
|-------|-------|-------|--------------------------|-----------------|
| 7  | 23.84  | 32.69  | 72.880 | 4  |
| 8  | 17.63  | 36.978 | 72.747 | 6  |
| 9  | 19.91  | 34.75  | 72.491 | 5  |
| 10 | 18.99  | 35.85  | 72.478 | 7  |
| 11 | 21.61  | 33.96  | 72.497 | 8  |
| 12 | 19.25  | 35.73  | 72.437 | 3  |
| 13 | 19.02  | 35.77  | 72.484 | 11 |
| 14 | 18.80  | 36.11  | 72.483 | 9  |
| 15 | 18.94  | 35.94  | 72.475 | 13 |
| 16 | 19.245 | 35.62  | 72.459 | 14 |
| 17 | 19.476 | 35.46  | 72.437 | 10 |
| 18 | 20.021 | 35.44  | 72.349 | 15 |
| 19 | 20.021 | 35.44  | 72.349 | 16 |
| 20 | 20.842 | 35.08  | 72.331 | 12 |
| 21 | 21.359 | 35.14  | 72.309 | 17 |
| 22 | 21.603 | 35.08  | 72.321 | 18 |
| 23 | 21.471 | 35.04  | 72.323 | 19 |
| 24 | 21.684 | 35.105 | 72.319 | 20 |
| 25 | 21.650 | 35.197 | 72.206 | 23 |
| 26 | 21.625 | 35.299 | 72.293 | 22 |

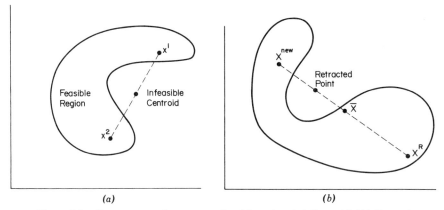

(a)                    (b)

**Figure 7.6.**  Consequence of nonconvex feasible region. ($a$) Centroid. ($b$) Retraction.

in Figure 7.6a, this need not be the case for a nonconvex feasible region. Similarly, with a nonconvex region, a retracted point need not be feasible even if the centroid is feasible (see Fig. 7.6b). Consequently, for nonconvex regions the method could fail to converge satisfactorily. In practice, the method is widely used and has successfully solved numerous nonconvex problems; thus, the above situations must arise with low probability.

As shown in Example 7.5, the complex method can slow down considerably once the trial points distribute along a constraint. In fact, in Example 7.5, with any reasonable termination criteria the algorithm might have terminated well short of the optimum. This behavior is not at all atypical, and consequently many implementations of this method will allow several terminations and subsequent restarts to occur before actually terminating the search. Thus, for the first several times the termination criterion is met the current best point is saved, and the entire search is restarted from this point. The search is terminated only after a specified number of restarts. This device seems to improve the overall convergence rate of the algorithm. Of course, neither this nor any other heuristic construction can guarantee that the method will always converge to a local minimum. Although the method generally is slow, it will reliably yield some improvement over the supplied starting point and thus has become widely used in engineering applications of modest variable dimensions.

As reviewed by Swann [reference 6, Chapter 7], numerous authors have presented enhancements to Box's basic algorithm. These have included more elaborate expansion/contraction rules, use of weighted centroids, modifications to accommodate infeasible centroids, simplex generation constructions that do not use random sampling, and use of quadratic interpolation constructions [4]. In general, these enhancements of the rudimentary algorithm have not resulted in significant improvement. If a more sophisticated algorithm is appropriate or needed, then the methods of Chapters 9 or 10 are clearly to be preferred to an elaborate heuristic method.

## 7.3 RANDOM SEARCH METHODS

Given that random sampling procedures have been applied to locate feasible starting points and to initiate the complex method, it is reasonable to consider the development of direct-search procedures based wholly or largely on random sampling. In fact, since the work reported by Brooks [9] in 1958, proposals employing random point-generation strategies have proliferated in the engineering literature. These proposals range from simple simultaneous sampling techniques to more elaborate procedures that involve a coupling of sequential sampling with heuristic hill-climbing methods. In this section we consider briefly several examples of search methods based on random sampling. We find that while sampling techniques can be effective in locating the vicinity of the optimum, they are quite inefficient if a closer estimate of the solution is required.

### 7.3.1 Direct Sampling Procedures

The most elementary form of sampling procedure is one in which trial points are generated via Eq. (7.1). Each generated trial point is tested by evaluating the inequality constraints, and if it is found to be feasible its objective value is compared to the best value currently available. If the current point yields a better value, it is retained; if not, it is rejected. The process continues until a specified number of points have been generated or the allocated CPU time has been exhausted. This procedure is referred to as *simultaneous* search and is known to be quite inefficient in its use of function evaluations. An analysis given by Spang [10] suggests that in order to attain 90% confidence that the range of uncertainty for each variable $x_1$ is reduced from $x_1^{(U)} - x_1^{(L)}$ to some specified value $\varepsilon_i(x_i^{(U)} - x_i^{(L)})$ $(0 < \varepsilon_i < 1)$, a simultaneous random search will require on the order of

$$2.3 \prod_{i=1}^{N} \varepsilon_i^{-1}$$

trial values. If $N = 5$ and $\varepsilon_i = 0.01$, this estimate indicates that the number of trial points required will be about $2.3 \times 10^{10}$. A purely simultaneous search is thus clearly unsatisfactory.

A somewhat more efficient approach is to divide the sampling into a series of simultaneous blocks. The best point of each block is used to initiate the next block, which is executed with a reduced variable sampling range. In the variant of this strategy proposed in reference 11, the user selects the number of points per block $P$ and the number of blocks $Q$ and proceeds as outlined below.

**Sampling with Interval Reduction.** Given an initial feasible solution $x^{(0)}$, an initial range estimate $z^{(0)}$, where $(x^0 - \frac{1}{2}z^{(0)}) \leq x < (x^{(0)} + \frac{1}{2}z^{(0)})$, and a range reduction factor $\varepsilon$, $0 < \varepsilon < 1$. For each of $Q$ blocks, and initially with $q = 1$, we execute the following steps:

**Step 1.** For $i = 1,\ldots, N$, sample the uniform distribution on $(-\frac{1}{2}, \frac{1}{2})$ and calculate $x_i^p = x_i^{q-1} + rz_i^{q-1}$.

**Step 2.** If $x^p$ is infeasible and $p < P$, repeat step 1. If $x^p$ is feasible, save $x^p$ and $f(x^p)$, and, if $p < P$, repeat step 1. If $p = P$, then set $x^q$ to be the point that has the lowest $f(x^p)$ over all feasible $x^p$.

**Step 3.** Reduce the range via $z_i^q = (1 - \varepsilon)z_i^{q-1}$.

**Step 4.** If $q > Q$, terminate the search. Otherwise increment $q$ and continue with step 1.

The authors suggest a value of $\varepsilon = 0.05$ and blocks of 100 points each. The number of blocks is related to the desired reduction in the variable uncertainty.

With 200 blocks, the initial range will be reduced by

$$(1 - 0.05)^{200} \simeq 3.5 \times 10^{-5}$$

at the cost of $2 \times 10^4$ trial evaluations. For simple low-dimensionality problems in which each evaluation of the set of constraints and the objective function may require only $10^{-3}$ s or less CPU time, the total computational burden may well be acceptable. For more realistic models—for instance, those requiring execution of a simulation—this type of sampling is prohibitively expensive.

### Example 7.6  Fuel Allocation in Power Plants [11]

Each of two electric power generators can be fired with either fuel oil or fuel gas or any combination of the two. The combined power that must be produced by the generators is 50 MW. The fuel gas availability is limited to 10 units/ph. It is desired to select the fuel mix for each generator so as to minimize the fuel oil utilization.

From the curve fitting of operating data, the fuel requirements of generator 1 to attain a power output of $x_1$ MW can be expressed as

$$w_1 = 1.4609 + 0.15186x_1 + 0.00145x_1^2 \tag{7.2}$$

$$w_2 = 1.5742 + 0.1631x_1 + 0.001358x_1^2 \tag{7.3}$$

where $w_1$ and $w_2$ are, respectively, fuel oil and fuel gas requirements in units per hour.

Similarly for generator 2, to attain $x_2$ MW of power, the fuel requirements are

$$y_1 = 0.8008 + 0.2031x_2 + 0.000916x_2^2 \tag{7.4}$$

$$y_2 = 0.7266 + 0.2256x_2 + 0.000778x_2^2 \tag{7.5}$$

where $y_1$ and $y_2$ are, respectively, fuel oil and fuel gas requirements in units per hour. The operating range of generator 1 is restricted to

$$18 \leqslant x_1 \leqslant 30 \tag{7.6}$$

and that of generator 2 to

$$14 \leqslant x_2 \leqslant 25 \tag{7.7}$$

Suppose the fuels can be combined in an additive fashion. That is, for a given value of power output $x_1^*$, any linear combination of fuel-utilization rates

$$\lambda_1 w_1(x_1^*) + (1 - \lambda_1)w_2(x_1^*)$$

where $0 \leqslant \lambda_1 \leqslant 1$, will also produce $x_1^*$ of power. A similar assumption holds for the second generator.

The problem thus can be posed as follows: Determine the output rates of each generator $x_1$ and $x_2$ and the fuel-mix fractions $\lambda_1$ and $\lambda_2$ so as to minimize the total oil consumption:

$$f = \lambda_1 w_1(x_1) + \lambda_2 y_1(x_1)$$

subject to the constraints on the availability of gas,

$$(1 - \lambda_1)w_2(x_1) + (1 - \lambda_2)y_2(x_2) \leqslant 10 \qquad (7.8)$$

the total requirement,

$$x_1 + x_2 = 50 \qquad (7.9)$$

and the variable bounds (7.6) and (7.7) as well as

$$0 \leqslant \lambda_1 \leqslant 1 \quad \text{and} \quad 0 \leqslant \lambda_2 \leqslant 1 \qquad (7.10)$$

Equality constraint (7.9) can clearly be used to solve for $x_2$, so that $x_2$ can be eliminated from the formulation. The resulting problem involves three variables, one nonlinear inequality (7.8), and the variable upper and lower bounds (7.10), as well as

$$25 \leqslant x_1 \leqslant 30$$

which results when (7.6), (7.7), and (7.9) are combined.

It is shown in reference 11 that with an initial solution estimate of

$$x_1 = 20 \qquad \lambda_1 = \tfrac{1}{2} \qquad \lambda_2 = \tfrac{1}{2}$$

and initial range estimate

$$z_1 = 20 \qquad z_2 = 1.0 \qquad z_3 = 1.0$$

a solution within 0.1% of the optimum objective function value is obtained in 55 blocks of 100 trial points each. To obtain a solution within 0.01% of the optimum, a total of 86 blocks were required, or *8600 trial points*. The optimum solution obtained is

$$f = 3.05 \text{ fuel oil units/hr}$$

$$x_1 = 30.00 \ (x_2 = 20.00)$$

$$\lambda_1 = 0.00 \quad \text{and} \quad \lambda_2 = 0.58$$

Obviously, the requirement of 5500 trial points to locate the optimum of a problem involving a cubic objective function, one cubic inequality, and three bounded variables is quite excessive. Moreover, the requirement of 3100 additional trial points to improve the solution accuracy by one additional significant figure is unsatisfactory.

While the above heuristic sampling method is based on the direct reduction of the variable range with each fixed block of trials, it is possible to force convergence more effectively by modifying the distribution function of the random variable. For instance, Gaddy and co-workers [12, 13] have advocated the use of the formula

$$x_i^p = \bar{x}_i + z_i(2r - 1)^k$$

where $\bar{x}$ is the current best solution, $z_i$ is the allowable range of variable $i$, and the random variable $r$ is uniformly distributed on the interval 0 to 1.

When $k = 1$, then in effect $x_i^p$ becomes a random variable uniformly distributed on the interval $(\bar{x}_i - z_i, \bar{x}_i + z_i)$. When $k = 3, 5, 7, \ldots$, and so on, the distribution of $x_i^p$ becomes more and more concentrated around $\bar{x}_i$. Thus $k$ can serve as an adaptive parameter whose value will regulate the contraction or expansion of the search region. In reference 12 an adjustment procedure is recommended in which $k$ increased by 2 whenever a specified number of improved points is found and decreased by 2 when no improvement is found after a certain number of trials. An advantage of this approach is that the number of trials between adjustments is flexible, varying with the progress of the search.

Further refinements of this type of search are reported in reference 14, in which new points are generated by sampling from a normal distribution centered at the current best point $\bar{x}$ with the variance of the distribution as the heuristically adjusted parameter. For instance, the variance is reduced by one-third after each improved point and perhaps increased by a factor of 2 after, say, 100 unsuccessful trial points. Empirical evidence given in both references 12 and 14 indicates that these methods are reasonably efficient for solving low-dimensionality problems, provided that accuracies within 1% of the optimum solution value are acceptable. However, the efficiency deteriorates significantly if more accurate solutions are sought. For instance, to increase the solution accuracy by one significant figure, the number of iterations must be tripled [14]. Use of such strategies in optimizing a six-variable chemical process simulation model is reported by Doering and Gaddy [13]. Typically, several hundred trial simulations were required to reach an optimum from fairly close initial solution estimates. On the whole, the body of available evidence indicates that the above simple techniques based on random sampling should not be used when problems of larger dimensions must be solved or more accurate solutions are required.

### 7.3.2   Combined Heuristic Procedures

In order to improve the performance of methods based on random sampling, various authors have proposed the incorporation of more complex adaptive or heuristic rules that serve to inject some hill-climbing moves into the overall search. These methods range from simple randomized descent schemes [11], to more complex adaptive range-reduction methods [15], to rather elaborate partial-enumeration strategies [16]. In this section we briefly consider only two representative strategies. The literature of the various engineering disciplines contains numerous reports of similar types of search schemes, most presented without extensive comparative testing to verify their relative effectiveness (see, for example, the review given in ref. 17). The two variants discussed here were selected primarily because they illustrate a certain type of strategy and not necessarily because they are numerically superior to other similar search methods based on sampling.

In the *adaptive step-size random search* [16], originally posed for unconstrained problems, sampling is used to generate a search direction, while the step length in that direction is determined by the previous history of successes and failures. Specifically, if two successive steps lead to improvements, the step size is increased by a factor $\alpha_s$. If $M$ successive steps do not result in improvement, the step size is reduced by the factor $\alpha_f$. In outline form, and with suitable modifications to accommodate constraints, the algorithm proceeds as follows.

**Adaptive Step-Size Random Search.**   Given parameters $\alpha_s$, $\alpha_f$, and $M$ and an initial feasible point $x^0$ set the initial step size $\alpha$ to 1.0 and the failure counter $m$ to zero.

**Step 1.**   Generate a random-direction vector $d$ of unit length, and calculate
$$x^{(1)} = x^{(0)} + \alpha d$$

**Step 2.**   If $x^{(1)}$ is feasible and $f(x^{(1)}) < f(x^{(0)})$, set $y = x^{(0)} + \alpha_s(x^{(1)} - x^{(0)})$ and go to step 3. Otherwise, set $m = m + 1$ and go to step 4.

**Step 3.**   If $y$ is feasible and $f(y) < f(x^{(0)})$, then set $\alpha = \alpha_s \alpha$, $x = y$ and go to step 5. Otherwise set $x^{(0)} = x^{(1)}$ and go to step 1.

**Step 4.**   If $m > M$, set $\alpha = \alpha_f \alpha$, $m = 0$, and go to step 5. Otherwise, continue with step 5.

**Step 5.**   Check for termination. If termination rule is not met, go to step 1.

Recommended values of the parameters of the method [16] are $\alpha_s = 1.618$, $\alpha_f = 0.618$, and $M = 3N$, where $N$ is the problem dimension.

A numerical investigation of several random searches [18], including the above, which used only unconstrained test problems indicated that the adaptive step-size random-search procedure was quite effective in reducing the objective function during the initial search phases with problems up to 10 variables. For more precise solutions, the average linear convergence rate was rather slow. Although comparisons with the strategies of references 11, 12, and 14 are not available, the more sequential nature of the adaptive direction-based strategy would appear to make it preferable.

By way of contrast, the *combinatorial heuristic method* [19] advanced for the solution of complex optimal mechanism-design problems takes the approach of discretizing the range of the independent variables and then carrying out a randomized search over the resulting finite grid of possible solutions. The discretized grid of solutions is constructed by simply dividing the expected range of each independent variable into a suitable number of fixed variable values. The search over this fixed grid of variable values is based on a semirandomized one-variable-at-a-time strategy. The strategy is given here in outline form,

## Combinatorial Heuristic Method

**Step 1.** Generate a random feasible starting point $x^0$ and set $F_{min} = f(x^0)$. For each variable $i$, $i = 1, 2, \ldots, N$, execute the following loop of steps.

**Step 2.** Optimize the $i$th variable, holding the others fixed.

(a) Conduct a randomized selection of values of the variable $i$ to determine $q$ additional feasible values that yield a better objective function value than the current base point. If none can be found, reinitiate step 2 with variable $i + 1$.

(b) Identify the best of the $q$ feasible solutions and set its objective function value to $T_{min}$.

(c) Conduct a look-ahead search:

(i) For each of the $q$ feasible solutions identified in step 2(a), conduct a randomized selection of $q$ values of the $(i + 1)$st variable to determine the feasible value of that variable that yields an objective function improvement over $T_{min}$.

(ii) Select the best of these $q$ feasible points. The value of the $i$th variable of that point is fixed as being optimum.

(d) If $i = N$, go to step 3. Otherwise, go to step 2(a) and consider the $(i + 1)$st variable.

**Step 3.** Perform a randomized search to identify an improved value of the $N$th variable, holding all other variables at their current base-point values. The result becomes the new base point, and its function value is the new $F_{min}$.

**Step 4.** Check for termination. If the termination criteria are not met, re-initiate step 2 with $i = 1$.

Several termination criteria could be employed in step 4, including the normal one of insufficient improvement in the objective function value. The authors of the method suggest that $q$, the number of look-ahead points, should be between 3 and 5.

The reader should note that once the random sampling to determine the starting point is completed, the remaining sampling involves only the selection of discrete values of a given variable. The sampling in this case only amounts to an unbiased way of choosing which discrete value to test next and is carried out so as to avoid an exhaustive search over all discrete values of the variable in question. The normal technique for carrying out this type of randomized selection is to assign an equal subinterval of the unit interval to each discrete value of the variable, to sample the uniform distribution on the interval $(0, 1)$, and then to select the discrete variable value based on the subinterval within which the sampled random variable value is located. Further enhancements of the combinatorial heuristic method reported by Datseris [20] include variable reordering, more elaborate look-ahead procedures, and variable reduction strategies.

The proponents of the combinatorial heuristic method have reported on the application of this technique to several mechanism-design problems, including the design of a variable-speed drive, a recording mechanism, a twin crank drive, and a two-mesh gear train [20, 21]. In these cases the method yield improved "optimal" solutions over designs previously reported in the literature. However, although some limited evidence is available that this technique is more efficient than a direct simultaneous random search, extensive comparative results on the basis of which an evaluation could be made are not available.

### 7.3.3 Discussion

The available fragmentary evidence indicates that random-sampling-based search methods ought to be used either with low-dimensionality problems or as devices for generating good starting points for more sophisticated optimization methods. It is true, however, that any sampling strategy that covers most of the feasible region inherently offers enhanced possibilities of locating the vicinity of the global optimum. In fact, research is under way in the unconstrained case to exploit this feature in solving multimodal optimization problems [22–24]. However, this work also indicates that sampling methods should be used only to identify the approximate neighborhood of the minimum and that theoretically based methods should be employed to refine this estimate.

Of the random search methods described in this section, those using some adaptive heuristics are likely to be more effective because they minimize the use of simultaneous samples. In general, a sequential sampling procedure will be more efficient than a simultaneous sampling procedure, because it can make use of the results of previous test points. Moreover, sampling procedures such as that of Luus and Jaakola [11], which successively force a reduction of the feasible region, must be used with some caution to ensure that the reduction heuristic does not eliminate the optimal solution itself. It is quite possible that many or most of the trial points employed per block can be infeasible, in which case the base point for the next block may be selected from very few feasible points and thus may direct the search toward suboptimal regions. With reduction-based strategies, the number of feasible points found per block must be monitored and the total number of trials per block increased if the count of the former is low.

Although systematic evidence is not available, the general impression gained from the literature is that random search methods can be quite effective for severely nonlinear problems that involve multiple local minima but are of low dimensionality. However, for problems of more dimensions with many inequality constraints, the methods of the next three chapters generally are much more efficient.

## 7.4  SUMMARY

This chapter has focused on techniques for locating constrained optima that rely only upon the use of objective and constraint function values. Because of the difficulty of directly satisfying nonlinear equality constraints, this class of methods requires that the problem be reformulated so that equality constraints are explicitly or implicitly eliminated. Moreover, these methods normally require that a feasible starting point be available. We found that random sampling was one possible way of generating such points. The methods themselves can be divided into adaptations of unconstrained direct-search methods and strategies based on random search procedures. We found that the algorithms that employ searches along a fixed set of directions invariably encounter situations that lead to premature termination. Thus, such methods necessarily require modifications that redirect the set of directions to take the constraint normals into account. The extension of the simplex search to accommodate constraints, on the other hand, requires only minor revisions, but, theoretically, these extensions could fail if the feasible region is not convex.

Random search procedures range from direct sampling and rejection methods to more complex strategies that employ sampling together with various heuristic or adaptive devices. We found that sequential sampling is always preferable to simultaneous sampling and that with the sequential procedures some mechanism of range reduction is quite expedient. In general, simple random-sampling-based techniques are best applied to low-dimensionality problems and in applications in which highly accurate solutions are not required. The chapter concluded with a brief examination of combined sampling and heuristic procedures. The adaptive step-size random search uses sampling to define a search direction but heuristics to adjust the step size. The combinatorial heuristic method is based on a discretization of the independent variable ranges and a randomized search of the resulting grid of trial points. Available fragmentary evidence suggests that such more elaborate sampling procedures can be quite effective for low-dimensionality, severely nonlinear problems. Random-search-based methods are probably best used to generate good feasible starting points and to identify the vicinity of global solutions for problems with multiple local optima.

# REFERENCES

1. Wait, R. *The Numerical Solution of Algebraic Equations*, Wiley, New York, 1979.
2. Westerberg, A. W., H. P. Hutchinson, R. L. Motard, and P. Winter, *Process Flowsheeting*, Cambridge University Press, Cambridge, England, 1979.
3. Sargent, R. W. H., "A Review of Methods for Solving Nonlinear Algebraic Equations," Proceedings Conference in *Foundations of Computer Aided Process Design*, (R. S. H. Mah and W. D. Seider, eds.) Vol. 1, Engineering Foundation, New York, pp. 27–76, 1981.
4. Box, M. J., D. Davies, and W. H. Swann, *Nonlinear Optimization Techniques*, ICI Monograph No. 5, Oliver & Boyd, London, 1969.
5. Klingman, W. R., and D. M. Himmelblau, "Nonlinear Programming with the Aid of a Multiple Gradient Summation Technique," *JACM*, **11**, 400–415 (1964).
6. Gill, P. E., and W. Murray (Eds.), *Numerical Methods for Constrained Optimization*, Academic Press, London, 1974.
7. Box, M. J., "A New Method of Constrained Optimization and a Comparison with Other Methods," *Computer J.*, **8**, 42–52 (1965).
8. Biles, W. E., "Optimization of Multiple Response Simulation Models," University of Notre Dame, Final Report, ONR-Contract N00014-76-C-1021, Oct. 31, 1978.
9. Brooks, S. H., "A Discussion of Random Methods for Seeking Maxima," *Oper. Res.*, **6**, 244–253 (1958).
10. Spang, H. A., "A Review of Minimization Techniques for Nonlinear Function," *SIAM Rev.*, **4**, 343–365 (1962).
11. Luus, R., and T. H. I. Jaakola, "Optimization by Direct Search and Systematic Reduction of the Size of Search Region," *AIChE J.*, **19**, 760–766 (1973).
12. Heuckroth, M. W., J. L. Gaddy, and L. D. Gaines, "An Examination of the Adaptive Random Search Technique," *AIChE J.*, **22**, 744–750 (1976).
13. Doering, F. J., and J. L. Gaddy, "Optimization of the Sulfuric Acid Process with a Flowsheet Simulator," *Computers Chem. Eng.*, **4**, 113–122 (1980).
14. Goulcher, R., and J. J. Casares Long, "The Solution of Steady State Chemical Engineering Optimization Problems using a Random Search Algorithm," *Computers Chem. Eng.*, **2**, 33–36 (1978).
15. Martin, D. L., and J. L. Gaddy, "Process Optimization with the Adaptive Randomly Directed Search," in *Selected Topics in Computer Aided Process Design and Analysis*, (R. S. H. Mah and G. V. Reklaitis, Eds.) CEP Symp. Ser., Vol. 214, American Institute of Chemical Engineers, New York, 1982.
16. Schumer, M. A., and K. Steiglitz, "Adaptive Step Size Random Search," *IEEE Trans.* **AC-13**, 270–276 (1968).
17. Jarvis, R. A., "Optimization Strategies in Adaptive Control: A Selective Survey," *IEEE Trans. Syst., Man, and Cybern.* **5**, 83–94 (1975).
18. Schrack, G., and N. Borowski, "An Experimental Comparison of Three Random Searches," in *Numerical Methods for Nonlinear Optimization* (F. A. Lootsma, Ed), Academic Press, New York, 1972.

19. Lee, T. W., and F. Freudenstein, "Heuristic Combinatorial Optimization in the Kinematic Design of Mechanisms: Part 1: Theory," *J. Eng. Ind. Trans. ASME*, 1277–1280 (1976).

20. Datseris, P., "Principles of a Heuristic Optimization Technique as an Aid in Design—An Overview," *ASME Tech. Paper* **80-DET-44**, Sept. 1980.

21. Lee, T. W., and R. Freudenstein, "Heuristic Combinatorial Optimization in the Kinematic Design of Mechanisms, Part 2: Applications," *J. Eng. Ind. Trans. ASME*, 1281–1284 (1976).

22. Mockus, J., "On Baysian Methods of Optimization," in *Towards Global Optimization*, (L. C. W. Dixon and G. P. Szego, Eds.), North-Holland, Amsterdam, 1975.

23. Zilinskas, A., "On Statistical Models for Multi-Modal Optimization," *Math. Operationsforsh. Stat. Ser. Stat.*, **9**, 255–266 (1978).

24. Devroye, L. P., "Progressive Global Random Search of Continuous Functions," *Math Programming*, **15**, 330–342 (1978).

# PROBLEMS

**7.1.** Why must equality constraints be eliminated before applying direct-search optimization methods?

**7.2.** Why might a penalty function construction be used to help generate a feasible starting point for an inequality-constrained problem?

**7.3.** What is the reason for using more than $N + 1$ points in the complex search method?

**7.4.** Why is convexity necessary to guarantee the successful operation of the complex method?

**7.5.** Why is restarting used in the context of the complex method?

**7.6.** In random sampling with interval reduction in blocks, what are the competing factors that must be considered in the choice of block size?

**7.7.** In interval-reduction sampling procedures, is it possible to discard the minimum solution during the reduction process? If so, how can this occur, and how can it be avoided?

**7.8.** Which is the adaptive component of the adaptive step-size random search procedure?

**7.9.** Show that the constraint $g_1(x) = x_1^2 - x_2 \leqslant 0$ is convex. Suppose the constraint is rewritten as an equality by adding a slack variable. Is the constraint set

$$g_1(x) = x_1^2 - x_2 + x_3 = 0$$

$$g_2(x) = x_3 \geqslant 0$$

convex?

**7.10.** (a) Is the set defined by the constraints

$$g_1(x) = x_1 + x_2 - 1 = 0$$

$$g_2(x) = 2 - x_1^2 - x_2^2 \geqslant 0$$

convex?

(b) Is the set defined by the constraints

$$g_1(x) = x_1 - 1 \leqslant 0$$

$$g_2(x) = x_1 + x_2^2 - 2 \geqslant 0$$

convex?

**7.11.** Is the feasible region

$$x_1 + x_2 + x_3 = 1$$

$$x_1^2 + x_2^2 \leqslant 1$$

$$x_3 + x_2^2 \geqslant 1$$

a convex region? Prove your answer.

**7.12.** Given the nonlinear program

$$\text{Minimize} \quad f(x) = x_1^2 + x_2^2 + x_3^2$$

$$\text{Subject to} \quad g_1(x) = 1 - x_2^{-1}x_3 \geqslant 0$$

$$g_2(x) = x_1 - x_3 \geqslant 0$$

$$h_1(x) = x_1 - x_2^2 + x_2x_3 - 4 = 0$$

$$0 \leqslant x_1 \leqslant 5$$

$$0 \leqslant x_2 \leqslant 3$$

$$0 \leqslant x_3 \leqslant 3$$

What transformations are necessary in order to use the complex method? Give the final transformed form.

**7.13.** Given the problem

$$\text{Minimize} \quad f(x) = -x_1x_2^2x_3^3/81$$

$$\text{Subject to} \quad h_1(x) = x_1^3 + x_2^2 + x_3 - 13 = 0$$

$$h_2(x) = x_2^2x_3^{-1/2} - 1 = 0$$

$$\text{All} \quad x_i \geqslant 0$$

(a) Reformulate the problem for solution using the complex method. Is the resulting feasible region convex?

(b) Estimate upper and lower bounds on the independent variables for use in the complex method.

(c) Find a starting feasible solution.

**7.14.** In solving the problem

$$\text{Minimize} \quad 3(x_2 - 4)^2 - 2x_1$$

$$\text{Subject to} \quad g_1(x) = 10 - x_1^2 - x_2^2 \geqslant 0$$

$$g_2(x) = 9 - x_1^2 - (x_2 - 4)^2 \geqslant$$

$$0 \leqslant (x_1, x_2) \leqslant 4$$

the following points are generated by the complex method:

|   | $x_1$ | $x_2$ | $f(x)$ |
|---|---|---|---|
| 1 | 2.0 | 2.0 | 8 |
| 2 | 0.5 | 3.0 | 2 |
| 3 | 0.6 | 1.1 | 24.03 |
| 4 | 2.3 | 2.1 | 6.23 |

Assuming $\alpha = 1.3$, carry out two iterations of the complex method. Construct a plot of the results.

**7.15.** Given the problem

$$\text{Minimize} \quad f(x) = 3x_1^2 - 2x_2$$

$$\text{Subject to} \quad g_1(x) = 2x_1 + x_2 \geqslant 4$$

$$g_2(x) = x_1^2 + x_2^2 \leqslant 40$$

(a) Estimate upper and lower bounds on the variables, and obtain a starting feasible solution.

(b) Write a program to carry out sampling with interval reduction, and use it to solve the problem. Use $\varepsilon = 0.05$ and 100 blocks of 50 points each.

**7.16.** Solve the problem

$$\text{Minimize} \quad f(x) = (x_1 - 1)^2 + (x_1 - x_2)^2 + (x_2 - x_3)^2$$

$$\text{Subject to} \quad h_1(x) = x_1(1 + x_2^2) + x_2^4 - 4 - 3\sqrt{2} = 0$$

$$\text{All} \quad x_i, 0 \leqslant x_i \leqslant 3$$

using a computer implementation of the complex method. [$f^* = 0.032567$; see Miele et al., *J. Optim. Theory Appl.*, **10**, 1 (1972).]

**7.17.** Solve the problem

$$\text{Minimize} \quad f(x) = (x_1 - 1)^2 + (x_1 - x_2)^2 + (x_3 - 1)^2$$

$$+ (x_4 - 1)^4 + (x_5 - 1)^6$$

$$\text{Subject to} \quad h_1(x) = x_1^2 x_4 + \sin(x_4 - x_5) - 2\sqrt{2} = 0$$

$$h_2(x) = x_2 + x_3^4 x_4^2 - 8 - \sqrt{2} = 0$$

using an adaptive random search procedure. [$f^* = 0.24149$; see Miele et al., *J. Optim. Theory Appl.*, **10**, 1 (1972).]

**7.18.** The minimum-weight design of the corrugated bulkheads for a tanker requires the solution of the following optimization problem (Kvalie et al., *Tech. Rep.*, Norwegian Technical Institute, Trondheim, Norway, 1966).

$$\text{Minimize} \quad f(x) = \frac{5.885 x_4 (x_1 + x_3)}{x_1 + (x_3^2 - x_2^2)^{1/2}}$$

$$\text{Subject to} \quad g_1(x) = x_2 x_4 (0.4 x_1 + \tfrac{1}{6} x_3)$$

$$- 8.94[x_1 + (x_3^2 - x_2^2)^{1/2}] \geqslant 0$$

$$g_2(x) = x_2^2 x_4 (0.2 x_1 + \tfrac{1}{12} x_3)$$

$$- 2.2\{8.94[x_1 + (x_3^2 - x_2^2)^{1/2}]\}^{4/3} \geqslant 0$$

$$g_3(x) = x_4 - 0.0156 x_1 - 0.15 \geqslant 0$$

$$g_4(x) = x_4 - 0.0156 x_3 - 0.15 \geqslant 0$$

$$g_5(x) = x_4 - 1.05 \geqslant 0$$

$$g_6(x) = x_3 - x_2 \geqslant 0$$

The variables $x$ correspond to the following corrugation dimension: $x_1$ width, $x_2$ depth, $x_3$ length, and $x_4$ plate thickness.

(a) Obtain a starting feasible solution be sampling, using the bounds

$$0 \leqslant x_1 \leqslant 100, \quad 0 \leqslant x_2 \leqslant 100, \quad 0 \leqslant x_3 \leqslant 100, \quad \text{and} \quad 0 \leqslant x_4 \leqslant 5$$

(b) Solve the problem using a random search method ($f^* = 6.84241$).

# Chapter 8

# Linearization Methods
# for Constrained Problems

In the previous chapters we considered in some detail the two problem classes that optimization theory has treated most thoroughly: the unconstrained problem and the completely linear constrained problem. We found that for these two special cases very efficient algorithms can be devised that have a high degree of reliability and for which a rather complete theoretical analysis can be presented. In view of these successes, it should not be surprising that most approaches to the general constrained nonlinear problem have involved devices by means of which the general problem can be solved by exploiting the techniques developed for these two special cases. As discussed in Chapter 6, one successful approach to solving the general problem has been to reformulate it as a parameterized unconstrained problem. In this way, the constrained problem can be solved as a sequence of unconstrained problems using well-developed existing algorithms. In this and the next chapter we consider the use of a second device, linearization, by means of which the general problem can be converted to a constrained but linear problem. The use of linearization approximations allows us to solve the general problem by employing linear programming (LP) methodology either by solving one or more linear programs or else by employing elements of the linear simplex methodology in a recursive fashion.

All of the methods discussed in this section employ the fact that a general nonlinear function $f(x)$ can be approximated in the vicinity of a point $x^0$ by Taylor's expansion,

$$f(x) = f(x^0) + \nabla f(x^0)(x - x^0) + O(\|x - x^0\|)^2$$

Almost without exception, the higher order terms $O(\|x - x^0\|^2)$ are ignored, and hence the function $f(x)$ is approximated by the *linearization* of $f$ at $x^0$ denoted by

$$\tilde{f}(x; x^0) = f(x^0) + \nabla f(x^0)(x - x^0)$$

We will call the point $x^0$ the *linearization point*. It should be recognized that the linearization is in most instances a very gross approximation, which must be used with great caution. Nonetheless, it is an approximation that is very commonly used not only in optimization but in virtually all of engineering. The representative optimization techniques discussed in this chapter differ primarily in the manner and frequency with which the linearizations are updated and in the devices that are employed to account for the errors introduced by employing such approximations.

## 8.1  DIRECT USE OF SUCCESSIVE LINEAR PROGRAMS

The simplest and most direct use of the linearization construction is to replace the general nonlinear problem with a complete linearization of all problem functions at some selected estimate of the solution. Since in this manner all problem functions are replaced by linear approximations, the resulting problem takes the form of a linear program and can be solved as such. The LP solution obtained will no doubt only be an approximation to the solution of the original problem; however, if adequate precautions are taken it ought to be an improvement over the linearization point. We first consider the case of the linearly constrained problem and discuss the supplementary calculations required to make direct linearization effective. Then we consider the general nonlinear programming (NLP) problem.

### 8.1.1  The Linearly Constrained Case

The linearly constrained NLP problem is that of

$$\text{Minimize} \quad f(x)$$

$$\text{Subject to} \quad \mathbf{A}x \leqslant b$$

$$x \geqslant 0$$

This differs from the standard LP problem only in the presence of a nonlinear objective function $f(x)$. As in the LP case, the feasible region remains a polyhedron, that is, a geometric entity formed by the intersection of a set of planes. Because $f(x)$ is nonlinear, the optimal solution no longer needs to be confined to vertices or corner points of the feasible region but can instead lie anywhere within the region. Moreover, if $f(x)$ is nonconvex, the linearly constrained NLP problem may have multiple local minima. Given these problem features, let us consider the consequences of formulating and solving the linearization of the linearly constrained nonlinear program constructed at some feasible point $x^0$.

Clearly, the problem

$$\text{Minimize} \quad \tilde{f}(x; x^0)$$

$$\text{Subject to} \quad \mathbf{A}x \leqslant b$$

$$x \geqslant 0$$

is an LP problem and as such, assuming the feasible region is bounded, will possess an optimal solution that is a feasible corner point. The key question is, How is $\tilde{x}^*$, the solution of the approximate problem, related to $x^*$, the solution of the original problem? First of all, ignoring for the moment the case of an infinite number of LP solutions, we have reduced the possible multiple local minima to a single approximate solution. To which of the local minima of the NLP the point $\tilde{x}^*$ will be closest is totally determined by our choice of linearization point. Hence, in the nonconvex case, we cannot be assured of finding the global minimum. However, even if $f(x)$ is convex, we have no real assurance that $\tilde{x}^*$ will in fact even be close to $x^*$. While $\tilde{x}^*$ will always be a corner point, $x^*$ can lie anywhere; it may even be an interior point. Quite obviously, even if $f(x)$ is convex, some further adjustment of $\tilde{x}^*$ will be required to ensure that we can draw nearer to $x^*$.

Note that both $x^0$ and $\tilde{x}^*$ are feasible points and that by virtue of the minimization of $\tilde{f}(x; x^0)$ it must be true that

$$\tilde{f}(x^0; x^0) > \tilde{f}(\tilde{x}^*, x^0)$$

Consequently, if we substitute, using the linearization formula, it follows that

$$f(x^0) > f(x^0) + \nabla f(x^0)(\tilde{x}^* - x^0)$$

or

$$\nabla f(x^0)(\tilde{x}^* - x^0) < 0$$

Evidently the vector $(\tilde{x}^* - x^0)$ is a *descent* direction. Recall from our discussion of the unconstrained case that while a descent direction is a desirable choice for a search direction, it can lead to an improved point only if it is coupled with a step-adjustment procedure or, better yet, a line search. Note that a line search from $x^0$ in the direction $(\tilde{x}^* - x^0)$ will eventually lead to the point $\tilde{x}^*$. Since $\tilde{x}^*$ is a corner point of the feasible region, and since $x^0$ is feasible, all points on the line between them will be feasible (since the feasible region is convex). Moreover, since $\tilde{x}^*$ is a corner point, points on the line beyond $\tilde{x}^*$ will lie outside the feasible region. The line search can thus be confined to the bounded line segment

$$x = x^0 + \alpha(\tilde{x}^* - x^0) \qquad 0 \leqslant \alpha \leqslant 1$$

Thus while the LP solution does not directly yield an improved estimate of the optimum, it does accomplish two important functions: It defines a search direction, and it defines the point at which the feasible region will be intersected.

As a consequence of solving the problem

$$\min f\left(x^0 + \alpha(\tilde{x}^* - x^0)\right) \qquad 0 \leqslant \alpha \leqslant 1$$

a feasible point $x^1$ will be found with the property

$$f(x^1) < f(x^0)$$

Since $\nabla f(x^1)$ will in general not be equal to the zero vector, the improved point $x^1$ will serve as a linearization point for the next approximating linear program. The sequence of alternate LP solutions followed by line searching must be continued until successive line-search optima $x^{(t)}$ differ by less than some selected convergence tolerance. The composite algorithm, attributed to Frank and Wolfe [1], is summarized as follows.

### Frank-Wolfe Algorithm

Given $x^0$, line search, and overall convergence tolerances $\varepsilon > 0$ and $\delta > 0$.

**Step 1.** Calculate $\nabla f(x^{(t)})$. If $\|\nabla f(x^{(t)})\| \leqslant \varepsilon$, stop. Otherwise, go to step 2.

**Step 2.** Solve the LP subproblem,

$$\text{Minimize} \quad \nabla f(x^{(t)}) y$$

$$\text{Subject to} \quad \mathbf{A} y \leqslant b$$

$$y \geqslant 0$$

Let $y^{(t)}$ be the optimal solution to the LP problem.

**Step 3.** Find $\alpha^{(t)}$ which solves the problem

$$\min \quad f\left(x^{(t)} + \alpha(y^{(t)} - x^{(t)})\right)$$

$$0 \leqslant \alpha \leqslant 1$$

**Step 4.** Calculate

$$x^{(t+1)} = x^{(t)} + \alpha^{(t)}(y^{(t)} - x^{(t)})$$

**Step 5.** Convergence check. If

$$\|x^{(t+1)} - x^{(t)}\| < \delta\|x^{(t+1)}\|$$

and if

$$\|f(x^{(t+1)}) - f(x^{(t)})\| \leq \varepsilon\|f(x^{(t+1)})\|$$

then terminate. Otherwise, go to step 1.

Note that in the statement of the LP subproblem we have deleted the constant terms of the linear objective function $\tilde{f}(x; x^{(t)})$ and retained only the variable portion.

### Example 8.1   Multistage Compressor Optimization

A gas flowing at a rate of $N$ moles/hr at 1 atm pressure is to be compressed to 64 atm, using the three-stage compressor shown in Figure 8.1. Assume that compression occurs reversibly and adiabatically and that after each stage of compression the gas is cooled back to its initial temperature $T$. Choose the interstage pressures so as to minimize the energy consumption.

**Solution.**   For reversible adiabatic compression with cooling to inlet temperature $T$, the work is given by

$$W = NRT\left(\frac{k}{k-1}\right)\left(\frac{P_{\text{out}}}{P_{\text{in}}}\right)^{(k-1)/k} - NRT\left(\frac{k}{k-1}\right)$$

where  $k = C_p/C_v$, the ratio of the gas heat capacities
       $R$ = the ideal gas constant

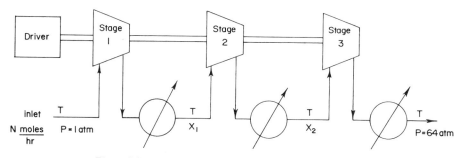

**Figure 8.1.**   Multistage compressor schematic, Example 8.1.

For three-stage compression, the total work is given by

$$W_{total} = NRT\left(\frac{k}{k-1}\right)\left\{\left(\frac{x_1}{1}\right)^\alpha + \left(\frac{x_2}{x_1}\right)^\alpha + \left(\frac{64}{x_2}\right)^\alpha - 3\right\}$$

where  $\alpha = k - 1/k$

$x_1$ = outlet pressure from the first stage

$x_2$ = outlet pressure from the second stage

If for the gas in question $\alpha = \frac{1}{2}$, then for fixed $T$ and $N$ the optimal pressures $x_1$ and $x_2$ will be obtained as the solution of the problem

$$\text{Minimize} \quad f(x) = x_1^{1/4} + \left(\frac{x_2}{x_1}\right)^{1/4} + \left(\frac{64}{x_2}\right)^{1/4}$$

$$\text{Subject to} \quad x_1 \geqslant 1$$

$$x_2 \geqslant x_1$$

$$64 \geqslant x_2$$

where the constraints are imposed to ensure that the gas pressures selected will be monotonically increasing from inlet to outlet.

This is a linearly constrained NLP problem and can be solved using the Frank-Wolfe algorithm. Suppose as our initial estimate we use $x_1^0 = 2$ and $x_2^0 = 10$, a feasible solution. The gradient of $f(x)$ at $x^0$ is

$$\frac{\partial f}{\partial x_1} = 0.25x_1^{-3/4}\left(1 - x_2^{1/4}x_1^{-1/2}\right) = -3.83 \times 10^{-2}$$

$$\frac{\partial f}{\partial x_2} = 0.25x_2^{-3/4}\left[x_1^{-1/4} - (64)^{1/4}x_2^{-1/2}\right] = -2.38 \times 10^{-3}$$

Assuming that these values are not sufficiently close to zero, we continue with step 2: Solve the LP problem

$$\text{Minimize} \quad -3.83 \times 10^{-2}y_1 - 2.38 \times 10^{-3}y_2$$

$$\text{Subject to} \quad y_1 \geqslant 1$$

$$y_2 \geqslant y_1$$

$$64 \geqslant y_2$$

The minimum will clearly occur when $y_1$ and $y_2$ are as large as possible, namely $y_1^0 = y_2^0 = 64$.

**Step 3.** Search the line

$$\begin{pmatrix} x_1 \\ x_2 \end{pmatrix} = \begin{pmatrix} 2 \\ 10 \end{pmatrix} + \alpha\left(\begin{pmatrix} 64 \\ 64 \end{pmatrix} - \begin{pmatrix} 2 \\ 10 \end{pmatrix}\right)$$

for $0 \leqslant \alpha \leqslant 1$. Using, say, an interpolation method, we find that the optimum occurs at $\alpha = 0.02732$.

**Step 4.** The new point is

$$\begin{pmatrix} x_1^{(1)} \\ x_2^{(1)} \end{pmatrix} = \begin{pmatrix} 2 \\ 10 \end{pmatrix} + 0.02732\begin{pmatrix} 62 \\ 54 \end{pmatrix} = \begin{pmatrix} 3.694 \\ 11.475 \end{pmatrix}$$

As shown in Figure 8.2, the point $x^{(1)}$ is an interior point of the feasible region. Since the search clearly has not yet converged, the iterations resume with step 1.

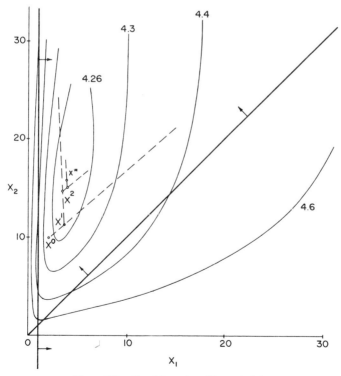

**Figure 8.2.** Feasible region, Example 8.1.

**Step 1**

$$\nabla f(x^{(1)}) = (3.97 \times 10^{-3}, -4.56 \times 10^{-3})$$

**Step 2.**  The LP subproblem becomes

$$\text{Minimize} \quad +3.97 \times 10^{-3}y_1 - 4.56 \times 10^{-3}y_2$$

$$\text{Subject to} \quad y_1 \geqslant 1$$

$$y_2 \geqslant y_1$$

$$64 \geqslant y_2$$

The minimum is attained at the corner point at which $y_1$ is as small and $y_2$ is as large as possible; thus,

$$y_1^{(1)} = 1 \quad \text{and} \quad y_2^{(1)} = 64$$

**Step 3.**  Search the line

$$\begin{pmatrix} x_1 \\ x_2 \end{pmatrix} = \begin{pmatrix} 3.694 \\ 11.475 \end{pmatrix} + \alpha\left(\begin{pmatrix} 1 \\ 64 \end{pmatrix} - \begin{pmatrix} 3.694 \\ 11.475 \end{pmatrix}\right)$$

for $0 \leqslant \alpha \leqslant 1$ such that $f(x(\alpha))$ is minimized. The optimum occurs at $\alpha = 0.06225$.

**Step 4.**  The new point is

$$\begin{pmatrix} x_1^{(2)} \\ x_2^{(2)} \end{pmatrix} = \begin{pmatrix} 3.694 \\ 11.475 \end{pmatrix} + 0.06225 \begin{pmatrix} -2.694 \\ 52.525 \end{pmatrix} = \begin{pmatrix} 3.526 \\ 14.745 \end{pmatrix}$$

From Figure 8.2, this point is also an interior point. Moreover, it is still sufficiently distant from the optimum. The process continues with the points

$$x^{(3)} = (3.924, 15.069)$$

$$x^{(4)} = (3.886, 15.710)$$

and so on, until suitably tight convergence tolerances are met.

From Figure 8.2 it is evident that with an interior optimum, the Frank-Wolfe algorithm advances very similarly to the unconstrained gradient method. In fact, the zigzagging along a limited set of directions is typical of the ordinary

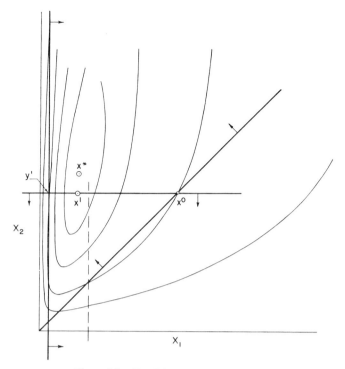

**Figure 8.3.** Feasible region, Example 8.2.

gradient method. If both the starting point and the solution are boundary points, then the search will eventually reduce to a sequence of one-parameter searches on the constraints.

### Example 8.2

Suppose that the compressor system of the previous example is constrained to satisfy the additional condition

$$x_2 \leqslant 14 \text{ atm}$$

Obtain the new optimum using as initial estimate $x^0 = (14, 14)$.

**Solution.** The given initial point is a corner point of the modified feasible region, as shown in Figure 8.3. At $x^0$,

$$\nabla f(x^0) = \left(1.67 \times 10^{-2}, -8.25 \times 10^{-3}\right)$$

The corresponding LP subproblem is

$$\text{Minimize} \quad 1.67 \times 10^{-2} y_1 - 8.25 \times 10^{-3} y_2$$

$$\text{Subject to} \quad y_1 \geqslant 1$$

$$y_2 \geqslant y_1$$

$$14 \geqslant y_2$$

The solution is obtained when $y_1$ is as small and $y_2$ is as large as possible. Thus,

$$y_1^0 = 1 \quad \text{and} \quad y_2^0 = 14$$

Step 3, the line search, involves finding $\alpha$, $0 \leqslant \alpha \leqslant 1$, such that

$$\begin{pmatrix} x_1 \\ x_2 \end{pmatrix} = \begin{pmatrix} 14 \\ 14 \end{pmatrix} + \alpha \left( \begin{pmatrix} 1 \\ 14 \end{pmatrix} - \begin{pmatrix} 14 \\ 14 \end{pmatrix} \right)$$

and $f(x(\alpha))$ is minimized. This amounts to a search on the constraint $y_2 = 14$. The solution is

$$\alpha = 0.7891$$

and the new point is

$$\begin{pmatrix} x_1^{(1)} \\ x_2^{(1)} \end{pmatrix} = \begin{pmatrix} 3.7417 \\ 14.0 \end{pmatrix}$$

At this point, $\nabla f(x^{(1)}) = (0, -1.275 \times 10^{-3})$. This is the optimal solution and, as shown in Figure 8.3, it lies in the middle of the constraint $y_2 = 14$. Note that if the additional constraint $x_1 \geqslant 5$ were imposed (dashed line in Fig. 8.3), then the optimal solution would be the corner point $(5, 14)$.

From the example, it is clear that even for boundary points, the Frank-Wolfe algorithm amounts to a gradient search projected onto the constraint surfaces. Convergence to a Kuhn-Tucker point from any feasible starting point can readily be proved (see Zangwill [2] or Wolfe [3]) under reasonable assumptions: continuously differentiable $f(x)$, bounded feasible region, and accurate LP and line-search subproblem solutions. If $f(x)$ is convex, clearly the solution will be the global minimum. It is awkward to analyze the rate of convergence because of the involvement of the vertex points in the definition of the descent direction. However, it is sufficiently clear that there are no grounds to expect a better rate than that of the ordinary gradient method.

Finally, it is interesting to note that with the Frank-Wolfe algorithms (in fact with any algorithm that uses a subproblem with a linearized objective function), if the objective function is convex, then after each subproblem solution it is possible to obtain a very convenient estimate of the remaining improvement yet to be achieved in the value of the objective function. Recall that for a convex function linearized at any point $x^{(t)}$,

$$f(x) \geqslant f(x^{(t)}) + \nabla f(x^{(t)})(x - x^{(t)}) \equiv \tilde{f}(x; x^{(t)})$$

for all points $x$. But this inequality implies that over all feasible points,

$$\min f(x) \geqslant \min \tilde{f}(x; x^{(t)}) \equiv \tilde{f}(y^{(t)}; x^{(t)})$$

Hence, the value of $\tilde{f}(y^{(t)}; x^{(t)})$ is a lower-bound estimate of the value of $f(x)$ at the minimum point. On the other hand, $f(y^{(t)})$, or better yet $f(x^{(t+1)})$, is an upper-bound estimate of the minimum value of $f(x)$. Thus, after each cycle of the algorithm, the difference

$$f(x^{(t+1)}) - \tilde{f}(y^{(t)}; x^{(t)})$$

gives a conservative estimate of the attainable improvement in the objective function value. This difference can thus be used as a good termination criterion, supplementing those defined in step 5 of the Frank-Wolfe algorithm.

With this basic introduction to the use of direct linearization in solving the special case of linearly constrained NLP problems, we continue with a discussion of the general case.

### 8.1.2   The General Nonlinear Programming Case

Consider the general NLP problem

$$\text{Minimize} \quad f(x)$$

$$\text{Subject to} \quad g_j(x) \geqslant 0 \qquad j = 1,\ldots, J$$

$$h_k(x) = 0 \qquad k = 1,\ldots, K$$

$$x_i^{(U)} \geqslant x_i \geqslant x_i^{(L)} \qquad i = 1,\ldots, N$$

Given an estimate $x^{(t)}$, the direct linear approximation problem that can be constructed at $x^{(t)}$ is

Minimize $\quad f(x^{(t)}) + \nabla f(x^{(t)})(x - x^{(t)})$

Subject to $\quad g_j(x^{(t)}) + \nabla g_j(x^{(t)})(x - x^{(t)}) \geqslant 0 \qquad j = 1, \dots, J$

$\qquad\qquad\quad h_k(x^{(t)}) + \nabla h_k(x^{(t)})(x - x^{(t)}) = 0 \qquad k = 1, \dots, K$

$\qquad\qquad\quad x_i^{(U)} \geqslant x_i \geqslant x_i^{(L)} \qquad\qquad\qquad\qquad i = 1, \dots, N$

Clearly this is an LP problem and as such can be solved using LP methods to give a new point $x^{(t+1)}$. Note, however, that even if $x^{(t)}$ is a feasible point of the original nonlinear problem, there is no assurance that $x^{(t+1)}$ will be feasible. In fact, if the constraint functions are nonlinear, the point $x^{(t+1)}$ will almost assuredly be infeasible. But if $x^{(t+1)}$ is infeasible, then the attainment of an improvement in the value of the approximate objective function, that is,

$$\tilde{f}(x^{(t+1)}; x^{(t)}) < \tilde{f}(x^{(t)}; x^{(t)})$$

is no assurance that an improved estimate of the true optimum solution has been attained. In general, if $\{x^{(t)}\}$ is a series of points each of which is the solution of the LP problem obtained by using the previous LP solution as linearization point, then in order to attain convergence to the true optimum solution it is sufficient that at each point $x^{(t)}$ an improvement be made in both the objective function value and the constraint infeasibility. As shown in the next example, these desirable conditions can occur in direct applications of the successive linear approximation strategy without further embellishments or safeguards.

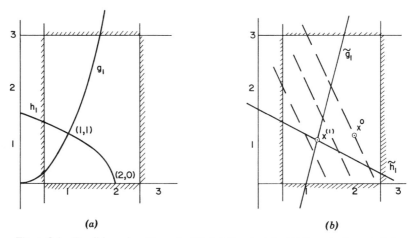

(a)                                                  (b)

**Figure 8.4.** Feasible region, Example 8.3. (a) Constraints; (b) linearized constraints.

*Example 8.3*

Consider the problem

$$\text{Minimize} \quad f(x) = x_1^2 + x_2^2$$

$$\text{Subject to} \quad g_1(x) = x_1^2 - x_2 \geq 0$$

$$h_1(x) = 2 - x_1 - x_2^2 = 0$$

$$\tfrac{5}{2} \geq x_1 \geq \tfrac{1}{2}$$

$$3 \geq x_2 \geq 0$$

As shown in Figure 8.4a, the points satisfying the problem constraints consist of all points on the curve $h_1(x) = 0$ between the point (2, 0) determined by the linear bound $x_2 \geq 0$ and the point (1, 1) determined by constraint $g_1(x) \geq 0$. Suppose we construct the linearized approximation to this problem at the point $x^0 = (2, 1)$. The result is

$$\text{Minimize} \quad \tilde{f}(x; x^0) = 5 + 4(x_1 - 2) + 2(x_2 - 1)$$

$$\text{Subject to} \quad \tilde{g}_1(x; x^0) = 3 + 4(x_1 - 2) - (x_2 - 1) \geq 0$$

$$\tilde{h}_1(x; x^0) = -1 - (x_1 - 2) - 2(x_2 - 1) = 0$$

$$\tfrac{5}{2} \geq x_1 \geq \tfrac{1}{2}$$

$$3 \geq x_2 \geq 0$$

As shown in Figure 8.4b, all points satisfying the linearized constraints will lie on the line $3 - x_1 - 2x_2 = 0$ between the point (2.5, 0.25) determined by the bound $\tfrac{2}{2} \geq x_1$ and the point $(\tfrac{11}{9}, \tfrac{8}{9})$ determined by the intersection with $\tilde{g}_1(x; x^0)$. While the original objective function contours are circles with center at the origin, the contours of the linearized objective functions are straight lines of slope $-2$. Clearly, the solution to the linear approximating problem is $x^{(1)} = (\tfrac{11}{9}, \tfrac{8}{9})$.
At this point,

$$g_1(x^{(1)}) = 0.6049 > 0$$

$$h_1(x^{(1)}) = -0.0123 \neq 0$$

that is, the equality constraint is violated. Consequently, the problem is relinearized at $x^{(1)}$. The result is

Minimize     $\tilde{f}(x; x^{(1)}) = 2.284 + 2.444(x_1 - \frac{11}{9}) + 1.778(x_2 - \frac{8}{9})$

Subject to   $\tilde{g}_1(x; x^{(1)}) = +0.6049 + 2.444(x_1 - \frac{11}{9}) - (x_2 - \frac{8}{9}) \geqslant 0$

$\tilde{h}_1(x; x^{(1)}) = -0.0123 - (x_1 - \frac{11}{9}) - 1.778(x_2 - \frac{8}{9}) = 0$

$\frac{5}{2} \geqslant x_1 \geqslant \frac{1}{2}$

$3 \geqslant x_2 \geqslant 0$

The new solution point lies at the intersection of $\tilde{g}_1(x; x^{(1)}) = 0$ and $\tilde{h}_1(x; x^{(1)}) = 0$ and is $x^{(2)} = (1.0187, 0.9965)$. The equality constraint is still violated $[h_1(x^{(2)}) = -0.0117]$, but to a lesser extent. If we continue in this fashion for two iterations, the solution $x^* = (1, 1)$ is reached with high accuracy. From the following tabulation of the successive function values

| Iteration | $f$ | $g$ | $h$ |
|-----------|-----|-----|-----|
| 0 | 5 | 3 | $-1.0$ |
| 1 | 2.284 | 0.605 | $-0.0123$ |
| 3 | 2.00015 | $3.44 \times 10^{-4}$ | $-1.18 \times 10^{-5}$ |
| optimum | 2.00 | 0.0 | 0.0 |

it is evident that both the objective function and the equality constraint value improve monotonically. The inequality constraint $g(x)$ remains feasible at all intermediate points.

This type of monotonic behavior will occur if the problem functions are mildly nonlinear, as in the flatter portions of the parabolic constraints of Example 8.3. However, if more pronounced nonlinearities are encountered, then the monotonic improvement of the objective and constraint function values breaks down. Typically, the linearization of the objective function proves such a poor approximation that when the LP solution is used to evaluate the true objective function, a worse rather than an improved value is obtained. Alternatively, the linearization of the nonlinear constraints is such a poor approximation that when the LP solution is used to evaluate the actual constraints, the current LP solution is further outside the feasible region than its predecessor. If either or both of these situations occur during the iterations, the algorithm may fail to converge. This type of behavior is illustrated in the next example.

*Example 8.4*

Suppose we begin the solution of the problem given in Example 8.3 at the point $x^0 = (1.9999, 0.01)$. At this point,

$$f(x^0) = 3.9997$$

$$g_1(x^0) = 3.9995 > 0$$

$$h_1(x^0) = 0.0$$

That is, the point is feasible. The linearized subproblem in this instance becomes, after simplification,

$$\text{Minimize} \quad -3.9997 + 3.9998x_1 + 0.02x_2$$

$$\text{Subject to} \quad 3.9998x_1 - x_2 - 3.9996 \geqslant 0$$

$$-x_1 - 0.02x_2 + 2.0001 = 0$$

$$\tfrac{5}{2} \geqslant x_1 \geqslant \tfrac{1}{2}$$

$$3 \geqslant x_2 \geqslant 0$$

The linearized constraints are shown in Figure 8.5, and the solution to the linearized subproblem is $x^{(1)} = (1.9401, 3)$. In going from $x^0$ to $x^{(1)}$ the linearized objective function does actually improve from $\tilde{f}(x^0; x^0) = 3.9997$ to $\tilde{f}(x^{(1)}; x^0) = 3.8203$. However, the *actual* objective function does *not* improve;

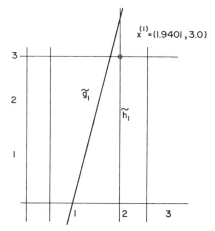

**Figure 8.5.** Linearized problem feasible region, Example 8.4.

that is,

$$f(x^0) = 3.9997 \qquad \text{but } f(x^{(1)}) = 12.764$$

Moreover, the new point turns out to be substantially infeasible,

$$g_1(x^{(1)}) = 0.764 > 0$$

$$h_1(x^{(1)}) = -8.9401 \neq 0$$

Thus, in this case, the iteration actually shows divergence.

**Step-Size Adjustment.** From Example 8.4 it is evident that the linearization cannot be relied upon to yield uniformly good results. Although the linearization is always a good approximation within some suitably small neighborhood of any given base point, it can be spectacularly bad farther away from the point, especially if the function undergoes significant changes of slope within the neighborhood of the base point. One way to ensure that the linearization is used only within the immediate vicinity of the base point is to impose limits on the allowable increments in the variables, that is, for each subproblem constructed around base point $x^{(t)}$, we impose the bounds

$$-\delta_i \leqslant x_i - x_i^{(t)} \leqslant \delta_i \qquad i = 1, \ldots, N$$

where $\delta_i$ is some suitably chosen positive step-size parameter. The parameters $\delta_i$ must be chosen small enough so that convergence is forced, that is, so that the original objective function is improved and also that the constraint infeasibilities are reduced in proceeding from $x^{(t)}$ to the solution $x^{(t+1)}$ of the current subproblem. Unfortunately, such a choice frequently leads to very slow overall convergence of the iterations.

*Example 8.5*

Suppose the bounds $\delta_1 = \delta_2 = 0.5$ are imposed upon the subproblem generated in Example 8.4. This in effect adds the constraints

$$1.9999 - 0.5 \leqslant x_1 \leqslant 0.5 + 1.9999$$

$$0.01 \quad -0.5 \leqslant x_2 \leqslant 0.5 + 0.01$$

to the problem, which, as shown in Figure 8.6, amounts to constructing a hypercube around the base point $x^0$. The solution will now lie at the point $\bar{x}$ at which the line $\tilde{h}(x; x^0) = 0$ intersects the upper bound $x_2 = 0.51$. The solution to the restricted subproblem is thus

$$\bar{x} = (1.9899, 0.51)$$

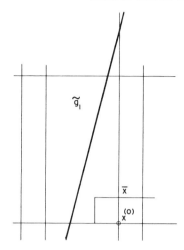

**Figure 8.6.**   Feasible region, Example 8.5.

with objective function value

$$f(\bar{x}) = 4.2198$$

and constraint values

$$g_1(\bar{x}) = 3.450 > 0 \qquad h_1(\bar{x}) = -0.25 \neq 0$$

Note that although the constraint infeasibility is improved, the objective function value remains worse than at the starting point. This indicates that the $\delta$'s chosen are still too large and should be reduced further. It turns out, as can be verified by trial-and-error evaluations, that the $\delta$'s will have to be reduced to 0.05 before improvement is observed. With $\delta_1 = \delta_2 = 0.05$, the solution will be

$$\bar{x} = (1.9989, 0.05) \qquad f(\bar{x}) = 3.9992$$

$$g_1(\bar{x}) = 3.936 > 0 \qquad h_1(\bar{x}) = -0.0025$$

But that in turn represents only a very small improvement over the base point $x^0$.

The difficulties with this modification of the direct approach become rather obvious from this example. To achieve solutions that are feasible or near feasible and at the same time yield improvement in the objective function value, the allowable increment may, in the presence of strong nonlinearities, have to be decreased to such an extent that only very slow progress in the iterations is achieved. Since the effort required to construct the linearized subproblems and to solve the resulting linear program is appreciable, the overall computational efficiency diminishes dramatically.

An algorithm employing basically the solution strategy of Example 8.5 was first proposed by R. E. Griffith and R. A. Stewart of the Shell Oil Company [4]. Subsequent implementations were reported by H. V. Smith of IBM [5], by G. W. Graves of Aerospace corporation ([6], cited in [7]), and others. The program developed by Griffith and Stewart, called MAP, includes special heuristics for adjusting the magnitudes of the $\delta$'s as the iterations progress. The IBM code, called P$\phi$P II, adjusts the $\delta$'s by using estimates of the local linearization error, but this restricts its applicability to inequality-constrained problems. Graves's version, also applicable only to inequality-constrained problems, differs from the other two primarily in incorporating estimates of the remainder term of the first-order Taylor's expansion of the constraints into the linearization formulas. Moreover, a scaling parameter is introduced to regulate the size of the correction to the successive LP solutions. Convergence of Graves's algorithm can be proved, provided the constraints $g_j(x) \geq 0$ are concave *and* the objective function is convex.

**Use of Penalty Functions.** Observe that the step-bounding logic of these strategies can be viewed as a one-parameter search of some penalty function

$$P(x, p) = f(x) + p\Omega(g, h, x)$$

over the direction defined by the vector $x^{(t+1)} - x^{(t)}$. Thus, a conceptual two-step algorithm analogous to the Frank-Wolfe algorithm could be formulated. In the first step the LP is constructed and solved to yield a new point $\bar{x}$. In the second step, for a suitable choice of parameter $p$, the line-search problem,

$$\text{Minimize} \quad P(x(\alpha), p)$$

where

$$x(\alpha) = x^{(t)} + \alpha(\bar{x} - x^{(t)}) \qquad \alpha \geq 0$$

would be solved to yield the new point $x^{(t+1)} = x^{(t)} + \alpha^{(t+1)}(\bar{x} - x^{(t)})$. Presumably some adjustment of the penalty parameter ought to be incorporated, and possibly the rigorous minimization of $P$ could be relaxed to seeking "sufficient" improvement.

Strategies of this general type were investigated by Palacios-Gomez [8]. A representative version of such a successive approach would consist of two primary phases as outlined below for a given specified tolerance $\varepsilon > 0$ and step size parameter $\delta$:

*Phase Check*

If at $x^{(t)}$, the exterior penalty term $\Omega$ satisfies $\Omega(g, h, x) > \varepsilon$, then go to phase I. Otherwise go to phase II.

*Phase I*

**Step 1.**  Linearize at $x^{(t)}$ and solve the LP problem to obtain $\bar{x}$.

**Step 2.**  If $\Omega(g, h, \bar{x}) < \Omega(g, h, x^{(t)})$, then $x^{(t+1)} = \bar{x}$ and proceed to the phase check. Otherwise reduce $\delta$ (e.g., $\delta = \delta/2$) and go to step 1.

*Phase II*

**Step 1.**  Linearize and solve the LP to obtain $\bar{x}$.

**Step 2.**  If $P(p, \bar{x}) < P(p, x^{(t)})$, set $x^{(t+1)} = \bar{x}$, increase the penalty parameter $p$, and go to the phase check. Otherwise reduce $\delta$ and go to step 1.

Termination occurs when both ($\|\bar{x} - x^{(t)}\|$) and $|P(p, \bar{x}) - P(p, x^{(t)})|$ are sufficiently small.

*Example 8.6*

Suppose a Palacios-type strategy is applied to the problem of Example 8.3 at the point $x^{(1)} = (1.9899, 0.51)$ obtained in Example 8.5 with $\delta = \frac{1}{2}$. For illustrative purposes we choose $\varepsilon = 10^{-2}$, and

$$\Omega(g, h, x) = h_1^2 + \left[\min(g_1(x), 0)\right]^2$$

At $x^{(1)}$, we have $\Omega = h_1^2 = \frac{1}{16} > 10^{-2}$. Therefore, the phase check directs the calculations to phase I.

**Step 1.**  The LP constructed at $x^{(1)}$ is

$$\text{Minimize} \quad 3.9798\Delta x_1 + 1.02\Delta x_2$$

$$\text{Subject to} \quad 3.45 + 3.9798\Delta x_1 - \Delta x_2 \geqslant 0$$

$$-0.25 - \Delta x_1 - 1.02\Delta x_2 = 0$$

$$-\tfrac{1}{2} \leqslant (\Delta x_1, \Delta x_2) \leqslant \tfrac{1}{2}$$

The solution is $\bar{x} = (1.4899, 0.7551)$.

**Step 2.**  At $\bar{x}$, $g(\bar{x}) = 1.4647$ and $h_1(\bar{x}) = -6.007 \times 10^{-2}$. Thus

$$\Omega(g, h, \bar{x}) = 3.6 \times 10^{-3} < \Omega\left(g, h, x^{(1)}\right) = \tfrac{1}{16}$$

Therefore,

$$x^{(2)} = \bar{x}.$$

The phase check indicates that $\Omega(g, h, \bar{x}) < 10^{-2}$ and thus directs the calculations to phase II.

**Step 1.**   The LP at $x^{(2)}$ is

$$\text{Minimize}\quad 2.9798\Delta x_1 + 1.5102\Delta x_2$$

$$\text{Subject to}\quad 1.4647 + 2.9798\Delta x_1 - \Delta x_2 \geqslant 0$$

$$-6.007 \times 10^{-2} - \Delta x_1 - 1.5102\Delta x_2 = 0$$

$$-\tfrac{1}{2} \leqslant (\Delta x_1, \Delta x_2) \leqslant \tfrac{1}{2}$$

The solution is $\bar{x} = (1.1369, 0.9889)$.

**Step 2.**   At this solution,

$$f(\bar{x}) = 2.2703$$

$$g_1(\bar{x}) = 0.3036$$

$$h_1(x) = -0.1147$$

Hence, $P(10^2, \bar{x}) = 3.586 > P(10^2, x^{(2)}) = 3.150$. The step size $\delta$ must be reduced, say by $\tfrac{1}{2}$, and step 1 is repeated with $-\tfrac{1}{4} \leqslant (\Delta x_1, \Delta x_2) \leqslant \tfrac{1}{4}$. At the solution, $\bar{x} = (1.2399, 0.8809)$, the function values are $f(\bar{x}) = 2.31327$, $g_1(\bar{x}) = 0.6565$, and $h_1(\bar{x}) = -0.01582$. Thus

$$P(10^2, \bar{x}) = 2.338 < P(10^2, x^{(2)})$$

and we set $x^{(3)} = \bar{x}$, increase $p$ to, say, 1000, and go to the phase check.

At the cost of one more LP solution, the point $\bar{x} = (1.0215, 0.9958)$ is obtained with

$$P(10^3, \bar{x}) = 2.210 < P(10^2, x^{(3)}) < P(10^3, x^{(3)})$$

Thus, $x^{(4)} = \bar{x}$. With one further iteration, the very close optimal solution estimate, $x^{(5)} = (1.00019, 0.99991)$ is calculated.

Note that by making use of the penalty function and penalty term as measures of progress in the solution process, drastic reductions in the step-size bound $\delta$ were avoided and thus good progress could be made. Unfortunately, although the use of the penalty function enhances the performance of the SLP strategy, it does not guarantee convergence in the general case. In fact, convergence results are available only in the linearly constrained case [8]. Of course, in all successive LP variants, if the solution of the LP approximation is such that $x^{(t+1)} - x^{(t)} = 0$, then it follows that there is no $\Delta x \neq 0$ such that

$$\nabla f \cdot \Delta x < 0$$

$$\nabla g_j \cdot \Delta x > 0 \qquad j = 1, \ldots, J$$

$$\nabla h_k \cdot \Delta x = 0 \qquad k = 1, \ldots, K$$

and consequently the Kuhn-Tucker conditions must hold [8].

### 8.1.3  Discussion and Applications

In spite of the computational enhancements of the basic successive LP strategy that are obtained by Palacios-Gomez [8] and others [9], the key limitation is the need to resort to the solution of a sequence of linear programs. If the problem is largely linear with only modest nonlinear contributions, then this overhead is acceptable. In this context, a good measure of nonlinearity is given by the number of *nonlinear variables*, that is, variables that appear in a nonlinear form in at least one problem function. For problems with a low relative number of nonlinear variables, the use of the successive LP strategy has found wide acceptance in various industries. For instance, Lasdon and Warren [10] report on the use of this approach to solve large oil field development problems by both Gulf Oil and the Kuwait Oil Company. Problems involving up to 4392 linear and 400 nonlinear variables are reported to have been solved. Marathon Oil is reported to be using PϕP II as part of a hierarchical model for optimizing a complete refinery. Coefficients of the PϕP II model are obtained from up to 20 different process-simulation models [10]. Similar specialized applications are reported by Chevron for solving refinery blending problems and by Union Carbide to solve multi-time-period process-optimization problems. In the latter case, the intermediate linearized subproblems are reported to involve up to 10,000 variables. However, the number of nonlinear variables is small, so that the successive LP subproblems are each solved in an average of only 23 simplex iterations [10]. Similar results are reported for Exxon applications by Baker and Lasdon [11]. A typical refinery application involving a small number of nonlinear elements is presented in the next example.

### *Example 8.7   The Pooling Problem*

In refinery and other processing applications such as coal-blending plants, the need frequently arises to mix (or pool) several raw products to form an intermediate mixture, which is subsequently used as one of the components in a final blended product. The need for pooling usually occurs because of limited storage tank availability or because the raw products must be shipped in a single tank car, tanker, or pipeline. A simplified system requiring pooling is shown in Figure 8.7. In this system [12], two raw product streams with flows $V_1$ and $V_2$ and sulfur compositions 3% and 1%, respectively, must be pooled and then blended with a third raw product stream with flow $F$ and 2% sulfur composition to produce two final blends.

The first blend must have maximum sulfur composition of 2.5% and maximum demand of 100 bbl/hr; the second blend has maximum sulfur composition of 1.5% and maximum demand of 200 bbl/hr. The costs of the three raw streams are 6, 16, and 10 \$/bbl, and the selling prices of the product blends are 9 and 10 \$/bbl. Determine the flows and pool composition so as to maximize profit.

Let $P_i$ = flow of blend $i$, bbl/hr
    $F_i$ = flow of the third raw product used in blend $i$
    $V_i$ = flow of pooled intermediate used in blend $i$
    $\bar{x}$ = the pool sulfur composition

From total flow balances,

$$V_1 + V_2 = \bar{V}_1 + \bar{V}_2$$

$$F_1 + \bar{V}_1 = P_1$$

$$F_2 + \bar{V}_2 = P_2$$

From sulfur balances,

$$0.3V_1 + 0.01V_2 = \bar{x}\left(\bar{V}_1 + \bar{V}_2\right)$$

$$0.02F_1 + \bar{x}\bar{V}_1 \leqslant 0.025P_1$$

$$0.02F_2 + \bar{x}\bar{V}_2 \leqslant 0.015P_2$$

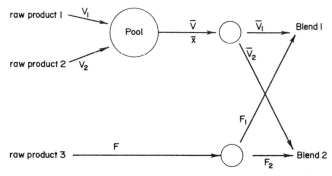

**Figure 8.7.** Schematic of pooling problem, Example 8.7.

In addition, the maximum demands reduce to the bounds

$$P_1 \leqslant 100 \qquad P_2 \leqslant 200$$

and the pool composition must be bounded by the raw stream compositions,

$$0.01 \leqslant \bar{x} \leqslant 0.03$$

The objective function in this case is simply the linear function

$$9P_1 + 10P_2 - 6V_1 - 16V_2 - 10F$$

The optimization problem thus reduces to

$$\text{Maximize} \quad f = 9P_1 + 10P_2 - 6V_1 - 16V_2 - 10F$$

$$\text{Subject to} \quad h_1 = V_1 + V_2 - \bar{V}_1 - \bar{V}_2 = 0$$

$$h_2 = 0.03V_1 + 0.01V_2 - \bar{x}(\bar{V}_1 + \bar{V}_2) = 0$$

$$h_3 = F_1 + \bar{V}_1 - P_1 = 0$$

$$h_4 = F_2 + \bar{V}_2 - P_2 = 0$$

$$g_1 = 0.025P_1 - 0.02F_1 - \bar{x}\bar{V}_1 \geqslant 0$$

$$g_2 = 0.015P_2 - 0.02F_2 - \bar{x}\bar{V}_2 \geqslant 0$$

and

$$P_1 \leqslant 100 \qquad P_2 \leqslant 200$$

$$0.01 \leqslant x \leqslant 0.03$$

Note that this problem would be an LP problem except for the product terms $\bar{x}\bar{V}_1$ and $\bar{x}\bar{V}_2$ that arise in $h_2$, $g_1$, and $g_2$. If an initial guess were made of $\bar{x}$, then the problem could be solved as a linear program. The solution could be used to update $\bar{x}$, and the process would be continued until convergence was obtained. However, as observed by Haverly [12], such a recursive solution procedure is not reliable. Since equality constraint $h_2$ is nonlinear, the feasible region for this problem is nonconvex; hence, the problem admits multiple local solutions. As pointed out by Lasdon et al. [12], this problem has three Kuhn-Tucker points:

1. A saddlepoint with $\bar{x} = 0.02$ and all flows zero
2. A local maximum with $\bar{x} = 0.03$, $P_1 = 100$, $F = V = 50$, all other flows zero, and a profit of $100/hr

**3.** The global maximum with $\bar{x} = 0.01$, $V_2 = F_2 = 100$, $P_2 = 200$, all other flows zero, and a profit of \$400/hr

The recursive solution strategy leads to one of these three points depending upon the initial value of $\bar{x}$ selected. However, as reported by Lasdon et al. [13], solution using the successive LP strategy devised by Palacios-Gomez [8] leads to the global maximum regardless of the initial value of $\bar{x}$ selected, requiring between two and three LP subproblem solutions. It is not clear why the successive LP strategy avoids the local solutions; however, it certainly is a welcome result. Since in pooling applications of practical scope, the number of pools, raw streams, product blends, and blend characteristics or compositions would be much larger, the possibility of local maxima increases, and hence the use of successive linear programming is clearly to be preferred.

In summary, the successive LP strategy, particularly with the penalty function modifications of Palacios-Gomez, is a useful and practical approach for solving large NLP problems with a moderate degree of nonlinearity. For highly nonlinear problems, however, this solution strategy will require substantial reductions in the step size in order to ensure progress in the search, thus resulting in slow convergence. In fact, the typically slow convergence with highly nonlinear problems has led to the use of the name *small step gradient methods* for this family of NLP techniques.

Improvements in the direct linear approximation methods have taken three alternative forms:

**1.** Global approximation of the problem nonlinearities by means of piecewise linear functions
**2.** Accumulation of successive linearizations from iteration to iteration so as to better define the feasible region
**3.** Use of the local linear approximation to generate only a direction for movement rather than a complete step

In the following sections we discuss the ideas involved in each of these constructions. As will become apparent, the first two approaches are applicable to only a limited class of problems, while the third has led to very efficient and powerful general-purpose algorithms.

## 8.2  SEPARABLE PROGRAMMING

Separable programming is a technique first developed by C. E. Miller [14] in 1963 by means of which certain types of nonlinear constrained optimization problems can be reformulated into equivalent problems involving only linear functions. The resulting approximating problems are then solved using a specially modified simplex algorithm. The motivation for this technique stems from the observation that a good way of improving the linear approximation to a general nonlinear function over a large interval is to partition the interval into subintervals and construct individual linear approximations over each subinterval. This is known as constructing a piecewise linear approximation.

In this section, we first consider the use of this construction for single-variable functions. Then we extend the construction to multivariable functions. We also examine the modifications to standard LP methodology that must be made to effectively solve the resulting linear subproblems. Finally, we discuss some practical applications of the separable programming construction.

### 8.2.1  Single-Variable Functions

In the special case of single-variable functions, the development of a piecewise linear approximation is quite straightforward. Consider an arbitrary continuous function $f(x)$ of a single variable defined over an interval $0 \leqslant x \leqslant a$. Suppose we arbitrarily choose a grid of $K$ points spaced over the interval of interest and denote the points $x^{(k)}$, $k = 1,\ldots, K$. For each $x^{(k)}$, we evaluate $f^k = f(x^{(k)})$, and between every pair of adjacent grid points $x^{(k)}$, $x^{(k+1)}$ we draw a straight line connecting the points $(x^k, f^k)$ and $(x^{k+1}, f^{k+1})$. As shown in Figure 8.8, we will have formed an approximating function $\tilde{f}(x)$ that is a piecewise linear function.

The equation of the line connecting points $(x^{(k)}, f^{(k)})$ and $(x^{(k+1)}, f^{(k+1)})$ is given by

$$\tilde{f}(x) = f^{(k)} + \frac{f^{(k+1)} - f^{(k)}}{x^{(k+1)} - x^{(k)}}(x - x^{(k)}) \tag{8.1}$$

An equation of this form will be associated with each subinterval $x^{(k)} \leqslant x \leqslant$

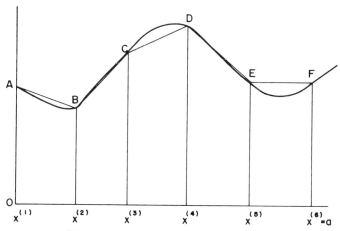

**Figure 8.8.**  Piecewise linear approximation.

$x^{(k+1)}$. Consequently, the entire approximation to $f(x)$ will consist of $K - 1$ such linear equations, with each one applicable over its restricted subinterval of validity.

It is convenient to condense this collection of linear equations into a more general form. First of all, observe that any $x$ in the range $x^{(k)} \leqslant x \leqslant x^{(k+1)}$ can be written as

$$x = \lambda^{(k)}x^{(k)} + \lambda^{(k+1)}x^{(k+1)} \tag{8.2}$$

where $\lambda^{(k)}$ and $\lambda^{(k+1)}$ are two nonnegative parameters normalized to one, that is,

$$\lambda^{(k)} + \lambda^{(k+1)} = 1$$

Suppose we substitute this expression for $x$ into Eq. (8.1):

$$\tilde{f}(x) = f(x^{(k)}) + \frac{f^{(k+1)} - f^{(k)}}{x^{(k+1)} - x^{(k)}} \left[ \lambda^{(k+1)}x^{(k+1)} - (1 - \lambda^{(k)})x^{(k)} \right]$$

Since $1 - \lambda^{(k)} = \lambda^{(k+1)}$, it follows that

$$\tilde{f}(x) = f(x^{(k)}) + \lambda^{(k+1)}(f^{(k+1)} - f^{(k)})$$

$$= \lambda^{(k)}f^{(k)} + \lambda^{(k+1)}f^{(k+1)} \tag{8.3}$$

Thus in the interval $x^{(1)} \leqslant x \leqslant x^{(K)}$, each $x$ and its approximate function value $\tilde{f}(x)$ can be generated by assigning appropriate values to the two $\lambda$ variables that correspond to the subinterval within which $x$ lies. Since any given value of $x$ can lie in only one subinterval, it further follows that only the two $\lambda$'s associated with that subinterval will have nonzero values. The other $\lambda$'s must all be equal to zero. Consequently, the $K - 1$ sets of Eqs. (8.2) and (8.3) can be combined conveniently into the two equations

$$x = \sum_{k=1}^{K} \lambda^{(k)}x^{(k)} \qquad \tilde{f}(x) = \sum_{k=1}^{K} \lambda^{(k)}f^{(k)} \tag{8.4}$$

where

(i)
$$\sum_{k=1}^{K} \lambda^{(k)} = 1$$

(ii)
$$\lambda^{(k)} \geqslant 0 \qquad k = 1, \ldots, K$$

and

(iii)
$$\lambda^{(i)}\lambda^{(j)} = 0 \qquad \text{if} \quad j > i + 1; i = 1, K - 1$$

Condition (iii) merely requires that no more than two of the $\lambda^{(i)}$'s be positive and if two are positive, say $\lambda^{(j)}$ and $\lambda^{(i)}$ with $j > i$, then it must be true that $j = i + 1$, that is, the $\lambda$'s must be adjacent. This restriction ensures that only points lying on the piecewise linear segments are considered part of the approximating function. To see this, note from Figure 8.8 that if $\lambda^{(1)}$ and $\lambda^{(2)}$ were positive and $\lambda^{(3)}$ to $\lambda^{(6)}$ were zero, then the resulting point would lie on the line segment joining points $A$ and $B$, a portion of the approximating function. However, if, say $\lambda^{(1)}$ and $\lambda^{(3)}$ were allowed to be positive and the other $\lambda$'s were all equal to zero, then a point on the line connecting $A$ and $C$, which is not part of the approximating function, would be generated. Finally if, say, $\lambda^{(3)}$ alone were positive, then from condition (i) it would have to be equal to 1, and the point $C$ lying on both $f(x)$ and $\tilde{f}(x)$ would be generated.

The piecewise linear construction does require an increase in problem size. In Figure 8.8 the nonlinear function $f(x)$ is converted into a linear function $\tilde{f}(x)$ at the cost of introducing six new variables (the $\lambda$'s) and two linear side conditions (8.4) and (i). Moreover, if we wanted a closer approximation of $f(x)$ we could introduce additional grid points; however, this would introduce additional variables $\lambda^{(k)}$.

The accuracy of the approximation is also strongly dependent on a judicious selection of the grid-point locations. The best way to select grid-point locations is to work from a plot of the function over the variable range of interest. Grid points should normally be located in the vicinity of inflection points (such as $C$ in Fig. 8.8) and should be concentrated in the regions in which the function changes values rapidly, that is, regions with large function derivative values. This means that in general the grid-point locations will be spaced nonuniformly.

### 8.2.2 Multivariable Separable Functions

The foregoing construction can be generalized to multivariable functions only if the multivariable functions are *separable*.

#### Definition

A function $f(x)$ of the $N$-component vector variable $x$ is said to be *separable* if it can be expressed as the sum of single-variable functions that each involve only one of the $N$ variables. Thus,

$$f(x) = \sum_{i=1}^{N} f_i(x_i)$$

For example, the function

$$f(x) = x_1^2 + e^{x_2} + x_2^{2/3}$$

is separable, but

$$f(x) = x_1\sin(x_2 + x_3) + x_2 e^{x_3}$$

is not.

The restriction of the approximation technique to separable functions may seem to be quite restrictive and it is. However, the class of separable nonlinear functions is nonetheless quite large. Moreover, there are various devices by means of which some functions that are not directly separable can be rewritten in separable form.

For instance, the product $x_1 x_2$ can be rewritten as $x_3^2 - x_4^2$ with the side conditions

$$x_3 = \tfrac{1}{2}(x_1 + x_2) \quad \text{and} \quad x_4 = \tfrac{1}{2}(x_1 - x_2) \tag{8.5}$$

The generalized product

$$c \prod_{i=1}^{N} x_i^{a_i}$$

can be replaced by $y$ with the side conditions

$$\ln(y) = \ln(c) + \sum_{i=1}^{N} a_i \ln(x_i)$$

provided the $x_i$ are restricted to be positive.

Under the restriction that the multivariable function $f(x)$ is separable and continuous, the construction of a piecewise linear approximation to $f(x)$ proceeds as follows: Subdivide the interval of values on each variable $x_i$ with $K_i$ grid points. Thus, each variable will be associated with a sequence of points

$$x_i^{(L)} = x_i^{(1)} < x_i^{(2)} < \cdots < x_i^{(j)} < \cdots < x_i^{(K_i)} = x_i^{(U)}$$

where $x_i^{(L)}$ and $x_i^{(U)}$ are the given or assumed values of the lower and upper bounds on variable $x_i$.

If we adopt the shorthand notation $f_i^{(k)} = f_i(x_i^{(k)})$, then the approximating function $\tilde{f}(x)$ will be given by

$$\tilde{f}(x) = \sum_{k=1}^{K_1} \lambda_1^{(k)} f_1^{(k)} + \sum_{k=1}^{K_2} \lambda_2^{(k)} f_2^{(k)} + \cdots + \sum_{k=1}^{K_N} \lambda_N^{(k)} f_N^{(k)}$$

where

$$x_i = \sum_{k=1}^{K_i} \lambda_i^{(k)} x_i^{(k)} \qquad i = 1, \ldots, N$$

given that for $i = 1, \ldots, N$

(i) $$\sum_{k=1}^{K_i} \lambda_i^{(k)} = 1$$

(ii) $$\lambda_i^{(k)} \geqslant 0 \qquad k = 1, \ldots, K_i$$

(iii) $$\lambda_i^{(i)} \lambda_i^{(j)} = 0 \qquad \text{if } j > i + 1; i = 1, 2, \ldots, K_i - 1$$

In this manner each nonlinear function $f(x)$ is replaced by a function $\tilde{f}(x)$ that is linear in the variables. Of course, if a very accurate approximation to $f(x)$ is required, then the number of grid points and hence variables $\lambda$ can become quite large. Again, linearity is attained at the price of substantially increased dimensionality.

**Example 8.8**

Consider the function

$$f(x) = f_1(x_1) + f_2(x_2) = \left(1 - (1 - x_1)^3\right) + \left(0.1x_2^4\right)$$

over the hypercube $0 \leqslant x_1 \leqslant 2$ and $0 \leqslant x_2 \leqslant 3$.

Let the approximating function to $f_1(x_1)$ be constructed by subdividing the interval $0 \leqslant x_1 \leqslant 2$ into three sections by using the four grid points 0, 0.5, 1.5, and 2.0. Thus,

$$x_1 = 0\lambda_1^{(1)} + 0.5\lambda_1^{(2)} + 1.5\lambda_1^{(3)} + 2.0\lambda_1^{(4)}$$

and

$$\tilde{f}_1(x_1) = f_1(0)\lambda_1^{(1)} + f_1(0.5)\lambda_1^{(2)} + f_1(1.5)\lambda_1^{(3)} + f_1(2.0)\lambda_1^{(4)}$$

$$= 0\lambda_1^{(1)} + 0.875\lambda_1^{(2)} + 1.125\lambda_1^{(3)} + 2.0\lambda_1^{(4)}$$

Similarly, to construct the approximation to $f_2(x_2)$, suppose we choose five grid points, 0, 1, 2, 2.6, and 3 with corresponding $f_2$ values of 0, 0.1, 1.6, 4.57, and 8.1. Then,

$$\tilde{f}(x_1, x_2) = 0\lambda_1^{(1)} + 0.875\lambda_1^{(2)} + 1.125\lambda_1^{(3)} + 2.0\lambda_1^{(4)} + 0\lambda_2^{(1)}$$

$$+ 0.1\lambda_2^{(2)} + 1.6\lambda_2^{(3)} + 4.57\lambda_2^{(4)} + 8.1\lambda_2^{(5)}$$

with

$$x_1 = 0\lambda_1^{(1)} + 0.5\lambda_1^{(2)} + 1.5\lambda_1^{(3)} + 2.0\lambda_1^{(4)}$$

$$x_2 = 0\lambda_2^{(1)} + 1\lambda_2^{(2)} + 2\lambda_2^{(3)} + 2.6\lambda_2^{(4)} + 3\lambda_2^{(5)}$$

The $\lambda_i^{(k)}$ are restricted to satisfy

$$\sum_{k=1}^{4} \lambda_1^{(k)} = 1 \qquad \sum_{k=1}^{5} \lambda_2^{(k)} = 1 \qquad \lambda_i^{(k)} \geqslant 0$$

and only two adjacent $\lambda_1^{(k)}$'s and two adjacent $\lambda_2^{(k)}$'s can be positive. For example,

$$\lambda_1^{(2)} = 0.4 \qquad \lambda_1^{(3)} = 0.6 \qquad \lambda_1^{(1)} = \lambda_1^{(4)} = 0$$

$$\lambda_2^{(3)} = 0.5 \qquad \lambda_2^{(4)} = 0.5 \qquad \lambda_2^{(1)} = \lambda_2^{(2)} = \lambda_2^{(5)} = 0$$

will generate the point $x = (1.1, 2.3)$ and the approximate function value $\tilde{f}(1.1, 2.3) = 4.11$. The exact function value is $f(1.1, 2.3) = 3.8$.

### 8.2.3   Linear Programming Solutions of Separable Problems

By using piecewise linear approximations of the form shown in Example 8.8, any NLP problem in which both the objective function and the constraints are separable can be reformulated as a large linear program in the variables $\lambda$ that can be solved by conventional LP codes. The only feature requiring special attention is condition (iii). This restriction can, however, be readily accommodated, since in the ordinary simplex method the basic variables are the only ones that can be nonzero. Thus, prior to entering one of the $\lambda$'s into the basis (which will make it nonzero), a check is made to ensure that no more than one other $\lambda$ associated with the corresponding variable $x_i$ is in the basis (is nonzero) and, if it is, that the $\lambda$'s are adjacent. If these checks are not satisfied, then the $\lambda$ to be entered into the basis is rejected and another is selected. This modification to the normal simplex rules required to ensure that condition (iii) is always met is known as *restricted basis entry*. Recall that the reason for imposing condition (iii) and for restricted basis entry is to ensure that the points generated lie on the piecewise linear segments rather than between them.

In the following example, we illustrate the use of separable programming constructions to solve a nonlinear program via an approximating linear program.

*Example 8.9*

$$\text{Maximize} \quad f(x) = x_1^4 + x_2$$

$$\text{Subject to} \quad g_1(x) = 9 - 2x_1^2 - 3x_2 \geqslant 0$$

$$x_1 \geqslant 0 \qquad x_2 \geqslant 0$$

Since both problem functions are separable, we can immediately identify

$$f(x) = f_1(x_1) + f_2(x_2)$$

$$g_1(x) = g_{11}(x_1) + g_{12}(x_2)$$

where

$$f_1(x_1) = x_1^4 \qquad f_2(x_2) = x_2$$

$$g_{11}(x_1) = -2x_1^2 \qquad g_{12}(x_2) = -3x_2$$

Since $f_2$ and $g_{12}$ are linear functions, they need not be replaced by piecewise linear functions. However, the other two, $f_1$ and $g_{11}$, will need to be approximated. Suppose we choose to construct our approximation over the interval $0 \leqslant x_1 \leqslant 3$, and suppose we use four equidistant grid points for this purpose. The data required for the piecewise approximation are tabulated below:

| $k$ | $x_1^{(k)}$ | $f_1(x_1^{(k)})$ | $g_{11}(x_1^{(k)})$ |
|---|---|---|---|
| 1 | 0 | 0 | 0 |
| 2 | 1 | 1 | $-2$ |
| 3 | 2 | 16 | $-8$ |
| 4 | 3 | 81 | $-18$ |

Thus the approximation to $f_1(x_1)$ becomes

$$\tilde{f}_1(x_1) = (0)\lambda_1^{(1)} + (1)\lambda_1^{(2)} + (16)\lambda_1^{(3)} + (81)\lambda_1^{(4)}$$

and the approximation to $g_{11}(x_1)$ becomes

$$\tilde{g}_{11}(x_1) = (0)\lambda_1^{(1)} - 2\lambda_1^{(2)} - 8\lambda_1^{(3)} - 18\lambda_1^{(4)}$$

with the restriction

$$\lambda_1^{(1)} + \lambda_1^{(2)} + \lambda_1^{(3)} + \lambda_1^{(4)} = 1$$

The resulting linear approximating problem thus becomes

$$\text{Maximize} \quad f(x) = \lambda_1^{(2)} + 16\lambda_1^{(3)} + 81\lambda_1^{(4)} + x_2$$

$$\text{Subject to} \quad 2\lambda_1^{(2)} + 8\lambda_1^{(3)} + 18\lambda_1^{(4)} + 3x_2 + x_3 = 9$$

$$\lambda_1^{(1)} + \lambda_1^{(2)} + \lambda_1^{(3)} + \lambda_1^{(4)} = 1$$

$$\text{all } \lambda_1^{(k)}, x_2, x_3 \geqslant 0$$

Variable $x_3$ is a slack variable inserted to convert the constraint to an equality. Suppose we convert the LP problem to tableau form and execute the usual simplex calculations.

| | | $C_J$ | | | | | | |
|---|---|---|---|---|---|---|---|---|
| | | 1 | 16 | 81 | 1 | 0 | 0 | |
| $C_B$ | Basis | $\lambda_1^{(2)}$ | $\lambda_1^{(3)}$ | $\lambda_1^{(4)}$ | $x_2$ | $x_3$ | $\lambda_1^{(1)}$ | |
| 0 | $x_3$ | 2 | 8 | 18 | 3 | 1 | 0 | 9 |
| 0 | $\lambda_1^{(1)}$ | 1 | ① | 1 | 0 | 0 | 1 | 1 |
| | $\overline{C}_J$ | 1 | 16 | 81 | 1 | 0 | 0 | 0 |

$\uparrow$

From the $\overline{C}_J$ row, we see that $\lambda_1^{(4)}$ is the entering variable, and from the minimum ratio rule it follows that the outgoing variable is $x_3$. But $\lambda_1^{(4)}$ cannot be brought into the basis, since $\lambda_1^{(4)}$ and $\lambda_1^{(1)}$ are not adjacent. Therefore, $\lambda_1^{(4)}$ is rejected and the next best value in the $\overline{C}_J$ row is sought. From the $\overline{C}_J$ row, $\lambda_1^{(3)}$ is the next candidate to enter the basis, and from the minimum ratio rule, $\lambda_1^{(1)}$ will leave the basis.

Continuing in this fashion, we eventually achieve the optimal solution with basic variables $\lambda_1^{(3)} = 0.9$ and $\lambda_1^{(4)} = 0.1$. The corresponding optimal value of $x_1$ is

$$x_1 = 0(0) + 1(0) + 2(0.9) + 3(0.1) = 2.1$$

Thus the optimal solution to the separable program is $\tilde{x}^* = (2.1, 0)$ with $\tilde{f}(x^*) = 22.5$. As can be readily verified, the exact optimum solution is

$$x^* = (2.12, 0) \qquad f(x^*) = 20.25$$

Note that if, in the tableau, the restricted basis entry rule had been neglected, then $\lambda_1^{(4)}$ would have entered the basis and $x_3$ would have left. The resulting next tableau would then have been

| | | $C_J$ | | | | | | |
|---|---|---|---|---|---|---|---|---|
| | | 1 | 16 | 81 | 1 | 0 | 0 | |
| $C_B$ | Basis | $\lambda_1^{(2)}$ | $\lambda_1^{(3)}$ | $\lambda_1^{(4)}$ | $x_2$ | $x_3$ | $x_1^{(1)}$ | |
| 81 | $\lambda_1^{(4)}$ | $\frac{1}{9}$ | $\frac{4}{9}$ | 1 | $\frac{1}{6}$ | $\frac{1}{18}$ | 0 | $\frac{1}{2}$ |
| 0 | $\lambda_1^{(1)}$ | $\frac{8}{9}$ | $\frac{5}{9}$ | 0 | $-\frac{1}{6}$ | $-\frac{1}{18}$ | 1 | $\frac{1}{2}$ |
| | $\overline{C}_J$ | $-8$ | $-20$ | 0 | $-12.5$ | $-4.5$ | 0 | 40.5 |

Apparently the optimal solution is $\lambda_1^{(1)} = \lambda_1^{(4)} = \frac{1}{2}$, with $\tilde{f} = \frac{81}{2} = 40.5$. However, this is clearly not correct, since these values of the $\lambda$'s give

$$x_1 = 0\left(\tfrac{1}{2}\right) + 1(0) + 2(0) + 3\left(\tfrac{1}{2}\right) = 1.5$$

and at the point $(1.5, 0)$ the actual objective function value is

$$f(1.5, 0) = (1.5)^4 + 0 = 5.0625$$

Neglect of the basis entry rule can therefore lead to points that do not lie on the approximating segments and hence may yield meaningless LP solutions.

For a special class of problems the restricted basis rule can, however, be relaxed. As proven in reference 15 (theorem 11.3.1), given a separable nonlinear program of the form

$$\text{Minimize} \quad \sum_{i=1}^{N} f_i(x_i)$$

$$\text{Subject to} \quad \sum_{i=1}^{N} g_{ji}(x_i) \leqslant b_j \quad j = 1, \ldots, J$$

$$x_i \geqslant 0 \quad i = 1, \ldots, N$$

the restricted basis rule will always be satisfied at the optimum of the separable programming LP provided

1. For all $i$, $f_i(x_i)$ is either strictly convex or it is linear.
2. For all $i$ and $j$, $g_{ji}$ is either concave or it is linear.

Since the conditions involve the convexity of the single-variable functions, they are easily checked and quite useful in practice.

### 8.2.4 Discussion and Applications

The solution obtained via separable programming is only as accurate as the underlying approximating problem. In fact, only if the components of the separable functions $g_{ji}$ are all either linear or concave can it be guaranteed that the solution will even be a feasible solution of the original problem. Solutions of improved feasibility and accuracy can be generated by following either of two strategies:

1. Increase the overall accuracy of each approximating function to the desired level by adding more grid points. However, this increases the

number of variables and thus, because of the restricted basis requirement, considerably increases the number of iterations required to solve
the approximating problem.

2.  Increase the accuracy of each approximating function locally in an
    iterative fashion. The problem is solved initially via a rather coarse grid.
    The grid is then refined only in the neighborhood of the resulting
    solution. The problem is solved again, and the refinement process
    continues until a sufficiently accurate solution is attained. In this
    manner the number of grid points can be kept constant at the price of
    repeated solution of the intermediate LP problems.

The latter strategy is usually preferable. However, it should be noted that local
iterative refinement can, in some cases, lead to convergence to false optima and
hence must be used with care. In general, separable programming solutions will
be only local minima. Attainment of global minima is guaranteed only in the
case of convex objective functions and convex feasible regions.

Clearly, separable programming is not suitable for general NLP problems.
It is appropriate for applications that are largely linear with some nonlinear
elements that can be expressed as separable functions. If highly accurate
solutions are not required, then, by use of the piecewise linear construction, the
approximate optimum will be obtained in a single solution of the LP subproblem. This feature has led some practitioners to rank separable programming
very highly as a useful and practical tool [16]. Separable programming has
found numerous applications in refinery scheduling and planning models [17].
We will consider one such application in the next example.

### *Example 8.10   Computing Tetraethyllead Requirements [17, 18]*

One very important nonlinearity encountered in refinery modeling arises in
computing the amount of tetraethyllead (TEL) that must be added to a
gasoline blend to increase the gasoline octane number to required levels. The
octane number is not a linear function of TEL. Instead, the exact relationship
is defined using TEL susceptibility charts, such as Figure 8.9. The assumption
used in the industry is that the octane number of a blend can be estimated as
the volumetric average of the octane numbers of its components all measured
at the *same* TEL concentration as that of the blend.

Suppose $N$ components plus TEL are to be blended to yield a gasoline with
specified minimum research and motor octane number. The TEL concentration of the gasoline is not to exceed 3 cm$^3$ TEL per U.S. gallon. For each of the
components, the relationship between the octane numbers and the TEL
concentration will be given by a chart similar to Figure 8.9.

Let $x_n$ = flow of component $n$, gal/day
$y$ = flow of blended gasoline, gal/day

$z$ = concentration of TEL in the blend, cm³/gal

$r_n(z)$ = research octane number of component $n$ when it contains TEL at concentration $z$

$m_n(z)$ = motor octane number of component $n$ when it contains TEL at concentration $z$

$r, m$ = minimum acceptable research (motor) octane number of the blend

$w$ = total usage of TEL, cm³/day

From a flow balance,

$$\sum_n x_n = y$$

Using the octane number blending assumption,

$$\sum_n r_n(z)x_n \geqslant ry \qquad \sum_n m_n(z)x_n \geqslant my \qquad (8.7)$$

Finally, $w = zy$, where $w$ would presumably be used in the objective function. Note that since $r_n$ and $m_n$ are functions of $z$, the two blending constraints are actually nonlinear functions. However, by choosing, say, four grid points, the curves of Figure 8.9 can be approximated by piecewise linear functions, as shown in Figure 8.10 and given in equation form below:

$$z = a_1\lambda_1 + a_2\lambda_2 + a_3\lambda_3 + a_4\lambda_4 \qquad \sum_{i=1}^{4} \lambda_i = 1 \qquad (8.8)$$

$$r_n(z) = r_{n1}\lambda_1 + r_{n2}\lambda_2 + r_{n3}\lambda_3 + r_{n4}\lambda_4$$

$$m_n(z) = m_{n1}\lambda_1 + m_{n2}\lambda_2 + m_{n3}\lambda_3 + m_{n4}\lambda_4 \qquad (8.9)$$

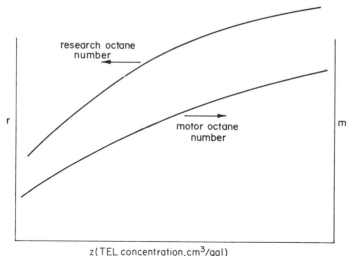

Figure 8.9. TEL susceptibility chart, Example 8.10.

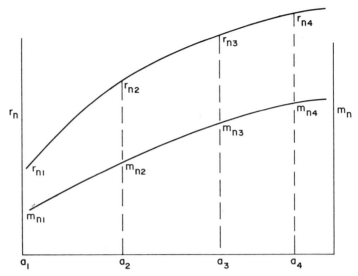

**Figure 8.10.**   Piecewise linear approximation to susceptibility chart.

The blending constraint thus takes the form

$$\sum_{n=1}^{N} x_n \left( r_{n1}\lambda_1 + r_{n2}\lambda_2 + r_{n3}\lambda_3 + r_{n4}\lambda_4 \right) \geqslant ry \qquad (8.10)$$

which, however, includes the nonlinear variable products $x_n \lambda_i$.

Let $r^{(u)} = \max_{n,i}(r_{ni})$ and $y^{(u)}$ be some reasonable upper bound on $y$, the gasoline production. Then, using substitution (8.5) discussed in Section 8.2, we can define

$$u_i = \frac{1}{2} \left[ \frac{1}{r^{(u)}y^{(u)}} \sum_{n=1}^{N} r_{ni}x_n + \lambda_i \right] \qquad i = 1,\ldots,4$$

$$v_i = \frac{1}{2} \left[ \frac{1}{r^{(u)}y^{(u)}} \sum_{n=1}^{N} r_{ni}x_n - \lambda_i \right] \qquad (8.11)$$

Note that since

$$0 \leqslant \lambda_i \leqslant 1 \quad \text{and} \quad 0 \leqslant \frac{1}{r^{(u)}y^{(u)}} \sum_{n} r_{ni}x_n \leqslant 1$$

it follows that

$$0 \leqslant u_i \leqslant 1 \qquad -\tfrac{1}{2} \leqslant v_i \leqslant \tfrac{1}{2} \qquad (8.12)$$

Thus, the blending constraint (8.10) takes the separable form,

$$\sum_{i=1}^{4} u_i^2 - \sum_{i=1}^{4} v_i^2 \geq \frac{ry}{r^{(u)}y^{(u)}} \tag{8.13}$$

Finally, each $u_i$, $v_i$ and its square is approximated by piecewise linear functions over the range given by Eqs. (8.12). For instance, with five equidistant grid points,

$$u_i = 0\beta_{i1} + \tfrac{1}{4}\beta_{i2} + \tfrac{1}{2}\beta_{i3} + \tfrac{3}{4}\beta_{i4} + \beta_{i5} \tag{8.14}$$

$$u_i^2 = 0\beta_{i1} + \tfrac{1}{16}\beta_{i2} + \tfrac{1}{4}\beta_{i3} + \tfrac{9}{16}\beta_{i4} + \beta_{i5} \tag{8.15}$$

and

$$\sum_{j=1}^{5} \beta_{ij} = 1 \tag{8.16}$$

Similar variables $\gamma_{ij}$ would be required for the $v_i$'s.

Thus, by using four grid points for the TEL concentration and five grid points for each $u_i$, $v_i$, we have introduced the following variables:

$$u_i, v_i, \lambda_i \qquad i = 1,\ldots,4$$

$$\beta_{ij}, \gamma_{ij} \qquad i = 1,\ldots,4; j = 1,\ldots,5$$

or a total of 52. We have introduced the following equations: Eq. (8.8) and the normalization condition on the $\lambda$'s, $N$ equations of type (8.9), eight equations of type (8.11), and 24 equations of the type (8.14)–(8.16). The total is $34 + N$ equations plus bounds (8.12). This may well seem like an enormous penalty to pay for adapting the octane number constraints (8.7) for LP solution. However, since these constraints normally constitute only a very small portion of a large otherwise linear refinery model, the grid-point constructions prove quite practical and are widely used in the industry [17, 18].

## 8.3 CUTTING PLANE METHODS

The two approaches considered in Section 8.2 represent two extremes in strategies for utilizing linear approximations to nonlinear functions. In the direct linearization method, we used a single linearization to approximate each nonlinear constraint. We found that this type of approximation is adequate in the vicinity of the base point but is generally unreliable if further excursions from the base point are made. Consequently, to ensure convergence, the step length allowed per iteration must be carefully constrained, leading to generally slow convergence rates. The piecewise linear approximations used in separable programming, on the other hand, yield good gross approximations over the entire range of each constraint but are generally restricted to separable functions and are limited in the accuracy that can be attained at any specific point by the density of grid points selected. Consequently, to improve local accuracy we found that we must either use a larger number of grid points or else use iterative refinement.

It seems apparent that it would be preferable to employ a hybrid algorithm that would employ rough approximations to the constraints to reach the immediate neighborhood of the optimum point and could then build up a more accurate local approximation to the constraints that would permit an accurate determination of the optimum point itself. In this way the labor of constructing good approximations would not be wasted while the iterates were still far from the optimum. This is in large part the motivation behind the algorithms that we consider in this section.

### 8.3.1  The Basic Cutting Plane Algorithm

For the purpose of the present discussion we restrict our attention to the problem

(I)                Minimize   $f(x)$

Subject to   $g_j(x) \geqslant 0$                $j = 1, \ldots, J$

$x_i^{(L)} \leqslant x_i \leqslant x_i^{(U)}$        $i = 1, \ldots, N$

The set of inequality constraints in problem (I) defines the region $F$ in $R^N$ within which all feasible solutions must lie. We assume that $F$ is nonempty and bounded. Following the general strategy suggested in the introduction to this section, we seek to devise an algorithm that will solve this problem by solving a sequence of intermediate problems constructed by starting out with a rough approximation to the feasible region and successively improving the approximation by adding constraint estimates updated at the intermediate solutions.

If the set $F$ is nonempty and bounded, then the simplest approximation to $F$ will consist of a hypercube containing $F$, that is, a set $Z^0$ given by

$$Z^0 = \{x: x_i^{(L)} \leqslant x_i \leqslant x_i^{(U)}, i = 1, \ldots, N\}$$

where, if they have not been previously defined, the bounds $x_i^{(L)}$ and $x_i^{(U)}$ are any estimates that will ensure that $Z^0 \supset F$. This approximation to $F$ can serve as the feasible region for the first subproblem:

Minimize   $f(x)$

Subject to   $x \in Z^0$

Suppose $x^{(1)}$ is the solution to this simplified problem, then two situations can arise: Either $x^{(1)}$ is feasible, that is, $x^{(1)} \in F$ or else $x^{(1)}$ is infeasible, that is, $x^{(1)} \in Z^0$ but $x^{(1)} \notin F$. In the first instance, the problem is solved, since $x^{(1)} \in F$ and $F \subset Z^0$ implies that $x^{(1)}$ must be the minimum point of the original problem. If the second case occurs, then we need to update our approximation to $F$ so that eventually a feasible point will be obtained. To accomplish this we observe that although $x^{(1)}$ is not feasible, it does give us some indication of the portion of the boundary of the feasible region within which the true constrained optimum is likely to be found, namely, the portion of $F$ closest to $x^{(1)}$. Clearly, if we are to expend effort to improve our approximation of $F$, then it is this portion of $F$ that should receive our attention.

In order to improve our approximation to $F$ in the vicinity of $x^{(1)}$, we will need to modify the boundaries of $Z^0$ near $x^{(1)}$ so that they will more closely approach the actual boundaries of $F$. This can be achieved by imposing on $Z^0$ additional constraints that will exclude from $Z^0$ the region in the vicinity of $x^{(1)}$. The effect of adding such "cutting" constraints will be to ensure that the minimum point of the updated approximate feasible region will be closer to the actual feasible region $F$ than $x^{(1)}$ was. Figure 8.11$a$ illustrates this idea graphically. In the figure the original feasible region $F$ is the cross-hatched area, while the set $Z^0$ is the cube within the limits $a_1 \leqslant x_1 \leqslant b_1$, $a_2 \leqslant x_2 \leqslant b_2$. If $x^{(1)}$ is the optimum for some objective function $f(x)$, for example, $f(x) = -x_1 - x_2$, over the set $Z^0$, then we can improve the current approximation to the boundary of $F$ in the neighborhood of $x^{(1)}$ by imposing one or more linear inequalities $p_i^{(1)}(x) \geqslant 0$ that will "cut off" some portion of $Z^0$ containing $x^{(1)}$, as shown in Figure 8.11$b$. If we let $H^{(1)}$ be the set

$$H^{(1)} = \left\{ x: \, p_i^{(1)}(x) \geqslant 0 \right\}$$

then the new approximation to the feasible region, designated as $Z^{(1)}$, will be given by

$$Z^{(1)} \equiv Z^{(0)} \cap H^{(1)}$$

Quite obviously, $Z^{(1)}$ will be contained in $Z^0$.

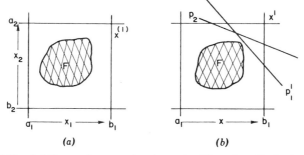

$(a)$                                     $(b)$

**Figure 8.11.** ($a$) Hypercube approximation to feasible region. ($b$) Cutting planes.

The smaller region $Z^{(1)}$ can next serve as the feasible region for the subproblem

$$\text{Minimize} \quad f(x)$$

$$\text{Subject to} \quad x \in Z^{(1)}$$

As before, the solution $x^{(2)}$ to this intermediate problem will either be a solution to the original problem or else help to locate the next set of linear inequalities that will be imposed to produce the still smaller region $Z^{(2)}$. If the computations are continued in this fashion, and if with each set of cuts we can be sure that a nonempty remaining portion of $Z^0$ is eliminated, then it seems reasonable that eventually a point $x$ will be reached that is feasible and that consequently will be the minimum of $f(x)$ over $F$. This is, in effect, the basic strategy of the family of methods known as *cutting plane methods*.

In outlining the general strategy employed in this family of algorithms, we have left unspecified both the numerical procedure to be used in solving the intermediate linearly constrained subproblems and the exact manner in which the cutting planes themselves are to be constructed. Alternative specifications of these two steps within the basic strategy lead to different versions of cutting plane algorithms. In the following discussion we consider the version reported by Kelley in 1960 [19] (and independently by Cheney and Goldstein [20]). This algorithm, known appropriately as *Kelley's cutting plane method*, must be considered one of the classic techniques of nonlinear programming.

### 8.3.2 Kelley's Algorithm

Kelley's cutting plane algorithm is designed to solve problems in the form of problem (II) in which the $g_j$ are concave functions:

(II) $$\text{Minimize} \quad f(x) = \sum_{i=1}^{N} c_i x_i$$

$$\text{Subject to} \quad g_j(x) \geqslant 0 \qquad j = 1, \ldots, J$$

$$x_i^{(L)} \leqslant x_i \leqslant x_i^{(U)} \qquad i = 1, \ldots, N$$

This formulation of the inequality-constrained nonlinear problem with linear objective function is entirely equivalent to problem (I) because if $f(x)$ is nonlinear it can always be reduced to the above form. This can be accomplished by introducing a new variable $x_0$ and the constraint

$$x_0 \geqslant f(x)$$

The reformulated problem becomes

$$\text{Minimize} \quad x_0$$

$$\text{Subject to} \quad g_j(x) \geq 0 \qquad j = 1,\ldots, J$$

$$x_i^{(L)} \leq x_i \leq x_i^{(U)} \qquad i = 1,\ldots, N$$

$$x_0 - f(x) \geq 0$$

which is just a special case of problem (II). The obvious advantage of this formulation is that, provided only linear cutting planes are generated during the course of the computations, the resulting intermediate optimization problems will always be *linear programming problems*. Consequently, the intermediate problems in Kelley's method are solved using standard LP algorithms.

The cuts used in the algorithm are constructed from the local linearization of the violated constraints in the following fashion. Suppose $x^{(1)}$ is the solution to

$$\text{Minimize} \quad f(x)$$

$$\text{Subject to} \quad x \in Z^0$$

and $x^{(1)} \notin F$. Since $x^{(1)}$ is not in $F$, there must be at least one constraint $k$ that is violated at $x^{(1)}$, that is,

$$g_k(x^{(1)}) < 0$$

The local linearization of each violated constraint at $x^{(1)}$,

$$\tilde{g}_j(x; x^{(1)}) \equiv g_j(x^{(1)}) + \nabla g_j(x^{(1)})(x - x^{(1)})$$

will serve to exclude the point $x^{(1)}$ from the remainder of the set $Z^0$, provided the linearization is written as a "greater than" inequality:

$$\tilde{g}_j(x; x^{(1)}) \geq 0$$

To verify this, note that since $g_j(x^{(1)}) < 0$, the point $x^{(1)}$ will not satisfy the above linearized inequality. Thus, in principle, the set $H^{(1)}$ generated at the subproblem optimum $x^{(1)}$ could consist of the linearizations of all constraints violated at $x^{(1)}$. In practice, however, this leads to the generation of an excessive number of linear constraints, and consequently Kelley proposed that only the linearization of the most violated constraint be used to construct the set $H^{(1)}$. In this fashion, with each subproblem only one additional linear inequality is added to update the current approximation to the feasible region.

All of the essential elements of the algorithm have now been established and can be assembled as summarized in the following.

### Kelley's Cutting Plane Method

Given a problem in the form of problem (II), a suitable linear programming technique, a constraint tolerance $\varepsilon > 0$, and an initial bound on the feasible region

$$Z^0 = \left\{x: x_i^{(L)} \leqslant x_i \leqslant x_i^{(U)}, i = 1, \ldots, N\right\}$$

such that

$$Z^0 \supset F$$

**Step 1.**  Solve the linear program

$$\text{Minimize} \quad \Sigma c_i x_i$$

$$\text{Subject to} \quad x \in Z^0$$

and designate the solution as $x^{(1)}$.

For $k = 1, 2, \ldots$, carry out the following series of steps.

**Step 2.**  Find $m$ such that

$$-g_m(x^{(k)}) = \max\left[-g_j(x^{(k)}), 0: j = 1, \ldots, J\right]$$

If $g_m(x^{(k)}) > -\varepsilon$, terminate. Otherwise go to step 3.

**Step 3.**  Construct the cutting plane,

$$p^{(k)}(x) \equiv \tilde{g}_m(x; x^{(k)}) = g_m(x^{(k)}) + \nabla g_m(x^{(k)})(x - x^{(k)})$$

and let $H^{(k)}$ be the half-space $H^{(k)} = \{x: \ p^{(k)}(x) \geqslant 0\}$. Solve the linear program:

$$\text{Minimize} \quad \Sigma c_i x_i$$

$$\text{Subject to} \quad x \in Z^{(k-1)} \cap H^{(k)}$$

Designate the solution $x^{(k+1)}$.

**Step 4.**  Set $Z^{(k)} = Z^{(k-1)} \cap H^{(k)}$ and $k = k + 1$. Go to step 2.

*Example 8.11*

Determine the solution to the following problem using Kelley's method.

$$\text{Minimize} \quad f(x) = x_1 - x_2$$

$$\text{Subject to} \quad g_1(x) = 2x_1 - x_2^2 - 1 \geq 0$$

$$g_2(x) = 9 - 0.8x_1^2 - 2x_2 \geq 0$$

$$x_1 \geq 0 \qquad x_2 \geq 0$$

From Figure 8.12 it is apparent that the bounds $0 \leq x_1 \leq 5$ and $0 \leq x_2 \leq 4$ adequately bracket the feasible region for this problem. Note that $g_1$ and $g_2$ are concave functions.

Step 1 can be initiated with the set $Z^0 = [x: 0 \leq x_1 \leq 5 \text{ and } 0 \leq x_2 \leq 4]$ and some suitable small $\varepsilon > 0$.

**Step 1.**  Find the minimum of

$$f(x) = -x_1 - x_2$$

$$\text{Subject to} \quad 0 \leq x_1 \leq 5 \quad \text{and} \quad 0 \leq x_2 \leq 4$$

The solution quite obviously is $x^{(1)} = (5, 4)$.

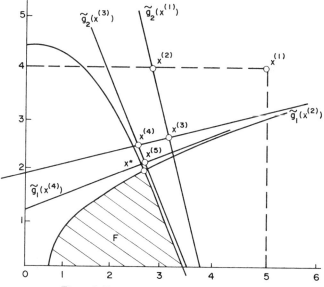

**Figure 8.12.**  Feasible region, Example 8.11.

**Step 2.** Since $g_1(x^{(1)}) = -7$ and $g_2(x^{(1)}) = -19$, the most violated constraint is $g_2(x)$. We continue with step 3.

**Step 3.** The cutting plane is calculated as the linearization of $g_2(x)$ at $x^{(1)}$, or

$$p^{(1)}(x) \equiv \tilde{g}_2(x; x^{(1)}) = -19 - 8(x_1 - 5) - 2(x_2 - 4)$$

The new inequality to be added to $Z^0$ is thus

$$-19 - 8(x_1 - 5) - 2(x_2 - 4) \geqslant 0$$

Note that the point $x^{(1)}$ does *not* satisfy this inequality. The second subproblem to be solved now is

**Step 1**

$$\text{Minimize} \quad -x_1 - x_2$$

$$\text{Subject to} \quad 29 - 8x_1 - 2x_2 \geqslant 0$$

$$0 \leqslant x_1 \leqslant 5 \qquad 0 \leqslant x_2 \leqslant 4$$

and its solution is $x^{(2)} = (2.625, 4)$.

**Step 2**

$$g_1(x^{(1)}) = -11.75 \qquad \text{and} \qquad g_2(x^{(2)}) = -4.5125$$

Clearly, $g_1(x)$ is most violated. Moreover, the violation is large, and consequently the algorithm does not terminate.

**Step 3.** The equation of the cutting plane is

$$p^{(2)}(x) \equiv g_1(x; x^{(2)})$$

$$= -11.75 + 2(x_1 - 2.625) - 8(x_2 - 4) = 0$$

and again the cut clearly separates $x^{(2)}$ from the feasible region $F$.

The third intermediate linear program is

$$\text{Minimize} \quad -x_1 - x_2$$

$$\text{Subject to} \quad 29 - 8x_1 - 2x_2 \geqslant 0$$

$$15 + 2x_1 - 8x_2 \geqslant 0$$

$$0 \leqslant x_1 \leqslant 5 \qquad 0 \leqslant x_2 \leqslant 9$$

The solution is $x^{(3)} = (2.971, 2.618)$.

The iterations continue in this fashion until the constraints are satisfied within the specified tolerance $\varepsilon$. The final solution is $x^* = (2.5, 2.0)$.

The history of the iterations is summarized as follows:

| Subproblem Index | $x_1^{(k)}$ | $x_2^{(k)}$ | $f(x^{(k)})$ |
|---|---|---|---|
| 1 | 5 | 4 | $-9.000$ |
| 2 | 2.625 | 4 | $-6.625$ |
| 3 | 2.971 | 2.618 | $-5.589$ |
| 4 | 2.343 | 2.461 | $-4.804$ |
| 5 | 2.516 | 2.050 | $-4.566$ |
| $\infty$ | 2.500 | 2.000 | $-4.500$ |

From the tabulation it is apparent that the optimum objective function value is approached from below. This comes about because the feasible regions for the subproblems are continually becoming smaller. Hence the subproblem minima must increase monotonically.

### 8.3.3 Computational Aspects and Properties

The basic Kelley algorithm has been used with some success in various applications. As we will note in Chapter 11, it has proved to be a very effective means of solving a special class of nonlinear optimization problems known as generalized geometric programs. The method has the advantage common to all linearization techniques that any linearity or near linearity in the original problem is preserved and directly utilized. It has the further obvious advantage that the subproblem to be solved at each major iteration is one for which the powerful techniques of linear programming are applicable. Specifically, since each successive LP subproblem differs from the preceding subproblem by only a single constraint, the dual simplex method mentioned in Chapter 4 is particularly effective. In particular, the addition of a constraint to the previous optimal subproblem tableau is accommodated with at most a few reduction steps to eliminate any basic variable coefficients. Moreover, the resulting dual feasible solution can be used immediately to initiate iterations to determine the next subproblem optimum.

Although the successive LP subproblem solutions can be obtained very efficiently, the basic technique does suffer from several theoretical and practical disadvantages, some of which fortunately can be ameliorated. The first and major disadvantage is that the algorithm generates a sequence of infeasible points. Thus, it cannot be terminated early with a "good" but perhaps not optimal solution. A second and theoretically important disadvantage is that the feasible region $F$ must be a *convex set*. Although many nonconvex problems have been solved using the technique, in principle convergence to an optimum point can be guaranteed only if $F$ is convex. Convexity of $F$ is, of course, equivalent to requiring that each of the constraints $g_i(x)$ be *concave* functions over $F$. Recall that for a concave function a local linearization will always overestimate the function value; that is, given a point $w$ and any other point $x$,

$$g(x) \leqslant g(w) + \nabla g(w)(x - w)$$

Thus, if $x$ is any point satisfying $g_j(x) \geqslant 0$, then it will also satisfy the inequality

$$0 \leqslant g_j(w) + \nabla g_j(w)(x - w)$$

for any $w$. In other words, if the constraints $g_j(x)$ are all concave functions, then the linearization will always include the original feasible region $F$. Consequently, the sequence of approximate feasible regions $Z^{(k)}$ generated by Kelley's algorithm will be a family of nested sets with each set containing the feasible equation $F$,

$$Z^0 \supset Z^{(1)} \supset Z^{(2)} \supset \cdots \supset Z^{(k)} \supset F$$

This feature is necessary to guarantee that in the process of generating cutting planes we do not generate planes that actually will eliminate a portion of the feasible region $F$. In the absence of convexity it is theoretically possible that this will happen.

A third practical disadvantage of Kelley's method is that the size of the LP subproblem grows continuously with each iteration. Moreover, as the optimum point is approached closely, the successive cutting planes begin to be very nearly identical. This leads to numerical problems in solving the linear programs, because the tableau rows will become very nearly linearly dependent. An obvious way of reducing the number of cuts that must be handled is to institute procedures for dropping old cutting planes from the subproblems whenever possible. However, care must be exercised when this is done to ensure that instabilities are not introduced into the iterations. In particular, as shown by Topkis [21] and Eaves and Zangwill [22], old cuts should be deleted only if they are not binding at the optimum of the current subproblem. These authors have proposed the following deletion procedure.

### Cut-Deletion Procedure

Let $I^t$ be the set of indices of the cutting planes accumulated but not yet deleted by iteration $t$.

Replace Step 4 of Kelley's method with the following:

**Step 4a.**   Determine the set $D$ of cutting planes to be deleted. For each $i \in I^t \cup \{t\}$, if both

$$p^{(i)}(x^{(t+1)}) > 0 \; (> \varepsilon)$$

and

$$f(x^{(t+1)}) \geq f(x^{(i)}) + p^{(i)}(x^{(i)})$$

then include $i$ in the set $D$. Let $I^{t+1} = I^t \cup \{t\} - D$.

**Step 4b.** Set

$$Z^{t+1} = Z^0 \cap \{x: p^{(i)}(x) \geq 0, i \in I^{t+1}\}$$

$$t = t + 1$$

and go to step 2.

This procedure deletes constraints generated prior to subproblem $t$ provided they are both nonbinding at $x^{(t+1)}$ and that the improvement in $f(x)$ since the old constraint was generated is large enough. The second criterion in effect guarantees that monotonicity of function values is maintained. As shown in references 21 and 22, this procedure will ensure convergence in the case of convex constraint sets. From an implementation point of view, cut deletion can readily be accommodated within the dual simplex method normally used for LP subproblem solution.

### Example 8.12

Consider the use of the modified Kelley method to solve the problem

$$\text{Minimize} \quad 3x_1 + x_2$$

$$\text{Subject to} \quad g(x) = 1 - (x_1 - 2)^2 - (x_2 - 5)^2 \geq 0$$

and suppose the set

$$Z^0 = \{x: 0 \leq x_1 \leq 5, 0 \leq x_2 \leq 7\}$$

is given as the first approximation to $F$. The solution to the first subproblem is obviously $x^{(1)} = 0$ with $f(x^{(1)}) = 0$. The second subproblem will include the constraint

$$-28 + 4x_1 + 10x_2 \geq 0$$

And has the solution

$$x^{(2)} = (0, 2.8) \qquad \text{with } f(x^{(1)}) = 2.8$$

Suppose we continue for three more iterations, retaining all previous cutting planes. The fifth subproblem will thus take the form:

$$\text{Minimize} \quad 3x_1 + x_2$$

$$\text{Subject to} \quad p^{(1)}(x) = -28 + 4x_1 + 10x_2 \geqslant 0$$

$$p^{(2)}(x) = -20.16 + 4x_1 + 4.4x_2 \geqslant 0$$

$$p^{(3)}(x) = -7.024 + 4x_1 + 0.84x_2 \geqslant 0$$

$$p^{(4)}(x) = -13.422 + 2.038x_1 + 2.62x_2 \geqslant 0$$

$$0 \leqslant x_1 \leqslant 5 \qquad 0 \leqslant x_2 \leqslant 7$$

Let us apply the deletion procedure to this subproblem. Since all previous cuts have been retained, $I = [1, 2, 3]$. The solution to the above linear program is $x^{(5)} = (0.813, 4.491)$ with $f(x^{(5)}) = 6.930$. At this point, we are ready for **step 4a:**

For $i = 1, 2, 3, 4$, we check whether the two conditions given are satisfied:

$i = 1$:

$$p^{(1)}(x^{(5)}) = 20.162 > 0$$

and

$$f(x^{(5)}) = 6.903 > f(x^{(1)}) + p^{(1)}(x^{(1)}) = 0 + (-28)$$

therefore, $D = \{1\}$.

$i = 2$:

$$p^{(2)}(x^{(5)}) = 2.852 > 0$$

and

$$f(x^{(5)}) = 6.930 > f(x^{(2)}) + p^{(2)}(x^{(2)}) = 2.8 + (-784)$$

therefore, $D = \{1, 2\}$.

$i = 3$:

$$p^{(3)}(x^{(5)}) = 0.0$$

$i = 4$:

$$p^{(4)}(x^{(5)}) = 0.0$$

Thus,

$$I^{(5)} = \{1, 2, 3\} \cup \{4\} - \{1, 2\} = \{3, 4\}$$

**Step 4b.**

$$Z^{(5)} = \{x: \ 0 \leqslant x_1 < 5, 0 \leqslant x_2 \leqslant 7\}$$

$$\cap \{x: \ p^{(3)}(x) \geqslant 0 \text{ and } p^{(4)}(x) \geqslant 0\}$$

As a result of step 4, two of the old planes have been deleted. The iteration returns to steps 2 and 3. The next cutting plane is

$$p^{(5)}(x) = -7.170 + 2.374x_1 + 1.018x_2 = 0$$

and the next subproblem becomes

$$\text{Minimize} \quad 3x_1 + x_2$$

$$\text{Subject to} \quad p^{(3)}(x) \geqslant 0$$

$$p^{(4)}(x) \geqslant 0$$

$$p^{(5)}(x) \geqslant 0$$

$$0 \leqslant x_1 \leqslant 5 \qquad 0 \leqslant x_2 \leqslant 7$$

The solution is $x^{(6)} = (0.5427, 5.778)$, with $f(x^{(6)}) = 7.406$. Again, step 4 has been reached. After checking the planes $\{3, 4, 5\}$, the iterations continue.

The results of the next six iterations are summarized below and are depicted in Figure 8.13.

| Subproblem Index | $x_1^{(t)}$ | $x_2^{(t)}$ | $f(x^{(t)})$ | $I^t$ |
|---|---|---|---|---|
| 6 | 0.5427 | 5.778 | 7.406 | 3, 5 |
| 7 | 0.8069 | 5.162 | 7.582 | 5, 6 |
| 8 | 0.9501 | 4.828 | 7.678 | 5, 7 |
| 9 | 1.052 | 4.590 | 7.746 | 5, 8 |
| 10 | 1.030 | 4.721 | 7.813 | 8, 9 |
| 11 | 1.058 | 4.658 | 7.831 | 9, 10 |
| 12 | 1.048 | 4.692 | 7.836 | 10, 11 |
| ∞ | 1.051 | 4.684 | 7.838 | — |

As can be seen from the summary, the deletion criterion does ensure that no more than three planes need be included in every subproblem. In the unmodified Kelley's algorithm, since all old planes would have to be retained, the twelfth subproblem would have to involve all 11 previously generated cutting plane inequalities.

Note from Figure 8.13 that although by virtue of the cut-deletion procedure the number of planes that have to be considered is reduced, the problem of dealing with nearly parallel planes nonetheless persists. As the cuts approach the feasible region, the constraints do become more and more parallel, for example, $p^{(8)}$ and $p^{(5)}$. Consequently, their accurate numerical solution to determine the point of intersection becomes more and more problematic. This is one difficulty of cutting plane methods that remains unresolved.

### 8.3.4    Discussion

The convergence of the Kelley cutting plane algorithm can be proven under general conditions that include the convexity of the constraint functions both with [21] and without [23] cut deletion. However, as evident from the examples presented in this section, the progress of cutting plane iterations is quite reminiscent of the trajectory of the gradient method of unconstrained optimi-

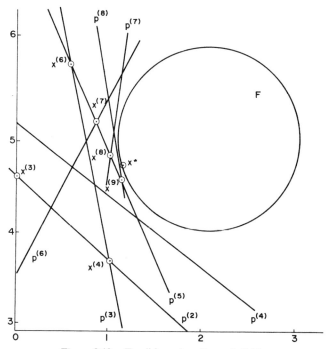

**Figure 8.13.**   Feasible region, Example 8.12.

zation. In fact, for special cases of cutting plane algorithms it has been shown that the rates of convergence of both the objective function values and the intermediate solution values are linear [3]. A generalized rate of convergence analysis of the entire family of cutting plane algorithms cannot be given, because the rate depends very directly on the details of the construction of the cutting planes.

Other variations of the cutting plane strategy have been reported by Veinott [24] and Elzinga and Moore [25]. These methods basically attempt to yield "deeper" cuts so as to obtain a better definition of the feasible region. The reader is invited to consult these references and the work of Eaves and Zangwill [22] for further details.

## 8.4 SUMMARY

This chapter explored three basic strategies for employing linear approximations in solving nonlinearly constrained problems: the direct successive LP approach, global approximation via piecewise linear functions, and outer approximations using linearizations of the most violated constraints. In the direct linearization approach we found that the LP solution can be used only to define a direction that must be searched using the methods of Chapter 2. The Frank-Wolfe algorithm, applicable to linearly constrained problems, can thus be viewed as an alternating series of LP and line-search subproblems. The application of the successive LP approach to nonlinear constraints introduced the need to adjust the LP solution to maintain constraint feasibility. The Griffith and Stewart method of bounding the step to both maintain feasibility and descent leads to very slow progress; the use of a penalty function is much preferred. Nonetheless, even with these enhancements, we conclude that the successive LP approach is most appropriate for problems with few nonlinear variables.

The separable programming technique involves the use of piecewise linear constructions to approximate separable nonlinear functions over the full range of the problem variables. This construction does substantially increase problem dimensionality and normally requires grid readjustment if more accurate solutions are required. For nonconvex feasible regions, the resulting linear program must be solved with restrictions in the choice of basic variables to ensure that only points on the approximating segments are generated. The separable programming construction is most appropriate for problems that would be LP problems except for the presence of a few separable nonlinear elements. Such situations occur widely in planning and scheduling applications.

Finally, we considered the class of cutting plane methods. This family of methods is based on the concept of successively improving the linear approximations to the constraint boundary in the vicinity of the solution as that solution is approached from outside the feasible region. In Kelley's classical algorithm, the linear cutting planes serve to successively eliminate portions of the previous approximation of the feasible region. This strategy is applicable only to inequality-constrained problems and, in principle, requires the feasible region to be convex. The series of LP subproblems can be solved efficiently using specialized LP methods, particularly if cut-deletion procedures are employed to reject old cutting planes that have been superseded. Cutting plane methods are susceptible to limitations on numerical accuracy that arise because near the solution successive planes become nearly parallel. In spite of these difficulties, this family of methods has been used effectively to solve difficult specially structured problems and continues to be the subject of further research.

## REFERENCES

1. Frank, N., and P. Wolfe, "An Algorithm for Quadratic Programming," *Naval Res. Log. Quart.*, **3**, 95–110 (1956).
2. Zangwill, W. I., *Nonlinear Programming*, Prentice-Hall, Englewood Cliffs, NJ, 1969.
3. Wolfe, P., "Convergence Theory in Nonlinear Programming," in *Integer and Nonlinear Programming* (J. Abadie, Ed.), North-Holland, Amsterdam, 1970.
4. Griffith, R. E., and R. A. Stewart, "A Nonlinear Programming Technique for Optimization of Continuous Processing Systems," *Manage. Sci.*, **7**, 379–391 (1961).
5. Smith, H. V., "A Process Optimization Program for Nonlinear Systems, Pop II," 7090 H9 IBM UD21, SHARE General Program Library, 1965.
6. Graves, G. W., "Development and Testing of a Nonlinear Programming Algorithm," Aerospace Corporation, Rep. No. ATR-64 (7040)-2, June 1964.
7. Clasen, R. J., G. W. Graves, and J. Y. Lu, "Sortie Allocation by a Nonlinear Programming Model for Determining a Munitions Mix," Rand Corporation Rep. R-1411-DDPAE, March 1974.
8. Palacios-Gomez, F. E., "The Solution of Nonlinear Optimization Problems Using Successive Linear Programming," Ph.D. dissertation, School Business Administration, The University of Texas, Austin, TX, 1980.
9. Beale, E. M. L., "Nonlinear Programming Using a General Mathematical Programming System," in *Design and Implementation of Optimization Software* (H. J. Greenberg, Ed.), Sijthoff and Noordhoff, Netherlands, 1978.
10. Lasdon, L. S., and A. D. Warren, "Survey of Nonlinear Programming Applications," *Oper. Res.*, **28**, 1029–1073 (1980).
11. Baker, T. E., L. S. Lasdon, "Successive Linear Programming at Exxon," Working Paper 81/82-2-8, Graduate School of Business, The University of Texas, Austin, TX, July 1982.
12. Haverly, C. A., "Studies of the Behavior of Recursion for the Pooling Problem," *SIGMAP Bull.*, **25**, 19–28 (Dec. 1978).
13. Lasdon, L. S., A. D. Warren, S. Sarkar, and F. E. Palacios-Gomez, "Solving the Pooling Problem using Generalized Reduced Gradient and Successive Linear Programming Algorithms," *SIGMAP Bull.*, **77**, 9–15 (July 1979).
14. Miller, C. E., "The Simplex Method for Local Separable Programming," in *Recent Advances in Mathematical Programming* (R. Graves and P. Wolfe, Eds.), McGraw-Hill, New York, 1963.

15. Bazaraa, M. S., and C. M. Shetty, *Nonlinear Programming: Theory and Algorithms*, Wiley, New York, 1979.

16. Beale, E. M. L., "Advanced Algorithmic Features for General Mathematical Programming Systems," in *Integer and Nonlinear Programming* (J. Abadie, Ed.), North-Holland, Amsterdam, 1970.

17. El Agizy, "Applications of Separable Programming to Refinery Models," ORSA 36th National Meeting, Nov. 1969.

18. Kawaratoni, T. K., R. J. Ullman, and G. B. Dantzig, "Computing Tetraethyl-lead Requirements in a Linear Programming Format," *Oper. Res.*, **8**, 24–29 (1960).

19. Kelley, J. E., "The Cutting Plane Method for Solving Convex Programs," *SIAM J.*, **8**, 703–712 (1960).

20. Cheney, E. W., and A. A. Goldstein, "Newton's Method of Convex Programming and Tchebycheff Approximation," *Numer. Math.*, **1**, 253–268 (1959).

21. Topkis, D. M., "Cutting Plane Methods without Nested Constraint Sets," *Oper. Res.*, **18**, 404–413 (1975).

22. Eaves, B. C., and Zangwill, W. I., "Generalized Cutting Plane Algorithms," *SIAM. J. Control*, **9**, 529–542 (1971).

23. Zangwill, W. I., "Nonlinear Programming: A Unified Approach," Prentice-Hall, Englewood Cliffs, NJ, 1969, chap. 9.

24. Veinott, A. F., Jr., "The Supporting Hyperplane Method for Unimodal Programming," *Oper. Res.*, **15**, 147–152 (1967).

25. Elzinga, J., and T. G. Moore, "A Central Cutting Plane Algorithm for the Convex Programming Problem," *Math. Programming*, **8**, 134–145 (1975).

## PROBLEMS

**8.1.** Why is a line search necessary in the Frank-Wolfe algorithm? Why must the algorithm be initiated with a feasible point?

**8.2.** What is the fundamental reason for placing bounds on the changes in the variables in the Griffith/Stewart direct linearization method?

**8.3.** Suggest a method by means of which a feasible starting point can be computed for the Frank-Wolfe algorithm.

**8.4.** Since the LP subproblem of the Frank-Wolfe algorithm always yields corner points as a solution, explain how the algorithm would approach a point that is strictly in the interior of the feasible region.

**8.5.** Recall that in the successive LP strategy using the penalty function, only a rough minimization of the penalty function is carried out. Discuss the merits of also solving the LP subproblem only approximately. Do you think just one LP iteration per subproblem might suffice? Explain.

**8.6.** Outline a penalty-type successive LP strategy that would exclude linear constraints and variable bounds from the penalty-function construction. Discuss the merits of such a strategy.

**8.7.** Under what problem conditions is separable programming likely to be preferable to successive LP methods?

**8.8.** What is restricted basis entry, why is it necessary, and under what conditions can it be disregarded?

**8.9.** If a grid is selected for a given variable $x_n$ and used to approximate a function $f(x_n)$, must the same grid be used for all other functions involving $x_n$? Explain.

**8.10.** Outline two strategies for accommodating nonlinear equality constraints in the cutting plane algorithm. Discuss their relative merits.

**8.11.** Explain how linear constraints should be treated in the cutting plane method. Can linear equalities be accommodated?

**8.12.** What is the purpose of adding a new constraint at each iteration of Kelley's algorithm?

**8.13.** Discuss the merits of using cut deletion together with the addition of more than one new cutting plane per subproblem.

**8.14.** Prove or disprove that Kelley's method will yield monotonic increases in the objective function as well as monotonic decreases in the constraint infeasibilities for convex problems.

**8.15.** Form a cutting plane constraint at the point $\bar{x} = (1, 2)$ for the following nonlinear inequality:

$$g(x) = x_1^2 x_2 + x_2^2 x_1 \leqslant 10$$

**8.16.** Consider the problem

$$\text{Minimize} \quad f(x) = \tfrac{1}{3}(x_1 + 3)^3 + x_2$$

$$\text{Subject to} \quad 3 \geqslant x_1 \geqslant 1$$

$$2 \geqslant x_2 \geqslant 0$$

(a) Construct a piecewise linear approximation to the cubic term over the range of $x_1$ using four grid points.

(b) Will restricted basis entry be necessary in solving the resulting LP problem? Explain.

(c) Solve the approximating linear program. Compare the solution to the actual minimum.

**8.17.** Consider the problem:

$$\text{Minimize} \quad f(x) = (x_1 - 2)^2 + (x_2 - 2)^2$$

$$\text{Subject to} \quad g_1(x) = 3 - x_1 - x_2 \geqslant 0$$

$$g_2(x) = 10x_1 - x_2 - 2 \geqslant 0$$

$$x_2 \geqslant 0$$

(a) Can $x^0 = (0, 0)$ be used as starting point for the Frank-Wolfe algorithm? Why?

(b) Suppose $x_0 = (0.2, 0)$ is used. Solve the problem using the Frank-Wolfe algorithm.

**8.18.** Consider the problem

$$\text{Maximize} \quad f(x) = 2x_1 - 3(x_2 - 4)^2$$

$$\text{Subject to} \quad g_1(x) = 10 - x_1^2 - x_2^2 \geqslant 0$$

$$g_2(x) = 9 - x_1^2 - (x_2 - 4)^2 \geqslant 0$$

$$x_1 \geqslant 0$$

(a) Given the starting point $x^0 = (2, 2)$, calculate the next iteration point that would result if a small step gradient method were applied with $\delta = 1$.

(b) Repeat the calculation with the point $x^0 = (0, 1)$ and with the point $(1, 3)$.

**8.19.** Replace the objective function of Example 8.16 with the new objective function,

$$\max f(x) = x_1^{1/4} + x_2$$

(a) Show that for separable programming purposes, restricted basis entry will not be required.

(b) Using the same grid of points as used in Example 8.16, solve the resulting linear program.

**8.20.** Given the constraint set

$$g(x) = x_1^2 + x_2^2 - 1 \geqslant 0$$

$$1.5 \geqslant x_1 \geqslant 0$$

and the point $x_0 = (\sqrt{3}/3, \sqrt{3}/3)$, construct the cutting plane at $x^0$, and show that the cut will eliminate a portion of the feasible region. [*Hint:* Use the trial point $(1, 0)$.]

**8.21.** Continue the solution of the problem of Example 8.12 from the fifth subproblem with a cut-deletion procedure in which a cut is deleted if it is loose at the new point. Make a graph of the results of the next three or four iterations, and explain what you observe.

**8.22.** (a) Construct a piecewise linear approximation to the separable function

$$f(x_1, x_2) = x_1^2 + x_1^{-1} + 2(x_2 + 2)^{-2}$$

over the intervals $\frac{1}{4} \leqslant x_1 \leqslant \frac{3}{2}$ and $-1 \leqslant x_2 \leqslant 1$. Use five grid points for $x_1$ and four grid points for $x_2$.

(b) Use the simplex method to minimize the approximating function over the region in $x_1$ and $x_2$.

**8.23.** Given the problem

$$\text{Maximize} \quad f(x) = x_1 + x_2^2$$

$$\text{Subject to} \quad g_1(x) = 25 - x_1^2 - x_2^2 \geqslant 0$$

$$g_2(x) = 9 - x_1^2 - x_2 \geqslant 0$$

$$5 \geqslant x_1 \geqslant 0 \qquad 10 \geqslant x_2 \geqslant 0$$

(a) Set up and solve the first subproblem for Kelley's cutting plane method.

(b) Calculate the cutting plane at the resulting point.

(c) Set up and solve the second subproblem, and generate the next cut.

**8.24.** Suppose the problem

$$\text{Minimize} \quad f(x) = -x_1 - x_2$$

$$\text{Subject to} \quad g_1(x) = 25 - x_1^2 - x_2^2 \geqslant 0$$

$$g_2(x) = 9 - 2x_1 - x_2 \geqslant 0$$

$$x_1 \geqslant 0 \qquad x_2 \geqslant 0$$

is solved using Kelley's method.

(a) Should constraints $g_1$ and $g_2$ be treated differently in the algorithm?

(b) Can this observation be generalized?

**8.25.** Show that the problem

$$\text{Minimize} \quad f(x) = x_1 + x_2^2 + x_3^3$$

$$\text{Subject to} \quad 64 x_1^{-2} x_2^{-4} x_3^{-6} \leqslant 1$$

$$\text{all } x_i \geqslant 0$$

can be solved using separable programming. Assuming an upper bound of 10 for all variables, formulate the separable program using 5 grid points for each variable.

**8.26.** Given the nonlinear program

$$\text{Minimize} \quad f(x) = x_1^2 + x_2^2 + x_3^2$$

$$\text{Subject to} \quad g_1(x) = 1 - x_2^{-1}x_3 \geqslant 0$$

$$g_2(x) = x_1 - x_3 \geqslant 0$$

$$h_1(x) = x_1 - x_2^2 + x_2 x_3 - 4 = 0$$

$$0 \leqslant x_1 \leqslant 5$$

$$0 \leqslant x_2 \leqslant 3$$

$$0 \leqslant x_3 \leqslant 3$$

Carry out one iteration of the modified successive LP algorithm of Palacios-Gomez with $x^0 = (4, 2, 2)$, $\delta = 1$, and $\varepsilon = 10^{-3}$.

**8.27.** Can the NLP of problem 8.26 be solved using the cutting plane method? If not, reformulate it so that it can be solved. Is the resulting problem convex? Make a plot of the feasible region to confirm.

**8.28.** Reformulate the pooling problem discussion in Example 8.7 for solution using the separable programming strategy. Use five grid points for flows and four for the composition. Can the resulting problem be solved using a library LP routine? Are any modifications required?

**8.29.** (a) Construct a piecewise linear approximation to the function

$$f(x) = 10 \sin x$$

over the interval $0 \leqslant x \leqslant 2\pi$ rad using nine grid points.

(b) What is the maximum error in your approximating function?

(c) Show that if the simplex LP method is applied to minimize the approximating function, then the minimum must occur at a grid point.

**8.30.** Given the problem

$$\text{Minimize} \quad f(x) = x_1^2 + x_2^2 + x_3^2$$

$$\text{Subject to} \quad x_1 x_2 x_3 = 9$$

$$x_1 + x_2^2 \leqslant 6$$

$$x_2^2 + x_1 \leqslant 4 \qquad x_1, x_2 \geqslant 0$$

How would you set up the problem for solution using Kelley's cutting plane method? Can the problem be solved using this method?

**8.31.** Explain what happens when Kelley's cutting plane method is used to solve the problem.

$$\text{Minimize} \quad f(x) = (x_1 - \tfrac{1}{2})^2 + (x_2 - 1)^2$$

$$\text{Subject to} \quad 4 - x_1^2 - 2x_2^2 \geqslant 0$$

$$-2 \leqslant (x_1, x_2) \leqslant 2$$

[*Hint*: First construct a rough sketch.]

**8.32.** Consider the problem

$$\text{Minimize} \quad f(x) = (x_1 - x_2)^2 + x_2$$

$$\text{Subject to} \quad x_1 + x_2 \geqslant 1$$

$$x_1 + 2x_2 \leqslant 3$$

$$x_1 \geqslant 0 \qquad x_2 \geqslant 0$$

with initial point $x^0 = (3, 0)$.

(a) Carry out three iterations of the Frank-Wolfe algorithm.

(b) Reformulate the problem for solution using the cutting plane method. Carry out three iterations.

(c) Use a sketch of the feasible region to explain differences in the iteration trajectories.

# Chapter 9

# Direction-Generation Methods Based on Linearization

The linearization-based algorithms that we examined in Chapter 8 have the common characteristic of using LP solution techniques to specify the sequence of intermediate solution points. Given a point $x^{(t)}$, the linearized subproblem at $x^{(t)}$ is updated in some prescribed fashion, and then the exact location of the next iterate $x^{(t+1)}$ is determined by the LP procedure. Yet we know that away from the base point for the linearization, the linearized subproblem cannot be expected to give a very good estimate of either the boundaries of the feasible region or the contours of the objective function.

Rather than relying on the admittedly inaccurate linearization to define the precise location of a point, perhaps it is more realistic to utilize the linear approximations only to determine a locally good *direction* for search. The location of the optimum point along this direction could then be established by direct examination of values of the original objective and constraint functions along that direction rather than by recourse to the old linearized constraints, which lose their accuracy once a departure from the base point is made. This strategy is analogous to that employed in gradient-based unconstrained search methods: A linear approximation (the gradient) is used to determine a direction, but the actual function values are used to guide the search along this direction. Of course, in the constrained case the linearization will involve the objective function as well as the constraints, and the directions generated will have to be chosen so that they lead to feasible points. This is the motivation for the methods to be discussed in this chapter.

## 9.1. THE METHOD OF FEASIBLE DIRECTIONS

Chronologically, the first of the direction-generating methods based on linearization have been the family of *feasible direction* methods developed by Zoutendijk [1].

Consider the inequality-constrained problem

$$\text{Minimize} \quad f(x)$$

$$\text{Subject to} \quad g_j(x) \geq 0 \quad j = 1, \ldots, J$$

Suppose that $x^{(1)}$ is a starting point that satisfies all the constraints, that is, $g_j(x^{(1)}) \geq 0$; $j = 1, \ldots, J$, and suppose that a certain subset of these constraints are binding at $x^{(1)}$. Zoutendijk postulated that a direction $d$ would be a good direction for search if it were a descent direction, that is,

$$\nabla f(x^{(1)}) \cdot d < 0$$

and if, for at least a small distance along the ray,

$$x(\alpha) = x^{(1)} + \alpha d \quad \text{with } \alpha \geq 0$$

the points $x$ along $d$ were feasible. To a first-order approximation, the points $x(\alpha)$ along $d$ will be feasible if, for all constraints that are binding at $x^{(1)}$,

$$\tilde{g}_j(x; x^{(1)}) \equiv g_j(x^{(1)}) + \nabla g_j(x^{(1)})(x - x^{(1)}) \geq 0$$

Since, by assumption,

$$g_j(x^{(1)}) = 0$$

and

$$x - x^{(1)} = \alpha d \quad \text{with } \alpha \geq 0$$

it follows that this requirement is equivalent to requiring that the directions $d$ satisfy

$$\nabla g_j(x^{(1)})d \geq 0$$

for all constraints $g_j(x)$ that are binding at $x^{(1)}$. A direction $d$ that satisfies the above inequalities at $x^{(1)}$ is appropriately called a *feasible direction*. Zoutendijk's basic idea, then, is at each stage of the iteration to determine a vector $d$ that will be both a feasible direction and a descent direction. This is accomplished numerically by finding a normalized direction vector $d$ and a scalar parameter $\theta > 0$ such that

$$\nabla f(x^{(1)})d \leq -\theta$$

and

$$\nabla g_j(x^{(1)})d \geq \theta$$

and $\theta$ is as large as possible. In carrying out the numerical solution of this direction-finding problem, the candidate $d$'s can be normalized conveniently by imposing the bounds $-1 \leqslant d_i \leqslant +1$, $i = 1,\ldots, N$. The vector $d$ that is determined in this manner will be a reasonable compromise between choosing a direction of maximum descent and following a path into the interior of the feasible region that avoids the constraints.

Once the direction vector is calculated, the next intermediate point can be found by searching on $\alpha$ along the line

$$x = x^{(1)} + \alpha d^{(1)}$$

until either the optimum of $f(x)$ is reached or else some constraint is encountered, whichever occurs first. Typically one would first conduct a line search to find the value $\bar{\alpha}$ at which some constraint $g_j(x)$ first becomes binding. That is, for each constraint $g_j(x)$ we find the value $\alpha_j > 0$ at which $g_j(x^{(1)} + \alpha d^{(1)}) = 0$, and then we choose $\bar{\alpha}$ to be the smallest of these $\alpha_j$'s. Numerically $\bar{\alpha}$ can be determined by first using a bracketing search to find a value of $x$ at which some $g_j(x) < 0$, followed by a root-finding scheme such as bisection. With $\bar{\alpha}$ known, we could then use any line search to find the $\alpha$ in the range $0 \leqslant \alpha \leqslant \bar{\alpha}$ that would minimize

$$f(x^{(1)} + \alpha d^{(1)})$$

### 9.1.1 The Basic Algorithm

At a given feasible point $x^{(t)}$, let $I^{(t)}$ be the set of indices of those constraints that are active at $x^{(t)}$, within some tolerance $\varepsilon$, that is,

$$I^{(t)} = \left\{ j: 0 \leqslant g_j(x^{(t)}) \leqslant \varepsilon, j = 1,\ldots, J \right\}$$

for some small $\varepsilon > 0$. A complete iteration of a feasible direction method then consists of the following three steps:

**Step 1.** Solve the linear programming problem

$$\text{Maximize} \quad \theta$$

$$\text{Subject to} \quad \nabla f(x^{(t)})d \leqslant -\theta$$

$$\nabla g_j(x^{(t)})d \geqslant \theta \qquad j \varepsilon I^{(t)}$$

$$-1 \leqslant d_i \leqslant 1 \qquad i = 1,\ldots, N$$

Label the solution $d^{(t)}$ and $\theta^{(t)}$.

**Step 2.** If $\theta^{(t)} \leqslant 0$, the iteration terminates, since no further improvement is possible. Otherwise, determine

$$\bar{\alpha} = \min\{\alpha: g_j(x^{(t)} + \alpha d^{(t)}) = 0, j = 1,\ldots, J \text{ and } \alpha \geqslant 0\}$$

If no $\bar{\alpha} > 0$ exists, set $\bar{\alpha} = \infty$.

**Step 3.** Find $\alpha^{(t)}$ such that

$$f(x^{(t)} + \alpha^{(t)}d^{(t)}) = \min\{f(x^{(t)} + \alpha(d^{(t)})): 0 \leqslant \alpha \leqslant \bar{\alpha}\}$$

Set $x^{(t+1)} = x^{(t)} + \alpha^{(t)}d^{(t)}$ and continue.

Note that the definition of the set $I^t$ of active constraints at $x^{(t)}$ is given using a tolerance parameter $\varepsilon > 0$. We will discuss the importance of this parameter in the next section.

### Example 9.1

$$\text{Minimize} \quad f(x) = (x_1 - 3)^2 + (x_2 - 3)^2$$

$$\text{Subject to} \quad g_1(x) = 2x_1 - x_2^2 - 1 \geqslant 0$$

$$g_2(x) = 9 - 0.8x_1^2 - 2x_2 \geqslant 0$$

The gradients of the problem functions are given by

$$\nabla f = [2(x_1 - 3), 2(x_2 - 3)]$$

$$\nabla g_1 = [2, -2x_2] \qquad \nabla g_2 = [-1.6x_1, -2]$$

Suppose the feasible starting point $x^{(1)} = (1, 1)$ is given. At this point,

$$g_1(x^{(1)}) = 0.0 \qquad g_2(x^{(1)}) = 6.2 > 0$$

Thus, $g_1$ is the only binding constraint, $I^{(1)} = \{1\}$, and the first subproblem to be solved becomes

$$\text{Maximize} \quad \theta$$

$$\text{Subject to} \quad -4d_1 - 4d_2 + \theta \leqslant 0$$

$$2d_1 - 2d_2 - \theta \geqslant 0$$

$$-1 \leqslant (d_1, d_2) \leqslant 1$$

The solution to this linear program is

$$d^{(1)} = (1,0) \qquad \text{with } \theta^{(1)} = 2$$

We must now search along the ray

$$x = \begin{pmatrix} 1 \\ 1 \end{pmatrix} + \begin{pmatrix} 1 \\ 0 \end{pmatrix} \alpha = \begin{pmatrix} 1 + \alpha \\ 1 \end{pmatrix} \quad \text{and} \quad \alpha \geqslant 0$$

to find the point at which the boundary of the feasible region is intersected. Since

$$g_1(\alpha) = 2(1 + \alpha) - 1 - 1 = 2\alpha$$

$g_1(x)$ is positive for all $\alpha \geqslant 0$ and is not violated as $\alpha$ is increased. To determine the point at which $g_2(x)$ will be intersected, we solve

$$g_2(\alpha) = 9 - 0.8(1 + \alpha)^2 - 2 = 0$$

to obtain $\bar{\alpha} = 1.958$. Finally, we search on $\alpha$ over the range $0 \leqslant \alpha \leqslant 1.958$ to determine the optimum of

$$f(\alpha) = [(1 + \alpha) - 3]^2 + (1 - 3)^2 = (\alpha - 2)^2 + 4$$

The minimum clearly occurs at the upper bound of $\alpha$, namely, at $\alpha^{(1)} = 1.958$. Thus, the iteration terminates with $x^{(2)} = (2.958, 1)$, at which point only $g_2(x)$ is binding.

Note that in any computer code for this algorithm, the search for the boundary as well as the line search for the optimum would have been carried out using iterative methods such as those discussed in Chapter 2.

The second iteration commencing at $x^{(2)}$ would first involve solving the subproblem

$$\text{Maximize} \quad \theta$$

$$\text{Subject to} \quad -0.084d_1 - 4d_2 + \theta \leqslant 0$$

$$-4.733d_1 - 2d_2 - \theta \geqslant 0$$

$$-1 \leqslant (d_1, d_2) \leqslant 1$$

constructed using the gradient of $g_2(x)$. The solution obtained by solving the

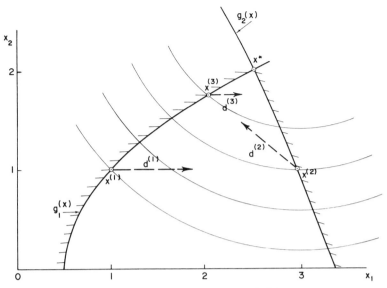

**Figure 9.1.** Feasible region, Example 9.1.

small linear program is

$$d^{(2)} = (-1, 0.8028) \qquad \text{with } \theta^{(2)} = 3.127$$

The search along the ray

$$x = \begin{pmatrix} 2.958 \\ 1 \end{pmatrix} + \alpha \begin{pmatrix} -1 \\ 0.8028 \end{pmatrix} \quad \text{and} \quad \alpha \geqslant 0$$

will yield $\bar{\alpha} = 0.9311$, at which point constraint $g_1(x)$ is intersected. The minimum value of $f(x)$ along the ray occurs at $\bar{\alpha}$; therefore, $\alpha^{(3)} = 0.9311$ and $x^{(3)} = (2.027, 1.748)$. The iterations can be continued in this fashion until the minimum point $x^* = (2.5, 2.0)$ is reached within specified accuracy. The iterations carried out above are shown in Figure 9.1.

### 9.1.2  Active Constraint Sets and Jamming

From Example 9.1 it is evident that the successive direction-generating subproblems of the feasible directions method differ in the active constraint set $I^{(t)}$ used at each iteration. Clearly, there is a computational gain in employing only a subset of all constraints to define the subproblem, since then the direction-generating LP is smaller. However, by considering only the constraints that are actually binding at the current feasible point, a zigzag iteration pattern results, which unfortunately slows down the progress of the iterations.

**Example 9.2**

Minimize    $f(x) = -x_1$

Subject to    $g_1(x) = x_2 \geqslant 0$         $g_2(x) = 1 - x_2 \geqslant 0$

$g_3(x) = x_1 \geqslant 0$         $g_4(x) = 8 - x_1 \geqslant 0$

Suppose the feasible point $x^{(1)} = (1,0)$ is given. Then, since $g_1(x)$ is tight, the direction-finding subproblem is

Maximize    $\theta$

Subject to    $(-1, 0)d \leqslant -\theta$

$(0, 1)d \geqslant \theta$

$+1 \geqslant d \geqslant -1$

Equivalently, we have

$d_1 \geqslant \theta \qquad d_2 \geqslant \theta$

$1 \geqslant (d_1, d_2) \geqslant -1$

The solution is $d^{(1)} = (1, 1)$, with $\theta = 1$. Moreover, the search of the line

$$x = \begin{pmatrix} 1 \\ 0 \end{pmatrix} + \alpha \begin{pmatrix} 1 \\ 1 \end{pmatrix}$$

will clearly lead to $\bar{\alpha} = 1$. The minimization of $f(x)$ on the interval $0 \leqslant \alpha \leqslant 1$ then gives $x^{(2)} = (2, 1)$. At $x^{(2)}$, constraint $g_2(x)$ is tight, and thus the direction subproblem becomes

Maximize    $\theta$

Subject to    $d_1 \geqslant \theta \qquad -d_2 \geqslant \theta$

$+1 \geqslant (d_1, d_2) \geqslant -1$

The resulting direction is $d^{(2)} = (1, -1)$. Moreover, $\bar{\alpha} = 1$ and $x^{(3)} = (3, 0)$. Evidently the iterations will zigzag across the feasible region as shown in Figure 9.2 until the upper bound on $x_1$ is reached. Obviously, if the proximity of both bounds on $x_2$ were taken into account at each step, a flatter search direction might have been selected and the optimum found more expeditiously.

In fact, the simple definition of the active constraint set used in the basic form of the feasible direction algorithm, namely,

$$I^{(t)} = \left\{ j: \ 0 \leqslant g_j(x^{(t)}) \leqslant \varepsilon, j = 1, \ldots, J \right\}$$

can not only slow down the progress of the iterations but can also lead to convergence to points that are not Kuhn-Tucker points [2, chap. 13]. Two-dimensional examples of this phenomena are not known, but a three-dimensional case is given by Wolfe [3]. This type of false convergence is known as *jamming* and, loosely speaking, occurs because the steps that the method takes become shorter and shorter as the direction vectors alternate between closely adjacent boundaries. The steps become shorter not because line search optima are reached but because a nearby constraint not considered in the direction-generation subproblem is encountered. Various modifications have been presented that can guarantee the elimination of jamming. We will briefly consider two of these: the $\varepsilon$ perturbation method [1] and the method of Topkis and Veinott [4].

**The $\varepsilon$-Perturbation Method.**    The definition of the active constraint set $I^{(t)}$ given above involves a small, positive tolerance parameter $\varepsilon$. As shown in reference 2, the following iterative adjustment of $\varepsilon$ will lead to guaranteed convergence:

1. At iteration point $x^{(t)}$ and with given $\varepsilon^{(t)} > 0$, define $I^{(t)}$ and carry out step 1 of the basic algorithm.
2. Modify step 2 with the following: If $\theta^{(t)} \geqslant \varepsilon^{(t)}$, set $\varepsilon^{(t+1)} = \varepsilon^{(t)}$ and continue. However, if $\theta^{(t)} < \varepsilon^{(t)}$, set $\varepsilon^{(t+1)} = \frac{1}{2}\varepsilon^{(t)}$ and proceed with the line search of the basic method. If $\theta^{(t)} < 0$, then a Kuhn-Tucker point has been found.

With this modification, it is efficient to set $\varepsilon$ rather loosely initially so as to include the constraints in a larger neighborhood of the point $x^{(t)}$. Then, as the iterations proceed, the size of the neighborhood will be reduced only when it is found to be necessary.

**The Topkis-Veinott Variant.**    An alternative approach to the problem is to simply dispense with the active constraint concept altogether and redefine the

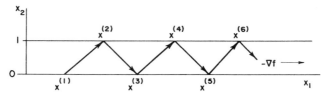

**Figure 9.2.**   Plot of Example 9.2.

direction-finding subproblem as follows:

Maximize  $\theta$

Subject to  $\nabla f(x^{(t)})d \leqslant -\theta$

$$g_j(x^{(t)}) + \nabla g_j(x^{(t)})d \geqslant \theta \qquad j = 1, \ldots, J$$

$$1 \geqslant d \geqslant -1$$

The remainder of the basic algorithm is unchanged. Note that the primary difference in the subproblem definition is the inclusion of the constraint value at $x^{(t)}$ in the inequality associated with each constraint. If the constraint is loose at $x^{(t)}$, that is, $g_j(x^{(t)}) > 0$, then the selection of $d$ is less affected by constraint $j$, because the positive constraint value will counterbalance the effect of the gradient term. As shown in reference 4, this definition of the subproblem ensures that no sudden changes are introduced in the search direction as constraints are approached, and thus jamming is avoided.

### Example 9.3

Consider the solution of Example 9.1 using the Topkis-Veinott variant. At $x^{(1)} = (1, 1)$, the direction-finding subproblem becomes

Maximize  $\theta$

Subject to  $-4d_1 - 4d_2 + \theta \leqslant 0$

$$2d_1 - 2d_2 \geqslant \theta$$

$$6.2 - 1.6d_1 - 2d_2 \geqslant \theta$$

$$1 \geqslant (d_1, d_2) \geqslant -1$$

Note that the third inequality corresponds to

$$g_2(x^{(1)}) + \nabla g_2(x^{(1)})d \geqslant \theta$$

The direction obtained in Example 9.1, namely $d^{(1)} = (1, 0)$, is the solution in this case also. The positive value of $g_2(x^{(1)})$ is sufficiently large to cancel out the gradient term, $-1.6(1) - 2(0) = -1.6$, so that the inequality corresponding to loose constraint $g_2$ becomes redundant.

At $x^{(2)} = (2.958, 1.0)$, the subproblem is

$$\text{Maximize} \quad \theta$$

$$\text{Subject to} \quad -0.084d_1 - 4d_2 + \theta \leqslant 0$$

$$-4.733d_1 - 2d_2 \geqslant \theta$$

$$3.916 + 2d_1 - 2d_2 \geqslant \theta$$

$$1 \geqslant (d_1, d_2) \geqslant -1$$

The third inequality corresponds to $g_1$, which is loose at $x^{(2)}$. In this case, the optimal solution is affected by the presence of the $g_1$ inequality. The direction obtained is

$$d^{(2)} = (-0.5816, 0.4669)$$

with $\theta^{(2)} = 1.8189$. The search of the line

$$x = \begin{pmatrix} 2.598 \\ 1.0 \end{pmatrix} + \alpha \begin{pmatrix} -0.5816 \\ 0.4669 \end{pmatrix}$$

leads to an optimum point $x^{(3)} = (2.0269, 1.7475)$, which is very slightly different from the point obtained in Example 9.1. Further iterations will show that as the iterates approach the vicinity of the optimum, the inclusion of the loose constraints serves to improve the progress toward the optimum, since it leads to somewhat steeper search directions.

### 9.1.3   Discussion

The basic algorithm in either its $\varepsilon$ form or the Topkis-Veinott form can be shown to converge to Kuhn-Tucker points under the reasonable assumptions that the problem functions are continuously differentiable [2, 4]. In the case of nonconvex feasible regions, the line search must, of course, be carried out carefully, since it is possible that the direction $d$, while initially feasible, may leave the feasible region and then reenter. Normally, only the closest feasible segment would be considered. Similarly, with a nonconvex objective function, multiple local minima could be encountered in the line search. Normally, the first local minimum encountered would be accepted. Under these provisions, the method has general applicability and has the important advantage of generating only *feasible* iteration points. That is an important consideration in engineering applications in which the system model employed may simply be invalid or meaningless outside the feasible region. These two features, guaranteed convergence for nonconvex problems and generation of feasible points, make feasible direction methods preferable to cutting plane methods. However, feasible direction methods share with the latter techniques the disadvantages of slow rate of convergence and inability to satisfactorily accommodate equality constraints.

From the definition of the direction-generation problem, the directions generated are gradient descent directions. Moreover, as we noted in Example 9.2, the requirement that the directions be strictly feasible causes a deflection from the steepest descent direction, which can potentially retard the rate of progress toward the solution. Thus, it can be expected that the rate of convergence will be no better than that attained by the unconstrained gradient method. No procedure has yet been proposed for accelerating the slow gradient-type convergence. Such procedures are not likely to be found, because of the propensity of the method for leaving the active constraint surface and thus complicating the possibilities of accumulating information about the constraint surface for use in accelerated steps.

The feasible direction machinery cannot directly handle nonlinear equality constraints, because there is no feasible interior to such constraints. The linearization of a nonlinear equality constraint will unavoidably generate a direction that leaves the constraint surface and thus will lead to infeasible points. In order to accommodate equalities, each constraint $h_k(x) = 0$ must be relaxed to allow limited excursion beyond the constraint surface, that is,

$$-\varepsilon \leqslant h_k(x) \leqslant \varepsilon$$

However, if $\varepsilon$ is small, then the line search will lead to very short steps and very slow progress. On the other hand, if the feasible direction constructions are relaxed to allow limited excursion beyond the constraint surface, then iterative solution of the constraint equations will be required to return to the feasible region, as shown in Figure 9.3.

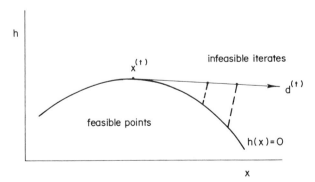

**Figure 9.3.** Iteration to reach the equality constraint surface.

A further major disadvantage of feasible direction methods is the continued necessity to solve LP subproblems. From our experience with the optimum gradient method for unconstrained problems, we recall that the locally best direction is not necessarily best from a global point of view. Thus, it might be that the effort involved in calculating the locally best compromise between descent and feasibility is not really warranted. Perhaps the computational savings achieved in selecting a descent direction without solving a linear program will more than offset the reduction in the rate of decrease achieved per iteration. The methods we discuss in the next two sections address these weaknesses of feasible direction methods.

## 9.2  SIMPLEX EXTENSIONS FOR LINEARLY CONSTRAINED PROBLEMS

The motivation for the methods considered in this section stems from the observation that since at a given point the number of search directions that are both descent directions and feasible directions is generally infinite, one ought to select candidate directions based on ease of computation. In the case of linear programs, the generation of search directions was simplified by changing only one variable at a time; feasibility was ensured by checking sign restrictions, and descent was ensured by selecting a variable with negative relative cost coefficient. In this section we consider the possibilities for modifying the very convenient simplex direction-generation machinery to accommodate nonlinear functions. We restrict our attention to *linearly constrained NLP problems* in order to simplify the constructions. The extension to the full NLPs will be made in Section 9.3.

### 9.2.1  The Convex Simplex Method

We begin with a brief review of the essential steps of the (linear) simplex method expressed in vector-matrix form and applied to the standard form of the problem:

$$\text{Minimize} \quad f(x) = cx$$

$$\text{Subject to} \quad \mathbf{A}x = b \quad x \geqslant 0$$

Given a feasible corner point $x^0$, the $x$ variable vector is partitioned into two sets: the basic variables $\hat{x}$, which are all positive, and the nonbasic variables $\bar{x}$, which are all zero. Note that if $\mathbf{A}$ has $M$ rows and $x$ has $N$ components, then $\hat{x}$ is an $M$ vector and $\bar{x}$ an $N - M$ vector. Corresponding to this partition of the variables, the columns of the coefficient matrix $\mathbf{A}$ are partitioned into $\mathbf{B}$ and $\bar{\mathbf{A}}$ and the objective function coefficient vector $c$ into $\hat{c}$

and $\bar{c}$. The first step of the algorithm then consists of expressing the basic variables in terms of the nonbasic variables. Thus, $\mathbf{B}\hat{x} + \overline{\mathbf{A}}\bar{x} = b$ is written as

$$\hat{x} + \mathbf{B}^{-1}\overline{\mathbf{A}}\bar{x} = \mathbf{B}^{-1}b \qquad (9.1)$$

by reducing the coefficients of the $\hat{x}$ variables to row-echelon form. In this form, the values of the current basic variables will be given by

$$\hat{x} = \mathbf{B}^{-1}b$$

while the current nonbasic variables will all be equal to zero:

$$\bar{x} = 0$$

The next step of the simplex method is to calculate the vector of relative cost coefficients by forming the inner product of the basic variable cost coefficients with the nonbasic variable tableau column, and subtracting the result from the cost coefficient of the nonbasic variable. Equivalently,

$$\tilde{c} = \bar{c} - \hat{c}\mathbf{B}^{-1}\overline{\mathbf{A}} \qquad (9.2)$$

The nonbasic variable to enter is selected by finding $\tilde{c}_s$ such that

$$\tilde{c}_s = \min\{\tilde{c}_i \colon \tilde{c}_i < 0, i = 1, \ldots, N - M\}$$

The basic variable $\hat{x}_r$ to leave the basis is selected using the minimum ratio rule. That is, we find $r$ such that

$$\Delta_r \equiv \frac{\hat{x}_r}{p_{rs}} = \min\left\{\frac{\hat{x}_j}{p_{js}} \colon p_{js} > 0, j = 1, \ldots, M\right\} \qquad (9.3)$$

where $p_{jk}$ are the elements of the matrix $\mathbf{B}^{-1}\overline{\mathbf{A}}$. The new basic feasible solution is thus

$$\hat{x}_j^{(1)} = \begin{cases} \Delta_r & \text{if } j = r \\ \hat{x}_j^0 - p_{js}\Delta_r & \text{if } j \neq r \end{cases} \qquad (9.4)$$

and all other variables are zero. At this point, the variables $\hat{x}$ and $\bar{x}$ are relabeled. Since an exchange will have occurred, $\mathbf{B}$, $\overline{\mathbf{A}}$, $\hat{c}$, and $\bar{c}$ will be redefined. The matrix $\mathbf{B}^{-1}$ is recomputed, and another cycle of iterations is begun.

Suppose we apply the same algorithm but replace the linear objective function coefficients $c$ with the linearized form of a nonlinear objective function $f(x)$. Thus, instead of $c$ we have $\nabla f(x^0)$ evaluated at some initial point $x^0$. If $x^0$ is a basic feasible solution and $x$ is again partitioned into $\hat{x}$ and

$\bar{x}$, then we can calculate the relative cost factor,

$$\nabla \tilde{f}(x^0) = \nabla \bar{f}(x^0) - \nabla \hat{f}(x^0)\mathbf{B}^{-1}\overline{\mathbf{A}} \tag{9.5}$$

in a fashion analogous to the linear case [Eq. (9.2)]. Here $\nabla \hat{f}$ denotes the partial derivatives of $f$ with respect to the $\hat{x}$ variables, and $\nabla \bar{f}$ those with respect to the $\bar{x}$ variables. In this expression, $\nabla \tilde{f}$ gives the relative change of the objective function $f(x)$ at $x^0$, taking into account the linear constraints. This fact can be verified as follows. Since $\hat{x}$ and $\bar{x}$ are related by Eq. (9.1), we can eliminate $\hat{x}$ from $f(x)$, to obtain

$$f(x) = f(\hat{x}, \bar{x}) = f(\mathbf{B}^{-1}b - \mathbf{B}^{-1}\overline{\mathbf{A}}\bar{x}, \bar{x})$$

Since $f(x)$ can be expressed as a function of $\bar{x}$ alone, we can apply the chain rule to calculate the derivative of $f$ with respect to $\bar{x}$ and thus obtain Eq. (9.5).

If any component of $\nabla \tilde{f}$ is negative, then, since to a first-order approximation

$$f(x) - f(x^0) = \nabla \tilde{f}(x^0)(\bar{x} - \bar{x}^0) \tag{9.6}$$

increasing the corresponding $\bar{x}$ variable will lead to a reduction in the objective function. As in the linear case, when we continue to increase that nonbasic variable, eventually a point is reached at which a basic variable is driven to zero. However, contrary to the linear case, once we depart from the point $x^0$, $\nabla \tilde{f}$ will change values, and hence the possibility exists of finding a minimum of $f(x)$ before reaching the adjacent corner point. Evidently, we will need to conduct a line search to find such a minimum. The endpoint of the line search will, however, be established using the minimum ratio rule as in the purely linear case.

In summary, then, the introduction of a nonlinear objective function in the simplex machinery requires:

1.  Evaluation of the function gradient
2.  Calculation of the modified relative cost factor via Eq. (9.5)
3.  Execution of a line search between the current point and the selected adjacent corner point

Before proceeding with further developments, let us consider a small example.

***Example 9.4***

$$\text{Minimize} \quad f(x) = (x_1 - 1)^2 + (x_2 - 2)^2$$

$$\text{Subject to} \quad 2x_2 - x_1 \leqslant 2$$

$$x_1 + x_2 \leqslant 4 \qquad x \geqslant 0$$

As shown in Figure 9.4, the unconstrained optimum to the problem lies at the point $(1, 2)$, which is infeasible. The constrained optimum lies on the face of the first constraint.

Suppose we rewrite the constraint set in canonical form by introducing slack variables $x_3$ and $x_4$. Thus,

$$-x_1 + 2x_2 + x_3 = 2$$

$$x_1 + x_2 + x_4 = 4$$

The initial basis will consist of $x_3$ and $x_4$; hence, $x_1 = x_2 = 0$, and our starting point lies at the origin of Figure 9.4. Corresponding to $\hat{x} = (x_3, x_4)$ and $\bar{x} = (x_1, x_2)$, we have

$$\mathbf{B} = \begin{pmatrix} 1 & 0 \\ 0 & 1 \end{pmatrix} \quad \text{and} \quad \bar{\mathbf{A}} = \begin{pmatrix} -1 & 2 \\ 1 & 1 \end{pmatrix}$$

and

$$\nabla \bar{f} = (-2, -4) \qquad \nabla \hat{f} = (0, 0)$$

Note that $\mathbf{B}^{-1} = \mathbf{B}$. The relative cost vector calculated using Eq. (9.5) is

$$\nabla \tilde{f} = (-2, -4) - (0, 0)\begin{pmatrix} 1 & 0 \\ 0 & 1 \end{pmatrix}\begin{pmatrix} -1 & 2 \\ 1 & 1 \end{pmatrix} = (-2, -4)$$

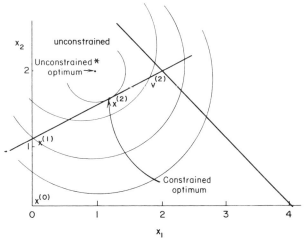

Figure 9.4. Feasible region, Example 9.4.

The nonbasic variable to enter will be $\bar{x}_2 = x_2$, since

$$\left| \frac{\partial \tilde{f}}{\partial \bar{x}_1} \right| < \left| \frac{\partial \tilde{f}}{\partial \bar{x}_2} \right|$$

The basic variable to leave will be $\hat{x}_1 = x_3$, since

$$\min\left(\tfrac{2}{2}, \tfrac{4}{1}\right) = \tfrac{2}{2} = 1 = \Delta_1$$

From Eq. (9.4), the new point is thus

$$\hat{x}_1^{(1)} = x_2 = \Delta_1 = 1$$

$$\hat{x}_2^{(1)} = x_4 = 4 - (1)\Delta_1 = 3$$

A line search between $x^0$ and $x^{(1)}$ is now required to locate the minimum of $f(x)$. Note that $x_1$ remains at 0, while $x_2$ changes as given by

$$x_2 = 0 + \alpha(1) = \alpha$$

Therefore, $f(\alpha) = (\alpha - 2)^2$. Clearly, $f(\alpha)$ is monotonically decreasing in $\alpha$ and the optimum $\alpha$ is $\alpha = 1$. The search thus advances to the adjacent corner point $x^{(1)}$.

At $x^{(1)}$, we have

$$\hat{x} = (x_2, x_4) = (1, 3)$$

$$\bar{x} = (x_1, x_3) = (0, 0)$$

and

$$\mathbf{B} = \begin{pmatrix} 2 & 0 \\ 1 & 1 \end{pmatrix} \qquad \bar{\mathbf{A}} = \begin{pmatrix} -1 & 1 \\ 1 & 0 \end{pmatrix} \qquad \mathbf{B}^{-1} = \begin{pmatrix} \tfrac{1}{2} & 0 \\ -\tfrac{1}{2} & 1 \end{pmatrix}$$

Moreover,

$$\nabla \hat{f}(x^{(1)}) = (-2, 0) \qquad \nabla \bar{f}(x^{(1)}) = (-2, 0)$$

Using Eq. (9.5),

$$\tilde{\nabla} f(x^{(1)}) = (-2, 0) - (-2, 0)\begin{pmatrix} \tfrac{1}{2} & 0 \\ -\tfrac{1}{2} & 1 \end{pmatrix}\begin{pmatrix} -1 & 1 \\ 1 & 0 \end{pmatrix} = (-3, 1)$$

Clearly, $\bar{x}_1 = x_1$ should be increased. Since the nonbasic variable coefficients are given by

$$\mathbf{B}^{-1}\bar{\mathbf{A}} = \begin{pmatrix} -\tfrac{1}{2} & \tfrac{1}{2} \\ +\tfrac{3}{2} & -\tfrac{1}{2} \end{pmatrix} \qquad \mathbf{B}^{-1}b = \begin{pmatrix} 1 \\ 3 \end{pmatrix}$$

from Eq. (9.3),

$$\Delta_2 = \min\left\{\begin{array}{c} 3 \\ \frac{3}{2} \end{array}\right\} = 2$$

Thus, $\hat{x}_2 = x_4$ leaves the basis; and the new corner point is

$$\hat{x}_1^{(2)} = x_2 = 1 + \tfrac{1}{2}\Delta = 2$$

$$\hat{x}_2^{(2)} = x_1 = \Delta = 2$$

A line search must now be executed on the line

$$x = \begin{pmatrix} x_1 \\ x_2 \\ x_3 \\ x_4 \end{pmatrix} = \begin{pmatrix} 0 \\ 1 \\ 0 \\ 3 \end{pmatrix} + \alpha\left\{\begin{pmatrix} 2 \\ 2 \\ 0 \\ 0 \end{pmatrix} - \begin{pmatrix} 0 \\ 1 \\ 0 \\ 3 \end{pmatrix}\right\}$$

Substituting into $f(x)$, we have

$$f(\alpha) = (2\alpha - 1)^2 + (\alpha - 1)^2$$

Setting $\partial f/\partial\alpha = 0$, it follows that $\alpha^* = 0.6$. Thus, the optimum point lies at

$$x^{(2)} = (1.2, 1.6, 0, 1.2)$$

Since the search stops short of the next corner point, the old basic variable $x_4$ will not go to zero, and hence a change of basis is not required. Suppose we retain the old basis of $(x_2, x_4)$. Then

$$\nabla\hat{f} = (-0.8, 0) \qquad \nabla\bar{f} = (0.4, 0)$$

Therefore,

$$\nabla\tilde{f}(x^{(2)}) = (0.4, 0) - (-0.8, 0)\begin{pmatrix} -\frac{1}{2} & \frac{1}{2} \\ \frac{3}{2} & -\frac{1}{2} \end{pmatrix} = (0, 0.4)$$

Since $\nabla\tilde{f} \geqslant 0$, it appears that $x^{(2)}$ is in fact the optimum solution to the problem. Note, however, that $\bar{x}^{(2)} = (x_1, x_3) = (1.2, 0)$; that is, we have a *nonzero* nonbasic variable or, equivalently, $x^{(2)}$ is not a corner point. This situation never occurs in the LP case but clearly must be handled in the extension to nonlinear objective functions.

This example, in addition to illustrating the main features of the extension to nonlinear objective functions, points out another situation that must be accommodated in extending the linear simplex method, namely, the occurrence of positive nonbasic variables. Evidently, positive nonbasic variables will be introduced whenever the line search yields an optimum that is not a corner point. To accommodate this case, we first of all need to modify our criterion for selecting the nonbasic variable that is to enter the basis. Then, we will need to introduce appropriate corrections in our formulas for calculating the next corner point (or target point).

At a point $x^{(t)}$, given a choice of basis and the $\nabla \tilde{f}(x^{(t)})$ vector associated with the nonbasic variables, consider the following two situations:

1. $\nabla \tilde{f}_i$ is negative for some $\bar{x}_i$, where $\bar{x}_i$ may be zero or positive, or
2. $\nabla \tilde{f}_i$ is positive for some $\bar{x}_i > 0$

From Eq. (9.6), we have, as a first-order approximation, that

$$f(x) - f(x^{(t)}) = \sum_{k=1}^{N-M} \nabla \tilde{f}_k (\bar{x}_k - \bar{x}_k^{(t)})$$

Now, if $\nabla \tilde{f}_i < 0$ for some $\bar{x}_i$, then by choosing $(\bar{x} - \bar{x}^{(t)})_i > 0$ and $(\bar{x} - \bar{x}^{(t)})_k = 0$ for all other variables, we have

$$f(x) - f(x^{(t)}) < 0$$

Thus by increasing $\bar{x}_i$, the objective function decreases. Similarly, if $\nabla \tilde{f}_i > 0$ for some $\bar{x}_i > 0$, then by choosing $(\bar{x} - \bar{x}^{(t)})_i < 0$ and $(\bar{x} - \bar{x}^{(t)})_k = 0$ for all other variables, we again have

$$f(x) - f(x^{(t)}) < 0$$

The choice $(\bar{x} - \bar{x}^{(t)})_i < 0$ is possible only if $\bar{x}_i^{(t)} > 0$.

Conversely, if at $x^{(t)}$, $\nabla \tilde{f}_i \geqslant 0$ for all $\bar{x}_i$ and $\nabla \tilde{f}_i = 0$ for all $\bar{x}_i > 0$, then no improvement is possible, and a local optimum has been obtained. We thus have deduced a rule for selecting nonbasic variables as well as a necessary condition for optimality. As will be shown in Section 9.3 in a more general setting, the conditions

$$\nabla \tilde{f} \geqslant 0 \qquad \text{(nonnegativity condition)} \qquad (9.7)$$

and

$$\bar{x}_i (\nabla \tilde{f}_i) = 0 \qquad \text{(complementary slackness condition)}$$

$$\text{for } i = 1, \ldots, N - M \quad (9.8)$$

are simply alternative ways of expressing the Kuhn-Tucker necessary conditions for a local minimum.

*Example 9.5*

Suppose we apply conditions (9.7) and (9.8) to the point $x^{(2)}$ at which the calculations of Example 9.4 terminated. At $x^{(2)}$, we had

$$\bar{x} = (1.2, 0) \quad \text{and} \quad \nabla \tilde{f} = (0, 0.4)$$

Clearly, (9.7) is satisfied, since $(0, 0.4) \geqslant 0$. Moreover,

$$\bar{x}_1 \nabla \tilde{f}_1 = 1.2(0) = 0$$

and

$$\bar{x}_2 \nabla \tilde{f}_2 = 0(0.4) = 0$$

Therefore, conditions (9.8) are satisfied and $x^{(2)}$ is thus a Kuhn-Tucker point. Since $f(x)$ is convex, $x^{(2)}$ is in fact the global constrained minimum.

Violation of either condition (9.7) or (9.8) provides an obvious construction for the selection of the nonbasic variable to enter the basis. In particular, let

$$\beta_s = \min\{0, \nabla \tilde{f}_i : i = 1, \ldots, N - M\} \tag{9.9}$$

$$\gamma_q = \max\{0, \nabla \tilde{f}_i \bar{x}_i : i = 1, \ldots, N - M\} \tag{9.10}$$

If $|\beta_s| > \gamma_q$, then $\bar{x}_s$ should be increased, adjusting only the basic variables until some $\hat{x}$ goes to zero. If $|\beta_s| < \gamma_q$, then $\bar{x}_q$ should be decreased, adjusting only the basic variables. In the latter case $\bar{x}_q$ should be decreased until either some $\hat{x}$ goes to zero or $\bar{x}_q$ itself goes to zero, whichever occurs first.

We are now in a position to outline the complete steps of the algorithm known as the *convex simplex method* ([2], chap. 8; see also [5]).

**Convex Simplex Algorithm**

Given a feasible point $x^0$, a partition $x = (\hat{x}, \bar{x})$ of the problem variables, and a convergence tolerance $\varepsilon > 0$.

**Step 1.** Compute $\nabla \tilde{f}(x^{(t)})$.

**Step 2.** Compute $\beta_s$ and $\gamma_q$ via Eqs. (9.9) and (9.10).

**Step 3.** If $|\beta_s| \leqslant \varepsilon$ and $\gamma_q \leqslant \varepsilon$, terminate. Otherwise, consider two cases:
   (a) If $|\beta_s| > \gamma_q$, determine

$$\Delta_r = \min\left\{ \frac{\hat{x}_j}{p_{js}} : p_{js} > 0, j = 1, \ldots, M \right\}$$

and set $\Delta = \Delta_r$.

(b) If $|\beta_s| < \gamma_q$, determine

$$\Delta_r = \min\left\{\frac{-\hat{x}_j}{p_{jq}} : p_{jq} < 0, j = 1, \ldots, M\right\}$$

and set $\Delta = -\min(\Delta_r, \bar{x}_q)$.

**Step 4.** Calculate the target point $v^{(t)}$:

$$\hat{v}_j^{(t)} = \hat{x}_j^{(t)} - p_{jk}\Delta$$

$$\bar{v}_i^{(t)} = \begin{cases} \bar{x}_i^{(t)} + \Delta & \text{if } i = k \\ 0 & \text{otherwise} \end{cases}$$

where $k$ is equal to $s$ or $q$ depending upon whether 3(a) or 3(b) occurs

**Step 5.** Find $\alpha^*$ such that

$$f\left(x^{(t)} + \alpha^*(v^{(t)} - x^{(t)})\right)$$

$$= \min_\alpha \left\{f\left(x^{(t)} + \alpha(v^{(t)} - x^{(t)})\right): 0 \leqslant \alpha \leqslant 1\right\}$$

**Step 6.** Set $x^{(t+1)} = x^{(t)} + \alpha^*(v^{(t)} - x^{(t)})$. If $\alpha^* = 1$ and $\Delta = \Delta_r$, update the basis and the basis inverse. Otherwise, retain the same basis and go to step 1.

*Example 9.6  [4]*

$$\text{Minimize} \quad f(x) = x_1^2 + x_2 x_3 + x_4^2$$

$$\text{Subject to} \quad \tfrac{1}{4}x_1 + x_2 + x_3 + \tfrac{1}{2}x_4 - x_5 = 4$$

$$\tfrac{1}{4}x_1 + x_2 - x_6 \qquad\quad = 2$$

$$x_i \geqslant 0 \qquad i = 1, \ldots, 6$$

Given the initial point $x^0 = (0, 2, 3, 0, 1, 0)$, at which $f(x^0) = 6.0$. Choose

$$\hat{x} = (x_2, x_3) \quad \text{and} \quad \bar{x} = (x_1, x_4, x_5, x_6)$$

Thus,

$$\mathbf{B}^{-1} = \begin{pmatrix} 0 & 1 \\ 1 & -1 \end{pmatrix} \qquad \bar{\mathbf{A}} = \begin{pmatrix} \tfrac{1}{4} & \tfrac{1}{2} & -1 & 0 \\ \tfrac{1}{4} & 0 & 0 & -1 \end{pmatrix}$$

$$\nabla\hat{f} = (3, 2) \qquad \nabla\bar{f} = (0, 0, 0, 0)$$

*First Iteration*

**Step 1.**  $\nabla \tilde{f}(x^0) = (-\frac{3}{4}, -1, 2, 1)$

**Step 2.**

$$\beta_2 = \min(0, -\tfrac{3}{4}, -1, 2, 1) = -1$$

$$\gamma_3 = \max(0, 0, 0, 2, 0) = 2$$

**Step 3.**  Since $|\beta_2| < \gamma_3$, case (b) occurs, that is, the third nonbasic variable $(x_5)$ is to be decreased. Then, since

$$\begin{pmatrix} p_{13} \\ p_{23} \end{pmatrix} = \mathbf{B}^{-1}\overline{\mathbf{A}}_3 = \begin{pmatrix} 0 & 1 \\ 1 & -1 \end{pmatrix}\begin{pmatrix} -1 \\ 0 \end{pmatrix} = \begin{pmatrix} 0 \\ -1 \end{pmatrix}$$

where $\overline{\mathbf{A}}_3$ denotes the third column of $\overline{\mathbf{A}}$, we have

$$\Delta_2 = \min\{-3/-1\} = 3 \quad \text{and} \quad \Delta = -\min\{3, 1\} = -1$$

Thus $x_5$ will decrease to zero before any basic variable is driven to zero.

**Step 4.**  The target point will be

$$\hat{v}_1 = v_2 = 2 - (0)(-1) \quad = 2$$

$$\hat{v}_2 = v_3 = 3 - (-1)(-1) = 2$$

$$\bar{v}_3 = v_5 = 1 + (-1) \quad\quad = 0$$

all other $\bar{v}_i$ (i.e., $v_1, v_4, v_6$) will remain zero. Thus,

$$v^0 = (0, 2, 2, 0, 0, 0)$$

**Step 5.**  The one-parameter minimum on the line

$$x = x^0 + \alpha(v^0 - x^0) \qquad 0 \leqslant \alpha \leqslant 1$$

occurs at $\alpha^* = 1.0$ with $f(\alpha^*) - 4.0$.

**Step 6.**  The next iteration point is $x^{(1)} = v^0$. Since no basic variable was driven to zero, a basis change or update of $\mathbf{B}^{-1}$ is not required.

*Second Iteration*

**Step 1.**  $\nabla \tilde{f}(x^{(1)}) = (-\frac{1}{2}, -1, 2, 0)$

**Step 2.**

$$\beta_2 = \min(0, -\tfrac{1}{2}, -1, 2, 0) = -1$$

$$\gamma = \max(0, 0, 0, 0, 0) = 0$$

**Step 3.** Obviously, case 3(a) occurs. The second nonbasic variable $(x_4)$ is increased.

$$\begin{pmatrix} p_{12} \\ p_{22} \end{pmatrix} = \begin{pmatrix} 0 & 1 \\ 1 & -1 \end{pmatrix} \begin{pmatrix} \frac{1}{2} \\ 0 \end{pmatrix} = \begin{pmatrix} 0 \\ \frac{1}{2} \end{pmatrix}$$

$$\Delta_2 = \min\left(\frac{2}{\frac{1}{2}}\right) = 4 = \Delta$$

The second basic variable $x_3$ will be forced to zero.

**Step 4.** The next target will be

$$\hat{v}_1 = v_2 = 2 - 0(4) = 2$$

$$\hat{v}_2 = v_3 = 2 - \tfrac{1}{2}(4) = 0$$

$$\bar{v}_2 = v_4 = 0 + 4 = 4$$

and all other $\bar{v}_i = 0$. Thus

$$v^{(1)} = (0, 2, 0, 4, 0)$$

**Step 5.** The one-parameter minimum occurs at $\alpha^* = \frac{1}{8}$, with $f(\alpha^*) = 3.625$.

**Step 6.** The new point is

$$x^{(2)} = (0, 2, 1.75, 0.5, 0, 0)$$

Again, the basis need not be changed, since the line search stops short of the target point. The iterations will continue in this fashion until the convergence check in step 3 is satisfied.

The foregoing example illustrates one characteristic feature of the convex simplex method: the tendency for the basis to remain unchanged over a series of iterations. This occurs because the intermediate solutions are not corner points, but instead lie within the feasible region (nonzero nonbasic variables). Thus, the partition into basic and nonbasic variables essentially amounts to the selection of a reference coordinate system in the nonbasic variables. When viewed in this light, the convex simplex method essentially amounts to the one-variable-at-a-time search called *sectioning* in the unconstrained case. It has been found empirically that convergence can be enhanced by occasional changes of basis even if none is strictly required by the computations. Two rules for basis selection have been suggested:

1. After each iteration, order the variables by decreasing magnitude, and select as basic variables the $M$ variables largest in magnitude. In this case $\mathbf{B}^{-1}$ must be updated after each iteration.

2.  Periodically update the basis by using a random selection rule. Say, after four or five iterations involving no basic change, randomly select a positive nonbasic variable to enter.

Clear empirical evidence indicating which of the three basis selection strategies (the basic algorithm or the two above) is best is not available. Regardless of which selection strategy is used, convergence to a Kuhn-Tucker point can be proven for continuously differentiable $f(x)$ and bounded feasible regions under the *nondegeneracy assumption* (the basis variables are always positive) [2, Chapter 8]. The rate of convergence to non-corner point optima can be quite slow; however, the direction-finding machinery is very simple and rapid. Thus, one would expect the solution efficiency to be better than that of the Frank-Wolfe algorithm.

## 9.2.2  The Reduced Gradient Method

From the interpretation of the convex simplex method as a sectioning proce-dure within the space of the nonbasic variables, we might deduce that the method could potentially be improved by changing all nonbasic variables simultaneously. In this way the change in the nonbasic variables would more closely resemble the unconstrained gradient method. However, since both condition (9.7) and condition (9.8) must be satisfied at a Kuhn-Tucker point, the nonbasic variable direction vector must be defined as follows:

$$\bar{d}_i = \begin{cases} -\nabla \tilde{f}_i & \text{if } \nabla \tilde{f}_i \leqslant 0 \\ -\bar{x}_i \nabla \tilde{f}_i & \text{if } \nabla \tilde{f}_i \geqslant 0 \end{cases} \qquad (9.11)$$

where $i = 1, \ldots, N - M$.

Thus, if $\nabla \tilde{f}_i < 0$, nonbasic variable $i$ is increased; while if $\nabla \tilde{f}_i > 0$ and $\bar{x}_i > 0$, then nonbasic variable $i$ is decreased. This definition ensures that when $\bar{d}_i = 0$ for all $i$, the Kuhn-Tucker conditions will be satisfied. When $\bar{d}$, the change in the nonbasic variables, is calculated using (9.11), then the change in the basic variables must be calculated using

$$\hat{d} = -\mathbf{B}^{-1} \bar{\mathbf{A}} \bar{d} \qquad (9.12)$$

in order that the new basic variables satisfy the linear constraints.

In this fashion a direction vector is defined that both satisfies the con-straints and is a descent direction. However, the variable nonnegativity condi-tions must also be satisfied. Thus, the step in the direction $d$ must be such that

$$\hat{x} = \hat{x}^{(t)} + \alpha \hat{d} \geqslant 0$$

$$\bar{x} = \bar{x}^{(t)} + \alpha \bar{d} \geqslant 0$$

where $x^{(t)}$ is the current feasible point.

In the first case, the limiting $\alpha$ value will be given by

$$\alpha_1 = \min\left\{\frac{-\hat{x}_i^{(t)}}{\hat{d}_i} : \text{for all } \hat{d}_i < 0, i = 1,\ldots, M\right\}$$

If all $\hat{d}_i \geqslant 0$, then set $\alpha_1 = \infty$.

In the second case, the limiting $\alpha$ value will be given by

$$\alpha_2 = \min\left\{\frac{-\bar{x}_j^{(t)}}{\bar{d}_j} : \text{for all } \bar{d}_j < 0, j = 1,\ldots, N - M\right\}$$

If all $\bar{d}_j \geqslant 0$, then set $\alpha_2 = \infty$.

Let $\alpha_{max} = \min(\alpha_1, \alpha_2)$ and use $\alpha_{max}$ as the maximum step size in the line search. The overall algorithm can now be defined as follows.

### Reduced Gradient Algorithm

Given a feasible point $x^0$, a partition of the problem variables $x = (\hat{x}, \bar{x})$ and a convergence tolerance $\varepsilon > 0$.

**Step 1.**  Compute $\nabla\tilde{f}(x^{(t)})$.

**Step 2.**  Compute $\bar{d}$ via (9.11) and $\hat{d}$ via (9.12). If $\|d\| < \varepsilon$, terminate. Otherwise, continue.

**Step 3.**  Calculate the step-size limit $\alpha_{max}$ as defined above.

**Step 4.**  Find $\alpha^*$ such that

$$f(x^{(t)} + \alpha^* d) = \min_{\alpha}\{f(x^{(t)} + \alpha d) : 0 \leqslant \alpha \leqslant \alpha_{max}\}$$

**Step 5.**  Calculate the new point,

$$x^{(t+1)} = x^{(t)} + \alpha^* d$$

If $\alpha^* = \alpha_{max} = \alpha_1$, change the basis to avoid degeneracy. Otherwise, go to step 1.

### Example 9.7

To illustrate the steps of the reduced gradient algorithm, consider again the problem of Example 9.6 with the initial point $x^0 = (0, 2, 3, 0, 1, 0)$. With the same choice of basic and nonbasic variables, $\hat{x} = (x_2, x_3)$ and $\bar{x} = (x_1, x_4, x_5, x_6)$, the reduced cost vector will be

$$\nabla\tilde{f}(x^0) = \left(-\tfrac{3}{4}, -1, 2, 1\right)$$

and

$$\mathbf{B}^{-1}\mathbf{A} = \begin{pmatrix} \frac{1}{4} & 0 & 0 & -1 \\ 0 & \frac{1}{2} & -1 & 1 \end{pmatrix}$$

**Step 2**

$$\bar{d} = \begin{cases} +\frac{3}{4} & \text{since } \nabla \tilde{f}_1 < 0 \\ +1 & \text{since } \nabla \tilde{f}_2 < 0 \\ -(1)(2) & \text{since } \nabla \tilde{f}_3 > 0 \\ -(0)(1) & \text{since } \nabla \tilde{f}_4 > 0 \end{cases}$$

Then,

$$\hat{d} = - \begin{pmatrix} \frac{1}{4} & 0 & 0 & -1 \\ 0 & \frac{1}{2} & -1 & 1 \end{pmatrix} \begin{pmatrix} \frac{3}{4} \\ 1 \\ -2 \\ 0 \end{pmatrix} = \left( -\frac{3}{16}, -\frac{5}{2} \right)$$

Therefore, $d = \left( \frac{3}{4}, -\frac{3}{16}, -\frac{5}{2}, 1, -2, 0 \right)$.

**Step 3.** Step-size check:

$$\alpha_1 = \min \left\{ \frac{-2}{-\left( \frac{3}{16} \right)}, \frac{-3}{-\left( \frac{5}{2} \right)} \right\} = 1.2$$

$$\alpha_2 = \min \left\{ \frac{-1}{(-2)} \right\} = \frac{1}{2}$$

$$\alpha_{\max} = \min\left( 1.2, \frac{1}{2} \right) = \frac{1}{2}$$

Therefore, at the target point, $v^0 = x^0 + \alpha_{\max} d$, the third nonbasic variable ($x_5$) will be reduced to zero.

**Step 4.** Line search on $\alpha$, $0 \leqslant \alpha \leqslant \alpha_{\max}$. It can be shown that $f(\alpha)$ decreases monotonically to $\alpha_{\max}$. Thus, the optimum $\alpha^*$ is equal to the bound $\alpha_{\max}$.

**Step 5.** The new point is

$$x^{(1)} = x^0 + \frac{1}{2}d = \left( \frac{3}{8}, \frac{61}{32}, \frac{7}{4}, \frac{1}{2}, 0 \right)$$

at which point

$$f(x^{(1)}) = 3.727$$

A basis change is not required, since the basic variables remain positive. The iterations will continue in this fashion. Note that in this first iteration a lower objective function value was attained than was the case with the convex simplex method ($f(x^{(1)}) = 4.0$).

The reduced gradient method was first proposed by Wolfe [6], who suggested this name because $\nabla \tilde{f}$ can be viewed as the gradient of $\tilde{f}(x)$ in the reduced space of the $\bar{x}$ variables. Modifications of the method to avoid jamming were presented by McCormick [7] and are incorporated in the above outline. A convergence proof, again under the nondegeneracy assumption and the requirement that $f(x)$ be continuously differentiable, is given in reference 8, Section 10.4. Numerical comparisons [9] between this algorithm and the convex simplex method show that their performance is very similar, with the convex simplex method slightly superior in the final phase of the search. However, for larger scale implementations it seems that the reduced gradient strategy is preferable because the effort involved in conducting a line search is the same regardless of the form of the direction vector. The small additional effort required to calculate $\hat{d}$ thus appears justified, since, especially farther from the solution, a gradient direction ought to be more efficient than a sectioning step.

As in the convex simplex method, alternate basis selection strategies can be employed. Recent implementations [10] favor changing basis only when a basic variable reaches zero (or equivalently, its upper or lower bound), since this saves recomputation of $\mathbf{B}^{-1}$. The nonbasic variable selected to replace this basic variable is the one that is largest (furthest from its bounds). Further discussion of these matters is given in reference 10.

### 9.2.3  Convergence Acceleration

The convex simplex method and the reduced gradient method can be viewed as, respectively, the sectioning search and the ordinary gradient method applied in a reduced variable space, and thus both can be expected to share the generally slow final convergence rates of these unconstrained methods. Given the enhanced convergence rates attained in the unconstrained case by the use of conjugate direction and quasi-Newton constructions, it is reasonable to consider the application of these techniques in the linearly constrained case.

Recall that the computation of the $\nabla \tilde{f}$ vector essentially made use of the linear constraints to eliminate the $\hat{x}$ variables. The independent problem variables are thus the nonbasic variables $\bar{x}$. Once the neighborhood of the optimum is approached and the basis no longer changes, then the succeeding iterations will effectively involve only an unconstrained search over the $\bar{x}$ variables. Thus, the gradient vector $\nabla \tilde{f}$ can be used to calculate a modified search direction vector $\bar{d}$ using either the conjugate gradient formula or one of the quasi-Newton formulas, such as the DFP method. The latter is preferable but requires storage of the quasi-Newton matrix. Whichever update formula is used, the acceleration calculations will be effective only if the basis remains unchanged. Any change of basis will alter the set of independent variables and thus will effectively alter the shape of the reduced dimensionality objective function. As a result, the usefulness of the accumulated information about the shape of the function will be lost.

The only modification of the reduced gradient method required to implement the acceleration calculations occurs in step 2, the computation of the direction vector $\bar{d}$. For the first iteration or any iteration following a basis change, the modified step will consist of the definition of the modified "gradient" $g$,

$$g_i^{(t)} = \begin{cases} \nabla \tilde{f}_i^{(t)} & \text{if } \bar{x}_i^{(t)} > 0 \quad \text{or} \quad \nabla \tilde{f}_i^{(t)} < 0; \quad i = 1,\ldots, N - M \\ 0 & \text{otherwise} \end{cases}$$

and the search vector specification, $\bar{d}^{(t)} = -g^{(t)}$. For all other iterations, $g$ is again defined as above, but $\bar{d}$ is calculated using the selected conjugate gradient or quasi-Newton formula. For instance, if the conjugate gradient method is used, then

$$\bar{d}^{(t)} = -g^{(t)} + \frac{\|g^{(t)}\|^2}{\|g^{(t-1)}\|^2} \bar{d}^{(t-1)}$$

As in the unconstrained case, if exact line searches are performed, if the objective function is quadratic, and if no basis changes occur, the exact optimum will be found in $N - M$ iterations. For general functions, continued iterations will be required until convergence is attained. However, the linear convergence rate of the unmodified algorithm can now in principle be upgraded to be superlinear.

### Example 9.8

Consider the solution of the problem

$$\text{Minimize} \quad f(x) = x_1^2 - 2x_1x_2 + x_2^2 + x_3^2$$

$$\text{Subject to} \quad x_1 + 2x_2 + x_3 = 3$$

$$x_i \geq 0 \quad i = 1, 2, 3$$

starting with the point $x^0 = (1, \frac{1}{2}, 1)$ and using the reduced gradient method coupled with the conjugate gradient direction formula.

### First Iteration

At $x^0$, $\nabla f(x^0) = (1, -1, 2)$. Suppose we select $x_1$ as basic variable and $x_2, x_3$ as nonbasics. Then, $\mathbf{B} = (1)$, $\bar{\mathbf{A}} = (2, 1)$, and $\mathbf{B}^{-1} = (1)$.

**Step 1.**   $\nabla f^0 = (-1, 2) - (1)(1)(2, 1) = (-3, 1)$

**Step 2.**   Next, we have

$$\bar{d} = \begin{pmatrix} 3 \\ -1 \end{pmatrix} \quad \text{and} \quad \hat{d} = -(2, 1)\begin{pmatrix} 3 \\ -1 \end{pmatrix} = -5$$

**Step 3.**   The maximum step will be given by

$$\alpha_1 = \min\left(\frac{-1}{-5}\right) = \tfrac{1}{5}$$

$$\alpha_2 = \min\left(\frac{-1}{-1}\right) = 1$$

$$\alpha_{\max} = \min(\tfrac{1}{5}, 1) = \tfrac{1}{5}$$

**Step 4.**   The line search will yield the optimum step

$$\alpha^* = \tfrac{1}{13} < \alpha_{\max}$$

at which value, $f(\alpha^*) = 0.8654$.

**Step 5.**   The new point is

$$x^{(1)} = \left(\tfrac{8}{13}, \tfrac{19}{26}, \tfrac{12}{13}\right)$$

and a basis change is not required.

### *Second Iteration*

**Step 1.**   $\nabla f(x^{(1)}) = (-\tfrac{3}{13}, \tfrac{3}{13}, \tfrac{24}{13})$ and then

$$\nabla \tilde{f}(x^{(1)}) = \left(\tfrac{3}{13}, \tfrac{24}{13}\right) - \left(-\tfrac{3}{13}\right)(2, 1) = \left(\tfrac{9}{13}, \tfrac{27}{13}\right)$$

**Step 2.**   The search vector $\bar{d}$ is calculated using the conjugate gradient formula,

$$\bar{d}^{(1)} = -\nabla \tilde{f}^{(1)} + \frac{\|\nabla \tilde{f}^{(1)}\|}{\|\nabla \tilde{f}^0\|} d^0$$

$$= -\begin{pmatrix} \tfrac{9}{13} \\ \tfrac{27}{13} \end{pmatrix} + \frac{\left(\left(\tfrac{9}{13}\right)^2 + \left(\tfrac{27}{13}\right)^2\right)}{(3^2 + 1^2)}\begin{pmatrix} 3 \\ -1 \end{pmatrix}$$

$$= \begin{pmatrix} 0.74556 \\ -2.5562 \end{pmatrix}$$

Then

$$\hat{d} = \mathbf{B}^{-1}\overline{\mathbf{A}}\,\overline{d} = -(2,1)\overline{d} = 1.06509$$

**Step 3.** The step-size limit is given by

$$\alpha_{\max} = \min(\alpha_1, \alpha_2) = \min(\infty, 0.36111) = 0.36111$$

**Step 4.** The line search will yield the exact optimum $\alpha^* = 0.36111$.

**Step 5.** The new point is $x^{(2)} = (1.0, 1.0, 0.0)$ with $f(x^{(2)}) = 0.0$. This is the exact solution, since

$$\nabla f(x^{(2)}) = (0,0,0)$$

As expected, the solution was obtained in two iterations, since the objective function was quadratic and the problem involved two nonbasic variables.

As shown by Zangwill [2, sect. 9.1], acceleration can also be applied with the convex simplex method. However, since this strategy reduces to a sectioning procedure in the space of the nonbasic variables, it proves convenient to use the conjugate direction method. While this is in itself an interesting mating of the linear simplex method with a direct-search method, it is inherently unsatisfactory to have available the function gradient and then resort to a direct-search method with its increased number of line searches.

In summary, the adaptation of simplex-like constructions to linearly constrained NLPs that we explored in this section have led to several very useful consequences. First, the partition into basic/nonbasic variables and the associated array manipulations have provided a means of calculating a feasible descent direction that only requires solution of a set of linear equations. Second, the modification of the relative cost coefficient calculating to accommodate nonlinear objective functions has provided an optimality test that not only allows us to deduce whether a point satisfies the necessary conditions for a local minimum but also is constructive so that if it is violated it can be used directly to determine a direction for improvement. Finally, the direction-generating formulation readily permits the application of quasi-Newton type convergence acceleration constructions. In the next section, we consider the extension of these constructions and their very desirable properties to nonlinearly constrained problems.

## 9.3 THE GENERALIZED REDUCED GRADIENT METHOD

The basic reduced gradient strategy introduced in the previous section can be extended readily to solve the general NLP problem. In this section we discuss the constructions involved in considerable detail. For simplicity we initially consider the nonlinear equality-constrained form of the NLP problem, namely,

$$\text{Minimize} \quad f(x)$$

$$\text{Subject to} \quad h_k(x) = 0 \quad k = 1,\ldots, K \quad (9.13)$$

We begin with a generalization of the implicit variable-elimination procedure employed in defining the reduced gradient vector $\tilde{\nabla} f$. As part of this analysis we show that the optimality criterion stated in terms of this vector are equivalent to the Lagrangian conditions. Then we develop the basic form of the algorithm, followed by a discussion of the various extensions and enhancements that are incorporated in state-of-the-art implementations of this method.

### 9.3.1  Implicit Variable Elimination

As we observed in our discussion of direct-search methods for constrained problems, if a problem involves nonlinear equality constraints that are simple enough that they can be solved explicitly and used to eliminate variables, then it is always expedient to do so. From a computational point of view this device reduces not only the number of variables in the problem but also the number of constraints. For instance, if the constraint

$$h_l(x) = 0$$

can be solved to yield

$$x_k = \phi(x_1, x_2, \ldots, x_{k-1}, \ldots, x_N)$$

then it is advantageous to do so because by using the above equation to substitute out the variable $x_k$ from the objective and remaining constraint functions, both the variable $x_k$ and the constraint $h_l(x)$ are eliminated from the problem. Now, in most cases the equality constraints are algebraically complex so that such explicit elimination is not possible. In those cases a reduction in problem dimensionality may still be carried out implicitly, in a manner analogous to our treatment of linear constraints in the previous section.

Suppose $x^{(1)}$ is a point satisfying the constraints of the equality-constrained problem (9.13), and suppose we construct a linear approximation to the problem constraints at the point $x^{(1)}$. The result is

$$\tilde{h}_k(x; x^{(1)}) \equiv h_k(x^{(1)}) + \nabla h_k(x^{(1)})(x - x^{(1)}) \quad k = 1,\ldots, K$$

Suppose we wish to use these linearized equations to predict the location of another feasible point, that is, a point at which

$$\tilde{h}_k(x; x^{(1)}) = 0 \quad k = 1,\ldots, K$$

Then, since $h_k(x^{(1)}) = 0$, $k = 1, \ldots, K$, that point must satisfy the system of linear equations

$$\nabla h_k(x^{(1)})(x - x^{(1)}) = 0 \qquad k = 1, \ldots, K \qquad (9.14)$$

In general, $K < N$; consequently, this system of equations will have more unknowns then equations and cannot be solved to yield a unique solution. However, we can solve for $K$ of the $N$ variables in terms of the other $N - K$. For instance, suppose we select the first $K$ variables and label them $\hat{x}$ (basic) and label the remaining variables $\bar{x}$ (nonbasic). Corresponding to this partition of the $x$'s, suppose we partition the row vectors $\nabla h_k$ into $\nabla \hat{h}_k$ and $\nabla \bar{h}_k$ and accumulate these subvectors into two arrays $\mathbf{J}$ and $\mathbf{C}$. The matrix $\mathbf{J}$ will consist of elements

$$\mathbf{J} = \begin{pmatrix} \nabla_1 \hat{h} \\ \nabla_2 \hat{h} \\ \vdots \\ \nabla_k \hat{h} \end{pmatrix}$$

and the matrix $\mathbf{C}$ of the elements

$$\mathbf{C} = \begin{pmatrix} \nabla_1 \bar{h} \\ \nabla_2 \bar{h} \\ \vdots \\ \nabla_k \bar{h} \end{pmatrix}$$

We can now rewrite the system of Eqs. (9.14) using the matrix notation introduced above as

$$\mathbf{J}(\hat{x} - \hat{x}^{(1)}) + \mathbf{C}(\bar{x} - \bar{x}^{(1)}) = 0$$

Assuming that the square $K \times K$ matrix $\mathbf{J}$ has nonzero determinant, we can solve this set of equations for the $\hat{x}$ variables. Equivalently we can invert $\mathbf{J}$ to obtain

$$\hat{x} - \hat{x}^{(1)} = -\mathbf{J}^{-1}\mathbf{C}(\bar{x} - \bar{x}^{(1)}) \qquad (9.15)$$

To a first-order approximation, all points in the vicinity of $x^{(1)}$ satisfying the constraints $h_k(x) = 0$ will be given by matrix equation (9.15). Moreover, for any choice of the nonbasic variables $\bar{x}$, the matrix equation will calculate values of the basic variables $\hat{x}$ that will satisfy the linear approximations to the original constraints. In effect, the linearization has allowed us to solve the constraints for $K$ variables even though we could not accomplish this directly with the constraints themselves.

Once we have this result, we can, of course, do what we previously did in the linearly constrained case. Namely, we can use Eq. (9.15) to eliminate variables $\hat{x}$ from the objective function $f(x)$. Thus,

$$\tilde{f}(\hat{x}; \bar{x}) \simeq f\left(\hat{x}^{(1)} - \mathbf{J}^{-1}\mathbf{C}(\bar{x} - \bar{x}^{(1)}), \bar{x}\right) \tag{9.16}$$

and we have reduced $f$ to a function involving only the $N - K$ nonbasic variables $\bar{x}$.

Since $f$ appears to be an unconstrained function of the nonbasic variable $\bar{x}$, we apply the usual necessary conditions for $x^{(1)}$ to be a local minimum of $f$. As with any unconstrained problem, necessary conditions for $x^{(1)}$ to be a local minimum of $f$ are that the gradient of $f$ with respect to $\bar{x}$ be zero.

Using the chain rule, since $\tilde{f}(\bar{x}) = \hat{f}(\hat{x}(\bar{x}), \bar{x})$, then

$$\frac{\partial \tilde{f}}{\partial \bar{x}} = \frac{\partial f}{\partial \bar{x}} + \frac{\partial f}{\partial \hat{x}} \cdot \frac{\partial \hat{x}}{\partial \bar{x}}$$

Since, from Eq. (9.15),

$$\frac{\partial \hat{x}}{\partial \bar{x}} = -\mathbf{J}^{-1}\mathbf{C}$$

and if we write $\nabla \bar{f} = (\partial f / \partial \bar{x})$ and $\nabla \hat{f} = (\partial f / \partial \hat{x})$, then it follows that

$$\nabla \tilde{f}(x^{(1)}) = \nabla \bar{f}(x^{(1)}) - \nabla \hat{f}(x^{(1)})\mathbf{J}^{-1}\mathbf{C} \tag{9.17}$$

The first-order necessary conditions thus become

$$\nabla \bar{f}(x^{(1)}) - \nabla \hat{f}(x^{(1)})\mathbf{J}^{-1}\mathbf{C} = 0$$

The vector $\nabla \tilde{f}$ defined by (9.17) is called the *reduced gradient* (or con-strained derivative) of the equality-constrained problems. This, of course, is precisely the same construction that we obtained in the development of the convex simplex method. The first-order necessary conditions for a local minimum expressed in terms of the reduced gradient reduces simply to the statement that

$$\nabla \tilde{f}(x^{(1)}) = 0$$

It can readily be shown that the reduced gradient optimality criterion is equivalent to the Lagrangian optimality criterion. Thus, zeros of the reduced gradient are Lagrangian stationary points. To verify this, we merely need to write the Lagrangian necessary conditions for the equality-constrained prob-lem. These are

$$\nabla f(x^*) - (v^*)^T \nabla h(x^*) = 0$$

where $\nabla h$ is understood to be the matrix of constraint gradients. If we now introduce the definition of independent and dependent variables $\bar{x}$ and $\hat{x}$, then the above system of equations can be written as

$$\nabla \hat{f}(x^*) - (v^*)^T \mathbf{J} = 0$$

$$\nabla \bar{f}(x^*) - (v^*)^T \mathbf{C} = 0$$

The first of these equations can be solved for $v^*$ to yield

$$v^* = \left( \nabla \hat{f}(x^*) \mathbf{J}^{-1} \right)^T$$

which when substituted into the second equation yields

$$\nabla \bar{f}(x^*) - \nabla \hat{f}(x^*) \mathbf{J}^{-1} \mathbf{C} = 0$$

This is precisely the reduced gradient condition that we derived earlier. Note that the Lagrange multipliers will be calculated automatically whenever the reduced gradient is evaluated.

By using linear approximations to the equality constraints we have thus accomplished two things:

1. We have obtained a set of linear equations that allow us to estimate the values of $\hat{x}$ corresponding to any perturbation of the nonbasic variables $\bar{x}$ about the point $x^{(1)}$ such that the resulting point is feasible.

2. We have derived a direct first-order necessary condition for the equality-constrained problem that implicitly accounts for the equality constraints.

In the next section we will show how these results of the implicit variable-elimination construction can be used to develop a reduced gradient algorithm for the general nonlinear problem. The *generalized reduced gradient* (GRG) method is only one of a class of algorithms based on implicit variable elimination. Other prominent members of this class that we have already considered are the simplex method of linear programming, the convex simplex method, and the reduced gradient method.

### 9.3.2 The Basic GRG Algorithm

In the previous section, we used linearizations of the constraints to express the objective function as an unconstrained function of $N - K$ independent variables. As a consequence of this construction, the problem can be treated as simply an unconstrained optimization problem—at least for small excursions about the base point for the linearization—and presumably it can be solved using any of a variety of unconstrained gradient methods. Thus, at least in principle, the following analog to the reduced gradient algorithm suggests itself.

At iteration $t$, suppose the feasible point $x^{(t)}$ is available along with the partition $x = (\hat{x}, \bar{x})$ which has associated with it a constraint gradient submatrix $\mathbf{J}$ with nonzero determinant.

**Step 1.**   Calculate

$$\nabla \tilde{f} = \nabla \bar{f}(x^{(t)}) - \nabla \hat{f}(x^{(t)}) \mathbf{J}^{-1} \mathbf{C}$$

**Step 2.**   If $\|\nabla \tilde{f}\| \leqslant \varepsilon$, stop. Otherwise, set

$$\bar{d} = (-\nabla \tilde{f})^T$$

$$\hat{d} = -\mathbf{J}^{-1} \mathbf{C} \bar{d}$$

$$d = (\hat{d}, \bar{d})^T$$

**Step 3.**   Minimize $f(x^{(t)} + \alpha d)$ with respect to the scalar parameter $\alpha$. Let $\alpha^{(t)}$ be the optimizing $\alpha$, set

$$x^{(t+1)} = x^{(t)} + \alpha^{(t)} d$$

and go to step 1.

It is easy to verify that regardless of the degree of nonlinearity the direction $d$ generated in this fashion is a descent direction. From a first-order Taylor's expansion of Eq. (9.16), we have that

$$f(x) - f(x^{(t)}) \approx \tilde{f}(x) - \tilde{f}(x^{(t)}) = \nabla \tilde{f}(x^{(t)})(\bar{x} - \bar{x}^{(t)})$$

$$= \alpha \nabla \tilde{f}(x^{(t)}) \bar{d}$$

Thus, if $\bar{d}$ is chosen as

$$\bar{d} = - (\nabla \tilde{f}(x^{(t)}))^T$$

then it follows that

$$f(x) - f(x^{(t)}) \approx \alpha(\nabla \tilde{f})(-\nabla \tilde{f})^T = -\alpha \|\nabla \tilde{f}\|^2$$

The right-hand side of this expression is less than zero for all positive values of $\alpha$. Consequently, for all positive $\alpha$'s small enough that the linear approximations used are sufficiently accurate, it follows that

$$f(x) - f(x^{(t)}) < 0$$

and consequently $d$ is a descent direction. Note that the specification

$$\hat{d} = -\mathbf{J}^{-1}\mathbf{C}\bar{d}$$

is implicit in the above construction, since Eq. (9.16) is based on using the constraint linearizations to eliminate the $\hat{x}$ variables.

While it is thus evident that step 2 produces a descent direction, it is not at all clear that the points generated in following that direction will be feasible. In fact, for nonlinear constraints, $d$ will not be a feasible direction. Rather, because it is constructed using a linearization of the equality constraints, it is almost certain to lead to points away from the constraints. As can be seen from the next example, the linearization of an equality constraint at a feasible point is equivalent to the construction of a plane tangent to the constraint at the point of linearization. In general, virtually all points on the tangent plane will not satisfy the original equality.

### Example 9.9

Consider the linearization of the constraint

$$h(x) = (x_1 - 1)^3 - x_2 + 1 = 0$$

at the point $x^0 = (\tfrac{1}{2}, \tfrac{7}{8})$. The linear approximation is

$$\tilde{h}(x; x^0) = 0 + \tfrac{3}{4}\left(x_1 - \tfrac{1}{2}\right) - \left(x_2 - \tfrac{7}{8}\right)$$

If we insist that all points $x$ satisfy this linearization, that is, $\tilde{h}(x; x^0) = 0$, we obtain the linear equality

$$\tfrac{3}{4}x_1 - x_2 + 0.5 = 0$$

From Figure 9.5 it is evident that no point on this plane, with the exception of the point $(2, 2)$, will satisfy $h(x) = 0$.

Summarizing the preceding discussion, we have demonstrated first of all that $d$ is a descent direction and second that it will in general lead to infeasible points. More precisely, we have shown that the vector $\bar{d}$ is a descent direction in the space of the nonbasic variables $\bar{x}$ but that the composite direction vector

$$d = \begin{pmatrix} \hat{d} \\ \bar{d} \end{pmatrix}$$

where $\hat{d}$ is calculated via the linear equation

$$\hat{d} = -\mathbf{J}^{-1}\mathbf{C}\bar{d}$$

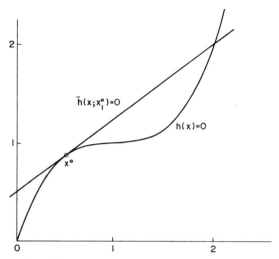

**Figure 9.5.**   Linearization of a nonlinear constraint.

yields infeasible points. Since $\bar{d}$ has desirable properties but the addition of $\hat{d}$ causes undesirable consequences, it seems reasonable that the computation of $\hat{d}$ should be revised. The question is how?

We can gain further insight into the situation by considering a graphical example. Consider the spherical constraint surface

$$h(x_1, x_2, x_3) = x_1^2 + x_2^2 + x_3^2 - 1 = 0$$

shown in Figure 9.6. Given the feasible point $x^0$, the linearization of $h(x) = 0$ at $x^0$ will yield a tangent plane. Suppose that $x_1$ and $x_2$ are chosen as nonbasic variables, that is, $\bar{x} = (x_1, x_2)$, and $x_3$ as a basic variable, that is, $\hat{x} = (x_3)$. Furthermore, suppose that for some objective function, step 2 of our prototype algorithm generates the direction $d$ shown in the figure. The subvector $\bar{d}$ will consist of the projection of $d$ on the $x_1 x_2$ plane, while $\hat{d}$ will simply be the projection of $d$ on the $x_3$ axis. Clearly the point $v^{(2)}$ that would be attained by following along $d$ from $x^0$ is not feasible, that is, $h(v^{(2)}) \neq 0$. The point $v^{(2)}$ is infeasible, because for the chosen change in the independent variables, $\bar{x}^{(2)} - \bar{x}^0$, the change in the dependent variable, $\hat{x}^{(2)} - \hat{x}^0$, calculated using the approximation

$$\hat{x}^{(2)} - \hat{x}^0 = -\mathbf{J}^{-1}\mathbf{C}(\bar{x}^{(2)} - \bar{x}^0)$$

is not accurate enough to yield a value of $\hat{x}$ that satisfies the equation

$$h(\bar{x}^{(2)}, \hat{x}) = 0$$

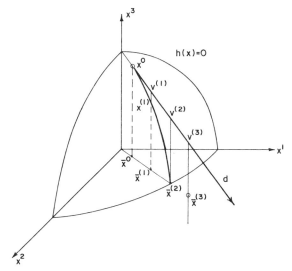

**Figure 9.6.**   The reduced gradient direction vector and curve of feasible points.

exactly. We should have expected this, because, as we observed previously, while the linearizations are good enough to determine local descent directions, they are not good enough to actually calculate feasible minimum points. What we really ought to do is to calculate $\bar{d}$ using the linearization, but then, rather than calculating $\hat{d}$ as in step 2 and minimizing $f$ along the *line* fixed by $d$, we should project $\bar{d}$ onto the constraint surface and minimize along the resulting *curve*. Thus, in effect we will search for the minimum of $f(x)$ along an implicitly defined curve that is given by the set of $\alpha$ and $\hat{x}$ values that satisfy the equation

$$h(\bar{x}^0 + \alpha\bar{d}, \hat{x}) = 0$$

This is the curve passing through points $\bar{x}^0$, $\bar{x}^{(1)}$, and $\bar{x}^{(2)}$ in Figure 9.6. If

$$\bar{d} = \begin{pmatrix} 1 \\ 1 \end{pmatrix} \quad \text{and} \quad x^0 = (0.1, 0.1)$$

then this curve will be given by all values of $\alpha$ and $x_3$ satisfying

$$(0.1 + \alpha)^2 + (0.1 + \alpha)^2 + x_3^2 - 1 = 0$$

In the general case with multiple constraints, this means that for every value of $\alpha$ that is selected as a trial, the constraint equation will have to be solved for the values of the dependent variables $\hat{x}$ that will cause the resulting point to be feasible.

Thus step 3 of our prototype algorithm must be replaced by an iterative procedure that will, for a given value of $\alpha$, iterate on the constraints to calculate a value of $\hat{x}$ that satisfies the constraints. This can be accomplished by using, for instance, Newton's method to solve the set of equations

$$h_k\left(\bar{x}^{(t)} + \alpha\bar{d}, \hat{x}\right) = 0 \qquad k = 1,\ldots, K$$

Recall from Chapter 3 that the iteration formula for Newton's method for solving a system of $K$ equations in $K$ unknowns $z_k$,

$$h_k(z) = 0 \qquad k = 1,\ldots, K$$

involves the gradients $\nabla_z h_k$ of the functions evaluated at the current best estimate $z^{(i)}$. If we denote the matrix with rows $\nabla_z h_k(z^{(i)})$ by $\nabla h^{(i)}$ and the vector of function values $h_k(z^{(i)})$ by $h^{(i)}$, then the iteration formula will be given by

$$z^{(i+1)} = z^{(i)} - \left(\nabla h^{(i)}\right)^{-1} h^{(i)}$$

In the case of our constraint iterations, we have

$$z^{(i)} = \hat{x}^{(i)}$$

$$\nabla h^{(i)} = \mathbf{J}\left(x^{(t)} + \alpha\bar{d}, \hat{x}^{(i)}\right)$$

and

$$h^{(i)} = h\left(x^{(t)} + \alpha\bar{d}, \hat{x}^{(i)}\right)$$

Therefore, the Newton iteration formula will be

$$\hat{x}^{(i+1)} = \hat{x}^{(i)} - \left[\mathbf{J}^{-1}\left(\bar{x}^{(t)} + \alpha\bar{d}, \hat{x}^{(i)}\right)\right] \cdot h\left(\bar{x}^{(t)} + \alpha\bar{d}, \hat{x}^{(i)}\right) \qquad (9.18)$$

where the notation $\mathbf{J}^{-1}(\bar{x}^{(t)} + \alpha\bar{d}, \hat{x}^{(i)})$ indicates that the elements of the inverse of $\mathbf{J}$ are functions of the step size $\alpha$ and of the basic variables $\hat{x}$.

Assuming that Newton's method will converge to an $\hat{x}$ that satisfies the constraints, then $f(x)$ can be evaluated at that point to determine if an improvement over $x^{(t)}$ has been achieved. If no improvement has been achieved, then $\alpha$ must be reduced using some search logic, and the Newton iterations must be repeated. If $f(x)$ is improved, then one can either continue with a new value of $\alpha$ until no further improvement in $f(x)$ can be obtained, or else, as is more typically done, the current point can be retained and a new direction vector calculated.

It is, of course, possible that for the $\alpha$ selected the Newton iterations may fail to find an $\hat{x}$ that satisfies the constraints. This indicates that $\alpha$ was chosen too large and should be reduced. In Figure 9.6 there is no value of $\hat{x}$ ($= x_3$)

corresponding to the point $\bar{x}^{(3)}$ that will intersect the constraint. If this occurs, then $\alpha$ must be reduced and another cycle of Newton iterations initiated.

With these additional steps, a composite *generalized reduced gradient* (GRG) algorithm can be formulated.

**A Generalized Reduced Gradient Algorithm.**   Given an initial feasible point $x^0$, a specified initial value of the search parameter $\alpha = \alpha^0$, termination parameters $\varepsilon_1$, $\varepsilon_2$, and $\varepsilon_3$, all $> 0$, and a reduction parameter $\gamma, 0 < \gamma < 1$.

**Step 1.**   Choose a partition of $x$ into $\hat{x}$ and $\bar{x}$ such that $\mathbf{J}$ has nonzero determinant, and calculate the reduced gradient;

$$\nabla \tilde{f}(x^{(t)})$$

**Step 2.**   If $\|\nabla \tilde{f}\| \leqslant \varepsilon_1$, stop. Otherwise, set

$$\bar{d} = -(\nabla \tilde{f})^T$$

$$\hat{d} = -\mathbf{J}^{-1}\mathbf{C}\bar{d}$$

and

$$d = (\hat{d}, \bar{d})^T$$

**Step 3.**   Set $\alpha = \alpha^0$. For $i = 1, 2, \ldots,$
(a) Calculate $v^{(i)} = x^{(t)} + \alpha d$. If $|h_k(v^{(i)})| \leqslant \varepsilon_2$; $k = 1, \ldots, K$, go to (d). Otherwise, continue.
(b) Let

$$\hat{v}^{(i+1)} = \hat{v}^{(i)} - \mathbf{J}^{-1}(v^{(i)}) \cdot h(v^{(i)})$$

$$\bar{v}^{(i+1)} = \bar{v}^{(i)}$$

(c) If $\|\hat{v}^{(i+1)} - \hat{v}^{(i)}\| > \varepsilon_3$, go to (b). Otherwise, if $|h_k(v^{(i)})| \leqslant \varepsilon_2$, $k = 1, \ldots, K$, to go step (d). Otherwise, set $\alpha = \gamma\alpha$ and go to (a).
(d) If $f(x^{(t)}) \leqslant f(v^{(i)})$, set $\alpha = \gamma\alpha$ and go to (a). Otherwise, set $x^{(t+1)} = v^{(i)}$ and go to step 1.

We illustrate the application of this algorithm with the following example.

*Example 9.10*

$$\text{Minimize} \quad f(x) = 4x_1 - x_2^2 + x_3^2 - 12$$

$$\text{Subject to} \quad h_1(x) = 20 - x_1^2 - x_2^2 = 0$$

$$h_2(x) = x_1 + x_3 - 7 = 0$$

Suppose we are given the feasible starting point $x^{(1)} = (2, 4, 5)$, and suppose we choose $x_1$ as nonbasic variable and $x_2$ and $x_3$ as basic variables. Thus,

$$\hat{x} = (x_2, x_3) \quad \text{and} \quad \bar{x} = (x_1)$$

Note that the number of nonbasic variables is equal to the total number of variables minus the number of equality constraints. The required function derivatives are $\nabla f = (4, -2x_2, 3x_3)$, $\nabla h_1 = (-2x_1, -2x_2, 0)$, and $\nabla h_2 = (1, 0, 1)$.

**Step 1.** Since

$$\nabla f = (4, -8, 10), \qquad \nabla h_1 = (-4, -8, 0), \qquad \nabla h_2 = (1, 0, 1)$$

it follows that

$$J = \begin{pmatrix} -8 & 0 \\ 0 & 1 \end{pmatrix} \qquad C = \begin{pmatrix} -4 \\ 1 \end{pmatrix} \qquad \nabla \hat{f} = (-8, 10) \qquad \nabla \bar{f} = (4)$$

Next, we calculate

$$J^{-1} = \begin{pmatrix} -\frac{1}{8} & 0 \\ 0 & 1 \end{pmatrix} \quad \text{and}$$

$$\nabla \tilde{f} = (4) - (-8, 10) \begin{pmatrix} -\frac{1}{8} & 0 \\ 0 & 1 \end{pmatrix} \begin{pmatrix} -4 \\ 1 \end{pmatrix} = -2$$

**Step 2.** The direction vector becomes

$$\bar{d} = -\nabla \tilde{f} = 2$$

$$\hat{d} = -\begin{pmatrix} -\frac{1}{8} & 0 \\ 0 & 1 \end{pmatrix} \begin{pmatrix} -4 \\ 1 \end{pmatrix} (2) = \begin{pmatrix} -1 \\ -2 \end{pmatrix}$$

**Step 3.** Set $\alpha^0 = 1$.

(a) $v^{(1)} = \begin{pmatrix} 2 \\ 4 \\ 5 \end{pmatrix} + \alpha \begin{pmatrix} 2 \\ -1 \\ -2 \end{pmatrix} = \begin{pmatrix} 4 \\ 3 \\ 3 \end{pmatrix}$

$$h_1(v^{(1)}) = -5 \quad \text{and} \quad h_2(v^{(1)}) = 0.0$$

Note that the linear constraint $h_2(x)$ is satisfied. This is to be expected, since a linear approximation to a linear constraint will be the constraint itself. Thus, linear constraints will always be satisfied automatically in this algorithm.

Since $h_1$ is not satisfied, Newton's method must be applied to the system of equations

$$h_1(4, x_2, x_3) = 0$$

$$h_2(4, x_2, x_3) = 0$$

to solve for $x_2$ and $x_3$ starting with the initial estimate $x_2 = x_3 = 3$.

(b) $\hat{v}^{(2)} = \begin{pmatrix} 3 \\ 3 \end{pmatrix} - \begin{pmatrix} -6 & 0 \\ 0 & 1 \end{pmatrix}^{-1} \begin{pmatrix} -5 \\ 0 \end{pmatrix} = \begin{pmatrix} 3 \\ 3 \end{pmatrix} + \frac{1}{6} \begin{pmatrix} 1 & 0 \\ 0 & -6 \end{pmatrix} \begin{pmatrix} -5 \\ 0 \end{pmatrix}$

$= \begin{pmatrix} 2.167 \\ 3 \end{pmatrix}$

(c) Since $\hat{v}^{(2)} \neq \hat{v}^{(1)}$, another Newton iteration must be carried out.

(b) $\hat{v}^{(3)} = \begin{pmatrix} \frac{13}{6} \\ 3 \end{pmatrix} - \begin{pmatrix} -\frac{13}{3} & 0 \\ 0 & 1 \end{pmatrix}^{-1} \begin{pmatrix} -0.69 \\ 0.0 \end{pmatrix} = \begin{pmatrix} 2.008 \\ 3 \end{pmatrix}$

If we accept this result as having converged, then we must test the constraint values [step (c)].

$$h_1(v^{(3)}) = -0.032 \quad \text{and} \quad h_2(v^{(3)}) = 0.0$$

Assuming that 0.032 is close enough to zero, we continue step (d).

(d) $v^{(3)} = (4, 2.008, 3)$

$$f(x^{(1)}) = 5 < f(v^{(3)}) = 9$$

No improvement in $f(x)$ is achieved. The initial $\alpha$ chosen was too large and must be reduced. Suppose we reduce $\alpha$ to 0.4.

(a) $\hat{v}^{(4)} = \begin{pmatrix} 2 \\ 4 \\ 5 \end{pmatrix} + 0.4 \begin{pmatrix} 2 \\ -1 \\ -2 \end{pmatrix} = \begin{pmatrix} 2.8 \\ 3.6 \\ 4.2 \end{pmatrix}$

$$h_1(v^{(4)}) = -0.80 \quad \text{and} \quad h_2(v^{(4)}) = 0.0$$

The point is infeasible and Newton iteration is required.

(b) $\hat{v}^{(5)} = \begin{pmatrix} 3.6 \\ 4.2 \end{pmatrix} \begin{pmatrix} -7.2 & 0 \\ 0 & 1 \end{pmatrix}^{-1} \begin{pmatrix} -0.8 \\ 0 \end{pmatrix} = \begin{pmatrix} 3.49 \\ 4.2 \end{pmatrix}$

(c) If we accept this result as having converged, then we must test

$$h_1(v^{(5)}) = -0.02 \quad \text{and} \quad h_2(v^{(5)}) = 0.0$$

Assuming that this is close enough to zero, we proceed to test if the objective function has improved.

(d) $f(x^{(1)}) = 5 > f(v^{(5)}) = 4.66$

Since improvement has been made, $x^{(2)}$ is set equal to $v^{(5)}$. The next iteration is begun with

$$x^{(2)} = (2.8, 3.49, 4.2)$$

The solution to the problem is $x^* = (2.5, 3.71, 4.5)$. Note that at that point the reduced gradient is zero:

$$\nabla \tilde{f} = (4) - (-7.42, 9.0) \begin{pmatrix} -7.42 & 0 \\ 0 & 1 \end{pmatrix}^{-1} \begin{pmatrix} -5 \\ 1 \end{pmatrix} = 4 - 4 = 0.0$$

As can be noted from the above example, most of the effort per iteration is involved in performing the Newton iterations. In order to keep these iterations to a low level, the one-parameter optimizations on $\alpha$ are normally not carried out very exactly. In fact, in our simplified version of the GRG algorithm the first feasible improved point is accepted without any further refinement. An alternative approach is to use an "inside-out" strategy that involves a bracketing search followed by a quadratic interpolation step. Thus, rather than choosing a larger initial $\alpha$ and then retracting the search if improvement is not achieved, a bracketing search is first carried out on $\alpha$ to bound the minimum. The bounding search is initiated with a small $\alpha$ so that the initial trials remain close to the linearization point $x^{(t)}$, and consequently the Newton correction iterations are less likely to encounter difficulties. As $\alpha$ is increased during bracketing, the previously calculated Newton result can always be used as a good estimate with which to start the Newton corrections at the next $\alpha$ value. Once a bracket is established, then a quadratic (or cubic) interpolation step is carried out to obtain an improved estimate of the line-search optimum $\alpha^*$. The resulting point may require further correction via Newton iterations but is not improved by further successive quadratic interpolations.

The extension of the reduced gradient method to nonlinearly constrained problems is generally credited to Abadie and Carpenter [11]. However, elements of the implicit variable-elimination strategy with nonlinear equalities appears in the early analytical-graphical optimal design work of Johnson as early as 1956 (see the review in ref. 12). The implicit variable-elimination strategy also underlies the differential algorithm of Beightler and Wilde [13] and the constrained derivative conditions of Wilde [14] and Reklaitis and Wilde [15]. The work of Abadie and Carpenter [11] did, however, lead to the first general GRG code.

### 9.3.3  Extensions of the Basic Method

The preceding development of the basic GRG algorithm has been restricted to equality constraints and has resulted in essentially a gradient method operating within the reduced space of the nonbasic variables. In this section we consider extensions of the basic algorithm to accommodate the general NLP problem with both upper and lower variable bounds as well as the inequality constraints:

$$\text{Minimize} \quad f(x)$$

$$\text{Subject to} \quad a_j \leqslant g_j(x) \leqslant b_j \qquad j = 1,\ldots,J$$

$$h_k(x) = 0 \qquad k = 1,\ldots,k$$

$$x_i^{(L)} \leqslant x_i \leqslant x_i^{(U)} \qquad i = 1,\ldots,N$$

In addition, we discuss the use of the conjugate direction or quasi-Newton formulas to modify the direction vector $\bar{d}$ generated by the basic GRG algorithm so as to attain improved terminal rates of convergence.

**Treatment of Bounds.** Variable upper and lower bounds can be accommodated either explicitly by treating each bound as an inequality constraint or implicitly by accounting for the bounds within the appropriate steps of the algorithm. The latter approach is clearly preferable, since it results in a smaller matrix **J** that must be inverted. To incorporate such constraints into the algorithm in an implicit fashion, three changes are necessary:

1. A check must be made to ensure that only variables that are not on or very near their bounds are labeled as basic variables. This check is necessary to ensure that some free adjustment of the basic variables can always be undertaken.

This can be accomplished by simply ordering the variables according to distance from their nearest bound. Thus, at the current feasible point $x^{(t)}$, let

$$z_i^{(t)} = \min\{(x_i^{(U)} - x_i^{(t)}), (x_i^{(t)} - x_i^{(L)})\} \tag{9.19}$$

The quantities $z_i^{(t)}$ can then be ordered by decreasing magnitude, and the variables associated with the first $K$ can be selected as dependent variables. Of course, this partition of the variables must also result in a nonsingular **J**. Therefore, one or more of the first $K$ variables may have to be rejected in favor of variables farther down the order before a nonsingular **J** results.

2. The direction vector $\bar{d}$ is modified to ensure that the bounds on the independent variables will not be violated if movement is undertaken in the $\bar{d}$ direction. This is accomplished by setting

$$\bar{d}_i = \begin{cases} 0 & \text{if } \bar{x}_i = \bar{x}_i^{(U)} \text{ and } (\nabla f)_i < 0, \\ 0 & \text{if } \bar{x}_i = \bar{x}_i^{(L)} \text{ and } (\nabla f)_i > 0 \\ -(\nabla f)_i & \text{otherwise} \end{cases}$$

This modification will also ensure that, if $\bar{d}_i = 0$ for all $i = 1, \ldots, N - K$, then the Kuhn-Tucker conditions will be satisfied at that point. This fact can be verified easily by applying the Kuhn-Tucker conditions to the problem with equality constraints and two types of inequalities:

$$g' \equiv x^{(U)} - x \geqslant 0$$

$$g'' \equiv x - x^{(L)} \geqslant 0$$

Verification is left as an exercise for the reader.

3.  Checks must be inserted in step 3 of the basic GRG algorithm to ensure that the bounds are not exceeded either during the search on $\alpha$ or during the Newton iterations.

As shown in Figure 9.7, this can be accomplished by simply checking if the result of step 3 satisfies the variable bounds. If at least one bound is violated, a linear interpolation can be made between $x^{(t)}$ and $v^{(i)}$ to estimate the point on this line segment at which a basic variable bound is first violated. Step (3) is then reinitiated at this point to obtain a $v^{(i)}$ at which $\hat{x}^{(L)} \leqslant \hat{v}^{(i)} \leqslant \hat{x}^{(U)}$.

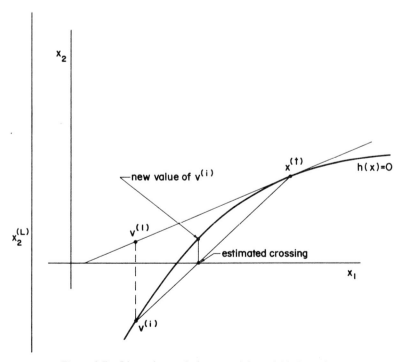

**Figure 9.7.** Linear interpolation to satisfy variable bounds.

**Treatment of Inequalities.** General inequality constraints can be handled within the GRG framework either by explicitly writing these constraints as equalities using slack variables or by implicitly using the concept of active constraint set as in feasible direction methods. In the former case, inequalities of the form

$$a_j \leqslant g_j(x) \leqslant b_j$$

are converted to equalities

$$h_{K+j}(x) = g_j(x) - x_{N+j} = 0$$

by the introduction of slack variables $x_{N+j}$, together with the bounds

$$a_j \leqslant x_{N+j} \leqslant b_j$$

This approach is attractive because it allows all constraints to be treated in a uniform fashion without further restructuring of the basic algorithm. The disadvantage is that $\mathbf{J}$ must include a row for each $g_j(x)$ even if the inequalities are not binding. In addition, the problem dimensionality is increased.

If an active constraint strategy is employed, then at each iteration point $x^{(t)}$ the active inequalities must be identified and their linearizations added to those of the equalities to define $\mathbf{J}$ and $\mathbf{C}$. The logic of step 3 of the basic algorithm must be modified to include tests for the violation of the previously inactive constraint. If at target point $v^{(i)}$ some previously inactive constraint is violated, then the step size $\alpha$ must be adjusted and the Newton calculations repeated until a point satisfying both the active and inactive constraints is found. At the resulting points, any tight but previously inactive constraints must be added to the active constraint set, and any active constraint now not binding must be deleted from the active set. These considerations will clearly complicate the step 3 computations and will result in the need to continually readjust the size of $\mathbf{J}$ and $\mathbf{C}$ and update $\mathbf{J}^{-1}$.

The two approaches to the treatment of inequalities are illustrated in the next example.

*Example 9.11*

Consider the problem

$$\text{Minimize} \quad f(x) = (x_1 - 3)^2 + (x_2 - 3)^2$$

$$\text{Subject to} \quad g_1(x) = 2x_1 - x_2^2 - 1 \geqslant 0$$

$$g_2(x) = 9 - 0.8x_1^2 - 2x_2 \geqslant 0$$

$$4 \geqslant (x_1, x_2) \geqslant 0$$

with initial feasible point $x^0 = (1, 1)^T$.

If the slack variable approach is used, then the constraints will be reformulated to

$$h_1(x) = g_1(x) - x_3 = 0$$

$$h_2(x) = g_2(x) - x_4 = 0$$

The expanded dimensionality starting point becomes

$$x^0 = (1,1,0,6.2)$$

At $x^0$,

$$\nabla h_1(x^0) = (2, -2, -1, 0)$$

$$\nabla h_2(x^0) = (-1.6, -2, 0, -1)$$

and

$$\nabla f(x^0) = (-4, -4, 0, 0)$$

Suppose we select $\hat{x} = (x_1, x_2)$ and $\bar{x} = (x_3, x_4)$. Then

$$J = \begin{pmatrix} 2 & -2 \\ -1.6 & -2 \end{pmatrix} \quad C = \begin{pmatrix} -1 & 0 \\ 0 & -1 \end{pmatrix}$$

**Step 1.**

$$\nabla \tilde{f} = (0,0) - \frac{(-4, -4)}{7.2} \begin{pmatrix} 2 & -2 \\ -1.6 & -2 \end{pmatrix} \begin{pmatrix} -1 & 0 \\ 0 & -1 \end{pmatrix}$$

$$= \frac{1}{7.2}(-1.6, 16)$$

**Step 2.** Set

$$\bar{d}_1 = \frac{1.6}{7.2}$$

since $\nabla \tilde{f}_1 < 0$ and $\bar{x}_1$ is at lower bound. Set

$$\bar{d}_2 = \frac{-1.6}{7.2}$$

since $\nabla \tilde{f}_2 > 0$ and $\bar{x}_2$ is not on its bounds. Then

$$\hat{d} = -J^{-1}Cd = \begin{pmatrix} 0.6790 \\ 0.5679 \end{pmatrix}$$

**Step 3.** As is evident from Figure 9.8, a larger initial step can easily be taken in this case. Suppose $\alpha^{(0)} = 2$ is employed. Then

$$v^0 = x^0 + 2d = (2.3580, 2.1358, 0.4444, 1.7555)$$

(a) $h_1(v^0) = -1.2900$ and $h_2(v^0) = -1.4752$

(b) $\hat{v}^{(1)} = \hat{v}^0 - J^{-1}h$

$$= \begin{pmatrix} 2.3580 \\ 2.1358 \end{pmatrix} - \frac{1}{7.2} \begin{pmatrix} 2 & -2 \\ -1.6 & -2 \end{pmatrix} \begin{pmatrix} -1.2900 \\ -1.4752 \end{pmatrix}$$

$$= (2.3066, 1.4393)^T$$

Steps 3b and 3c are repeated until the point $\hat{v} = (2.1859, 1.7110)^T$ is reached, at which

$$h_1 = -1 \times 10^{-4} \quad \text{and} \quad h_2 = -3 \times 10^{-5}$$

(d) At this point,

$$f(\hat{v}) = 2.324 < f(x^0) = 8.0$$

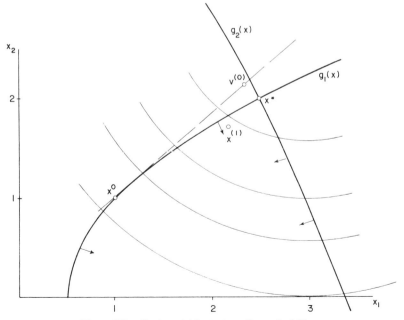

**Figure 9.8.** Slack variable strategy, Example 9.11.

Therefore,

$$x^{(1)} = v = (2.1859, 1.7110, 0.4444, 1.7555)^T$$

The iterations continue in this fashion until the optimum point $x^* = (2.5, 2, 0, 0)^T$ is reached.

Alternatively, if the active constraint strategy is used, then the constraints are first tested at $x^0$ to determine which are active. Since $g_1(x^0) = 0.0$ and $g_1(x^0) = 6.2$, constraint 1 is active and $g_2$ will be temporarily disregarded. If the partition $\hat{x} = (x_2)$, $\bar{x} = (x_1)$ is selected, then

$$\mathbf{J} = (-2) \qquad \mathbf{C} = (2)$$

$$\nabla \hat{f} = (-4) \qquad \nabla \bar{f} = (-4)$$

**Step 1.**  $\nabla \tilde{f} = (-4) - (-4)(-\frac{1}{2})(2) = -8$

**Step 2.**  Since $x_2$ is away from its bounds, we can set

$$\bar{d} = -(-8) = 8$$

$$\hat{d} = -\mathbf{J}^{-1}\mathbf{C}\bar{d} = -(-\frac{1}{2})(2)(8) = +8$$

**Step 3.**

$$v^0 = x^0 + \alpha d = \begin{pmatrix} 1 \\ 1 \end{pmatrix} + \alpha \begin{pmatrix} 8 \\ 8 \end{pmatrix}$$

Suppose we use $\alpha^0 = \frac{1}{4}$; then $v^0 = (3, 3)$ and $g_1(v^0) = -4$.

**Step 3b.**  $\hat{v}^{(1)} = 3 - (-\frac{1}{6})(-4) = 2.333$

Steps 3b and 3c will continue until the point $\hat{v}^{(3)} = 2.2361$ is reached, at which $g_1 \simeq -1 \times 10^{-4}$. At the point, $v^{(3)} = (3, 2.2361)$, the objective function value is $0.5835 < f(x^0)$. However, $g_2(v^{(3)}) = -2.672$, so that this constraint has become violated, as shown in Figure 9.9.

The step size $\alpha$ must therefore be reduced and step (3) repeated. For instance, if $\alpha^{(1)} = \frac{1}{6}$, we obtain

$$v^{(4)} = (2.3333, 2.3333)$$

A repeat of the Newton iterations will yield the new feasible point $x^{(1)} = (2.3333, 1.9149)$, at which

$$f(x^{(1)}) = 1.622 \quad \text{and} \quad g_2(x^{(1)}) = 0.815$$

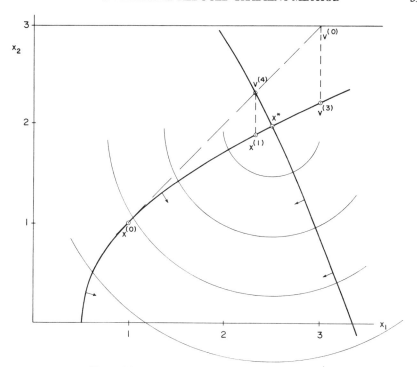

**Figure 9.9.**   Active constraint strategy, Example 9.11.

Because of the iterations required to correct the infeasibilities that can occur among the previously inactive constraints, the active constraint strategy is not used with the GRG implementations reported in the literature [11, 16, 17].

**Modification of the Direction Vector.**   Convergence proofs are not available for versions of GRG as it is typically implemented. Convergence has been proved [18] for an idealized version of GRG in which

1.  Exact searches on $\alpha$ are performed.
2.  An $\varepsilon$ perturbation procedure is used to avoid jamming.
3.  A special basis-updating procedure is introduced that may require basis changes at each iteration.
4.  It is assumed that at each iteration a nondegenerate, nonsingular basis can always be found.

Under these conditions it can be shown that the GRG calculations will converge to a Kuhn-Tucker point. An interesting terminal convergence rate result developed for the basic equality-constrained problem by Luenberger [19] indicates that the GRG algorithm can be expected to exhibit a performance analogous to the unconstrained gradient method. Specifically, it is shown that if $\{x^{(t)}\}$ is a sequence of iterates generated using the same variable partition and converging to $x^*$, then the sequence of objective function values $\{f(x^{(t)})\}$ converges to $f(x^*)$ *linearly* with a ratio bounded above by $[(B - b)/(B + b)]^2$. Quantities $b$ and $B$ are, respectively, the smallest and largest eigenvalues of the matrix,

$$Q = (-\mathbf{J}^{-1}\mathbf{C}^T, \mathbf{I})\mathbf{H}_L\begin{pmatrix} -\mathbf{J}^{-1}\mathbf{C} \\ \mathbf{I} \end{pmatrix}$$

and matrix $\mathbf{H}_L$ is the matrix of second derivatives of the Lagrangian function $L = f - v^T h$. This result indicates that the convergence rate is strongly influenced by the choice of basis (as reflected by $\mathbf{J}^{-1}\mathbf{C}$) as well as by the second derivatives of all problem functions. Note that this definition of convergence rate differs from the normal definition of the rate in terms of the error in $x^{(t)}$, that is, $\|x^{(t)} - x^*\|$. However, the ratio bound on the objective function values is similar to results available for the ordinary unconstrained gradient method. Luenberger's analysis thus clearly indicates the desirability of using convergence acceleration techniques.

As in the reduced gradient method for the linearly constrained case, the direction vector $\bar{d}$ can be updated using conjugate gradient or quasi-Newton methods. Provided that the variable partition is not altered, the use of such updates will substantially enhance the convergence rate. Of course, whenever a change is made in the partition, the updating formula must be restarted, since the nature of the search subspace is altered. Thus, the modification of direction vectors is most effective in the final stages of the iterations when no further significant partitioning changes are encountered. Different authors have, for one reason or another, selected different update methods. For instance, Lasdon et al. [16] recommend the use of the BFS formula and resort to conjugate gradient methods only if storage becomes limiting, while Abadie and Carpentier [11] use the Fletcher-Reeves conjugate gradient method. In view of the generally superior performance of the BFS update, it appears that it ought to be the preferred choice.

### 9.3.4   Computational Considerations

From the preceding discussion it is evident that implementation of an efficient GRG algorithm involves consideration of a variety of numerical issues: Newton iterations, linear equation solving, line searching, step-size adjustment, and so on. In this section we briefly consider the following computational enhancements to the GRG algorithm:

1. The use of approximate Newton iterations
2. The indirect calculation of $\mathbf{J}^{-1}$
3. The special treatment of linear problem elements
4. The use of numerical derivatives

**Approximate Newton Calculations.**   The main computational burden associated with the GRG algorithm arises from the Newton iterations. Various rules and approximation devices have been proposed to reduce this overhead. These are:

1.  Using the inverse of **J** evaluated at $x^{(t)}$ without subsequent updating.
2.  Using the approximation suggested by Abadie and Carpentier [11],

$$\mathbf{J}(v^{(k)})^{-1} = 2\mathbf{J}(x^{(t)})^{-1} - \mathbf{J}(x^{(t)})^{-1}\mathbf{J}(v^{(k)})\mathbf{J}(x^{(t)})$$

3.  Using a quasi-Newton equation-solving method to update $\mathbf{J}^{-1}$ without recomputing derivatives.

The first approach clearly is the simplest but can lead to substantially reduced rates of convergence. The second approach requires the recomputation of the constraint gradients at the intermediate iteration points $v^{(k)}$ but avoids recomputation of the inverse of **J**. The third approach appears very promising, since $\mathbf{J}^{-1}(x^{(k)})$ can be used as the initial value, and subsequent updates only require the constraint function values. The two latest GRG implementations [16, 17] use alternative 1. When coupled with an "inside-out" search strategy, this appears to be an acceptable simplification.

**Avoidance of Matrix Inverses.**   Note that although all the key formulas in terms of which the GRG algorithm is defined are expressed in terms of $\mathbf{J}^{-1}$, the inverse need not be explicitly calculated. For instance, to calculate the reduced gradient it is not necessary to use

$$\nabla \tilde{f} = \nabla \bar{f} - \nabla \hat{f} \mathbf{J}^{-1} \mathbf{C}$$

Instead, we first solve $\nabla \hat{f} - v^T \mathbf{J} = 0$ to obtain the multipliers $v$ and then evaluate the reduced gradient via

$$\nabla \tilde{f} = \nabla \bar{f} - v^T \mathbf{C}$$

Similarly, to calculate $\hat{d} = -\mathbf{J}^{-1}\mathbf{C}\bar{d}$, we first calculate $\mathbf{C}\bar{d}$ and then solve the linear equation set $\mathbf{J}\hat{d} = -\mathbf{C}\bar{d}$ for $\hat{d}$. Finally, the Newton calculations would be executed by solving the linear system

$$\mathbf{J}\Delta\hat{x} = -h$$

All these matrix calculations can be readily modified to take account of any special problem structure. In particular, since large problems normally tend to be sparse—that is, the matrix $(\mathbf{J}, \mathbf{C})$ tends to have relatively few nonzero elements—specialized techniques developed for the solution of large, sparse linear equation sets can be readily adopted [17]. Thus, the GRG method appears to be particularly suitable for the solution of large-scale nonlinear problems.

**Treatment of Linear Elements.**   Further significant computational savings can be achieved if, in implementing a GRG algorithm, special account is taken of linear elements in the problem formulations. Such savings are particularly significant in the case of large problems. As proposed by Murtagh and Saunders [10], each problem function can be subdivided into linear and nonlinear components,

$$h(x) = h^N(x^N) + a_k x^L + b_k$$

where the variables $x^L$ are those that do not appear in any nonlinear terms, while the variables $x^N$ are those that appear in at least one nonlinear term. The advantages of this classification are twofold. First, the columns of the matrix of constraint gradients corresponding to the $x^L$ variables will always remain constant and hence need not be recomputed. Second, the classification leads to an effective reduction in the dimensionality of the space of nonbasic variables.

Following Murtagh and Saunders [10], suppose the problem variables are divided into three sets of variables:

1.   The $K$ basic variables, which have values strictly between their upper and lower bounds
2.   A set of $S$ *superbasic* variables, which have values strictly between their upper and lower bounds but are not basic
3.   The remaining set of $N - K - S$ nonbasic variables whose values lie on one or the other of the bounds

The significance of this subdivision lies in the fact [a discovery attributed to A. Jain (see ref. 10)] that if a nonlinear program has NL nonlinear variables, then an optimal solution can be found at which the number of superbasic variables $S$ is less than or equal to NL. This result suggests that it is sufficient to form the search direction $\bar{d}$ only from the superbasic variables.

The concept of superbasic variable, therefore, effectively reduces the dimensionality of the search space $\bar{d}$ and, as a result, the size of any quasi-Newton update formula that is stored and used to modify $\bar{d}$. The nonbasic variables need in principle be considered only when no further improvement is possible within the space of the superbasics. Thus, if $\bar{d} = 0$ (within some tolerance), then the reduced gradient with respect to the nonbasic variables is computed to determine if further improvement is possible by allowing one (or more) of these variables to leave their bounds and thus to join the set of superbasics. This strategy has led to the very efficient reduced gradient implementation MINOS [20] for linearly constrained problems and is an important concept in GRG implementations for large-scale problems [21].

**Use of Numerical Derivatives.** The GRG algorithm, and in fact all of the linearization-based algorithms we have discussed, require values of the derivatives of the problem functions. As in the unconstrained case, these derivatives may be supplied in analytical form by the algorithm user or may be calculated numerically from the function values by forward or central difference formulas. Since forward differencing requires $N$ additional evaluations of each function, while central differencing requires $2N$, the former is normally selected because the higher accuracy offered by central differencing is not really necessary. To further save function evaluations, two devices have been reported. Gabriele [17] suggests the use of a test to determine whether a variable has changed sufficiently before the partial derivatives of the constraint functions with respect to that variable are computed. Given a differencing parameter $\delta$, the function values

$$\tilde{h}_k^{(i)} \equiv h_k(x_1, \ldots, x_{i-1}, x_i + \delta, x_{i+1}, \ldots, x_N) \qquad k = 1, \ldots, K$$

needed to estimate the partial derivatives $\partial h_k / \partial x_i$ by forward difference,

$$\frac{\partial h_k^{(t)}}{\partial x_i} \approx \frac{\tilde{h}_k^{(i)} - h_k^{(t)}}{\delta} \tag{9.20}$$

are evaluated only if

$$\left| \frac{x_i^{(t)} - x_i^{(t-1)}}{x_i^{(t-1)}} \right|$$

is greater than some selected tolerance $\varepsilon$. This, of course, assumes that the value of each partial derivative with respect to $x_i$ is dominated by the variable $x_i$. Such an approximation can be rather poor, for instance in the case $h(x) = x_1 x_2$. However, Gabriele reports good results with large-scale problems [17].

Objective function evaluations can be saved in calculating the reduced gradient by taking advantage of the fact that changes in the basic variables can be estimated from changes in the nonbasic variables by using the linearization

$$\hat{d} = -\mathbf{J}^{-1}\mathbf{C}\bar{d}$$

The procedure proposed by Gabriele and Ragsdell [22] calculates the reduced gradient at a point $x^{(t)}$ as follows.

Given the differencing increment $\delta$,

1. Calculate the matrix $\mathbf{J}^{-1}\mathbf{C}$ at $x^{(t)}$, and denote the columns of this array $D_i$.
2. Evaluate the objective function at $x^{(t)}$, that is, $f^{(t)} = f(x^{(t)})$.
3. For $i = 1, \ldots, N - K$, set

$$z_i = \bar{x}_i + \delta \tag{9.21}$$

Calculate

$$w = \hat{x} - \delta D_i \qquad (9.22)$$

Evaluate

$$\tilde{f}^{(i)} = f(w, \bar{x}_1, \ldots, \bar{x}_{i-1}, z_i, \bar{x}_{i+1}, \ldots, \bar{x}_{N-K}) \qquad (9.23)$$

Then estimate the reduced gradient with respect to $\bar{x}_i$, using

$$\left(\nabla \tilde{f}^{(t)}\right)_i = \frac{\tilde{f}^{(i)} - f^{(t)}}{\delta} \qquad (9.24)$$

The sequence of calculations (9.21)–(9.24) is executed $N - K$ times to obtain $\nabla \tilde{f}^{(t)}$. Thus, the number of objective function evaluations is $N - K + 1$ rather than the $N + 1$ that would be required if the complete gradient of $f$ were estimated. When $K$ is large or $f(x)$ is complex to evaluate, this can be a significant savings. A potential disadvantage is that the inverse $\mathbf{J}^{-1}$ and the product $\mathbf{J}^{-1}\mathbf{C}$ must be available explicitly.

## 9.4   GRADIENT PROJECTION METHODS

In spite of the power and versatility of the GRG algorithm, it does have several inherently unsatisfactory features, all of which are a result of the variable-partitioning strategy. First of all, the partitioning strategy imposes the need for checks of matrix singularity and basic variable degeneracy and for strategies for recovergy from these conditions. Second, the process of partition changes requires expensive numerical calculations to update $\mathbf{J}^{-1}$ and to reevaluate the reduced gradient. Moreover, if quasi-Newton techniques are used to modify search directions, then each partition change results in a loss of the update information accumulated up to that point. Finally, there is a conceptually unsatisfactory arbitrariness to the choice of partitions. We know from the convergence rate analysis that the choice of partition affects the terminal rate of convergence, yet we choose partitions without actually being able to take this into account.

Nonetheless, the main reason for using partitioning and the variable-elimination construction is to calculate a descent direction that takes into account the neighboring constraint surfaces, that is, to determine a constrained descent direction. The family of methods we study in this section, whose development actually predates that of the GRG algorithm, accomplish this objective without variable elimination by directly calculating a projection of the objective function gradient onto the constraint surfaces. We introduce the main ideas of the projection construction by first considering the linearly constrained case. Then we outline the modifications required to accommodate nonlinear constraints. Finally, we examine the rather close relationship between reduced gradient and projection methods and review their relative merits.

### 9.4.1 The Linearly Constrained Case

Consider first the rather simple case of minimizing $f(x)$ subject to a single linear equality constraint:

$$\sum a_i x_i = a^T x = b$$

Given a feasible point $x^{(t)}$ at which the gradient $\nabla f$ is not equal to zero, we seek to find a search direction that lies on the constraint and is a descent direction. As shown in Figure 9.10, such a direction could be obtained geometrically by projecting the negative of $\nabla f(x^{(t)})$ perpendicularly onto the constraint surface. This component of $-\nabla f$, denoted by $-\nabla f_c$, clearly is a feasible direction, since for all $\alpha \geqslant 0$ the points given by

$$x = x^{(t)} + \alpha(-\nabla f_c)$$

will satisfy the linear constraint. The direction is a descent direction because the angle between $\nabla f$ and $-\nabla f_c$ is greater than 90°. The geometric construction for determining the perpendicular projection of a vector onto a plane is obvious; the difficulty lies in carrying it out algebraically. Algebraically, the process of perpendicular projection amounts to a decomposition of the projected vector into two orthogonal components: one parallel to the constraint surface and one perpendicular to the surface. The parallel component will be the desired vector projection.

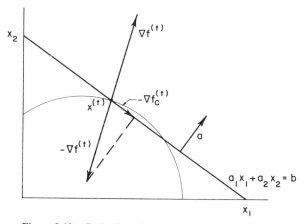

**Figure 9.10.** Projection of gradient onto a constraint.

As the first step in the development, observe that the statement $a^T s = 0$ implies that $s$ is a feasible direction (is parallel to the surface) and that the vector $a$ is perpendicular to the plane. In fact, if we scale the constraint so that $\|a\| = 1$, then the vector $a$ will be the normal to the constraint surface. Since $a$ is the normal to the constraint, all vectors perpendicular to the constraint surface must be parallel to $a$. Consequently, for any vector $s$, the component $s'$ that is perpendicular to the constraint surface must be equal to a constant times $a$. The component of $s$ that is parallel to the constraint surface $s''$ must satisfy $a^T s'' = 0$. Thus, any vector $s$ can be written as the vector sum

$$s = s' + s'' \tag{9.25}$$

where $s' = \gamma a$ for some suitable scalar $\gamma$, and $s''$ satisfies the equation $a^T s'' = 0$. It remains to determine how $\gamma$ is calculated.

Consider the product $a^T s$. By virtue of (9.25),

$$a^T s = a^T(\gamma a) + a^T s'' = \gamma(a^T a) + 0$$

Therefore, solving for $\gamma$, we obtain

$$\gamma = \frac{a^T s}{a^T a} = (a^T a)^{-1} a^T s \tag{9.26}$$

Given this expression for $\gamma$, we can now use it in (9.25) to calculate the desired component $s''$ of $s$ that is parallel to the constraint. Specifically, from (9.25) we have

$$s = s' + s'' = \gamma a + s'' = a(a^T a)^{-1} a^T s + s''$$

Solving for $s''$,

$$s'' = s - a(a^T a)^{-1} a^T s = \left(I - a(a^T a)^{-1} a^T\right) s \tag{9.27}$$

where $I$ is an identity matrix of order conformable with the dimension of $s$. Note that while $(a^T a)^{-1}$ is a scalar, the product $a(a^T a)^{-1} a^T$ is a matrix, since a column vector $a$ is multiplied by a row vector $a^T$. The composite matrix $P \equiv (I - a(a^T a)^{-1} a^T)$ is an example of a *projection matrix*. It serves to project the vector $s$ into the plane of the constraint along the direction of the normal $a$ to the constraint surface. Similarly, the matrix $a(a^T a)^{-1} a^T$ is a projection matrix that serves to project the vector $s$ along a direction parallel to the constraint surface onto the space spanned by the constraint normal.

Note that both these projection matrices are symmetric and positive semi-definite. Symmetry is obvious, since $(a^T a)^{-1}$ is a scalar and $(a^T a)^T = a^T a$. To demonstrate the positive semidefiniteness of $P$, we consider the product $y^T P y$

for any arbitrary but nonzero vector $y$. Thus,

$$y^T \mathbf{P} y = y^T \left[ \mathbf{I} - a(a^T a)^{-1} a^T \right] y = y^T y - \frac{(y^T a)(a^T y)}{(a^T a)}$$

$$= \frac{\|y\|^2 \|a\|^2 - |a^T y|^2}{\|a\|^2}$$

The numerator of the last expression is nonnegative, since the inner product of any two vectors always satisfies the (Schwartz) inequality,

$$|a^T y| \leq \|y\| \|a\|$$

Before examining extensions of these constructions and their utilization in formulating an algorithm, let us consider a small numerical example.

### Example 9.12

Calculate the projection of the vector $s = (0, 1)^T$ onto the constraint $2x_1 + x_2 = 3$ at the point $(\frac{3}{4}, \frac{3}{2})^T$.

The constraint normal is $(2, 1)^T$. The projection matrix $\mathbf{P}$ is given by

$$\mathbf{P} = \left( \mathbf{I} - a(a^T a)^{-1} a^T \right)$$

$$= \begin{pmatrix} 1 & 0 \\ 0 & 1 \end{pmatrix} - \begin{pmatrix} 2 \\ 1 \end{pmatrix} \left\{ (2, 1) \begin{pmatrix} 2 \\ 1 \end{pmatrix} \right\}^{-1} (2, 1)$$

$$= \begin{pmatrix} 1 & 0 \\ 0 & 1 \end{pmatrix} - \frac{1}{5} \begin{pmatrix} 4 & 2 \\ 2 & 1 \end{pmatrix} = \begin{pmatrix} 0.2 & -0.4 \\ -0.4 & 0.8 \end{pmatrix}$$

The projection of $s$ onto the constraint surface along the constraint normal is thus

$$s'' = \mathbf{P}s = \begin{pmatrix} 0.2 & -0.4 \\ -0.4 & 0.8 \end{pmatrix} \begin{pmatrix} 0 \\ 1 \end{pmatrix} = \begin{pmatrix} -0.4 \\ 0.8 \end{pmatrix}$$

The component $s'$ parallel to $a$ is given by

$$s' = s - s'' = (0, 1)^T - (-0.4, 0.8)^T = (0.4, 0.2)^T$$

Both are shown in Figure 9.11. Observe that $s'$ and $s''$ are clearly orthogonal,

$$(s')^T s'' = (0.4, 0.2) \begin{pmatrix} -0.4 \\ 0.8 \end{pmatrix} = 0$$

and $s''$ and $a$ are orthogonal,

$$a^T s'' = (2, 1)\begin{pmatrix} -0.4 \\ 0.8 \end{pmatrix} = 0$$

Furthermore, **P** is clearly symmetric and positive semidefinite.

The construction of a projection matrix can be easily extended to multiple constraints. Given a system with $K$ linear equalities,

$$a_k^T x = b_k \qquad k = 1, \ldots, K$$

the surface of feasible points will be given by the intersection of all these constraints. The subspace orthogonal to the feasible surface will be spanned by the set of constraint normals $a_k$. Thus, any vector $s$ can be decomposed into a component $s''$ that is parallel to the feasible surface and thus satisfies

$$a_k^T s'' = 0 \qquad k = 1, \ldots, K$$

and a component $s'$ that is orthogonal to the constraint surface and thus an be expressed as a linear combination of the $a_k$. Thus, $s$ can be written as

$$s = s' + s'' = A\gamma + s'' \tag{9.28}$$

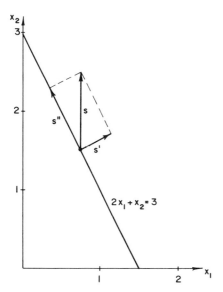

**Figure 9.11.**   Plot of Example 9.12.

where $\mathbf{A}$ is the matrix whose rows consist of the vectors $a_k^T$, and $\gamma$ is a vector of constants such that $s' = \mathbf{A}^T\gamma$. As before, if (9.28) is multiplied by $\mathbf{A}$, we obtain an expression for $\gamma$:

$$\mathbf{A}s = \mathbf{A}\mathbf{A}^T\gamma + \mathbf{A}s'' = \mathbf{A}\mathbf{A}^T\gamma$$

or

$$\gamma = (\mathbf{A}\mathbf{A}^T)^{-1}\mathbf{A}s \tag{9.29}$$

where the inverse of $\mathbf{A}\mathbf{A}^T$ exists only if the vectors $a_k$ are linearly independent. Substituting (9.29) into (9.28), we next obtain a formula for the projection matrix $\mathbf{P}$ analogous to (9.22). That is,

$$s'' = s - \mathbf{A}^T(\mathbf{A}\mathbf{A}^T)^{-1}\mathbf{A}s$$

$$= \left(\mathbf{I} - \mathbf{A}^T(\mathbf{A}\mathbf{A}^T)^{-1}\mathbf{A}\right)s \equiv \mathbf{P}s \tag{9.30}$$

Again, the projection matrix $\mathbf{P}$ is symmetric and positive semidefinite. Before defining an algorithm based on the use of $\mathbf{P}$, we need to establish three additional properties of the constructions involving $\mathbf{P}$ and the gradient vector $\nabla f$.

### Properties of Projection Constructions

(i)   $s = -\mathbf{P}\nabla f$ is a descent direction.
(ii)  If $s = 0$, the point $x^{(t)}$ satisfies the Lagrangian necessary conditions.
(iii) The vector of Lagrange multipliers is given by

$$v = (\mathbf{A}\mathbf{A}^T)^{-1}\mathbf{A}\nabla f$$

The first statement is a direct consequence of the fact that $\mathbf{P}$ is symmetric and positive semidefinite. Consider the product, $\nabla f^T \cdot s$. Since $s = -\mathbf{P}\nabla f$ and $\mathbf{P}$ is positive semidefinite, we have $\nabla f^T \cdot s = \nabla f^T(-\mathbf{P}\nabla f) \leqslant 0$.

If $s = -\mathbf{P}\nabla f = 0$, then $\nabla f$ must be orthogonal to the constraint surface. Thus, from Eq. (9.28), there must exist a vector $\gamma$ such that

$$\nabla f = \mathbf{A}^T\gamma \tag{9.31}$$

Since the rows of $\mathbf{A}$ are the coefficient vectors of the linear constraints, Eq. (9.31) is just a restatement of the Lagrangian necessary conditions,

$$\nabla f - \sum v_k a_k = 0$$

with $\gamma = v$. Moreover, from Eq. (9.29) the Lagrange multipliers will be given by

$$v = (\mathbf{A}\mathbf{A}^T)^{-1}\mathbf{A}\nabla f \qquad (9.32)$$

As in the reduced gradient method, the search-vector definition $s = -\mathbf{P}\nabla f$ thus also serves as a test for optimality. The complete gradient projection algorithm can now be stated.

### The Basic Gradient Projection Algorithm

Calculate **P**, assuming the vectors $a_k$ are linearly independent, and given a feasible point $x^{(t)}$ and a convergence parameter $\varepsilon > 0$:

**Step 1.**   Calculate $s^{(t)} = -\mathbf{P}\nabla f$.

**Step 2.**   If $\|s^{(t)}\| \leqslant \varepsilon$, calculate $v$ using Eq. (9.32) and terminate. Otherwise, continue.

**Step 3.**   Determine the maximum step size:

$$\alpha_{\max} = \min\left\{\max\left(0, \frac{b_k - a_k^T x^{(t)}}{a_k^T s^{(t)}} \text{ or } \infty \text{ if } a_k^T s^{(t)} = 0\right); k = 1, \dots, K\right\}$$

**Step 4.**   Solve the line-search problem

$$\text{Minimize } f(x^{(t)} + \alpha s^{(t)}) \qquad 0 \leqslant \alpha \leqslant \alpha_{\max}$$

**Step 5.**   Set $x^{(t+1)} = x^{(t)} + \alpha^* s^{(t)}$, and return to step 1.

Except for the definition of the search vector, the algorithm differs very little from the reduced gradient method. Note that in order to calculate **P** we require that the constraint coefficient vectors $a_k$ be linearly independent. This is no real restriction, because dependent linear constraints can simply be deleted from the problem definition since they will not affect the solution.

**Treatment of Inequalities and Bounds.**   While the basic algorithm is stated in terms of linear equalities, it can readily be extended to handle inequalities using either of the devices considered in Section 9.3.3: slack variables or an active constraint strategy. The latter is normally favored in the context of projection methods because of the reduction that results in the size of the **A** matrix [23]. As typically posed, the active constraint strategy makes use of the estimate of the multipliers obtained from (9.32) to delete constraints, one at a time, from the active constraint set. The modification proceeds as follows. At a given point $x^{(t)}$, the inequality constraints

$$a_j^T x \geqslant b_j \qquad j = 1, \dots, J$$

are tested to identify the active set. The constraint matrix $\mathbf{A}$ is then formed so that the first $K$ rows correspond to the equality constraints and the next, say $M$, correspond to the active inequalities. The projection operator is calculated as before. However, in step 2, if the termination criterion $\|s^{(t)}\| \leqslant \varepsilon$ is met, then after the multiplier vector

$$(v, u) = (\mathbf{A}\mathbf{A}^T)^{-1}\mathbf{A}\nabla f$$

is calculated, the signs of the last $M$ multipliers are checked. The last $M$ multipliers will correspond to the inequality constraints and must, from the Kuhn-Tucker conditions, be nonnegative at the optimum. If all are nonnega-tive, the process terminates. Otherwise, the constraint with the most negative multiplier is deleted from the active constraint set, the new projection matrix $\mathbf{P}$ is calculated, and step 1 is reinitiated. Note that in the active constraint strategy, bounds on the variables are treated as inequality constraints.

### Example 9.13

Consider the application of the gradient projection method to the problem of Example 9.4:

$$\text{Minimize} \quad f(x) = (x_1 - 1)^2 + (x_2 - 2)^2$$

$$\text{Subject to} \quad g_1(x) = x_1 - 2x_2 \geqslant -2$$

$$g_2(x) = -x_1 - x_2 \geqslant -4$$

$$g_3(x) = x_1 \geqslant 0 \qquad g_4(x) = x_2 \geqslant 0$$

with initial feasible point $x^0 = (0, 0)$.

At $x^0$, $g_1$ and $g_2$ are loose, while $g_3$ and $g_4$, the two nonnegativity conditions, are tight. Thus,

$$a_3 = (1, 0)^T \qquad a_4 = (0, 1)^T \qquad \text{and} \quad \mathbf{A} = \begin{pmatrix} 1 & 0 \\ 0 & 1 \end{pmatrix}$$

Clearly, $\mathbf{A}\mathbf{A}^T = \mathbf{I}$, $(\mathbf{A}\mathbf{A}^T)^{-1} = \mathbf{I}$, and therefore $\mathbf{P} = (\mathbf{I} - \mathbf{I}) = 0$.

**Step 1.** $s^0 = -\mathbf{P}\nabla f = -0\begin{pmatrix} -2 \\ -4 \end{pmatrix} = 0$. Therefore, the constraint multipliers must be checked.

$$u = (\mathbf{A}\mathbf{A}^T)^{-1}\mathbf{A}\nabla f = I\begin{pmatrix} -2 \\ -4 \end{pmatrix} = \begin{pmatrix} -2 \\ -4 \end{pmatrix}$$

Since $u_2$ is the most negative, the second tight constraint $g_4$ is deleted from the active constraint set. The active constraint set now reduces to a single constraint, $g_3$. Thus,

$$\mathbf{A} = (1,0) \quad \text{and} \quad \mathbf{A}\mathbf{A}^T = (1,0)\binom{1}{0} = 1$$

Therefore,

$$\mathbf{P} = \begin{pmatrix} 1 & 0 \\ 0 & 1 \end{pmatrix} - \binom{1}{0}(1 \, 0) = \begin{pmatrix} 0 & 0 \\ 0 & 1 \end{pmatrix}$$

**Step 1.**

$$s^0 = -\mathbf{P}\nabla f = - \begin{pmatrix} 0 & 0 \\ 0 & 1 \end{pmatrix}\begin{pmatrix} -2 \\ -4 \end{pmatrix} = \begin{pmatrix} 0 \\ 4 \end{pmatrix}$$

**Step 2.**   Since $\|s^{(0)}\| = 4 > 0$, we continue with step 3.

**Step 3.**

$$\alpha_{\max} = \min \begin{cases} \max\left(0, \dfrac{-2-0}{-8}\right) & \text{for } g_1 \\[2mm] \max\left(0, \dfrac{-4-0}{-4}\right) & \text{for } g_2 \\[2mm] \max(0, \infty) & \text{for } g_3 \text{ since } (1,0)\binom{0}{4} = 0 \\[2mm] \max\left(0, \dfrac{0-0}{4}\right) & \text{for } g_4 \end{cases} = \frac{1}{4}$$

**Step 4.**   We search the line,

$$x = \binom{0}{0} + \alpha\binom{0}{4} \qquad 0 \leqslant \alpha \leqslant \tfrac{1}{4}$$

The function $f(\alpha) = (-1)^2 + (4\alpha - 2)^2$ decreases monotonically for increasing $\alpha$ until $\alpha = \tfrac{1}{2}$ is reached. Therefore,

$$\alpha^* = \alpha_{\max} = \tfrac{1}{4}$$

**Step 5.**   $x^{(1)} = (0,1)^T$

At the new point, we again check for active constraints. Clearly, $g_1$ and $g_3$ are tight. Thus,

$$\mathbf{A} = \begin{pmatrix} 1 & -2 \\ 1 & 0 \end{pmatrix}$$

It is easy to verify that $\mathbf{P} = 0$ so that $s = -\mathbf{P}\nabla f = 0$. The multipliers are $u = (1, -3)^T$. Since $u_2 < 0$, the second active constraint $g_3$ must be dropped from the active constraint set. The active set now consists only of $g_1$. Recomputing $\mathbf{P}$ with $\mathbf{A} = (1, -2)$, we obtain

$$\mathbf{P} = \begin{pmatrix} .8 & .4 \\ .4 & .2 \end{pmatrix}$$

**Step 1.**

$$s^{(1)} = -\mathbf{P}\nabla f = \begin{pmatrix} .8 & .4 \\ .4 & .2 \end{pmatrix}\begin{pmatrix} -2 \\ -2 \end{pmatrix} = \begin{pmatrix} 2.4 \\ 1.2 \end{pmatrix}$$

**Step 2.** $\|s^{(1)}\| > 0$; therefore, we continue.

**Step 3.** The step size $\alpha_{max} = \frac{5}{6}$ is determined by $g_2$.

**Step 4.** We next seek the minimum of

$$f(\alpha) = (2.4\alpha - 1)^2 + (1 + 1.2\alpha - 2)^2$$

The minimum occurs at $\alpha^* = \frac{1}{2} < \alpha_{max}$.

**Step 5.** The new point is $x^{(2)} = (1.2, 1.6)^T$.

At this point only $g_1$ is tight. The active constraint set remains unchanged, and consequently $\mathbf{P}$ need not be updated. At $x^{(2)}$ we have

$$s^{(2)} = -\mathbf{P}\nabla f = -\begin{pmatrix} 0.8 & 0.4 \\ 0.4 & 0.2 \end{pmatrix}\begin{pmatrix} 0.4 \\ -0.8 \end{pmatrix} = \begin{pmatrix} 0 \\ 0 \end{pmatrix}$$

and

$$u^{(2)} = (\mathbf{A}\mathbf{A}^T)^{-1}\mathbf{A}\nabla f = \frac{1}{5}(1, -2)\begin{pmatrix} 0.4 \\ -0.8 \end{pmatrix} = 0.4 > 0$$

Therefore, the point $x^{(2)}$ satisfies the Kuhn-Tucker conditions.

As is evident from the example calculations, the major computational burden of this algorithm consists of the updates of $(\mathbf{A}\mathbf{A}^T)^{-1}$ and $\mathbf{P}$ required each time the active constraint set is modified. However, since constraints are normally deleted and added one at a time, it is possible to develop formulas for updating $\mathbf{P}$ and $(\mathbf{A}\mathbf{A}^T)^{-1}$ that make use of the prior values of these matrices [23]. The interested reader is directed to reference 23 for the algebraic details.

At present, no convergence proof is known for the gradient projection method, since technically it is subject to jamming [8, 18], although no practical instances of jamming have been reported.

### 9.4.2 The General Nonlinear Programming Case

The projection algorithm can clearly be extended to accommodate nonlinear constraints, equalities or inequalities, by simply using their linearizations within the projection formulas. However, as in the GRG method, the computed search direction will in general lead outside the feasible region and thus will require corrections to return to the constraint surface. The organization of the correction calculations must differ from that used in the GRG method because here variable partitioning is not used. However, the basic strategy remains the same.

Recall that in the linearly constrained case, the constraint normals $a_k$ could be used not only to define a projection parallel to the constraint surfaces but also to define a direction orthogonal to the constraint surfaces. Whereas in the linearly constrained case only the former type of projection is needed, in the nonlinear case both types of projections can be used. Projection parallel to the constraint surfaces will serve to define a search direction tangent to the constraint surface. Projection parallel to the constraint normals will serve to define a direction useful for returning to the nonlinear constraint surface. As shown in Figure 9.12 for a single constraint, at $x^{(t)}$ the constraint gradient vector $\nabla h(x^{(t)})$ does define a direction normal to the constraint surface. This direction could be used at the point $w^{(t)}$ obtained by stepping along $s^{(t)}$ from $x^{(t)}$. However, since the surface curvature changes away from $x^{(t)}$, $\nabla h(x^{(t)})$ may no longer point toward the surface, as is the case at point $z^{(t)}$. To avoid such problems, the gradient of $h$ could be recomputed at $w^{(t)}$ to define a direction $\nabla h(w^{(t)})$ that is more likely to return to the surface. This, however, requires recomputation of the gradients of all constraints that were active at $x^{(t)}$. The resulting direction can then serve as the basis of a Newton iteration to locate the precise intersection with the constraint surface. The details of the calculations can be developed as follows.

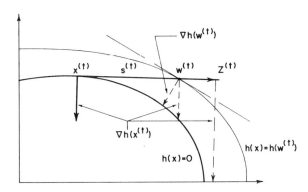

**Figure 9.12.** Projection to return to the constraint surface.

Let $I^{(t)}$ be the set of constraints active at $x^{(t)}$, and let $w^{(t)}$ be the infeasible point reached in direction $s^{(t)}$ from $x^{(t)}$. The direction normal to the constraint surfaces will be given by some linear combination of the normals of the linearized constraints, namely, the gradient vectors $\nabla h_k(w^{(t)})$, $k = 1, \ldots, K$ and $\nabla g_l(w^{(t)})$, $l \in I^{(t)}$. Thus,

$$s = A^T(w^{(t)})\gamma \qquad (9.33)$$

where the columns of $A^T(w^{(t)})$ comprise the gradient vectors $\nabla h_k(w^{(t)})$, $k = 1, \ldots, K$, followed by $\nabla g_l(w^{(t)})$, $l \in I^{(t)}$. We now apply Newton's method to determine $\gamma$. For simplicity, let us denote by $h$ the vector of equality constraints and active inequalities. Using the usual Taylor series approximation, we have

$$0 = h(w^{(t)} + A^T\gamma) = h(w^{(t)}) + A(w^{(t)}) \cdot A^T(w^{(t)})\gamma \qquad (9.34)$$

where the last term is a result of the chain rule differentiation

$$\nabla_\gamma h = \frac{dh}{dx} \cdot \frac{dx}{d\gamma}$$

Solving (9.34) for $\gamma$, we obtain

$$\gamma = -(AA^T)^{-1}h$$

which when substituted into (9.33) effectively yields an iteration formula for the corrected point,

$$w^{(t+1)} - w^{(t)} = s^{(t)} = -A^T(AA^T)^{-1}h \qquad (9.35)$$

Note that all terms in the right-hand matrix product are evaluated at $w^{(t)}$. Moreover, the construction requires that the gradient vectors $\nabla_k h(w^{(t)})$, $k = 1, \ldots, K$, and $\nabla g_l(w^{(t)})$, $l \in I^{(t)}$, be linearly independent.

*Example 9.14*

Consider the use of the above construction to determine the feasible projection of the point $w^0 = (1, 1)^T$ obtained by a step from the point $x^0 = (0, 1)^T$ onto the surface of the constraint

$$h(x) = x_1^2 + x_2^2 - 1 = 0$$

As shown in Figure 9.13, the gradient vector $\nabla h$ evaluated at $w^0$ is $\nabla h(w^0) = (2, 2)^T$, and thus $-\nabla h$ points toward the surface $h(x) = 0$. The application of (9.35) yields

$$w^{(1)} = w^0 - \binom{2}{2}\left\{(2, 2)\binom{2}{2}\right\}^{-1}(1)$$

$$= \binom{1}{1} - \binom{2}{2}\frac{1}{8} = \binom{0.75}{0.75}$$

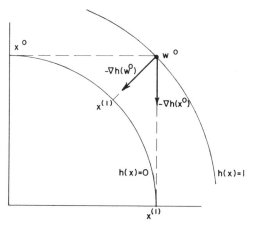

**Figure 9.13.**   Plot of Example 9.14.

Repeated application gives

$$w^{(2)} = \begin{pmatrix} 0.75 \\ 0.75 \end{pmatrix} - \begin{pmatrix} \frac{3}{2} \\ \frac{3}{2} \end{pmatrix} \left\{ \left( \frac{3}{2}, \frac{3}{2} \right) \begin{pmatrix} \frac{3}{2} \\ \frac{3}{2} \end{pmatrix} \right\}^{-1} \left( \frac{1}{8} \right)$$

$$= \begin{pmatrix} 0.75 \\ 0.75 \end{pmatrix} - \begin{pmatrix} \frac{3}{2} \\ \frac{3}{2} \end{pmatrix} \left( \frac{4}{18} \right) \left( \frac{1}{8} \right) = \begin{pmatrix} 0.7083 \\ 0.7083 \end{pmatrix}$$

$$w^{(3)} = \begin{pmatrix} 0.7071 \\ 0.7071 \end{pmatrix}$$

at which point $h(w^{(3)}) = 3 \times 10^{-6}$.

Note that if, as an approximation, $\nabla h(x^0)$ were used instead of $\nabla h(w^{(t)})$, then the updating of the term $\nabla h(\nabla h^T \nabla h)^{-1}$ could be saved. Convergence to the point $(1, 0)^T$ would be obtained, but convergence would be quite slow. The first few iterations are given below:

$$w^{(1)} = w^0 - \nabla h(x^0) \left( \nabla h(x^0)^T \nabla h(x^0) \right)^{-1} h(w^0)$$

$$= \begin{pmatrix} 1 \\ 1 \end{pmatrix} - \begin{pmatrix} 0 \\ 2 \end{pmatrix} \left\{ (0, 2) \begin{pmatrix} 0 \\ 2 \end{pmatrix} \right\}^{-1} (1)$$

$$= \begin{pmatrix} 1 \\ 1 \end{pmatrix} - \frac{1}{4} \begin{pmatrix} 0 \\ 2 \end{pmatrix} = \begin{pmatrix} 1 \\ 0.5 \end{pmatrix}$$

then,

$$w^{(2)} = \begin{pmatrix} 1 \\ 0.5 \end{pmatrix} - \begin{pmatrix} 0 \\ 0.5 \end{pmatrix} h(w^{(1)}) = \begin{pmatrix} 1 \\ 0.375 \end{pmatrix}$$

$$w^{(3)} = (1, 0.3046)^T$$

$$w^{(4)} = (1, 0.2580)^T \quad \text{and so forth}$$

Thus, with fixed gradient terms, at best a linear convergence rate is obtained.

The preceding example illustrates that, as in the GRG method, the attempt to save gradient evaluations and updates of the inverse can lead to a substantially reduced convergence rate. Clearly, a better approach is to use quasi-Newton equation-solving methods.

The correction calculations, of course, must be preceded by some strategy for adjusting the step length $\alpha$ that determines how far along the gradient projection direction

$$s = -\mathbf{P}\nabla f$$

the iterations proceed. The inside-out strategy described in Section 9.3.3 followed by quadratic interpolation for an estimate of the minimum appears to be quite appropriate. Of course, if an active constraint strategy is used, then the step parameter $\alpha$ must also be selected so as not to violate any previously inactive constraint.

### Gradient Projection Algorithm for a General Nonlinear Program

Given a feasible point $x^0$, convergence tolerance $\varepsilon_1 > 0$, and active set tolerance $\varepsilon_2 > 0$.

**Step 0.** Evaluate all inequality constraints at $x^{(t)}$ to identify the active set

$$I^{(t)} = \{ j: \; g_j(x^{(t)}) \leqslant \varepsilon_2, \; j = 1, \ldots, J \}$$

**Step 1.** Calculate $\mathbf{P}$ and $s = -\mathbf{P}\nabla f(x^{(t)})$.

**Step 2.** If $\|s^{(t)}\| > \varepsilon_1$, go to step 3. Otherwise, calculate the multipliers

$$(v, u) = (\mathbf{A}\mathbf{A}^T)^{-1}\mathbf{A}\nabla f$$

and find

$$u_m = \min\{ u_l: \; l \in I^{(t)} \}$$

If $|u_m| \leqslant \varepsilon_1$, terminate. Otherwise, delete inequality constraint $m$ from $I^{(t)}$ and go to step 1.

**Step 3.** Determine the maximum step size $\alpha_{\max}$ such that $g_l(w(\alpha)) \geqslant 0$ for all $l \notin I^{(t)}$.

For each $\alpha$, $w(\alpha)$ is the result of iterations (9.35).

**Step 4.**

    (a) Conduct a bracketing search on $\alpha, 0 \leqslant \alpha \leqslant \alpha_{\max}$ to determine $\alpha_1$, $\alpha_2$, $\alpha_3$ and $w(\alpha_1)$, $w(\alpha_2)$, $w(\alpha_3)$, which bracket the minimum of $f(x)$ on the curve $h(w(\alpha)) = 0$.

    (b) Use quadratic interpolation to estimate $\alpha^*$.

**Step 5.** Use (9.35) to determine $w(\alpha^*)$ and go to step 0.

Note that the logic of steps 3 and 4 are in general no less complex than the corresponding step correction calculations of the GRG method.

### Example 9.15

Consider the problem of Example 9.10:

$$\text{Minimize} \quad f(x) = 4x_1 - x_2^2 + x_3^2 - 12$$

$$\text{Subject to} \quad h_1(x) = 20 - x_1^2 - x_2^2 = 0$$

$$h_2(x) = x_1 + x_3 - 7 = 0$$

with $x^0 = (2, 4, 5)^T$.

**Step 0.** All constraints are equalities, and hence active.

**Step 1.** $\nabla h_1(x^0) = (-4, -8, 0)^T$
$\nabla h_2(x^0) = (1, 0, 1)^T$
and
$\nabla f(x^0) = (4, -8, 10)^T$

$$\mathbf{AA}^T = \begin{pmatrix} -4 & -8 & 0 \\ 1 & 0 & 1 \end{pmatrix} \begin{pmatrix} -4 & 1 \\ -8 & 0 \\ 0 & 1 \end{pmatrix} = \begin{pmatrix} 80 & -4 \\ -4 & 2 \end{pmatrix}$$

and

$$\mathbf{P} = \begin{pmatrix} 1 & 0 & 0 \\ 0 & 1 & 0 \\ 0 & 0 & 1 \end{pmatrix} - \frac{1}{144} \begin{pmatrix} -4 & 1 \\ -8 & 0 \\ 0 & 1 \end{pmatrix} \begin{pmatrix} 2 & 4 \\ 4 & 80 \end{pmatrix} \begin{pmatrix} -4 & -8 & 0 \\ 1 & 0 & 1 \end{pmatrix}$$

$$= \frac{1}{9} \begin{pmatrix} 4 & -2 & -4 \\ -2 & 1 & 2 \\ -4 & 2 & 4 \end{pmatrix}$$

Then,

$$s^0 = -\mathbf{P}\nabla f(x^0) = \left(\tfrac{8}{9}, \tfrac{4}{9}, -\tfrac{8}{9}\right)^T$$

**Step 2.** $\|s^0\| \gg 0$. Therefore, continue with step 3.

**Step 3.** All constraints are active; continue with step 4.

**Step 4.** Bracketing search on $\alpha$. Suppose $\alpha^0 = \tfrac{1}{2}$ is chosen. Then

$$w^0 = \begin{pmatrix} 2 \\ 4 \\ 5 \end{pmatrix} + \frac{1}{2} \begin{pmatrix} \tfrac{8}{9} \\ \tfrac{4}{9} \\ -\tfrac{8}{9} \end{pmatrix} = \begin{pmatrix} 2.444 \\ 4.222 \\ 4.555 \end{pmatrix}$$

At $w^0$, $h_1 = -3.800$ and (as expected because $h_2$ is linear) $h_2 = 0.00$. Application of (9.35) yields

$$w^{(1)} = w^0 - \begin{pmatrix} -4.888 & 1 \\ -8.444 & 0 \\ 0 & 1 \end{pmatrix}$$

$$\times \left\{ \begin{pmatrix} -4.888 & -8.444 & 0 \\ 1 & 0 & 1 \end{pmatrix} \begin{pmatrix} -4.888 & 1 \\ -8.444 & 0 \\ 0 & 1 \end{pmatrix} \right\}^{-1} \begin{pmatrix} -3.8 \\ 0 \end{pmatrix}$$

$$= (2.333, 3.837, 4.667)^T$$

At $w^{(1)}$, $h_1 = -0.1635$, and again $h_2 = 0.00$.

In an additional iteration, we obtain

$$w(\alpha^0) = (2.3274, 3.8188, 4.6726)^T$$

with $h_1 = -3 \times 10^{-4}$ and $f = 4.5543$. Since $f(x^0) = 5.0$, an improved point has been reached. To obtain a bracket, step 4a is repeated with $\alpha^{(1)} = \tfrac{3}{2}$. Thus,

$$w^0 = (3.333, 4.666, 3.666)^T \quad \text{and} \quad h_1 = -12.888$$

Successive application of (9.35) leads to the point

$$w(\alpha^{(1)}) = (2.8804, 3.4210, 4.1196)^T$$

with $h_1 = 10^{-6}$ and $f = 4.7893$. Since $f(w(\alpha^{(1)})) > f(w(\alpha^0))$, a bracket has been obtained.

**Step 4b.** Quadratic interpolation on $\alpha$. We have the three points $\alpha = 0$, $\alpha = \tfrac{1}{2}$, and $\alpha = \tfrac{3}{2}$, and the associated objective function values.

From the usual interpolation formulas, we readily obtain $\alpha^* = 0.8447$. This yields the point

$$w^0 = (2.7509, 4.3754, 4.2491)^T$$

at which $h_1 = -6.7119$. Repeated application of (9.35) yields the feasible point

$$w(\alpha^*) = (2.5329, 3.6861, 4.4671)^T$$

with $h_1 = 2 \times 10^{-3}$ and $f = 4.4995$.

Note that while the quadratic interpolation gives a reasonable estimate of the objective function value, it does not yield a feasible point, since the interpolation does not directly account for the variation in the values of $h_1(x)$.

This completes one full iteration of the generalized gradient projection method. Repeated iterations will yield the true optimum objective function value 4.4859.

The original form of the foregoing algorithm was proposed by Rosen [24] along with the appropriate formulas for updating the matrices $(AA^T)^{-1}$ and **P**. As in the linearly constrained case, a rigorous convergence proof is not available. However, Luenberger [19] presents a terminal convergence rate analysis that shows that once the active constraint set at the optimal solution is known, then for a convergent sequence of iterates $\{x^{(t)}\}$ the objective function values $f(x^{(t)})$ will converge to $f(x^*)$ linearly with a ratio bounded above by $(A - a)/(A + a)$. Here $A$ and $a > 0$ are the largest and smallest eigenvalues of the matrix of second derivatives of the Lagrangian function at $x^*$ over the subspace defined by $\nabla h_k^T(x^*)\Delta x = 0, k = 1, \ldots, K; \nabla g_j^T(x^*)\Delta x = 0, j \in I^*$. This very interesting result indicates quite clearly that the convergence rate is influenced equally by the constraint shapes and the objective function contours. The ratio bound parallels the bound known for the unconstrained optimal gradient method and clearly indicates that some quasi-Newton acceleration method should be employed.

In the linearly constrained case, the use of the DFP update within the framework of the gradient projection method was reported by Goldfarb [25]. Updates involving the BFS formula and a detailed treatment of numerical issues is given by Gill and Murray [26]. For the Goldfarb variant, Powell [27] has shown that for a positive definite quadratic objective function involving $P$ active constraint set changes, the exact minimum can be found by performing at most $N + P$ exact line searches. The use of quasi-Newton updates in the general nonlinear case was developed by Murtagh and Sargent [28]. If inequality constraints are treated using slack variables, then, in contrast to the GRG method, the full update to the second derivative matrix of the Lagrangian is obtained. If the active constraint strategy is used, then each active set change will result in some loss of efficiency. Limited computational results involving an improved projection method show the performance of quasi-Newton-based projection to be competitive with that of a GRG-based code [29].

Extensions and adaptations of projection methods for use within the context of mechanical design are reported by Haug and Arora [30]. Successful applications include the optimum design of planar frames and long-barrel cannon under dynamic loads [31]. Computational comparisons carried out with SUMT penalty-type methods indicate that the projection methods offer a factor of 5 reduction in computing time. Comparisons with GRG methods have not been reported.

### 9.4.3  Relationship between GRG and Projection Methods

From the preceding development it is apparent that there is considerable similarity between the reduced gradient and projection strategies. Both calculate a descent direction that lies in the tangent plane at the linearization point. Both in general require a correction to return to the constraint surface. Both methods can accommodate inequality constraints by using either slack variables or active constraint strategies. Both methods can and have been modified to incorporate quasi-Newton updates to accelerate convergence for general functions and obtain convergence in a fixed number of steps for quadratic functions. In view of these similarities it is thus not surprising that it can be shown that these two families of methods are merely different implementations of the same basic construction [32].

Consider the problem of minimizing $f(x)$ subject to a set of $M$ linear equality constraints $Ax = b$. The GRG strategy amounts to defining a new nonorthogonal coordinate system $\bar{y}$ and $\hat{y}$ such that

$$\hat{y} = Ax - b \qquad (9.36)$$

and

$$\bar{y} = \bar{x} \qquad (9.37)$$

The variables $\hat{y}$ measure the constraint satisfaction. Feasible points are those for which $\hat{y} = 0$. The variables $\bar{y}$, on the other hand, define the position of points on the $(N - M)$-dimensional constraint surface. Thus, the GRG method is characterized by a transformation of variables

$$y = \begin{pmatrix} \hat{y} \\ \bar{y} \end{pmatrix} = \begin{pmatrix} \hat{A} & \bar{A} \\ 0 & I \end{pmatrix} \begin{pmatrix} \hat{x} \\ \bar{x} \end{pmatrix} - \begin{pmatrix} b \\ 0 \end{pmatrix} \qquad (9.38)$$

The gradient projection method, on the other hand, is characterized by the use of a transformation

$$y = Tx - d \qquad (9.39)$$

in which the first $M$ variables $\hat{y}$ again measure the constraint satisfaction. However, instead of defining $\hat{y}$ via (9.36), a transformation matrix $\mathbf{C}$ is introduced so that

$$\hat{y} = \mathbf{C}(\mathbf{A}x - b)$$

and $\mathbf{T}$ and $\mathbf{C}$ are chosen so that $\hat{y}$ is *orthogonal* to the remaining variables $\bar{y}$.

A graphical representation of these constructions in the case of a single constraint and two dimensions is depicted in Figure 9.14 after Sargent [32]. The GRG variable partition essentially results in the nonorthogonal local coordinate choice $\bar{y}$ and $\hat{y}$ (Figure 9.14$a$), while the gradient projection method results in an orthogonal choice (Figure 9.14$b$). Neither method explicitly operates in the $y$-coordinate system. If explicit use were made of the coordinate transformation, then gradients and other problem information available in the $x$ coordinates would first have to be converted to the $y$ coordinates. The desired search directions would be computed in the $y$ variables, and then these $y$-search directions would have to be transformed back into the $x$ variables to allow evaluation of the problem functions. Instead of employing these cumbersome intermediate steps, in both algorithms the $y$-coordinate search directions are expressed directly in terms of the original $x$ variables. Thus, the reduced gradient of the GRG is computed as

$$\nabla \tilde{f}^T = \left( \begin{pmatrix} \hat{\mathbf{A}} & \bar{\mathbf{A}} \\ \mathbf{0} & \mathbf{I} \end{pmatrix}^{-1} \right)^T \begin{pmatrix} \mathbf{0} & \mathbf{0} \\ \mathbf{0} & \mathbf{I} \end{pmatrix} \begin{pmatrix} \hat{\mathbf{A}} & \bar{\mathbf{A}} \\ \mathbf{0} & \mathbf{I} \end{pmatrix}^{-1} \nabla f^T$$

where the matrix $\begin{pmatrix} \mathbf{0} & \mathbf{0} \\ \mathbf{0} & \mathbf{I} \end{pmatrix}$ restricts consideration to the $\bar{y}$ components.

The reduced gradient of the projection method is computed via

$$\nabla \tilde{f}^T = \left( \mathbf{I} - \mathbf{A}^T (\mathbf{A}\mathbf{A}^T)^{-1} \mathbf{A} \right) \nabla f^T$$

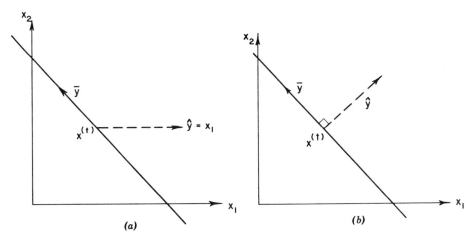

**Figure 9.14.** ($a$) GRG transformation. ($b$) Projection transformation.

In the former case, because of the particular form of the transformation, it is possible to operate directly on the reduced $\bar{y} = \bar{x}$ dimension. In the latter case, it is not possible to eliminate variables directly, and hence the algorithm must operate in the full original dimensions.

Given that the two algorithms are at root equivalent, the only basis for a choice between them are the numerical and implementation features. On that basis it appears that the GRG strategy is clearly to be preferred for large-scale applications in which model sparsity must be exploited. The GRG variable-partition (basis-change) strategy clearly lends itself to sparse matrix methods. The projection strategy, with its sparsity-destroying matrix products and need for the explicit projection matrix, does not. Storage considerations and number of arithmetic operations would also seem to strongly favor GRG. However, for small-dense equality-constrained problems, the projection method may be preferable, because it does not require the extraneous matrix calculations resulting from variable-partition changes. In fact, for linear equality-constrained problems, once the projection matrix is calculated it never needs to be updated. If a choice had to be made of a general purpose NLP code, then available computational evidence suggests that a competent GRG implementation should be preferred.

## 9.5   DESIGN APPLICATION

In this section we consider the least-cost design of multiproduct processes consisting of a series of batch and semicontinuous units. We first define the problem, then discuss its formulation, and finally consider its solution using a GRG code.

### 9.5.1   Problem Statement

The series process shown in Figure 9.15, consisting of three batch units (two reactors and a dryer) and five semicontinuous units (three pumps, one heat exchanger, and one centrifuge), is to process three products over a planning horizon of one year (8000 hr). The production requirements are

Production Requirements, lb/yr

| | |
|---|---|
| Product 1 | 400,000 |
| Product 2 | 300,000 |
| Product 3 | 100,000 |

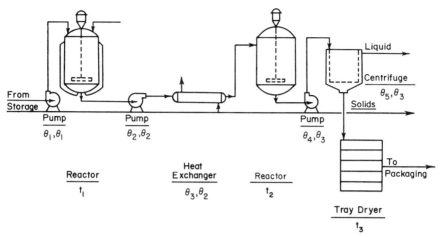

**Figure 9.15.**   Typical batch semicontinuous process.

The batch processing times of each product in each of the three batch units are:

Batch Processing Time, hr

|  | Reactor 1 | Reactor 2 | Dryer |
|---|---|---|---|
| Product 1 | 3 | 1 | 4 |
| Product 2 | 6 | — | 8 |
| Product 3 | 2 | 2 | 4 |

All three products are to be processed serially in all units, with the exception that product 2 will by-pass reactor 2 and be sent directly to the centrifuge.

The volume of material that must be processed at each stage to produce a pound of final product varies from product to product. The volume factors for the three products and eight units are given in Table 9.1.

The plant equipment is to be sized for minimum capital cost. For batch units the equipment cost is given by a power law correlation of the form $a_i V_i^{\alpha_i}$, where $V_i$ is the unit volume (ft$^3$) and $a_i$, $\alpha_i$ are empirical constants. Semicontinuous unit costs are given by $b_k R_k^{\beta_k}$, where $R_k$ is the production rate (ft$^3$/hr) of unit $k$. The correlation coefficients for the eight equipment types are given in Table 9.2.

### 9.5.2   General Formulation

Many fine chemicals, pharmaceuticals, polymers, and foods are produced in processes that consist of a series of batch and/or semicontinuous processing stages. A batch operation is simply one in which a specified quantity of

Table 9.1. Volume Factors, ft$^3$ / lb Final Product

| | Product 1 | Product 2 | Product 3 |
|---|---|---|---|
| Batch Unit Volume Factors $S_{ip}$ | | | |
| Reactor 1 | 1.2 | 1.5 | 1.1 |
| Reactor 2 | 1.4 | — | 1.2 |
| Dryer | 1.0 | 1.0 | 1.0 |
| Continuous Unit Volume Factors $\bar{S}_{kp}$ | | | |
| Pump 1 | 1.2 | 1.5 | 1.1 |
| Pump 2 | 1.2 | 1.5 | 1.1 |
| Heat exchanger | 1.2 | 1.5 | 1.1 |
| Pump 3 | 1.4 | — | 1.2 |
| Centrifuge | 1.4 | 1.5 | 1.2 |

material is loaded into the unit, the material is subjected to some chemical or physical treatment for a specified time period, and then the resulting product is discharged. A semicontinuous unit is one that operates with continuous feed and removal of material but is intermittently started up or shut down. Batch/semicontinuous plants are normally used to produce multiple products so that the products actually share in the utilization of the plant equipment. It is preferable to operate such plants in the *overlapping* mode. Under such an operating policy each product is produced in campaigns that each involve the production of a certain number of batches of that product. During a campaign the process is dedicated to the exclusive production of one product. The processing of successive batches of that product is timed so that a new batch is started up as soon as a unit becomes available.

Table 9.2. Cost Coefficients for Eight Equipment Types

| Equipment Type | Batch Unit | | Continuous Unit | |
|---|---|---|---|---|
| | $a_i$ | $\alpha_i$ | $b_i$ | $\beta_i$ |
| Reactor 1 | 592 | 0.65 | | |
| Reactor 2 | 582 | 0.39 | | |
| Dryer | 1200 | 0.52 | | |
| Pump 1 | | | 370 | 0.22 |
| Pump 2 | | | 250 | 0.40 |
| Heat exchanger | | | 210 | 0.62 |
| Pump 3 | | | 250 | 0.4 |
| Centrifuge | | | 200 | 0.83 |

Given a system consisting of three sequential batch stages with processing times $t_1$, $t_2$, $t_3$, the operation of the process can be depicted in a bar graph (or Gantt chart) as shown in Figure 9.16. In this chart a solid horizontal line indicates that that unit is busy during the corresponding time interval. Note that unit 2 is occupied continuously. Unit 1 is idle two time units and unit 3 four time units between successive batches because they must wait before batches of material can be transferred to or from unit 2. For fixed processing times, each product will have associated with it a characteristic time, called the *cycle* time, between the completion of batches of that material. The cycle time will be equal to the maximum of the individual unit processing times, or

$$T_p = \max_i (t_i)$$

For the operation depicted in Figure 9.16, the cycle time will be equal to

$$T_p = \max(t_1, t_2, t_3) = t_2$$

since $t_2$ is the longest processing time.

The product cycle time determines the batch size $B_p$ and the number of batches $N_p$ that must be processed to produce a given quantity of material $Q_p$ over the fixed total production time $T$. Thus, $N_p = T/T_p$ and $B_p = Q_p/N_p$. If $P$ products share in the use of the plant, then the total time $T$ must be divided

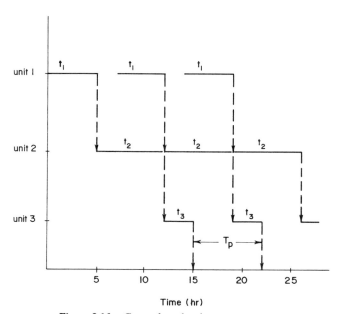

Time (hr)

**Figure 9.16.** Gantt chart for three-stage process.

among these products. Thus,

$$\sum_{p=1}^{p} \frac{Q_p}{B_p} T_p \leqslant T \qquad (9.40)$$

The processing time of the semicontinuous units also influences the cycle time. First, if a batch unit $i$ is preceded by semicontinuous unit $l$ and followed by unit $k$, and if $\theta_{kp}$ indicates the processing time of unit $k$, then the total processing time of batch unit $i$, $\bar{\theta}_{ip}$, for a given product $p$ becomes

$$\bar{\theta}_{ip} = \theta_{lp} + t_{ip} + \theta_{kp} \qquad (9.41)$$

Thus, the cycle time for product $p$ will be given by

$$T_p = \max\left\{ \max_i \left( \bar{\theta}_{ip} \right), \max_k \left( \theta_{kp} \right) \right\} \qquad (9.42)$$

The volume of material that must be processed at stage $i$ to produce a batch of $B_p$ pounds of product $p$ is equal to $S_{ip} B_p$. Since these volume will be different for different products, the unit size must be selected to be big enough for all products. Thus,

$$V_i = \max_p \left( S_{ip} B_p \right) \qquad (9.43)$$

The continuous unit processing times $\theta_{kp}$ will be related to the batch size and the unit rate $R_k$ (ft$^3$/hr) through the equation

$$\theta_{kp} = \frac{\bar{S}_{kp} B_p}{R_k} \qquad (9.44)$$

Finally, the operating times of consecutive semicontinuous units must be related to each other. If for product $p$, two semicontinuous units $k$ and $l$ are consecutive with $l$ following $k$, then

$$\theta_{kp} \geqslant \theta_{lp} \qquad (9.45)$$

The cost function for the plant will consist of the sum of the capital costs of the batch and semicontinuous units. Thus,

$$\text{Capital cost} = \sum_i a_i V_i^{\alpha_i} + \sum_k b_k R_k^{\beta_k} \qquad (9.46)$$

The design-optimization problem thus consists of minimizing (9.46) subject to Eqs. (9.40)–(9.45) plus nonnegativity conditions on the variables. This is an NLP problem with equality and inequality constraints. It could certainly be solved using a GRG algorithm. Note, however, that constraints (9.42) and (9.43) are nondifferentiable and consequently must be modified before a GRG algorithm can be employed. This is easily done by replacing (9.42) with the inequalities

$$T_p \geqslant \bar{\theta}_{ip} \tag{9.47}$$

and

$$T_p \geqslant \theta_{kp} \tag{9.48}$$

Similarly, (9.43) is replaced with the inequalities

$$V_i \geqslant S_{ip} B_p \tag{9.49}$$

As a further simplification we can use (9.41) and (9.44) to eliminate $\bar{\theta}_{ip}$ and $\theta_{kp}$ from the problem formulation, to result in a problem in the variables $V_i$, $R_k$, $T_p$, and $B_p$.

### 9.5.3 Model Reduction and Solution

For the process of Figure 9.15, the design-optimization problem thus takes the form

Minimize $\quad a_1 V_1^{\alpha_1} + a_2 V_2^{\alpha_2} + a_3 V_3^{\alpha_3} + b_1 R_1^{\beta_1} + b_2 R_2^{\beta_2} + b_3 R_3^{\beta_3} + b_4 R_4^{\beta_4} + b_5 R_5^{\beta_5}$

where the coefficients $a_i$, $\alpha_i$, $b_i$, $\beta_i$ are given

Subject to $\quad$ Total time constraint (9.40):

$$\frac{400,000}{B_1} T_1 + \frac{300,000}{B_2} T_2 + \frac{100,000}{B_3} T_3 \leqslant 8000 \tag{9.50}$$

Reactor 1 volume limits (9.49):

$$V_1 \geqslant 1.2 B_1 \qquad V_1 \geqslant 1.5 B_2 \qquad V_1 \geqslant 1.1 B_3$$

Reactor 2 volume limits (9.49):

$$V_2 \geqslant 1.4 B_1 \qquad V_2 \geqslant 1.2 B_3 \tag{9.51}$$

Dryer volume limits (9.49):

$$V_3 \geqslant 1.0 B_1 \qquad V_3 \geqslant 1.0 B_2 \qquad V_3 \geqslant 1.0 B_3$$

Cycle time of product 1, (9.40):

$$T_1 \geqslant \frac{1.2B_1}{R_1} \qquad T_1 \geqslant \frac{1.2B_1}{R_2} \qquad T_1 \geqslant \frac{1.2B_1}{R_3}$$

$$T_1 \geqslant \frac{1.4B_1}{R_4} \qquad T_1 \geqslant \frac{1.4B_1}{R_5} \qquad (9.52)$$

Cycle time of product 2 (9.48):

$$T_2 \geqslant \frac{1.5B_2}{R_1} \qquad T_2 \geqslant \frac{1.5B_2}{R_2} \qquad T_2 \geqslant \frac{1.5B_2}{R_3}$$

$$T_2 \geqslant \frac{1.5B_2}{R_5}$$

Cycle time of product 3, (9.48):

$$T_3 \geqslant \frac{1.1B_3}{R_1} \qquad T_3 \geqslant \frac{1.1B_3}{R_2} \qquad T_3 \geqslant \frac{1.1B_3}{R_3}$$

$$T_3 \geqslant \frac{1.2B_3}{R_4} \qquad T_3 \geqslant \frac{1.2B_3}{R_5}$$

Consecutive semicontinuous units 2 and 3 ((9.45) and (9.44)):

$$\text{Product 1.} \quad \frac{1.2B_1}{R_2} \geqslant \frac{1.2B_1}{R_3}$$

$$\text{Product 2.} \quad \frac{1.5B_2}{R_2} \geqslant \frac{1.5B_2}{R_3} \quad \left.\right\} \text{ reduces to } R_3 \geqslant R_2 \qquad (9.53)$$

$$\text{Product 3.} \quad \frac{1.1B_3}{R_2} \geqslant \frac{1.1B_3}{R_3}$$

Consecutive semicontinuous units 4 and 5 [(9.45) and (9.44)]:

$$\text{Product 1.} \quad \frac{1.4B_1}{R_4} \geqslant \frac{1.4B_1}{R_5}$$

$$\text{Product 3.} \quad \frac{1.2B_3}{R_4} \geqslant \frac{1.2B_3}{R_5} \quad \left.\right\} \text{ reduces to } R_5 \geqslant R_4 \qquad (9.54)$$

For product 2, the heat exchanger will feed directly to the centrifuge,

$$\frac{1.5B_2}{R_2} \geqslant \frac{1.5B_2}{R_5} \qquad \text{or} \qquad R_5 \geqslant R_2 \qquad (9.55)$$

Cycle time of product 1 [(9.47), (9.41), and (9.44)]:

$$\text{Reactor 1. } T_1 \geqslant \frac{1.2B_1}{R_1} + 3 + \frac{1.2B_1}{R_2} \qquad (9.56)$$

$$\text{Reactor 2. } T_1 \geqslant \frac{1.2B_1}{R_3} + 1 + \frac{1.4B_1}{R_4}$$

$$\text{Dryer.} \quad T_1 \geqslant \frac{1.4B_1}{R_5} + 4$$

Cycle time of product 2 [(9.47), (9.41), and (9.44)]:

$$\text{Reactor 1. } T_2 \geqslant \frac{1.5B_2}{R_1} + 6 + \frac{1.5B_2}{R_2}$$

$$\text{Dryer.} \quad T_2 \geqslant \frac{1.5B_2}{R_5} + 8$$

Cycle time of product 3 [(9.47), (9.41), and (9.44)]:

$$\text{Reactor 1. } T_3 \geqslant \frac{1.1B_3}{R_1} + 2 + \frac{1.1B_3}{R_2}$$

$$\text{Reactor 2. } T_3 \geqslant \frac{1.1B_3}{R_3} + 2 + \frac{1.2B_3}{R_4}$$

$$\text{Dryer.} \quad T_3 \geqslant \frac{1.2B_3}{R_5} + 4$$

The problem thus has the following variables: three each of $T_p$, $B_p$, and $V_i$ and five rates $R_k$. Since pump 2 and the heat exchanger are always operated together, there is no loss in setting $R_3 = R_2$ and eliminating (9.53) and $R_3$ from the problem. The problem is all inequality constrained, with the following breakdown:

1 constraint of type (9.50)
8 constraints of type (9.51)
9 constraints of type (9.56)
14 constraints of type (9.52)
1 constraint of type (9.54)
1 constraint of type (9.55)

Note, however, that the constraints of type (9.52) are really subsumed by the constraints of type (9.56). Thus, the problem can be reduced to 13 variables and 19 inequality constraints. If all inequalities are reformulated to equalities, the problem grows to 32 variables. Using the GRG code OPT [33], the minimum cost of $159,483 is obtained in about 200 CPU seconds on a CDC 6500. The initial solution estimate and the optimal solution are given in Table 9.3. At the optimum, the cycle time of product 1 is limited by reactor 1, the cycle time of product 2 is limited by the dryer, and the cycle time of product 3 is limited by reactor 2. The semicontinuous times constitute a substantial fraction of the cycle time: 57% for product 1, 26% for product 2, and 71% for product 3.

For a more detailed discussion of this class of design problems, the reader is directed to references 34 and 35.

## 9.6 SUMMARY

In this chapter we have focused on methods that use linearizations of the nonlinear problem functions to generate good search directions. We examined basically three types of algorithms: feasible direction methods; extensions of the linear simplex method, which eventually led to the GRG algorithm; and gradient projection methods. The family of feasible direction methods required the solution of an LP subproblem to determine a direction that was both a descent direction and strictly feasible. The formulation of the direction-finding subproblem uses an active constraint set strategy that could be susceptible to

Table 9.3.  Results

|  | Initial Estimate | Optimal Solution |
|---|---|---|
| Volume reactor 1, ft$^3$ | 2000 | 1181.4 |
| Volume reactor 2, ft$^3$ | 2000 | 1250.6 |
| Volume dryer, ft$^3$ | 2000 | 893.3 |
| Rate of pump 1, ft$^3$/hr | 1000 | 753.1 |
| Rate of pump 2, ft$^3$/hr | 1000 | 422.1 |
| Rate of pump 4, ft$^3$/hr | 1000 | 422.1 |
| Rate of centrifuge, ft$^3$/hr | 1000 | 422.1 |
| Batch size, lb |  |  |
| Product 1 | 2000 | 893.3 |
| Product 2 | 2000 | 787.6 |
| Product 3 | 2000 | 892.9 |
| Cycle time, hr |  |  |
| Product 1 | 5.0 | 6.963 |
| Product 2 | 5.0 | 10.799 |
| Product 3 | 5.0 | 6.865 |

jamming unless special precautions such as the ε-perturbation device were incorporated. This family of methods is suitable for nonconvex problems, generates a sequence of feasible points, but cannot directly accommodate nonlinear equality constraints. We found that the major disadvantage of feasible direction methods is the continued necessity of solving a series of LP subproblems.

The second family of methods considered in this chapter seeks to avoid the LP solution by merely solving a set of linear equations to determine a good descent direction. For linearly constrained problems, the directions obtained will also be feasible. When only one nonbasic variable is used to define the search direction, we obtain the convex simplex method, the direct analog to the linear simplex method. When all nonbasic variables are used to define the direction, the reduced gradient method is obtained. Both variants essentially possess the convergence properties of Cauchy's method. However, both can be accelerated by incorporating conjugate direction or quasi-Newton constructions. The reduced gradient strategy was then extended to accommodate nonlinear constraints. In this case the directions generated were no longer feasible, and hence the major revision required was the addition of a Newton iteration to adjust the basic variable to achieve feasibility. The resulting GRG algorithm could also be modified to incorporate quasi-Newton acceleration constructions and to treat variable bounds implicitly. The key assumptions underlying the GRG algorithm were continuous differentiability of the problem functions and the existence of nondegenerate, nonsingular bases throughout the iteration process. Implementation of the algorithm was found to involve a number of numerical issues whose efficient resolution could materially affect algorithm performance. The implementation of a GRG code, a number of which are in wide use, was thus seen to be a significant effort.

Finally we examined the family of gradient projection methods and developed the matrix constructions required to decompose the gradient vectors into a component parallel to the constraint surface and a component orthogonal to the constraint surface. In the linearly constrained case the former component produced a feasible descent direction that could also be subjected to quasi-Newton modifications. In the nonlinearly constrained case, the parallel component led to infeasible iterates, and hence the orthogonal component was used to return to the constraints. We then studied the relationship between projection and GRG methods. We found that the two algorithms are at root equivalent and can be expected to have essentially the same convergence behavior. We observed that the projection calculations avoided the basis exchange overheads of the GRG method but did not easily allow the incorporation of techniques for handling large sparse problems. We concluded the chapter with a process design application that has been solved by both GRG and projection methods. Without a doubt, the GRG and projection methods discussed in this chapter constitute the gradient-based NLP algorithms most widely used in engineering at the present time.

# REFERENCES

1. Zoutendijk, G., *Methods of Feasible Directions*, Elsevier, Amsterdam, 1960.

2. Zangwill, W. I., *Nonlinear Programming: A Unified Approach*, Prentice-Hall, Englewood Cliffs, NJ, 1969.

3. Wolfe, P., "On the Convergence of Gradient Methods Under Constraints," *IBM J. Res. Develop.* **16**, 407–411 (1972).

4. Topkis, D. M., and A. F. Veinott, "On the Convergence of Some Feasible Direction Algorithms for Nonlinear Programming," *SIAM J. Control*, **5**, 268–279 (1967).

5. Zangwill, W. I., "The Convex Simplex Method," *Manage. Sci., A* **14**, 221–238 (1967).

6. Wolfe, P., "Methods of Nonlinear Programming," in *Recent Advances in Mathematical Programming* (R. L. Graves and P. Wolfe, Eds.), pp. 76–77, 1963.

7. McCormick, G. P., "Anti-Zig-Zagging by Bending," *Manage. Sci.*, **15**, 315–320 (1969).

8. Bazaraa, M., and C. M. Shetty, *Nonlinear Programming: Theory and Algorithms*, Wiley, New York, 1979.

9. Hans-Tijan, T. Y., and W. I. Zangwill, "Analysis and Comparison of the Reduced Gradient and the Convex Simplex Method for Convex Programming," paper presented at ORSA 41st National Meeting, New Orleans, April 1972.

10. Murtagh, B. A., and M. A. Saunders, "Large Scale Linearly Constrained Optimization," *Math. Prog.*, **14**, 41–72 (1978).

11. Abadie, J., and J. Carpentier, "Generalization of the Wolfe Reduced Gradient Method to the Case of Nonlinear Constraints," in *Optimization* (R. Fletcher, Ed.), Academic Press, New York, Chapter 4, 1969.

12. Johnson, R. C., "A Method of Optimum Design," *ASME J. Mech. Des.*, **101**, 667–673 (1979).

13. Beightler, C., and D. J. Wilde, *Foundation of Optimization Theory*, Prentice-Hall, Englewood Cliffs, NJ, 1967.

14. Wilde, D. J., "Differential Calculus in Nonlinear Programming," *Oper. Res.*, **10**, 764–773 (1962). See also *Oper. Res.* **13**, 848–855 (1965).

15. Reklaitis, G. V., and D. J. Wilde, "Necessary Conditions for a Local Optimum Without Prior Constraint Qualifications," in *Optimizing Methods in Statistics* (J. S. Rustagi, Ed.), Academic Press, New York, pp. 339–368, 1971.

16. Lasdon, L. S., A. D. Waren, A. Jain, and M. Ratner, "Design and Testing of a Generalized Reduced Gradient Code for Nonlinear Programming," *ACM Trans. Math. Software*, **4**, 34–50 (1978).

17. Gabriele, G. A., "Large Scale Nonlinear Programming Using The Generalized Reduced Gradient Method," dissertation, Purdue University, W. Lafayette, IN, 1980.

18. Smeers, Y., "Generalized Reduced Gradient Method as an Extension of Feasible Direction Methods," *J. Opt. Theory Appl.*, **22**, 209–226 (1977).

19. Luenberger, D. G., *Introduction to Linear and Nonlinear Programming*, Addison-Wesley, Reading, MA, 1973.

20. Murtagh, B. A., and M. A. Saunders, "MINOS User's Guide," *Tech. Rep.* **77-9**, Systems Optimization Laboratory, Stanford University, Stanford, CA, Feb. 1977.

21. Lasdon, L. S., and A. D. Warren, "Generalized Reduced Gradient Software for Linearly and Nonlinearly Constrained Problems," in *Design and Implementation of Optimization Software* (H. Greenberg, Ed.), Sijthoff and Noordhoff, Holland, 1978.

22. Gabriele, G. A., and K. M. Ragsdell, "The Generalized Reduced Gradient Method: A Reliable Tool for Optimal Design," *ASME J. Eng. Ind.*, **99**, 384–400 (1977).

23. Rosen, J. B., "The Gradient Projection Method for Nonlinear Programming: Part I, Linear Constraints," *SIAM J. Appl. Math.*, **8**, 181–217 (1960).

24. Rosen, J. B., "The Gradient Projection Method for Nonlinear Programming: II. Nonlinear Constraints," *SIAM J. Appl. Math.*, **9**, 514–532 (1961).

25. Goldfarb, D., "Extension of Davidson's Variable Metric Method to Maximization under Linear Inequality and Equality Constraints," *SIAM J. Appl. Math.*, **17**, 739–764 (1969).

26. Gill, R. E., and W. Murray, "Quasi-Newton Methods for Linearly Constrained Optimization," in *Numerical Methods for Constrained Optimization* (P. E. Gill and W. Murray, Eds.), Academic Press, London, 1974.

27. Powell, M. J. D., "Quadratic Termination Properties of Minimization Algorithms I: Statements and Discussion of Results," *J. Inst. Math. Appl.*, **10**, 333–342 (1972).

28. Murtagh, B. A., and R. H. Sargent, "A Constrained Minimization Method with Quadratic Convergence," in *Optimization* (R. Fletcher, Ed.), Academic Press, New York, 1969.

29. Sargent, R. W. H., and B. A. Murtagh, "Projection Methods for Nonlinear Programming," *Math. Prog.*, **4**, 245–268 (1973).

30. Haug, E. J., and J. S. Arora, *Applied Optimal Design*, Wiley, New York, 1979.

31. Haug, E. J., J. S. Arora, and T. T. Feng, "Sensitivity Analysis and Optimization of Structures for Dynamic Response," *ASME J. Mech. Des.*, **100**, 311–318 (1979).

32. Sargent, R. W. H., "Reduced Gradient and Projection Methods for Nonlinear Programming," in *Numerical Methods for Constrained Optimization* (P. E. Gill and W. Murray, Eds.), Academic Press, London, England, 1974.

33. Gabriele, G. A., and K. M. Ragsdell, "OPT-A Nonlinear Programming Code in FORTRAN IV," Purdue Research Foundation, W. Lafayette, IN, 1976.

34. Knopf, F. C., M. R. Okos, and G. V. Reklaitis, "Optimal Design of Batch/Semi-Continuous Processes," *Ind. Eng. Chem. Proc. Des. Dev.*, **21**, 79–86 (1982).

35. Grossmann, I., and R. W. H. Sargent, "Optimum Design of Multipurpose Chemical Plants," *Ind. Eng. Chem. Proc. Des. Dev.*, **18**, 343–348 (1979).

# PROBLEMS

**9.1.** In the development of the feasible direction method, descent directions are employed that remain in the interior of the feasible region. Is this a good idea? Why or why not?

**9.2.** Can the feasible directions method be applied to equality-constrained problems? Explain.

**9.3.** (a) What is the major advantage of using an active constraint strategy? (b) How would such a strategy be used with GRG?

**9.4.** What are the major simplifications associated with a linear constraint set when using the GRG method?

**9.5.** The feasible direction method is defined with equal weighting of feasibility and descent. Formulate a variant that might weigh descent more heavily. (*Hint*: Multiply $\theta$ by suitable constants.)

**9.6.** What potential difficulties can arise if a step of the convex simplex method is executed with a degenerate basis?

**9.7.** Describe a procedure for estimating the remaining objective function improvement that could be carried out as part of the convex simplex algorithm for problems with convex objective functions.

**9.8.** Outline the way in which the *conjugate directions* method could be incorporated within the convex simplex method.

**9.9.** Contrast the relative advantages and disadvantages of using a reduced gradient and a projection method for a linearly constrained problem.

**9.10.** Discuss how bounds on the variables might best be accommodated in a projection method.

**9.11.** Outline an implementation of the generalized gradient projection method that would not require explicit calculation of the inverse of the matrix $AA^T$.

**9.12.** How might one take advantage of the fact that some of the constraints are linear in the generalized gradient projection method?

**9.13.** Outline a modification of the convex simplex method in which variable bounds $a_i \leqslant x_i \leqslant b_i$ would be handled without treating them as explicit inequality constraints.

**9.14.** Consider the problem

$$\text{Minimize} \quad f(x) = 100(x_2 - x_1^2)^2 + (1 - x_1^2)$$

$$\text{Subject to} \quad g_1(x) = x_1 + 1 \geqslant 0$$

$$g_1(x) = 1 - x_2 \geqslant 0$$

$$g_3(x) = 4x_2 - x_1 - 1 \geqslant 0$$

$$g_4(x) = 1 - 0.5x_1 - x_2 \geqslant 0$$

Given the point $x^0 = (-1, 1)$.

(a) Perform one full iteration of the convex simplex method

(b) Perform one full iteration of the reduced gradient method.

(c) Sketch the feasible region and explain the difference in the results.

**9.15.** Given the linearly constrained NLP problem

$$\text{Minimize} \quad f(x) = (x_1 - 2)^2 + (x_2 - 2)^2$$

$$\text{Subject to} \quad g_1(x) = 3 - x_1 - x_2 \geqslant 0$$

$$g_2(x) = 10x_1 - x_2 - 2 \geqslant 0$$

$$x_2 \geqslant 0$$

and the starting point $x^0 = (0.2, 0)$.

(a) Carry out three iterations of the method of feasible directions.

(b) Carry out three iterations of the reduced gradient method. In both cases, carry out subproblem solutions using graphical or algebraic shortcuts.

(c) Contrast the iteration paths of the two methods, using a graph.

**9.16.** Suppose the constraint $g_1$ of problem 9.15 is replaced by

$$g_1(x) = 4.5 - x_1^2 - x_2^2 \geqslant 0$$

(a) Discuss how this modification will affect the iteration path of problem 9.15(b).

(b) Calculate the reduced gradient at the feasible intersection of constraints $g_1$ and $g_2$. Is the point a local minimum?

(c) Calculate the search direction from the point of part (b), using the method of feasible directions. Contrast to the direction generated by the GRG method.

(d) Calculate the reduced gradient at the point $(\frac{3}{2}, \frac{3}{2})$. Is this point a local minimum?

**9.17.** Suppose the constraint $g_1$ of problem 9.15 is replaced by

$$g_1(x) = 2.25 - x_1 x_2 \geqslant 0$$

Reply to questions (b), (c), and (d) of problem 9.16 with this problem modification.

**9.18.** Consider the problem

$$\text{Minimize} \quad f(x) = (x_1 - x_2)^2 + x_2$$

$$\text{Subject to} \quad g_1(x) = x_1 + x_2 - 1 \geqslant 0$$

$$g_2(x) = 3 - x_1 + 2x_2 \geqslant 0$$

$$x_1 \geqslant 0 \qquad x_2 \geqslant 0$$

with initial point $x^0 = (3, 0)$.

(a) Calculate the search direction at $x^0$ using the convex simplex method, the reduced gradient method, and the method of feasible directions.

(b) Use a sketch to show the differences in the search directions.

**9.19.** Consider the problem

$$\text{Minimize} \quad f(x) = x_1 + x_2$$

$$\text{Subject to} \quad h(x) = x_1^2 + x_2^2 - 1 = 0$$

(a) Calculate the reduced gradient at the point $x^0 = (0, 1)$.

(b) Suppose the algorithm progresses from $x^0$ to the point $(-1, 0)$. Which variable should be selected to be independent and which dependent? Why?

**9.20.** Consider the problem

$$\text{Maximize} \quad f(x) = 2x_1 - 3(x_2 - 4)^2$$

$$\text{Subject to} \quad g_1(x) = 10 - x_1^2 - x_2^2 \geqslant 0$$

$$g_2(x) = 9 - x_1^2 - (x_2 - 4)^2 \geqslant 0$$

$$x_1 \geqslant 0$$

(a) Given the starting point $x^0 = (2, 2)$:

(1) Calculate the next iteration point that would result if a method of feasible direction were applied with $\varepsilon = 10^{-3}$.

(2) Calculate the next iteration point that would be attained if the generalized reduced gradient method were applied with $\varepsilon = 10^{-3}$.

(3) Exhibit these results graphically.

(b) Given the point $(0, 1)$, repeat the three parts outlined in (a).

(c) Use the reduced gradient to determine whether the point $(1, 3)$ is a local maximum. How can you prove if it is the global maximum?

**9.21.** Consider the problem

$$\text{Minimize} \quad f(x) = -x_1 x_2^2 x_3^3 / 81$$

$$\text{Subject to} \quad h_1(x) = x_1^3 + x_2^2 + x_3 - 13 = 0$$

$$h_2(x) = x_2^2 x_3^{-1/2} - 1 = 0$$

(a) Calculate the reduced gradient at the point $x = (1, \sqrt{3}, 9)$.

(b) Show that this point is a constrained local minimum.

(c) Confirm (b) by using the Lagrangian conditions.

**9.22.** Consider the problem

$$\text{Minimize} \quad f(x) = -x_1 - 2x_2 - 4x_1 x_2$$

$$\text{Subject to} \quad g_1(x) = 9 - (x_1 - 2)^2 - (x_2 - 1)^2 \geqslant 0$$

$$g_2(x) = x_1 \geqslant 0$$

$$g_3(x) = x_2 \geqslant 0$$

(a) Calculate the direction that would be selected by the feasible directions method at $x^0 = (0, 0)$.

(b) Repeat (a) with $x^0 = (5, 1)$.

(c) If the direction calculated in (a) were used, what would be the next iteration point?

**9.23.** Given the NLP problem,

$$\text{Minimize} \quad f(x) = (x_1 - 1)^2 + (x_2 - 1)^2 + (x_3 - 1)^2$$

$$\text{Subject to} \quad g_1(x) = x_1 + x_2 + x_3 - 2 \leqslant 0$$

$$h_1(x) = 1.5x_1^2 + x_2^2 + x_3^2 - 2 = 0$$

Consider a feasible point $\bar{x} = (1, \frac{1}{2}, \frac{1}{2})$.

(a) Calculate the reduced gradient at the point $\bar{x}$, and show that this corresponds to a Kuhn-Tucker point.

(b) Calculate the multipliers at $\bar{x}$.

(c) Suppose the constraint $h_1(x)$ is *deleted*. Calculate the reduced gradient and the new search direction. What would be the new point generated by the GRG algorithm?

**9.24.** (a) Your reduced gradient code used to solve the problem

$$\text{Minimize} \quad f(x) = 4x_1 - x_2^2 + x_3^2 - 12$$

$$\text{Subject to} \quad h_1(x) = 20 - x_1^2 - x_2^2 = 0$$

$$h_2(x) = x_1 + x_3 - 7 = 0$$

$$x_1, x_2, x_3, \geqslant 0$$

terminates at the point $(2.5, 3.71, 4.5)$. Is this the solution to the problem? Why?

(b) Upon checking the problem formulation, you discover that the constraint

$$h_3(x) = x_3 x_1^{-1} - 2 = 0$$

was left out. What is the solution now?

**9.25.** Consider the problem

$$\text{Minimize} \quad f(x) = \exp(x_2) + x_3 - \exp(x_4)$$

$$\text{Subject to} \quad g_1(x) = \exp(x_1 - 1) - x_2 + x_3 - 1 \geqslant 0$$

$$g_2(x) = x_2 - x_3^2 + x_4 \geqslant 0$$

$$\text{all } x_i \geqslant 0$$

Suppose the GRG method commences with the point $(1, 0, 0, 0)$.

(a) Show that all choices of basic variables will be either degenerate or degenerate and singular.

(b) Show that the choice $x = (x_1, x_2)$ will lead to a search direction that will force violation of variable nonnegativity.

(c) Suggest a way to calculate an acceptable descent direction.

**9.26.** Carry out three iterations of problems 9.15 using the gradient projection method. Show that the point $(\frac{3}{2}, \frac{3}{2})$ is a local minimum.

**9.27.** Calculate the gradient projection direction and the multipliers for problem 9.18 at the point $x^0 = (3, 0)$.

**9.28.** Consider the problem

$$\text{Minimize} \quad f(x) = x_1 + x_2$$

$$\text{Subject to} \quad h(x) = x_1^2 + x_2^2 - 1 = 0$$

at the point $(0, 1)$.

(a) Calculate the gradient projection direction.

(b) Use the direction of part (a) to take a step of length $\alpha = 0.1$. Compute the feasible projection of the resulting point.

**9.29.** Repeat the calculation of problem 9.23 using the equivalent generalized gradient projection constructions.

# Chapter 10

# Quadratic Approximation Methods for Constrained Problems

In the preceding two chapters we considered a number of alternative strategies for exploiting linear approximations to nonlinear problem functions. In general we found that, depending upon the strategy employed, linearizations would either lead to vertex points of the linearized constraint sets or generate descent directions for search. In either case, some type of line search was required in order to approach the solution of non–corner-point constrained problems. Based upon our experience with unconstrained problems, it is reasonable to consider the use of higher order approximating functions since these could lead directly to better approximations of non–corner-point solutions. For instance, in the single-variable case we found that a quadratic approximating function could be used to predict the location of optima lying in the interior of the search interval. In the multivariable case, the use of a quadratic approximation (e.g., in Newton's method) would yield good estimates of unconstrained minimum points. Furthermore, the family of quasi-Newton methods allowed us to reap some of the benefits of a quadratic approximation without explicitly developing a full second-order approximating function at each iteration. In fact, in the previous chapter we did to some extent exploit the acceleration capabilities of quasi-Newton methods by introducing their use within the direction-generation mechanics of the reduced gradient and gradient projection methods. Thus, much of the discussion of the previous chapters does point to the considerable algorithmic potential of using higher order, specifically quadratic, approximating functions for solving constrained problems.

In this chapter we examine in some detail various strategies for using quadratic approximations. We begin by briefly considering the consequence of direct quadratic approximation, the analog of the successive LP strategy. Then we investigate the use of second derivatives and Lagrangian constructions to formulate quadratic programming subproblems, the analog to Newton's method. Finally, we discuss the application of quasi-Newton formulas to generate updates of quadratic terms. We will find that the general NLP problem can be solved very efficiently via a series of subproblems consisting of a quadratic objective function and linear constraints, provided a suitable line search is carried out from the solution of each such subproblem. The resulting class of exterior point algorithms can be viewed as a natural extension of quasi-Newton methods to constrained problems.

## 10.1  DIRECT QUADRATIC APPROXIMATION

Analogous to the successive LP strategy of Section 8.1, we could approach the solution of the general NLP problem by simply replacing each nonlinear function by its local quadratic approximation at the solution estimate $x^0$ and solving the resulting series of approximating subproblems. If each function $f(x)$ is replaced by its quadratic approximation,

$$q(x; x^0) = f(x^0) + \nabla f(x^0)^T(x - x^0) + \tfrac{1}{2}(x - x^0)^T \nabla^2 f(x^0)(x - x^0)$$

then the subproblem becomes one of minimizing a quadratic function subject to quadratic equality and inequality constraints. While it seems that this subproblem structure ought to be amenable to efficient solution, in fact, it is not. To be sure, the previously discussed strategies for constrained problems can solve this subproblem but at no real gain over direct solution of the original problem. For a sequential strategy using subproblem solutions to be effective, the subproblem solutions must be substantially easier to obtain than the solution of the original problem.

   Recall from the discussion in Section 9.2.3 that problems with a quadratic positive-definite objective function and linear constraints can be solved in a *finite* number of reduced gradient iterations provided that quasi-Newton or conjugate gradient enhancement of the reduced gradient direction vector is used. Of course, while the number of iterations is finite, each iteration requires a line search, which strictly speaking is itself not a finite procedure. However, as we will discover in Chapter 11, there are specialized methods for these so-called *quadratic programming* (QP) problems that will obtain a solution in a finite number of iterations *without* line searching, using instead simplex-like pivot operations. Given that QP problems can be solved efficiently with truly finite procedures, it appears to be desirable to formulate our approximating subproblems as quadratic programs. Thus, assuming that the objective function is twice continuously differentiable, a plausible solution strategy would consist of the following steps:

### *Direct Successive Quadratic Programming Solution*

Given $x^0$, an initial solution estimate, and a suitable method for solving QP subproblems.

**Step 1.** Formulate the QP problem:

$$\text{Minimize} \quad \nabla f(x^{(t)})^T d + \tfrac{1}{2} d^T \nabla^2 f(x^{(t)}) d$$

$$\text{Subject to} \quad h_k(x^{(t)}) + \nabla h_k(x^{(t)})^T d = 0 \qquad k = 1,\ldots, K$$

$$g_j(x^{(t)}) + \nabla g_j(x^{(t)})^T d \geqslant 0 \qquad j = 1,\ldots, J$$

**Step 2.** Solve the QP problem, and set $x^{(t+1)} = x^{(t)} + d$.

**Step 3.** Check for convergence. If not converged, repeat step 1.

As shown in the following example, this approach can be quite effective.

### Example 10.1

Solve the problem

$$\text{Minimize} \quad f(x) = 6x_1 x_2^{-1} + x_2 x_1^{-2}$$

$$\text{Subject to} \quad h(x) = x_1 x_2 - 2 = 0$$

$$g(x) = x_1 + x_2 - 1 \geqslant 0$$

from the initial feasible estimate $x^0 = (2, 1)$, using the direct successive QP strategy.

At $x^0$, $f(x^0) = 12.25$, $h(x^0) = 0$, and $g(x^0) = 2 > 0$. The derivatives required to construct the QP subproblem are

$$\nabla f(x) = \left( (6x_2^{-1} - 2x_2 x_1^{-3}), (-6x_1 x_2^{-2} + x_1^{-2}) \right)^T$$

$$\nabla^2 f = \begin{pmatrix} (6x_2 x_1^{-4}) & (-6x_2^{-2} - 2x_1^{-3}) \\ (-6x_2^{-2} - 2x_1^{-3}) & (12x_1 x_2^{-3}) \end{pmatrix}$$

$$\nabla h(x) = (x_2, x_1)^T$$

Thus, the first QP subproblem will be

$$\text{Minimize} \quad (\tfrac{23}{4}, -\tfrac{47}{4})d + \tfrac{1}{2} d^T \begin{pmatrix} \tfrac{3}{8} & -\tfrac{25}{4} \\ -\tfrac{25}{4} & 24 \end{pmatrix} d$$

$$\text{Subject to} \quad (1, 2)d = 0$$

$$(1, 1)d + 2 \geqslant 0$$

Since the first constraint can be used to eliminate one of the variables, that is,

$$d_1 = -2d_2$$

the resulting single-variable problem can be solved easily analytically to give

$$d^0 = (-0.92079, 0.4604)$$

Thus, the new point becomes

$$x^{(1)} = x^0 + d^0 = (1.07921, 1.4604)$$

at which point

$$f(x^{(1)}) = 5.68779$$

$$h(x^{(1)}) = -0.42393$$

$$g(x^{(1)}) > 0$$

Note that the objective function value has improved substantially but that the equality constraint is violated. Suppose we continue with the solution procedure. The next subproblem is

Minimize   $(1.78475, -2.17750)d + \frac{1}{2}d^T \begin{pmatrix} 6.4595 & -4.4044 \\ -4.4044 & 4.1579 \end{pmatrix} d$

Subject to   $(1.4604, 1.07921)d - 0.42393 = 0$

$$(1, 1)d + 1.5396 \geqslant 0$$

The solution is $d^{(1)} = (-0.03043, 0.43400)$, resulting in the new point, with

$$x^{(2)} = (1.04878, 1.89440)$$

$$f(x^{(2)}) = 5.04401$$

$$h(x^{(2)}) = -0.013208$$

$$g(x^{(2)}) > 0$$

Note that both the objective function value and the equality constraint violation are reduced. The next two iterations produce the results

$$x^{(3)} = (1.00108, 1.99313) \qquad f(x^{(3)}) = 5.00457 \qquad h(x^{(3)}) = -4.7 \times 10^{-3}$$

$$x^{(4)} = (1.00014, 1.99971) \qquad f(x^{(4)}) = 5.00003 \qquad h(x^{(4)}) = -6.2 \times 10^{-6}$$

The exact optimum is $x^* = (1, 2)$ with $f(x^*) = 5.0$; a very accurate solution has been obtained in four iterations.

As is evident from Figure 10.1, the constraint linearizations help to define the search directions, while the quadratic objective function approximation

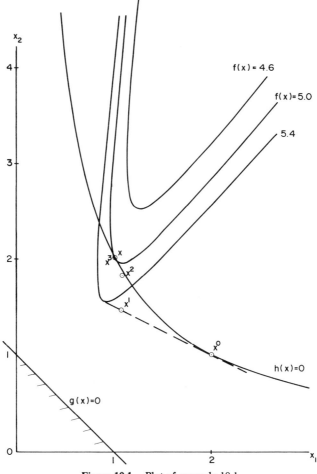

**Figure 10.1.**  Plot of example 10.1.

effectively fixes the step length along that direction. Thus, the use of a nonlinear objective function approximation does lead to nonvertex solutions. However, as shown in the next example, the self-bounding feature of the quadratic approximation does not always hold.

*Example 10.2*

Suppose the objective function and equality constraint of Example 10.1 are interchanged. Thus, consider the problem

$$\text{Minimize} \quad f(x) = x_1 x_2$$

$$\text{Subject to} \quad h(x) = 6x_1 x_2^{-1} + x_2 x_1^{-2} - 5 = 0$$

$$g(x) = x_1 + x_2 - 1 \geq 0$$

The optimal solution to this problem is identical to that of Example 10.1. With the starting point $x^0 = (2, 1)$ the first subproblem becomes

$$\text{Minimize} \quad (1, 2)d + \tfrac{1}{2}d^T \begin{pmatrix} 0 & 1 \\ 1 & 0 \end{pmatrix} d$$

$$\text{Subject to} \quad (\tfrac{23}{4}, -\tfrac{47}{4})d + \tfrac{29}{4} = 0$$

$$(1, 1)d + 2 \geq 0$$

The solution to this subproblem is $d^0 = (-1.7571, -0.24286)$, at which both constraints are tight. Thus, a subproblem corner point is reached. The resulting intermediate solution is

$$x^1 = x^0 + d^0 = (0.24286, 0.75714)$$

with

$$f(x^{(1)}) = 0.18388 \quad h(x^{(1)}) = 9.7619 \quad \text{and} \quad g(x^{(1)}) = 0$$

Although the objective function value decreases, the equality constraint violation is worse. The next subproblem becomes

$$\text{Minimize} \quad (0.75714, 0.24286)d + \tfrac{1}{2}d^T \begin{pmatrix} 0 & 1 \\ 1 & 0 \end{pmatrix} d$$

$$\text{Subject to} \quad -97.795d_1 + 14.413d_2 + 9.7619 = 0$$

$$d_1 + d_2 \geq 0$$

The resulting new point is

$$x^{(2)} = (0.32986, 0.67015)$$

with

$$f(x^{(2)}) = 0.2211 \quad h(x^{(2)}) = 4.1125 \quad g(x^{(2)}) = 0$$

Again, $g(x)$ is tight. The next few points obtained in this fashion are

$$x^{(3)} = (0.45383, 0.54618)$$

$$x^{(4)} = (-0.28459, 1.28459)$$

$$x^{(5)} = (-0.19183, 1.19183)$$

These and all subsequent iterates all lie on the constraint $g(x) = 0$. Since the objective function decreases toward the origin, all iterates will be given by the intersection of the local linearization with $g(x) = 0$. Since the slope of the linearization becomes larger or smaller than $-1$ depending upon which side of the constraint "elbow" the linearization point lies on, the successive iterates simply follow an oscillatory path up and down the surface $g(x) = 0$.

Evidently the problem arises because the linearization cannot take into account the sharp curvature of the constraint $h(x) = 0$ in the vicinity of the optimum. Since both the constraint and the objective function shapes serve to define the location of the optimum, both really ought to be taken into account. This can be accomplished implicitly by limiting the allowable step or by using a penalty function–based line search, as discussed in Section 8.1.2. Alternatively, the constraint curvature can be taken into account explicitly by incorporating it into a modified subproblem objective function, as discussed in the next section.

## 10.2  QUADRATIC APPROXIMATION OF THE LAGRANGIAN FUNCTION

The examples of the previous section suggest that it is desirable to incorporate into the subproblem definition not only the curvature of the objective function but also that of the constraints. However, based on computational considerations, we also noted that it is preferable to deal with linearly constrained rather than quadratically constrained subproblems. Fortunately, this can be accomplished by making use of the Lagrangian function, as will be shown below. For purposes of this discussion, we first consider only the equality-constrained problem. The extension to inequality constraints will follow in a straightforward fashion.

Consider the problem

$$\text{Minimize} \quad f(x)$$

$$\text{Subject to} \quad h(x) = 0$$

Recall that the necessary conditions for a point $x^*$ to be a local minimum are that there exist a multiplier $v^*$ such that

$$\nabla_x L(x^*, v^*) = \nabla f^* - v^{*T} \nabla h^* = 0 \quad \text{and} \quad h(x^*) = 0 \qquad (10.1)$$

Sufficient conditions for $x^*$ to be a local minimum (see Theorem 5.7 of Chapter 5) are that conditions (10.1) hold and that the Hessian of the Lagrangian function,

$$\nabla_x^2 L(x^*, v^*) = \nabla^2 f^* - (v)^{*T} \nabla^2 h^*$$

satisfy

$$d^T \nabla_x^2 L d > 0 \qquad \text{for all } d \text{ such that } (\nabla h)^{*T} d = 0 \qquad (10.2)$$

Given some point $(\bar{x}, \bar{v})$, we construct the following subproblem expressed in terms of the variables $d$:

$$\text{Minimize} \quad \nabla f(\bar{x})^T d + \tfrac{1}{2} d^T \nabla_x^2 L(\bar{x}, \bar{v}) d \qquad (10.3)$$

$$\text{Subject to} \quad h(\bar{x}) + \nabla h(\bar{x})^T d = 0 \qquad (10.4)$$

We now observe that if $d^* = 0$ is the solution to the problem consisting of (10.3) and (10.4), then $\bar{x}$ must satisfy the necessary conditions for a local minimum of the original problem. First note that if $d^* = 0$ solves the subproblem, then from (10.4) it follows that $h(\bar{x}) = 0$; in other words, $\bar{x}$ is a feasible point. Next, there must exist some $v^*$ such that the subproblem functions satisfy the Lagrangian necessary conditions at $d^* = 0$. Thus, since the gradient of the objective function (10.3) with respect to $d$ at $d^* = 0$ is $\nabla f(\bar{x})$ and that of (10.4) is $\nabla h(\bar{x})$, it follows that

$$\nabla f(\bar{x}) - (v)^{*T} \nabla h(\bar{x}) = 0 \qquad (10.5)$$

Clearly $v^*$ will serve as Lagrange multiplier for the original problem, and thus $\bar{x}$ satisfies the necessary conditions for a local minimum.

Now suppose we take the $v^*$ given above and reformulate the subproblem at $(\bar{x}, v^*)$. Clearly, (10.5) still holds so that $d = 0$ satisfies the necessary conditions. Suppose the subproblem also satisfies the second-order sufficient conditions at $d = 0$ with $v^*$. Then, it must be true that

$$d\{\nabla_x^2 L(\bar{x}, v^*) - v^*(0)\}d > 0 \qquad \text{for all } d \text{ such that } \nabla h(x)^T d = 0$$

Note that the second derivative with respect to $d$ of (10.4) is zero, since it is a linear function in $d$. Consequently, the above inequality implies that $d^T \nabla_x^2 L(\bar{x}, v^*)d$ is positive also. Therefore, the pair $(\bar{x}, v^*)$ satisfies the sufficient conditions for a local minimum of the original problem.

This demonstration indicates that the subproblem consisting of (10.3) and (10.4) has the following very interesting features:

1. If no further corrections can be found, that is, $d = 0$, then the local minimum of the original problem will have been obtained.
2. The Lagrange multipliers of the subproblem can be used conveniently as estimates of the multipliers used to formulate the next subproblem.
3. For points sufficiently close to the solution of the original problem the quadratic objective function is likely to be positive definite, and thus the solution of the QP subproblem will be well behaved.

By making use of the sufficient conditions stated for both equality and inequality constraints, it is easy to arrive at a QP subproblem formulation for the general case involving $K$ equality and $J$ inequality constraints. If we let

$$L(x, u, v) = f(x) - \sum v_k h_k(x) - \sum u_j g_j(x) \qquad (10.5)$$

then at some point $(\bar{x}, \bar{u}, \bar{v})$ the subproblem becomes

$$\text{Minimize} \quad q(d; \bar{x}) \equiv \nabla f(\bar{x})^T d + \tfrac{1}{2} d^T \nabla_x L(\bar{x}, \bar{u}, \bar{v})d \qquad (10.6)$$

$$\text{Subject to} \quad \tilde{h}_k(d; \bar{x}) \equiv h_k(\bar{x}) + \nabla h_k(\bar{x})^T d = 0 \qquad k = 1, \dots, K \qquad (10.7)$$

$$\tilde{g}_j(d; \bar{x}) \equiv g_j(\bar{x}) + \nabla g_j(\bar{x})^T d \geq 0 \qquad j = 1, \dots, J \qquad (10.8)$$

The algorithm retains the basic steps outlined for the direct QP case. Namely, given an initial estimate $x^0$ as well as $u^0$ and $v^0$ (the latter could be set equal to zero), we formulate the subproblem [Eqs. (10.6)–(10.8)]; solve it; set $x^{(t+1)} = x^{(t)} + d$; check for convergence; and repeat, using as next estimates of $u$ and $v$ the corresponding multipliers obtained at the solution of the subproblem.

### Example 10.3

Repeat the solution of the problem of Example 10.1 using the Lagrangian QP subproblem with initial estimates $x^0 = (2, 1)^T$, $u^0 = 0$, and $v^0 = 0$. The first subproblem becomes

$$\text{Minimize} \quad (\tfrac{23}{4}, -\tfrac{47}{4})d + \tfrac{1}{2}d^T \begin{pmatrix} \tfrac{3}{8} & -\tfrac{25}{4} \\ -\tfrac{25}{4} & 24 \end{pmatrix} d$$

$$\text{Subject to} \quad (1, 2)d = 0$$

$$(1, 1)d + 2 \geqslant 0$$

This is exactly the same as the first subproblem of Example 10.1, because with the initial zero estimates of the multipliers the constraint terms of the Lagrangian will vanish. The subproblem solution is thus, as before,

$$d^0 = (-0.92079, 0.4604)^T$$

Since the inequality constraint is loose at this solution, $u^{(1)}$ must equal zero. The equality constraint multiplier can be found from the solution of the Lagrangian necessity conditions for the subproblem. Namely,

$$\nabla q(d^0; x^0) = v \nabla \tilde{h}(d^0; x^0)$$

or

$$\left\{ (\tfrac{23}{4}, -\tfrac{47}{4}) + d^T \begin{pmatrix} -\tfrac{3}{8} & -\tfrac{25}{4} \\ -\tfrac{25}{4} & 24 \end{pmatrix} \right\} = v(1, 2)^T$$

Thus, $v^{(1)} = 2.52723$. Finally, the new estimate of the problem solution will be $x^{(1)} = x^0 + d^0$, or

$$x^{(1)} = (1.07921, 1.4604)^T \qquad f(x^{(1)}) = 5.68779 \qquad h(x^{(1)}) = -0.42393$$

as was the case before.

The second subproblem requires the gradients

$$\nabla f(x^{(1)}) = (1.78475, -2.17750)^T \qquad \nabla h(x^{(1)}) = (1.4604, 1.07921)^T$$

$$\nabla^2 f(x^{(1)}) = \begin{pmatrix} 6.45954 & -4.40442 \\ -4.40442 & 4.15790 \end{pmatrix}$$

$$\nabla^2 h(x^{(1)}) = \begin{pmatrix} 0 & 1 \\ 1 & 0 \end{pmatrix}$$

The quadratic term is therefore equal to

$$\nabla^2 L = \nabla^2 f - v \nabla^2 h$$

$$= \nabla^2 f - 2.52723 \begin{pmatrix} 0 & 1 \\ 1 & 0 \end{pmatrix} = \begin{pmatrix} 6.45924 & -6.93165 \\ -6.93165 & 4.15790 \end{pmatrix}$$

The complete subproblem becomes

$$\text{Minimize} \quad (1.78475, \, -2.17750)d + \tfrac{1}{2}d^T \begin{pmatrix} 6.45924 & -6.93165 \\ -6.93165 & 4.15790 \end{pmatrix} d$$

$$\text{Subject to} \quad 1.4604d_1 + 1.07921d_2 = 0.42393$$

$$d_1 + d_2 + 1.539604 \geqslant 0$$

The solution is $d^{(1)} = (0.00614, \, 0.38450)$. Again, since $\tilde{g}(d^{(1)}; x^{(1)}) > 0$, $u^{(2)} = 0$, and the remaining equality constraint multiplier can be obtained from

$$\nabla q(d^{(1)}; x^{(1)}) = v^T \nabla h(d^{(1)}; x^{(1)})$$

or

$$\begin{pmatrix} -0.84081 \\ -0.62135 \end{pmatrix} = v \begin{pmatrix} 1.46040 \\ 1.07921 \end{pmatrix}$$

Thus,

$$v^{(2)} = -0.57574 \quad \text{and} \quad x^{(2)} = (1.08535, \, 1.84490)^T$$

with

$$f(x^{(2)}) = 5.09594 \quad \text{and} \quad h(x^{(2)}) = 2.36 \times 10^{-3}$$

Continuing the calculations for a few more iterations, the results obtained are

$$x^{(3)} = (0.99266, \, 2.00463)^T \qquad v^{(3)} = -0.44046$$

$$f^{(3)} = 4.99056 \qquad h^{(3)} = -1.008 \times 10^{-2}$$

and

$$x^{(4)} = (0.99990, \, 2.00017)^T \qquad v^{(4)} = -0.49997$$

$$f^{(4)} = 5.00002 \qquad h^{(4)} = -3.23 \times 10^{-5}$$

It is interesting to note that these results are essentially comparable to those obtained in Example 10.1 without the inclusion of the constraint second derivative terms. This might well be expected, because at the optimal solution $(1, 2)$ the constraint contribution to the second derivative of the Lagrangian is small.

$$\nabla^2 f^* - v^* \nabla^2 h^* = \begin{pmatrix} 12 & -3.5 \\ -3.5 & 1.5 \end{pmatrix} - \left(-\tfrac{1}{2}\right) \begin{pmatrix} 0 & 1 \\ 1 & 0 \end{pmatrix} = \begin{pmatrix} 12 & -3.0 \\ -3.0 & 1.5 \end{pmatrix}$$

The basic algorithm illustrated in the preceding example can be viewed as an extension of Newton's method to accommodate constraints. Specifically, if no constraints are present, the subproblem reduces to

$$\text{Minimize} \quad \nabla f(\bar{x})^T d + \tfrac{1}{2} d^T \nabla^2 f(\bar{x}) d$$

which is the approximation used in Newton's method. Viewing it in this light, we might expect that some of the difficulties associated with Newton's method may also arise in the extension. Specifically, we might expect to generate nonminimum stationary points if the second derivative term is nonpositive definite and may expect failure to converge from poor initial solution estimates. As shown in the next example, this is in fact the case.

### Example 10.4

Consider the problem of Example 10.2 with the initial estimate $x^0 = (2, 2.789)$ and $u = v = 0$. The first subproblem will be given by

$$\text{Minimize} \quad (2.789, 2)d + \tfrac{1}{2} d^T \begin{pmatrix} 0 & 1 \\ 1 & 0 \end{pmatrix} d$$

$$\text{Subject to} \quad 1.4540 d_1 - 1.2927 d_2 = 1.3 \times 10^{-4}$$

$$d_1 + d_2 + 3.789 \geqslant 0$$

The solution is $d^0 = (-1.78316, -2.00583)$. The inequality constraint is tight, so both constraint multipliers must be computed. The result of solving the system

$$\nabla \tilde{q}(d^0; x^0) = v \nabla \tilde{h}(d^0; x^0) + u \nabla \tilde{g}(d^0; x^0)$$

$$\begin{pmatrix} 0.78317 \\ 0.21683 \end{pmatrix} = v \begin{pmatrix} -145.98 \\ 19.148 \end{pmatrix} + u \begin{pmatrix} 1 \\ 1 \end{pmatrix}$$

is $v^{(1)} = -0.00343$ and $u^{(1)} = 0.28251$.

At the corresponding intermediate point,

$$x^{(1)} = (0.21683, 0.78317)^T$$

we have

$$f(x^{(1)}) = 0.1698 \qquad h(x^{(1)}) = 13.318 \qquad g(x^{(1)}) = 0$$

Note that the objective function decreases substantially, but the equality constraint violation becomes very large. The next subproblem constructed at

$x^{(1)}$ with multiplier estimates $u^{(1)}$ and $v^{(1)}$ is

Minimize $\quad (0.78317, 0.21683)d$

$$+ \tfrac{1}{2}d^T\left\{\begin{pmatrix} 0 & 1 \\ 1 & 0 \end{pmatrix} - v^{(1)}\begin{pmatrix} 2125.68 & -205.95 \\ -205.95 & 5.4168 \end{pmatrix} - u^{(1)}(0)\right\}d$$

Subject to $\quad -145.98d_1 + 19.148d_2 = -13.318$

$$d_1 + d_2 \geqslant 0$$

The subproblem solution is $d^{(1)} = (+0.10434, -0.10434)^T$, and the multipliers are $v^{(2)} = -0.02497$, $u^{(2)} = 0.38822$. At the new point $x^{(2)} = (0.40183, 0.59817)^T$, the objective function value is 0.24036 and the constraint value is 2.7352. The results of the next iteration,

$$x^{(3)} = (0.74969, 0.25031)^T \qquad f(x^{(3)}) = 0.18766 \qquad h(x^{(3)}) = 13.416$$

indicate that the constraint violation has increased considerably while the objective function value has decreased somewhat. Comparing the status at $x^{(1)}$ and $x^{(3)}$, it is evident that the iterations show no real improvement. In fact, both the objective function value and the equality constraint violation have increased in proceeding from $x^{(1)}$ to $x^{(3)}$.

The solution to the problem of unsatisfactory convergence is, as in the unconstrained case, to perform a line search from the previous solution estimate in the direction obtained from the current QP subproblem solution. However, since in the constrained case both objective function improvement and reduction of the constraint infeasibilities need to be taken into account, the line search must be carried using some type of penalty function. For instance, as in the case of the successive LP strategy advanced by Palacios-Gomez (recall Section 8.1.2), an exterior penalty function of the form

$$P(x, R) = f(x) + R\left\{\sum_{k=1}^{K} [h_k(x)]^2 + \sum_{j=1}^{J} [\min(0, g_j(x))]^2\right\}$$

could be used along with some strategy for adjusting the penalty parameter $R$. This approach is illustrated in the next example.

### Example 10.5

Consider the application of the penalty function line search to the problem of Example 10.4 beginning with the point $x^{(2)}$ and the direction vector $d^{(2)} = (0.34786, -0.34786)^T$ which was previously used to compute the point $x^{(3)}$ directly.

Suppose we use the penalty function

$$P(x, R) = f(x) + 10\{h(x)^2 + [\min(0, g(x))]^2\}$$

and minimize it along the line

$$x = x^{(2)} + \alpha d^{(2)} = \begin{pmatrix} 0.40183 \\ 0.59817 \end{pmatrix} + \begin{pmatrix} 0.34786 \\ -0.34786 \end{pmatrix} \alpha$$

Note that at $\alpha = 0$, $P = 75.05$, while at $\alpha = 1$, $P = 1800.0$. Therefore, a minimum ought to be found in the range $0 \leqslant \alpha \leqslant 1$. Using any convenient line-search method, the approximate minimum value, $P = 68.11$, can be found with $\alpha = 0.1$. The resulting point will be

$$x^{(3)} = (0.43662, 0.56338)^T$$

with $f(x^{(3)}) = 0.24682$ and $h(x^{(3)}) = 2.6053$.

In order to continue the iterations, updated estimates of the multipliers are required. Since $\bar{d} = x^{(3)} - x^{(2)}$ is no longer the optimum solution of the previous subproblem, this value of $d$ cannot be used to estimate the multipliers. The only available updated multiplier values are those associated with $d^{(2)}$, namely $v^{(3)} = -0.005382$ and $u^{(3)} = 0.37291$.

The results of the next four iterations obtained using line searches of the penalty function after each subproblem solution are shown in Table 10.1. As is evident from the table, the use of the line search is successful in forcing convergence to the optimum from poor initial estimates.

The use of the quadratic approximation to the Lagrangian was proposed by Wilson [1]. Although the idea was pursued by Beale [2] and by Bard and Greenstadt [3], it has not been widely adopted in its direct form. As with Newton's method, the barriers to the adoption of this approach in engineering applications have been twofold: first, the need to provide second derivative values for all model functions, and second, the sensitivity of the method to

**Table 10.1   Results for Example 10.5**

| Iteration | $x_1$ | $x_2$ | $f$ | $h$ | $v$ |
|-----------|-------|-------|-----|-----|-----|
| 3 | 0.43662 | 0.56338 | 0.24682 | 2.6055 | −0.005382 |
| 4 | 0.48569 | 0.56825 | 0.27599 | 2.5372 | −0.3584 |
| 5 | 1.07687 | 1.8666 | 2.0101 | 0.07108 | −0.8044 |
| 6 | 0.96637 | 1.8652 | 1.8025 | 0.10589 | −1.6435 |
| 7 | 0.99752 | 1.99503 | 1.9901 | 0.00498 | −1.9755 |
| ∞ | 1.0 | 2.0 | 2.0 | 0.0 | −2.0 |

poor initial solution estimates. Far from the optimal solution, the function second derivatives and especially the multiplier estimates will really have little relevance to defining a good search direction. (For instance, in Table 10.1, $v^{(3)} = -5.38 \times 10^{-3}$, while $v^* = -2$.) Thus, during the initial block of iterations, the considerable computational burden of evaluating all second derivatives may be entirely wasted. A further untidy feature of the above algorithm involves the strategies required to adjust the penalty parameter of the line-search penalty function. First, a good initial estimate of the parameter must somehow be supplied, and second, to guarantee convergence the penalty parameter must in principle be increased to large values. The manner in which this is done requires heuristics of the type considered in Chapter 6. Recent work discussed in the next section has largely overcome these various shortcomings of the quadratic approximation approach.

## 10.3  VARIABLE METRIC METHODS FOR CONSTRAINED OPTIMIZATION

In Chapter 3 we noted that in the unconstrained case the desirable improved convergence rate of Newton's method could be approached by using suitable update formulas to approximate the matrix of second derivatives. Thus, with the wisdom of hindsight, it is not surprising that, as first shown by Garcia-Palomares and Mangasarian [4], similar constructions can be applied to approximate the quadratic portion of our Lagrangian subproblems. The idea of approximating $\nabla^2 L$ using quasi-Newton update formulas that only require differences of gradients of the Lagrangian function was further developed by Han [5, 6] and Powell [7, 8]. The basic variable metric strategy proceeds as follows.

### Constrained Variable Metric Method

Given initial estimates $x^0$, $u^0$, $v^0$, and a symmetric positive definite matrix $H^0$.

**Step 1.**  Solve the problem

$$\text{Minimize} \quad \nabla f(x^{(t)})^T d + \tfrac{1}{2} d^T \mathbf{H}^{(t)} d$$

$$\text{Subject to} \quad h_k(x^{(t)}) + \nabla h_k(x^{(t)})^T d = 0 \qquad k = 1, \ldots, K$$

$$g_j(x^{(t)}) + \nabla g_j(x^{(t)})^T d \geq 0 \qquad j = 1, \ldots, J$$

**Step 2.**  Select the step size $\alpha$ along $d^{(t)}$, and set $x^{(t+1)} = x^{(t)} + \alpha d^{(t)}$.

**Step 3.**  Check for convergence.

**Step 4.** Update $\mathbf{H}^{(t)}$, using the gradient difference

$$\nabla_x L\left(x^{(t+1)}, u^{(t+1)}, v^{(t+1)}\right) - \nabla_x L\left(x^{(t)}, u^{(t+1)}, v^{(t+1)}\right)$$

in such a way that $\mathbf{H}^{(t+1)}$ remains positive definite.

The key choices in the above procedure involve the update formula for $\mathbf{H}^{(t)}$ and the manner of selecting $\alpha$. Han [4, 5] considered the use of several well-known update formulas, particularly DFP. He also showed [4] that if the initial point is sufficiently close, then convergence will be achieved at a superlinear rate without a step-size procedure or line search by setting $\alpha = 1$. However, to assure convergence from arbitrary points, a line search is required. Specifically, Han [5] recommends the use of the penalty function

$$P(x, R) = f(x) + R\left\{\sum_{k=1}^{K} |h_k(x)| - \sum_{j=1}^{J} \min\left[0, g_j(x)\right]\right\}$$

to select $\alpha^*$ so that

$$P(x(\alpha^*)) = \min_{0 \leqslant \alpha \leqslant \delta} P\left(x^{(t)} + \alpha d^{(t)}, R\right)$$

where $R$ and $\delta$ are suitably selected positive numbers.

Powell [7], on the other hand, suggests the use of the BFS formula together with a conservative check that ensures that $\mathbf{H}^{(t)}$ remains positive definite. Thus, if

$$z = x^{(t+1)} - x^{(t)}$$

and

$$y = \nabla_x L\left(x^{(t+1)}, u^{(t+1)}, v^{(t+1)}\right) - \nabla_x L\left(x^{(t)}, u^{(t+1)} v^{(t+1)}\right)$$

Then define

$$\theta = \begin{cases} 1 & \text{if } z^T y \geqslant 0.2 z^T \mathbf{H}^{(t)} z \\ \dfrac{0.8 z^T \mathbf{H}^{(t)} z}{z^T \mathbf{H}^{(t)} z - z^T y} & \text{otherwise} \end{cases} \tag{10.9}$$

and calculate

$$w = \theta y + (1 - \theta)\mathbf{H}^{(t)} z \tag{10.10}$$

Finally, this value of $w$ is used in the BFS updating formula,

$$\mathbf{H}^{(t+1)} = \mathbf{H}^{(t)} - \frac{\mathbf{H}^{(t)} z z^T \mathbf{H}^{(t)}}{z^T \mathbf{H}^{(t)} z^T} + \frac{w w^T}{z^T w} \tag{10.11}$$

Note that the numerical value 0.2 is selected empirically and that the normal BFS update is usually stated in terms of $y$ rather than $w$.

On the basis of empirical testing, Powell [8] proposed that the step-size procedure be carried out using the penalty function

$$P(x, \mu, \sigma) = f(x) + \sum_{k=1}^{K} \mu_k |h_k(x)| - \sum_{j=1}^{J} \sigma_j \min(0, g_j(x)) \quad (10.12)$$

where for the first iteration

$$\mu_k = |v_k| \qquad \sigma_j = |u_j|$$

and for all subsequent iterations $t$,

$$\mu_k^{(t)} = \max\{|v_k^{(t)}|, \tfrac{1}{2}(\mu_k^{(t-1)} + |v_k^{(t)}|)\} \quad (10.13)$$

$$\sigma_j^{(t)} = \max\{|u_j^{(t)}|, \tfrac{1}{2}(\sigma_j^{(t+1)} + |u_j^{(t)}|)\} \quad (10.14)$$

The line search could be carried out by selecting the largest value of $\alpha$, $0 \leqslant \alpha \leqslant 1$, such that

$$P(x(\alpha)) < P(x(0)) \quad (10.15)$$

However, Powell [8] prefers the use of quadratic interpolation to generate a sequence of values of $\alpha_k$ until the more conservative condition

$$P(x(\alpha_k)) \leqslant P(x(0)) + 0.1\alpha_k \frac{dP}{d\alpha}(x(0)) \quad (10.16)$$

is met. It is interesting to note, however, that examples have been found for which the use of Powell's heuristics can lead to failure to converge [9]. Further refinements of the step-size procedure have been reported [10], but these details are beyond the scope of the present treatment.

We illustrate the use of a variant of the constrained variable metric (CVM) method using update (10.11), penalty function (10.12), and a simple quadratic interpolation-based step-size procedure.

### Example 10.6

Solve Example 10.1 using the CVM method with initial metric $\mathbf{H}^0 = \mathbf{I}$.

At the initial point $(2, 1)$, the function gradients are

$$\nabla f = \left(\tfrac{23}{4}, -\tfrac{47}{4}\right)^T \qquad \nabla h = (1, 2)^T \qquad \nabla g = (1, 1)^T$$

Therefore, the first subproblem will take the form

$$\text{Minimize} \quad (\tfrac{23}{4}, -\tfrac{47}{4})d + \tfrac{1}{2}d^T \mathbf{I} d$$

$$\text{Subject to} \quad (1, 2)d = 0$$

$$(1, 1)d + 2 \geq 0$$

It is easy to show that the problem solution lies at the intersection of the two constraints. Thus, $d^0 = (-4, 2)^T$, and the multipliers at this point are solutions of the system

$$\left\{ \begin{pmatrix} \tfrac{23}{4} \\ -\tfrac{47}{4} \end{pmatrix} + \begin{pmatrix} -4 \\ 2 \end{pmatrix} \right\} = v \begin{pmatrix} 1 \\ 2 \end{pmatrix} + u \begin{pmatrix} 1 \\ 1 \end{pmatrix}$$

or $v^{(1)} = -\tfrac{46}{4}$, $u^{(1)} = \tfrac{53}{4}$. For the first iteration, we use the penalty parameters

$$\mu^{(1)} = |-\tfrac{46}{4}| \quad \text{and} \quad \sigma^{(1)} = |\tfrac{53}{4}|$$

The penalty function (10.12) thus takes the form

$$P = 6x_1 x_2^{-1} + x_2 x_1^{-2} + \tfrac{46}{4}|x_1 x_2 - 2| - \tfrac{53}{4}\min(0, x_1 + x_2 - 1)$$

We now conduct a one-parameter search of $P$ on the line $x = (2, 1)^T + \alpha(-4, 2)^T$. At $\alpha = 0$, $P(0) = 12.25$. Suppose we conduct a bracketing search with $\Delta = 0.1$. Then,

$$P(0 + 0.1) = 9.38875 \quad \text{and} \quad P(0.1 + 2(0.1)) = 13.78$$

Clearly, the minimum on the line has been bounded. Using quadratic interpolation on the three trial points of $\alpha = 0$, $0.1$, and $0.3$, we obtain $\bar{\alpha} = 0.1348$ with $P(\alpha) = 9.1702$. Since this is a reasonable improvement over $P(0)$, the search is terminated with this value of $\alpha$. The new point is

$$x^{(1)} = (2, 1)^T + (0.1348)(-4, 2) = (1.46051, 1.26974)$$

We now must proceed to update the matrix $\mathbf{H}$. Following Powell, we calculate

$$z = x^{(1)} - x^0 = (-0.53949, 0.26974)^T$$

Then

$$\nabla_x L(x^0, u^{(1)}, v^{(1)}) = \begin{pmatrix} \tfrac{23}{4} \\ -\tfrac{47}{4} \end{pmatrix} - (-\tfrac{46}{4})\begin{pmatrix} 1 \\ 2 \end{pmatrix} - (\tfrac{53}{4})\begin{pmatrix} 1 \\ 1 \end{pmatrix} = \begin{pmatrix} 4 \\ -2 \end{pmatrix}$$

$$\nabla_x L(x^{(1)}, u^{(1)}, v^{(1)}) = \begin{pmatrix} 3.91022 \\ -4.96650 \end{pmatrix} - (-\tfrac{46}{4})\begin{pmatrix} 1.26974 \\ 1.46051 \end{pmatrix} - \tfrac{53}{4}\begin{pmatrix} 1 \\ 1 \end{pmatrix}$$

$$= \begin{pmatrix} 5.26228 \\ 4.81563 \end{pmatrix}$$

Note that both gradients are calculated using the same multiplier values $u^{(1)}$, $v^{(1)}$. By definition,

$$y = \nabla_x L(x^{(1)}) - \nabla_x L(x^0) = (1.26228, 6.81563)^T$$

Next, we check condition (10.9):

$$z^T y = 1.15749 > 0.2 z^T I z = 0.2(0.3638)$$

Therefore, $\theta = 1$ and $w = y$. Using (10.11), the update $H^{(1)}$ is

$$H^{(1)} = I - \frac{z z^T}{\|z\|^2} + \frac{y y^T}{z^T y}$$

$$= \begin{pmatrix} 1 & 0 \\ 0 & 1 \end{pmatrix} - (0.3638)^{-1} \begin{pmatrix} -0.53949 \\ +0.26974 \end{pmatrix} (-0.53949, 0.26974)$$

$$+ (1.15749)^{-1} \begin{pmatrix} 1.26228 \\ 1.26228 \end{pmatrix} (1.26228, 1.26228)$$

$$= \begin{pmatrix} 1.57656 & 7.83267 \\ 7.83267 & 40.9324 \end{pmatrix}$$

Note that $H^{(1)}$ is positive definite.

This completes one iteration. We will carry out the second in abbreviated form only.

The subproblem at $x^{(1)}$ is

$$\text{Minimize} \quad (3.91022, -4.96650)d + \tfrac{1}{2} d^T \begin{pmatrix} 1.57656 & 7.83267 \\ 7.83267 & 40.9324 \end{pmatrix} d$$

$$\text{Subject to} \quad (1.26974, 1.46051)d - 0.14552 = 0$$

$$d_1 + d_2 + 1.73026 \geqslant 0$$

The solution of the quadratic program is

$$d^{(1)} = (-0.28911, 0.35098)^T$$

At this solution, the inequality is loose, and hence $u^{(2)} = 0$. The other multiplier value is $v^{(2)} = 4.8857$.

The penalty function multipliers are updated using (10.13) and (10.14):

$$\mu^{(2)} = \max\left(|4.8857|, \frac{46}{4} + \frac{4.8857}{2}\right) = 8.19284$$

$$\sigma^{(2)} = \max\left(|0|, \frac{\frac{53}{4} + 0}{2}\right) = 6.625$$

The penalty function now becomes

$$P(x(\alpha)) = f(x) + 8.19284|x_1 x_2 - 2| - 6.625 \min(0, x_1 + x_2 - 1)$$

where

$$x(\alpha) = x^{(1)} + \alpha d^{(1)} \qquad 0 \leqslant \alpha \leqslant 1$$

At $\alpha = 0$, $P(0) = 8.68896$, and the minimum occurs at $\alpha = 1$, $P(1) = 6.34906$. The new point is

$$x^{(2)} = (1.17141, 1.62073)$$

with

$$f(x^{(2)}) = 5.5177 \quad \text{and} \quad h(x^{(2)}) = -0.10147$$

The iterations continue with an update of $H^{(1)}$. The details will not be elaborated since they are repetitious. The results of the next four iterations are summarized below.

| Iteration | $x_1$ | $x_2$ | $f$ | $h$ | $v$ |
|-----------|-------|-------|-----|-----|-----|
| 3 | 1.14602 | 1.74619 | 5.2674 | 0.001271 | -0.13036 |
| 4 | 1.04158 | 1.90479 | 5.03668 | -0.01603 | -0.17090 |
| 5 | 0.99886 | 1.99828 | 5.00200 | -0.003994 | -0.45151 |
| 6 | 1.00007 | 1.99986 | 5.00000 | $-1.9 \times 10^{-6}$ | -0.50128 |

Recall that in Example 10.3, in which analytical second derivatives were used to formulate the QP subproblems, comparable solution accuracy was attained in four iterations. Thus, the quasi-Newton result obtained using only first derivatives is quite satisfactory, especially in view of the fact that the line searches were all carried out only approximately.

It should be reemphasized that the available convergence results (superlinear rate) [6, 11] assume that the penalty function parameters remain unchanged and that exact line searches are used. Powell's modifications (10.13) and (10.14) and the use of approximate searches thus amount to useful heuristics justified solely by numerical experimentation.

Finally, it is noteworthy that an alternative formulation of the QP subproblem has been reported by Biggs, as early as 1972 [12]. The primary differences of that approach lie in the use of an active constraint strategy to select the inequality constraints that are linearized and the fact that the quadratic approximation appearing in the subproblem is that of a penalty function. In view of the overall similarity of that approach to the Lagrangian-based construction, we offer no elaboration here, but instead invite the interested reader to study the recent exposition of this approach offered in reference 13 and the references cited therein. It is of interest that as reported by Bartholomew-Biggs [13], a code implementing this approach (OPRQ) has been quite successful in solving a number of practical problems, including one with as many as 79 variables and 190 constraints. This gives further support to the still sparse but growing body of empirical evidence suggesting the power of constrained variable metric approaches.

## 10.4   DISCUSSION

Having outlined the general form of the family of CVM algorithms, we continue in this section to discuss some practical implementation features as well as to review the relationship between GRG and CVM methods.

### 10.4.1   Problem Scaling

First of all, as noted by Powell [7] and underscored by the experience of Berna et al. [14], the variable metric approach should be and apparently is insensitive to scaling of the constraints. However, although the method should similarly be insensitive to variable scaling, the experience reported in reference 14 indicates that scaling is in fact important. Berna et al. recommend scaling all variables so that they take on values between 0.1 and 10.

### 10.4.2   Constraint Inconsistency

As can be seen from Example 10.6 and is true in general, this family of methods will generate a series of infeasible points. This can lead to various difficulties ranging from model evaluation failures, because the model functions may be invalid for some variable values, to the generation of linearized constraint sets that are inconsistent. The former difficulty is a common feature of all exterior point methods and can be remedied in part by carefully formulating the model to avoid the possibility of obvious sources of numerical problems, such as nonpositive arguments of logarithms or power functions, or division by very small numbers or zero. The situation of inconsistent linearized constraint sets can in principle occur even though the original problem has a nonempty feasible region, particularly for very infeasible initial solution estimates. The device recommended by Powell [7] is to insert a dummy variable $\xi$, constrained to lie in the range $0 \leqslant \xi \leqslant 1$, into the subproblem formulation. All linearized equality constraints are written

$$\nabla h_k(\bar{x})d + \xi h_k(\bar{x}) = 0$$

and the linearizations of all inequalities with negative values are written

$$\nabla g_j(\bar{x})d + \xi g_j(\bar{x}) \geq 0$$

The QP objective function is modified by adding the term $(1 - \xi)M$, where $M$ is a large positive constant. When $\xi = 1$, the original subproblem is obtained. If no feasible $d$ can be found, then a value of $\xi < 1$ will allow additional freedom to find a $d$ satisfying the modified subproblem constraints. The use of this device is strongly endorsed by Berna et al. [14] for dealing with large-scale applications.

### 10.4.3  Modification of $\mathbf{H}^{(t)}$

As noted by Chao et al. [15] in the context of structural optimization problems, two additional numerical difficulties may arise in the use of the CVM algorithm: excessively large direction-vector values $d$ and nearly singular values of $\mathbf{H}^{(t)}$. The former may occur when at the optimal solution of the original problem the total number of active constraints (inequalities plus equalities) is less than the number of variables. The latter situation can occur because even though the updating procedure does maintain $\mathbf{H}^{(t)}$ symmetric and positive definite, the matrix may become nearly positive semidefinite. The correction suggested in both instances is to add an identity matrix weighted by an adjustable nonnegative parameter $\pi$ to the approximation $\mathbf{H}^{(t)}$. No systematic adjustment procedure is reported other than a heuristic adjustment based on the resulting magnitude of $d$. The excellent results obtained using the resulting modified CVM algorithm [15] with several large structural optimization problems when compared to other NLP approaches suggest that such heuristic adjustment can be satisfactory. Nonetheless, this issue merits further investigation, since in principle such modification of the $\mathbf{H}^{(t)}$ matrix could cause deterioration of the desirable CVM algorithm convergence properties.

### 10.4.4  Comparison of GRG with CVM

Since both the accelerated GRG method and the CVM method make use of quasi-Newton updates to improve the search directions deduced from linear approximations, it is reasonable to question whether the two approaches might not in fact be largely equivalent. In the following discussion we show via qualitative arguments that there are substantive differences between the two approaches. For purposes of this discussion we consider only the equality-constrained case. The inclusion of inequality constraints is merely a matter of detail that the reader might consider as an interesting exercise.

Given an equality-constrained problem, both methods, in principle, seek a point that satisfies the Lagrangian necessary conditions

$$r(x, v) \equiv \nabla f(x) - v^T \nabla h(x) = 0 \tag{10.17}$$

$$h(x) = 0 \tag{10.18}$$

where for simplicity we denote by $h$ the vector of equality constraint functions. The CVM approach can be viewed as an attempt to solve the above set of nonlinear equations by using Newton's method [16]. Thus, at some initial estimate $(x^0, v^0)$, the application of Newton's method requires the solution of the linearized equations

$$r(x^0, v^0) + \nabla_x r(x^0, v^0)^T \Delta x + \nabla_v r(x^0, v^0)^T \Delta v = 0$$

$$h(x^0) + \nabla h(x^0)^T \Delta x = 0$$

If we substitute for $r$ its definition in terms of the gradient of the Lagrangian, the result is

$$\nabla f(x^0)^T - v^T \nabla h(x^0) + d^T \nabla_x^2 L(x^0, v^0) = 0 \qquad (10.19)$$

$$h(x^0) + \nabla h(x^0)^T d = 0 \qquad (10.20)$$

As shown in Section 10.3, these equations are simply the necessary conditions for a minimum of the problem

$$\text{Minimize} \quad \nabla f(x^0)^T d + \tfrac{1}{2} d^T \nabla_x^2 L(x^0, v^0) d$$

$$\text{Subject to} \quad h(x^0) + \nabla h(x^0)^T d = 0$$

Evidently, the CVM strategy can thus be viewed as a process of solving equations (10.19) and (10.20) by using a quasi-Newton method to approximate $\nabla_x^2 L$. Notice that in general the CVM method does not require $x^0$ to satisfy $h(x^0) = 0$. Since neither the condition $h(x) = 0$ nor the requirement $\min f(x)$ are directly enforced in solving these equations, it proves necessary to execute a line search along $d$ to minimize a penalty function. Note that in Eq. (10.19) the term

$$\nabla f(x^0)^T - v^T \nabla h(x^0)$$

includes the new value of the multiplier $v$. Thus, the need for updating the approximation $\mathbf{H}$ of $\nabla_x^2 L$ by using the difference

$$\left[ \nabla f(x^{(1)})^T - (v^{(1)})^T \nabla h(x^{(1)}) \right] - \left[ \nabla f(x^0)^T - (v^{(1)})^T \nabla h(x^0) \right]$$

is apparent.

On the other hand, the accelerated GRG method approaches the solution of Eqs. (10.17) and (10.18) by first partitioning the variables and function gradients into

$$x = (\hat{x}, \bar{x}) \qquad \nabla h = (\mathbf{J}, \mathbf{C}) \qquad \nabla f = (\nabla \hat{f}, \nabla \bar{f})$$

to obtain the system

$$\nabla \hat{f}(x)^T - v^T \mathbf{J}(x) = 0 \tag{10.21}$$

$$\nabla \tilde{f}(x)^T - v^T \mathbf{C}(x) = 0 \tag{10.22}$$

$$h(x) = 0$$

Then condition (10.21) is forced by defining

$$v^T = \nabla \hat{f}(x)^T \mathbf{J}^{-1}(x) \tag{10.23}$$

From the implicit function theorem of calculus, given a system of equations $h(x) = 0$, if $\mathbf{J}(x)$ is nonsingular in the neighborhood of a point $x^0$, then there exists a set of functions $q$ such that

$$\hat{x} = q(\bar{x})$$

Consequently, the dependence on $x$ of all functions in (10.22) and (10.23) can be reduced to a dependence on $\bar{x}$ only. Condition (10.22) can thus be restated to require that

$$\nabla \tilde{f}(\bar{x}) = 0 \tag{10.24}$$

Again the application of Newton's method at a point $x^0$ that is *feasible*, that is, where $h(x^0) = 0$, results in the system

$$\nabla \tilde{f}(\bar{x}^0)^T + \bar{d}^T \nabla^2 \tilde{f}(\bar{x}^0) = 0 \quad \text{or} \quad \bar{d}^T = -\nabla \tilde{f}(\bar{x}^0)^T (\nabla^2 \tilde{f}(\bar{x}^0))^{-1} \tag{10.25}$$

where $\nabla^2 \tilde{f}$ is the "reduced" second derivative. An exact expression for this quantity can be derived in terms of the first and second derivatives of the problem functions as well as $v^0$ as given by Eq (10.23) [17]. For present purposes, that expression is not really essential, since $\nabla^2 \tilde{f}$ is approximated by a quasi-Newton matrix $\mathbf{H}$ that is updated by employing the reduced gradient differences

$$\nabla \tilde{f}(x^{(1)}) - \nabla \tilde{f}(x^0)$$

Note that, implicitly, $\nabla \tilde{f}(x^{(1)})$ uses $v^{(1)}$ and $\nabla \tilde{f}(x^0)$ employs $v^0$, calculated at $x^{(1)}$ and $x^0$, respectively, using Eq. (10.23). In the GRG method the correction $\hat{d}$ is calculated so that constraint feasibility is maintained. Therefore, the line search on $\bar{d}$ can be executed using $f(x)$ directly.

The conceptual differences between the GRG and CVM strategies are now quite evident. First, the multipliers for GRG are estimated via (10.23) using only first derivative information, whereas $v$ in the CVM approach satisfies (10.19) and thus does reflect second-order information to some degree. Second, the size of the quasi-Newton approximation matrix $\mathbf{H}$ is quite different in the two cases: $N \times N$ for CVM and $(N - K) \times (N - K)$ in the case of GRG. Third, GRG executes its line search by using the objective function directly, while CVM employs a penalty function. Finally, GRG repeatedly solves the constraint equations to maintain feasibility, while CVM merely deals with the constraint linearizations. While the effects of the differences in the multiplier estimates are not obvious, it is clear that for large problems the differences in the size of the matrix $\mathbf{H}$ that must be stored and updated can be quite significant. It is interesting to note that Berna et al. [14] have shown that the implicit variable-elimination strategy used in GRG can be employed within the CVM constructions to result in a QP subproblem of dimension $N - K$. However, in their scheme the full $\mathbf{H}$ matrix is always updated and then reduced to size $(N - K) \times (N - K)$ by implicit variable-elimination calculations. While the storage and updating of $\mathbf{H}$ represents a significant computational burden, quite clearly the major difference between GRG and CVM lies in the iterations to maintain constraint feasibility.

On balance, the above considerations clearly suggest that if the NLP model to be solved is sufficiently stable to allow substantial excursions outside the feasible region, then the CVM approach is to be preferred. However, if the model cannot be evaluated at points that are infeasible, then the accelerated GRG method is best employed. An interesting example of this situation is given in the study reported by Biegler and Hughes [18], who employed the CVM approach to optimize the linearized response surface models generated from rigorous chemical process simulation models. They found that substantial constraint violations could not be tolerated, and hence the CVM approach had to be modified to include iterations to satisfy constraints. Evidently, in that work an accelerated GRG approach could well have been used, avoiding the need to update the complete $\mathbf{H}$ matrix. Further discussion of issues arising in the adaptation of the CVM method to larger scale problems is given by Gill et al. [19]. It is clear from the above discussion that both approaches have an important role in engineering optimization and constitute the two best approaches to solving constrained nonlinear optimization problems presently available.

## 10.5 SUMMARY

In this chapter we have studied the consequences of attempting to use higher order, specifically quadratic, approximations to nonlinear problem functions. We first examined full quadratic approximation and found that a problem with quadratic constraints was no easier to solve than one with general nonlinear constraints. Next we considered the formulation of quadratic programming subproblems: quadratic objective functions and linear constraints.

We found that such subproblems were convenient to solve but because they involve only the original constraints in linearized form the result is not a significant improvement over the successive LP approaches of Chapter 8. We then examined the use of a subproblem objective function whose quadratic term was the second derivative of the Lagrangian function. We found that if a solution of the subproblem satisfies the necessary conditions and the second-order sufficient conditions for a local minimum, then the corresponding point will also satisfy the sufficient conditions for the original problem. Moreover, the Lagrangian multipliers of a given subproblem will conveniently serve as multipliers for the construction of the next subproblem. Thus, the basis was established for an efficient algorithm for generating good search directions. However, we also observed that the method does not necessarily generate feasible directions, and hence to ensure convergence we found it expedient to conduct a line search along the generated direction using an exterior penalty function.

The remaining major difficulty was the need to provide second derivatives of all problem functions. This difficulty was resolved by using quasi-Newton methods, which only require differences of gradients of the Lagrangian function to approximate and update the second derivative term. The result was a constrained variable metric (CVM) algorithm that can be shown to possess a local superlinear convergence rate even without line searching. Of course, for global convergence, line searching using a penalty function proved necessary. Our discussion of successive QP methods proceeded with a brief summary of enhancements that have been suggested by various investigators to treat cases of subproblem constraint inconsistency and ill-conditioning or near-singularity of the quasi-Newton matrix. Finally, we compared the relative merits of the GRG algorithm with quasi-Newton acceleration and the CVM algorithm. We found that the key differences are that the CVM algorithm generates infeasible intermediate points while GRG maintains feasibility, and that GRG uses first-order multiplier estimates while CVM uses some level of higher order information. Of course, from a storage viewpoint, the quasi-Newton matrix of the GRG method is, in general, smaller than that of the CVM algorithm. On balance, we found that both methods have features that are unique and hence both are important to engineering optimization.

## REFERENCES

1. Wilson, R. B., "A Simplicial Algorithm for Concave Programming," Ph.D. dissertation, Harvard University, Cambridge, MA, 1963.
2. Beale, E. M. L., "Numerical Methods," in *Nonlinear Programming* (J. Abadie, Ed.), North-Holland, Amsterdam, 1967.
3. Bard, Y., and Greenstadt, J. L., "A Modified Newton Method for Optimization with Equality Constraints," in *Optimization* (R. Fletcher, Ed.), Academic Press, New York, 1969.

4.  Garcia-Palomares, U. M., and O. L. Mangasarian, "Superlinearly Convergent Quasi-Newton Algorithms for Nonlinearly Constrained Optimization Problem," *Math. Programming*, **11**, 1–13 (1976).

5.  Han, S. P., "Superlinearly Convergent Variable Metric Algorithms for General Nonlinear Programming Problems," *Math. Programming*, **11**, 263–282 (1976).

6.  Han, S. P., "A Globally Convergent Method for Nonlinear Programming," *J. Opt. Theory Appl.* **22**, 297–309 (1977).

7.  Powell, M. J. D., "A Fast Algorithm for Nonlinearly Constrained Optimization Calculations," in *Numerical Analysis, Dundee 1977* (G. A. Watson, Ed.), Lecture Notes in Mathematics No. 630, Springer-Verlag, New York, 1978.

8.  Powell, M. J. D., "Algorithms for Nonlinear Functions that Use Lagrangian Functions," *Math. Programming*, **14**, 224–248 (1978).

9.  Chamberlain, R. M., "Some Examples of Cycling in Variable Metric Methods for Constrained Minimization," *Math. Programming*, **16**, 378–383 (1979).

10. Mayne, D. Q., "On the use of Exact Penalty Functions to Determine Step Length in Optimization Algorithms," in *Numerical Analysis, Dundee, 1979* (G. A. Watson, Ed.), Lecture Notes in Mathematics, Vol. 773, Springer-Verlag, New York 1980.

11. Powell, M. J. D., "The convergence of Variable Metric Methods for Nonlinearly Constrained Optimization Calculations," in *Nonlinear Programming 3*, (O. L. Mangasarian, R. R. Meyer, and S. M. Robinson, Eds.), Academic Press, New York, 1978.

12. Biggs, M. C., "Minimization Using Recursive Equality Quadratic Programming," in *Numerical Methods for Nonlinear Optimization* (F. A. Lootsma, Ed.), Academic Press, New York, 1972.

13. Bartholomew-Biggs, M. C., "Recursive Quadratic Programming Based on Penalty Functions for Constrained Minimization," in *Nonlinear Optimization: Theory and Algorithms* (L. C. W. Dixon, E. Spedicato, and G. P. Szego, Eds.), Birkhauser, Boston, 1980.

14. Berna, T. J., M. H. Locke, and A. W. Westerberg, "A New Approach to Optimization of Chemical Processes," *AICHE J.*, **26**, 37–43 (1980).

15. Chao, N. H., S. J. Fenves, and A. W. Westerberg, "Application of a Reduced Quadratic Programming Technique to Structural Design," Proc. Intern. Symp. Optimum Structural Design, Tuscon, AZ, Oct. 1981, pp. 8.5–8.12.

16. Powell, M. J. D., "Constrained Optimization Using Variable Metric Methods," in *Nonlinear Optimization: Theory and Algorithms* (L. C. W. Dixon, E. Spedicato, and G. P. Szego Eds.), Birkhauser, Boston 1980.

17. Reklaitis, G. V., and D. J. Wilde, "A Computationally Compact Sufficient Condition for Constrained Minima," *Oper. Res.*, **17**, 425–436 (1969).

18. Biegler, L. T., and R. R. Hughes, "Approximation Programming of Chemical Process with Q/LAP," *Chem. Eng. Prog.*, **77**, 76–83 (1981).

19. Gill, P. E., W. Murray, M. A. Saunders, and M. H. Wright, "QP-Based Methods for Large Scale Nonlinearly Constrained Optimization," in *Nonlinear Programming, 4*, (O. L. Mangasarian, R. R. Meyer, and S. M. Robinson, Eds.), Academic Press, New York, 1981.

## PROBLEMS

**10.1.** What is the major disadvantage associated with a solution technique based upon direct use of full quadratic approximations to all functions in the nonlinear program?

**10.2.** Consider the general NLP problem of Eqs. (10.6)–(10.8). Show that if $d = 0$ satisfies the second-order sufficient conditions for a local minimum, then the approximating point satisfies the sufficient conditions for the original problem.

**10.3.** Outline an implementation of a successive Lagrangian QP algorithm that would employ the more conservative step-adjustment strategy of the Griffith and Stewart successive LP algorithm. Discuss the advantages and disadvantages relative to the penalty function strategy.

**10.4.** It is possible to modify the step-size adjustment mechanism of the CVM method so as to generate feasible points as in the GRG method. Explain how this might be accomplished.

**10.5.** Compare the treatment of inequality constraints in the GRG and CVM algorithms. How do the methods of estimating multiplier values differ?

**10.6.** Suppose the CVM algorithm were employed with a problem involving a quadratic objective function and quadratic inequality constraints. How many iterations are likely to be required to solve the problem, assuming exact arithmetic? What assumptions about the problem are likely to be necessary in making this estimate?

**10.7.** Construct a full quadratic approximation to the problem

$$\text{Minimize} \quad f(x) = x_1 + x_2^2 + x_3^3$$

$$\text{Subject to} \quad g_1(x) = 1 - 64x_1^{-2}x_2^{-4}x_3^{-6} \geqslant 0$$

$$\text{all } x_i \geqslant 0$$

at the point $(\frac{1}{2}, \frac{1}{2}, \frac{1}{2})$. Is the resulting problem a convex problem?

**10.8.** Consider the NLP

$$\text{Minimize} \quad f(x) = x_1^{-1} + x_2^{-1}$$

$$\text{Subject to} \quad h(x) = \frac{1}{2}x_1^2 + x_2^2 - 1 = 0$$

$$x_1, x_2 \geqslant 0$$

(a) Construct a full quadratic approximation to the problem at the point $x^0 = (\frac{3}{4}, \frac{3}{4})$.

(b) Solve the resulting subproblem for the correction vector $d$, and set $x^1 = x^0 + d$.

(c) Is the resulting point an improvement over $x^0$? How might it be further improved?

**10.9.** Suppose the NLP problem of problem 10.8 is approximated at the same $x^0$ by the Lagrangian-based QP subproblem [Eqs (10.3) and (10.4)].

(a) Solve the resulting subproblem for the correction vector $d$, and set $x^1 = x^0 + d$.

(b) Calculate the new estimates of the multipliers.

(c) Formulate the next QP subproblem.

**10.10.** Consider the problem

$$\text{Minimize} \quad f(x) = \tfrac{1}{3}(x_1 + 3)^3 + x_2^2$$

$$\text{Subject to} \quad h(x) = x_2^3 - x_1 + 1 = 0$$

$$x_1 \geqslant 1$$

(a) Given the point $x^0 = (2, 1)$, construct the initial subproblem for the CVM algorithm.

(b) Solve the subproblem to obtain $d$.

(c) Calculate the step length using Powell's procedure, and calculate the next point.

(d) Construct the next subproblem of the CVM method.

**10.11.** Given the problem

$$\text{Minimize} \quad f(x) = 3x_1^2 - 4x_2$$

$$\text{Subject to} \quad h(x) = 2x_1 + x_2 - 4 = 0$$

$$g(x) = 37 - x_1^2 - x_2^2 \geqslant 0$$

the point $x^0 = (-1, 6)$, and the multiplier values $(v, u) = (-\tfrac{40}{13}, \tfrac{1}{13})$.

(a) Formulate the Lagrangian QP subproblem.

(b) Show that $d = 0$ is the subproblem solution.

(c) Show that the point satisfies the second-order conditions for the original problem.

**10.12.** Suppose that in problem 10.11 the multiplier estimates are $(v, u) = (-3, 0.1)$ at $x^0 = (-1, 6)$.

(a) Formulate the Lagrangian QP subproblem.

(b) Obtain the subproblem solution. Is it still $d = 0$?.

(c) Calculate the new multiplier values, and check the sufficient conditions for a local minimum.

**10.13.** Consider problem 10.11 with the given point $x^0 = (-0.9, 5.8)$ and the multiplier values $(v, u) = (-3, 0.1)$.

(a) Formulate and solve the Lagrangian QP subproblem to obtain the new $x$ and new values of the multipliers.

(b) At the new $x$, calculate the multiplier values as they would be obtained in the GRG method.

(c) Compare the results of (a) and (b) to the optimal multiplier values $(v, u) = (-\frac{40}{13}, \frac{1}{13})$.

**10.14.** (a) Carry out three iterations using the Powell variant of the CVM algorithm for solution of the problem

$$\text{Minimize} \quad f(x) = \log(1 + x_1^2) - x_2$$

$$\text{Subject to} \quad h(x) = (1 + x_1^2)^2 + x_2^2 - 4 = 0$$

using the starting point $(1, 1)$.

(b) Verify that the solution is $x^* = (0, \sqrt{3})$.

**10.15.** The optimal design of a through circulation system for drying catalyst pellets involves the choice of fluid velocity $x_1$ and bed depth $x_2$ so as to maximize the production rate. The resulting problem takes the form [see Thygeson, J. R., and E. D. Grossmann, *AICHE J.*, **16**, 749 (1970)]

$$\text{Minimize} \quad f(x) = 0.0064x_1[\exp(-0.184x_1^{0.3}x_2) - 1]$$

$$\text{Subject to} \quad g_1(x) = 1.2 \times 10^{13} - (3000 + x_1)x_1^2 x_2 \geq 0$$

$$g_2(x) = 4.1 - \exp(0.184x_1^{0.3}x_2) \geq 0$$

where $x_1$ and $x_2$ both must be nonnegative. A reasonable starting point is $x^0 = (30{,}000, 0.25)$.

(a) For solution via the CVM method, are function or variable scaling likely to be useful? Suggest a rescaled form of the problem.

(b) At $x^0$, both constraints are loose. Check whether $\nabla^2 f$ is positive definite at $x^0$. What problems are likely to occur in solving the first subproblem? How are they avoided in the CVM algorithm?

(c) Solve the problem using a successive QP or CVM program. [The solution is $x^* = (31{,}766, 0.342)$.]

# Chapter 11

# Structured Problems
# and Algorithms

This chapter is devoted to the study of optimization problems with special
structures. So far, we have discussed a number of methods for solving general
nonlinear programming (NLP) problems. When the objective function, constraints, and variables have certain forms (e.g., quadratic objective function,
linear constraints, integer variables), special-purpose algorithms can be developed to take advantage of these special forms. Often these algorithms are more
efficient than the general-purpose algorithms and are capable of solving larger
problems.

In this chapter, we study the following structured problems and their
algorithms:

1. *Integer programming*   Specialized LP problems with the added restriction that some or all of the design variables must be integer valued.
2. *Quadratic programming*   Optimization problems with a quadratic objective function and linear constraints.
3. *Geometric programming*   Optimization problems wherein the constraints and the objective function can be expressed as the sum of
   generalized polynomial functions and the design variables are restricted
   to be positive.

## 11.1   INTEGER PROGRAMMING

An *integer linear programming problem*, henceforth called an *integer program*,
is an LP problem wherein some or all of the decision variables are restricted to
be integer valued. A *pure integer program* is one in which all the variables are
restricted to be integers. A *mixed integer program* restricts some of the
variables to be integers while others can assume continuous (fractional) values.

The reason for considering integer programs is that many practical problems require integer solutions. To solve such problems, we could simply solve

the linear program, ignoring the integer restrictions, and then either round off or truncate the fractional values of the LP optimal solution to get an integer solution. Of course, while doing this, we have to be careful that the resulting solution stays feasible. Such an approach is frequently used in practice, especially when the values of the variables are so large that rounding or truncating produces negligible change. But in dealing with problems in which the integer variables assume small values, rounding and truncating may produce a solution far from the true optimal integer solution. In addition, for large problems such a procedure can become computationally expensive. For instance, if the optimal LP solution is $x_1 = 2.4$ and $x_2 = 3.5$, then we have to try four combinations of integer values of $x_1$ and $x_2$ that are closest to their continuous values, namely, $(2, 3)$ $(2, 4)$, $(3, 3)$, and $(3, 4)$. The one that is feasible and closest to the LP optimal value of the objective function will be an approximate integer solution. With just 10 integer variables, we have to try $2^{10} = 1024$ combinations of integer solutions! Even after examining all such combinations, we cannot guarantee that one of them is an optimal integer solution to the problem.

### 11.1.1 Formulation of Integer Programming Models

The formulation of integer programs is illustrated by the following examples.

#### Example 11.1 Fixed Charge Problem

Consider a production planning problem with $N$ products such that the $j$th product requires a fixed production or setup cost $K_j$, independent of the amount produced, and a variable cost $C_j$ per unit, proportional to the quantity produced. Assume that every unit of product $j$ requires $a_{ij}$ units of resource $i$ and there are $M$ resources. Given that product $j$, whose sales potential is $d_j$, sells for \$$p_j$ per unit and no more than $b_i$ units of resource $i$ are available $(i = 1, 2, \ldots, M)$, the problem is to determine the optimal product mix that maximizes the net profit.

**Formulation.** The total cost of production (fixed plus variable) is a nonlinear function of the quantity produced. But, with the help of binary (0-1) integer variables, the problem can be formulated as an integer linear program.

Let the binary integer variable $\delta_j$ denote the decision to produce or not to produce product $j$. In other words,

$$\delta_j = \begin{cases} 1 & \text{if product } j \text{ is produced} \\ 0 & \text{otherwise} \end{cases}$$

Let $x_j (\geq 0)$ denote the quantity of product $j$ produced. Then the cost of producing $x_j$ units of product $j$ is $K_j \delta_j + C_j x_j$, where $\delta_j = 1$ if $x_j > 0$ and $= 0$

if $x_j = 0$. Hence, the objective function is

$$\text{Maximize} \quad Z = \sum_{j=1}^{N} p_j x_j - \sum_{j=1}^{N} \left( K_j \delta_j + c_j x_j \right)$$

The supply constraint for the $i$th resource is given by

$$\sum_{j=1}^{N} a_{ij} x_j \leqslant b_i \qquad \text{for } i = 1, 2, \ldots, M$$

The demand constraint for the $j$th product is given by

$$x_j \leqslant d_j \delta_j \qquad \text{for } j = 1, 2, \ldots, N$$

$$x_j \geqslant 0 \quad \text{and} \quad \delta_j = 0 \text{ or } 1 \qquad \text{for all } j$$

Note that $x_j$ can be positive only when $\delta_j = 1$, in which case its production is limited by $d_j$ and the fixed production cost $K_j$ is included in the objective function.

### Example 11.2   Handling Nonlinear 0-1 Integer Problems

Consider a nonlinear (binary) integer programming problem:

$$\text{Maximize} \quad Z = x_1^2 + x_2 x_3 - x_3^3$$

$$\text{Subject to} \quad -2x_1 + 3x_2 + x_3 \leqslant 3$$

$$x_1, x_2, x_3 \in \{0, 1\}$$

This problem can be converted to a linear integer programming problem for solution by using Watters' transformation [1]. Since for any positive $k$ and a binary variable $x_j$, $x_j^k = x_j$, the objective function immediately reduces to

$$Z = x_1 + x_2 x_3 - x_3$$

Now consider the product term $x_2 x_3$. For binary values of $x_2$ and $x_3$, the product $x_2 x_3$ is always 0 or 1. Hence, introduce a binary variable $y_1$ such that $y_1 = x_2 x_3$. When $x_2 = x_3 = 1$, we want the value of $y_1$ to be 1, while for all other combinations $y_1$ should be zero. This can be achieved by introducing the following two constraints:

$$x_2 + x_3 - y_1 \leqslant 1$$

$$-x_2 - x_3 + 2y_1 \leqslant 0$$

Note that when $x_2 = x_3 = 1$, the above constraints reduce to $y_1 \geqslant 1$ and $y_1 \leqslant 1$, implying $y_1 = 1$. When $x_2 = 0$ or $x_3 = 0$ or both are zero, the second constraint,

$$y_1 \leqslant \frac{x_2 + x_3}{2}$$

forces $y_1$ to be zero. Thus the equivalent linear (binary) integer program becomes

$$\text{Maximize} \quad Z = x_1 + y_1 - x_3$$

$$\text{Subject to} \quad -2x_1 + 3x_2 + x_3 \leqslant 3$$

$$x_2 + x_3 - y_1 \leqslant 1$$

$$-x_2 - x_3 + 2y_1 \leqslant 0$$

$$x_1, x_2, x_3, y_1 \text{ are } (0, 1) \text{ variables}$$

### Remarks

1. Glover and Woolsey [2] have suggested an improvement to Watters' transformation that introduces a continuous variable rather than an integer variable.
2. The procedure for handling the product of two binary variables can be easily extended to the product of any number of variables.

For additional examples of integer programming formulations, the reader is referred to the texts by Phillips, Ravindran, and Solberg [3], and Plane and McMillan [4]. Chapter 14 describes a case study on facility location using a mixed integer programming model.

### 11.1.2 Solution of Integer Programming Problems

**The Branch-and-Bound Algorithm.** The *branch-and-bound algorithm* is the most widely used method for solving both pure and mixed integer programming problems in practice. Most commercial computer codes for solving integer programs are based on this approach. Basically the branch-and-bound algorithm is just an efficient enumeration procedure for examining all possible integer feasible solutions.

We discussed earlier that a practical approach to solving an integer program is to ignore the integer restrictions initially and solve the problem as a linear program. If the LP optimal solution contains fractional values for some integer variables, then by the use of truncation and rounding-off procedures, one can attempt to get an approximate optimal integer solution. For instance, if there are two integer variables $x_1$ and $x_2$ with fractional values 3.5 and 4.4, then we could examine the four possible integer solutions $(3, 4)$, $(4, 4)$, $(4, 5)$, $(3, 5)$ obtained by truncation and rounding methods. We also observe that the true optimal integer solution may not correspond to any of these integer solutions, since it is possible for $x_1$ to have an optimal (integer) value less than 3 or greater than 5. Hence, to obtain the true optimal integer solution we would have to consider all possible integer values of $x_1$ smaller and larger than 3.5. In other words, the optimal integer solution must satisfy either

$$x_1 \leqslant 3$$

or

$$x_1 \geqslant 4$$

When a problem contains a large number of integer variables, it is essential to have a systematic method that will examine all possible combinations of integer solutions obtained from the LP optimal solution. The branch-and-bound algorithm essentially does this in an efficient manner.

### Basic Principles

To illustrate the basic principles of the branch-and-bound method, consider the following mixed integer program.

### Example 11.3

$$\text{Maximize} \quad Z = 3x_1 + 2x_2$$

$$\text{Subject to} \quad x_1 \leqslant 2$$

$$x_2 \leqslant 2$$

$$x_1 + x_2 \leqslant 3.5$$

$$x_1, x_2 \geqslant 0 \text{ and integral}$$

The initial step is to solve the mixed integer program (MIP) problem as a linear program by ignoring the integer restrictions on $x_1$ and $x_2$. Call this linear program LP-1. Since we only have two variables, a graphical solution of LP-1 is presented in Figure 11.1. The LP optimal solution is $x_1 = 2$, $x_2 = 1.5$, and the maximum value of the objective function $Z_o = 9$. Since $x_2$ takes a fractional value, we do not have an optimal solution for the MIP problem. But observe that the optimal integer solution cannot have an objective function value larger than 9, since the imposition of integer restrictions on $x_2$ can only make the LP solution worse. Thus we have an *upper bound* on the maximum value of $Z$ for the integer program given by the optimal value of LP-1.

The next step of the branch-and-bound method is to examine other integer values of $x_2$ that are larger or smaller than 1.5. This is done by adding a new constraint, either $x_2 \leqslant 1$ or $x_2 \geqslant 2$, to the original linear program (LP-1). This creates two new linear programs (LP-2 and LP-3) as follows:

<table>
<tr><td align="center">LP-2</td><td align="center">LP-3</td></tr>
</table>

| | |
|---|---|
| Maximize $Z = 3x_1 + 2x_2$ | Maximize $Z = 3x_1 + 2x_2$ |
| Subject to $x_1 \leqslant 2$ | Subject to $x_1 \leqslant 2$ |
| $x_2 \leqslant 2$ | $x_2 \leqslant 2$ |
| $x_1 + x_2 \leqslant 3.5$ | $x_1 + x_2 \leqslant 3.5$ |
| (new constraint) $x_2 \leqslant 1$ | (new constraint) $x_2 \geqslant 2$ |
| $x_1, x_2 \geqslant 0$ | $x_1, x_2 \geqslant 0$ |

The feasible regions corresponding to LP-2 and LP-3 are shown graphically in Figures 11.2 and 11.3, respectively. (Note that the feasible region for LP-3 is just the straight line $AB$.) Observe also that the feasible regions of LP-2 and LP-3 satisfy the following:

1. The optimal solution to LP-1 ($x_1 = 2$, $x_2 = 1.5$) is infeasible to both LP-2 and LP-3. Thus the old fractional optimal solution will not be repeated.

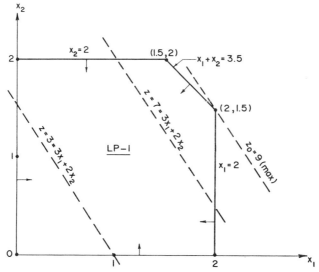

**Figure 11.1.** Solution to LP-1.

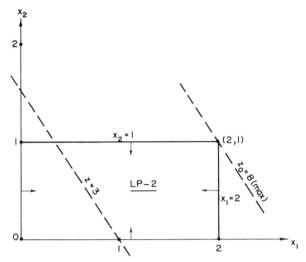

**Figure 11.2.** Solution to LP-2.

**2.** Every integer (feasible) solution to the original (MIP) problem is contained in either LP-2 or LP-3. Thus none of the feasible (integer) solutions to the MIP problem is lost due to the creation of two new linear programs.

The optimal solution to LP-2 (Figure 11.2) is $x_1 = 2$, $x_2 = 1$, and $Z_0 = 8$. Thus we have a feasible (integer) solution to the MIP problem. Even though LP-2

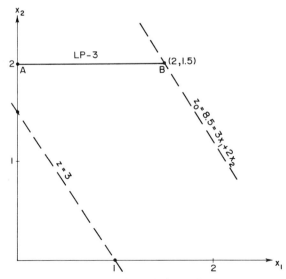

**Figure 11.3.** Solution to LP-3.

may contain other integer solutions, their objective function values cannot be larger than 8. Hence $Z_0 = 8$ is a *lower bound* on the maximum value of $Z$ for the mixed integer program. In other words, the optimal value of $Z$ for the MIP problem cannot be lower than 8. Since we had computed the upper bound earlier as 9, we cannot call the LP-2 solution the optimal integer solution without examining LP-3.

The optimal solution to LP-3 (Figure 11.3) is $x_1 = 1.5$, $x_2 = 2$, and $Z_0 = 8.5$. This is not feasible for the mixed integer program since $x_1$ is taking a fractional value. But the maximum $Z$ value (8.5) is larger than the lower bound (8). Hence it is necessary to examine whether there exists an integer solution in the feasible region of LP-3 whose value of $Z$ is larger than 8. To determine this we add the constraint either $x_1 \leqslant 1$ or $x_1 \geqslant 2$ to LP-3. This gives two new linear programs LP-4 and LP-5. The feasible region for LP-4 is the straight line $DE$ shown in Figure 11.4, while LP-5 becomes infeasible.

The optimal solution to LP-4 (Figure 11.4) is given by $x_1 = 1$, $x_2 = 2$, and $Z_0 = 7$. This implies that every integer solution in the feasible region of LP-3 cannot have an objective function value better than 7. Hence the integer solution obtained while solving LP-2, $x_1 = 2$, $x_2 = 1$, and $Z_0 = 8$, is the optimal integer solution to the MIP problem.

The sequence of LP problems solved under the branch-and-bound procedure for Example 11.3 may be represented in the form of a *network* or *tree* diagram as shown in Figure 11.5. A network or a tree consists of a set of *nodes*,

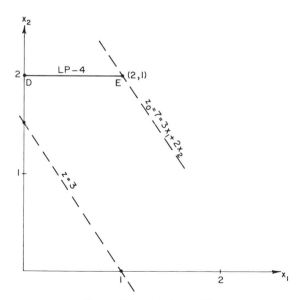

**Figure 11.4.** Solution to LP-4.

which are entry or exit points in the network, and a set of *arcs* or *branches*, which are the paths used to enter or exit a node. Node 1 represents the equivalent LP problem (LP-1) of the mixed integer program ignoring the integer restrictions. From node 1 we *branch* to node 2 (LP-2) with the help of the integer variable $x_2$ by adding the constraint $x_2 \leqslant 1$ to LP-1. Since we have an integer optimal solution for node 2, no further branching from node 2 is necessary. Once this type of decision can be made, we say that node 2 has been *fathomed*. Branching on $x_2 \geqslant 2$ from node 1 results in LP-3 (node 3). Since the optimal solution to LP-3 is fractional, we branch further from node 3 using the integer variable $x_1$. This results in the creation of nodes 4 and 5. Both have been fathomed, since LP-4 has an integer solution while LP-5 is infeasible. The best integer solution obtained at a fathomed node (in this case, node 2) becomes the optimal solution to the MIP problem.

### Details of the Algorithm

Consider an MIP problem of the following form:

$$\text{Maximize} \quad Z = cx$$

$$\text{Subject to} \quad Ax = b$$

$$x \geqslant 0$$

$$x_j \text{ is an integer for } j \in \mathbf{I}$$

where **I** is the set of all integer variables.

The first step is to solve the MIP problem as a linear program by ignoring the integer restrictions. Let us denote by LP-1 the linear program whose

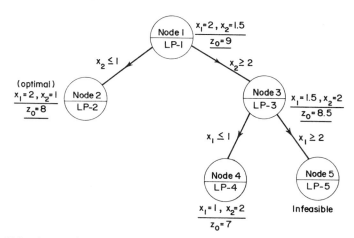

**Figure 11.5.** A network representation of the branch-and-bound method or Example 11.3

optimal value of the objective function is $Z_1$. Assume the optimal solution to LP-1 contains some integer variables at fractional values. Hence we do not have an optimal solution to the MIP problem. But $Z_1$ is an *upper bound* on the maximum value of $Z$ for the MIP problem.

The next step is to partition the feasible region of LP-1 by branching on one of the integer variables at a fractional value. A number of rules have been proposed to select the proper branching variable. They include:

1. Selecting the integer variable with the largest fractional value in the LP solution.

2. Assigning priorities to the integer variables such that we branch on the most important variable first. The importance of an integer variable may be based on one or more of the following criteria:

   (a) It represents an important decision in the model.

   (b) Its cost or profit coefficient in the objective function is very large compared to the others.

   (c) Its value is critical to the model based on the experience of the user.

3. Arbitrary selection rules; for instance, selecting the variable with the lowest index first.

Suppose that the integer variable $x_j$ is selected for further branching and its fractional value is $\beta_j$ in the LP-1 solution. Now we create two new LP problems LP-2 and LP-3 by introducing the constraints $x_j \leqslant \lfloor \beta_j \rfloor$ and $x_j \geqslant \lceil \beta_j \rceil$ respectively, where $\lfloor \beta_j \rfloor$ is the largest integer less than $\beta_j$, while $\lceil \beta_j \rceil$ is smallest integer greater than $\beta_j$. (See Figure 11.6.) In other words,

| LP-2 | | LP-3 | |
|---|---|---|---|
| Maximize | $Z = cx$ | Maximize | $Z = cx$ |
| Subject to | $Ax = b$ | Subject to | $Ax = b$ |
| | $x_j \leqslant \lfloor \beta_j \rfloor$ | | $x_j \geqslant \lceil \beta_j \rceil$ |
| | $x \geqslant 0$ | | $x \geqslant 0$ |

Assume that the optimal solutions to LP-2 and LP-3 are still fractional and hence are infeasible to the MIP problem with integer restrictions.

The next step is to select either LP-2 or LP-3 and branch from that by adding a new constraint. Here again a number of rules have been proposed for selecting the proper node (LP problem) to branch from. They include:

1. *Using the optimal value of the objective function.* Considering each of the nodes that can be selected for further branching, we choose the one whose LP optimal value is the largest (for a maximization problem). The rationale for this rule is that the LP feasible region with the largest $Z$ value may contain better integer solutions. For instance, any integer

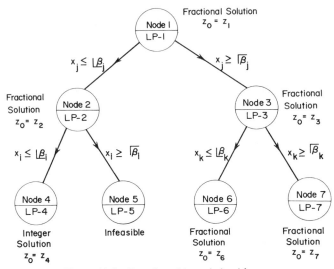

**Figure 11.6.** Branch-and-bound algorithm.

solution obtained by branching from LP-2 cannot have a $Z$ value better than the optimal value of $Z$ for LP-2.

2. *Last-in-first-out rule.* The LP problem that was solved most recently is selected (arbitrarily) for further branching.

Once the proper node (LP region) is selected for further branching, we branch out by choosing an integer variable with a fractional value. This process of branching and solving a sequence of linear programs is continued until an integer solution is obtained for one of the linear programs. The value of $Z$ for this integer solution becomes a *lower bound* on the maximum value of $Z$ for the MIP problem. At this point we can eliminate from consideration all those nodes (LP regions) whose values of $Z$ are not better than the lower bound. We say that these nodes have been *fathomed* because it is not possible to find a better integer solution from these LP regions than what we have now.

As an illustration, consider the tree diagram given in Figure 11.6. With the solution of LP-4 we have a lower bound on the maximum value of $Z$ for the MIP problem given by $Z_4$. In other words, the optimal solution to the MIP problem cannot have a $Z$ value smaller than $Z_4$. Further branching from node 4 is unnecessary, since any subsequent LP solution can only have a $Z$ value less than $Z_4$. In other words, node 4 has been fathomed. Node 5 has also been fathomed since the additional constraints render the LP problem infeasible. This only leaves nodes 6 and 7 for possible branching. Suppose $Z_6 < Z_4$ and $Z_7 > Z_4$. This means node 6 has also been fathomed (implicitly), since none of the integer solutions under node 4 can produce a better value than $Z_4$. However, it is possible for the LP region of node 7 to contain an integer solution better than that of node 4 since $Z_7 > Z_4$. Hence, we select node 7 for further branching and continue. In this manner, an intermediate node (LP problem) is explicitly or implicitly fathomed whenever it satisfies one of the following conditions:

1. The LP optimal solution of that node is integer valued; that is, it is feasible for the MIP problem.
2. The LP problem is infeasible.
3. The optimal value of $Z$ for the LP problem is not better than the current lower bound.

The branch-and-bound algorithm continues to select a node for further branching until all the nodes have been fathomed. The fathomed node with the largest value of $Z$ gives the optimal solution to the mixed integer program. Hence, the efficiency of the branch-and-bound algorithm depends on how soon the successive nodes are fathomed. The fathoming conditions (1) and (2) generally take considerable time to reach. Condition (3) cannot be used until a lower bound for the MIP problem is found. However, a lower bound is not available until a feasible (integer) solution to the MIP problem is obtained (condition 1). Hence it is always helpful if a feasible integer solution to the MIP problem can be found before the start of the branch-and-bound procedure. This will provide the initial lower bound to the MIP problem until a better lower bound is found by the branch-and-bound algorithm. In many practical problems, the present way of operating the system may provide an initial solution.

### 11.1.3 Guidelines on Problem Formulation and Solution

The solution time for solving the integer programming problem is very sensitive to the way the problem is formulated initially. From practical experience in solving a number of integer programs by the branch-and-bound method, the IBM research staff has come up with some suggestions on model formulations. We describe them briefly here. For more information, see the *IBM General Information Manual* [5]. The following guidelines should not be considered as restrictions but rather as suggestions that have often reduced computational time in practice.

1. Keep the number of integer variables as small as possible. One way to do this is to treat all integer variables whose values will be at least 20 as continuous variables.
2. Provide a good (tight) lower and upper bound on the integer variables when possible.
3. Unlike the general LP problem, the addition of new constraints to an MIP problem will generally reduce the computational time, especially when the new constraints contain integer variables.

4.  If there is no critical need to obtain an exact optimal integer solution, then considerable savings in computational time may be obtained by accepting the first integer solution that is 3% of the continuous optimum. In other words, for a maximization problem we can terminate the branch-and-bound procedure whenever

$$\frac{\text{Upper bound} - \text{lower bound}}{\text{Upper bound}} < 0.03$$

5.  The order in which the integer variables are chosen for branching affects the solution time. It is recommended that the integer variables be processed in a priority order based on their economic significance and user experience.

The branch-and-bound approach can also be used to solve nonlinear integer programming problems. Gupta and Ravindran [6] have recently implemented the branch-and-bound method with the generalized reduced gradient algorithm with some success.

**Solution of 0-1 Problems.**  For integer programming problems involving 0-1 integer variables, there exist special purpose algorithms for efficient solutions. These are generally based an implicit enumeration method. For a good discussion of these algorithms, the reader is referred to Plane and McMillan [4].

## 11.2  QUADRATIC PROGRAMMING

A quadratic programming problem is an optimization problem involving a quadratic objective function and linear constraints. It can be stated in a general form as

$$\text{Minimize} \quad f(x) = \sum_{j=1}^{n} c_j x_j + \sum_{i=1}^{n} \sum_{j=1}^{n} q_{ij} x_{ij}$$

$$\text{Subject to} \quad \sum_{j=1}^{n} a_{ij} x_j = b_i \quad \text{for } i = 1, \dots, m$$

$$x_j \geqslant 0$$

In vector-matrix notation, it may be written as

$$\text{Minimize} \quad f(x) = cx + x'Qx \tag{11.1}$$

$$\text{Subject to} \quad Ax = b \tag{11.2}$$

$$x \geqslant 0 \tag{11.3}$$

where $A$ is an $m \times n$ matrix of constraint coefficients, $b$ is an $m \times 1$ column vector, $c$ is a $1 \times n$ row vector, $Q$ is an $n \times n$ matrix of quadratic form and $x$ is an $n \times 1$ vector of design variables.

### 11.2.1  Applications of Quadratic Programming

We shall now illustrate some of the applications of quadratic programming models.

*Example 11.4  Portfolio Selection Problem*

An important problem faced by financial analysts in investment companies (banks, mutual funds, insurance companies) is the determination of an *optimal investment portfolio*. A *portfolio* specifies the amount invested in different securities which may include bonds, common stocks, bank CD's, treasury notes, and others. Because of the economic significance of the problem and its complexity, a number of mathematical models have been proposed for analyzing the portfolio selection problem.

Assume we have $N$ securities for possible investment and are interested in determining the portion of available capital $C$ that should be invested in each of the securities for the next investment period. Let the decision variables be denoted by $x_j$, $j = 1, 2, \ldots, N$, which represent the dollar amount of capital invested in security $j$. We then have the following system constraints on the decision variables:

$$x_1 + x_2 + \cdots + x_N \leqslant C$$

$$x_j \geqslant 0 \qquad j = 1, \ldots, N$$

Suppose we have historical data on each security for the past $T$ years, which give the price fluctuations and dividend payments. We can then estimate the return on investment from each security from past data. Let $r_j(t)$ denote the total return per dollar invested in security $j$ during year $t$. Then

$$r_j(t) = \frac{p_j(t + 1) - p_j(t) + d_j(t)}{p_j(t)}$$

where $p_j(t)$ is the price of security $j$ at the beginning of year $t$, and $d_j(t)$ is the total dividends received in year $t$.

Note that the values of $r_j(t)$ are not constants and can fluctuate widely from year to year. In addition, $r_j(t)$ may be positive, negative, or zero. Hence, in order to assess the investment potential of security $j$, we can compute the *average* or *expected return* from security $j$ per dollar invested, denoted by $\mu_j$ as

$$\mu_j = \frac{1}{T} \sum_{t=1}^{T} r_j(t)$$

The total expected return from the investment portfolio is then given by

$$E = \sum_{j=1}^{N} \mu_j x_j = \mu^T x$$

where $\mu^T = (\mu_1, \ldots, \mu_N)$ and $x = (x_1, \ldots, x_N)^T$.

**Model I.** A simple optimization model for the investor's problem is to maximize the total expected return subject to constraints on investment goals as follows:

$$\text{Maximize} \quad Z = \sum_{j=1}^{N} \mu_j x_j$$

$$\text{Subject to} \quad \sum_{j=1}^{N} x_j \leqslant C$$

$$x_j \geqslant 0$$

A number of policy constraints may also be imposed on the portfolio. Most investment companies limit the amount that can be invested in common stocks, whose returns are subject to wide variations. This can be expressed as

$$\sum_{j \in J_1} x_j \leqslant b_1$$

where $J_1$ represents the common stock securities and $b_1$ is the maximum investment (\$) in common stocks.

Many investment companies also require a certain amount in ready cash or fluid state to meet withdrawal requests from customers. This may be expressed as

$$\sum_{j \in J_2} x_j \geqslant b_2$$

where $J_2$ represents those securities in fluid state (e.g., savings accounts, checking accounts) and $b_2$ is the minimum cash reserve required. Similarly, one can introduce a number of such policy constraints on the investment portfolio.

A major drawback of this simple LP model is that it completely ignores the risk associated with the investment. Hence, it can lead to an investment portfolio that may have a very high average return but also has a very high risk associated with it. Because of the high risk, the actual return may be far less than the average value!

**Model II.**   In this model, we shall incorporate the "risk" associated with each security. Some securities, such as "speculative stocks," may have a larger price appreciation possibility and hence a greater average return, but they may also fluctuate more in value increasing its risk. On the other hand, "safe" investments, such as savings accounts and bank CD's, may have smaller returns on investments. We can measure the *investment risk* of a security by the amount of fluctuation in total return from its average value during the past $T$ years. Let $\sigma_{jj}^2$ denote the *investment risk* or *variance* of security $j$, which can be computed as

$$\sigma_{jj}^2 = \frac{1}{T} \sum_{t=1}^{T} \left[ r_j(t) - \mu_j \right]^2$$

In addition to the variance, a group of securities may be tied to a certain aspect of the economy, and a downturn in that economy would affect the prices of all those securities in that group. Automobile stocks, utility stocks, and oil company stocks are some examples of such security groups, which may rise or fall together. To avoid this kind of risk, we must diversify the investment portfolio so that funds are invested in several different security groups. This can be done by computing the relationships in rate of return between every pair of securities. Such a relationship, known as *covariance*, is denoted by $\sigma_{ij}^2$ and can be computed from past data.

$$\sigma_{ij}^2 = \frac{1}{T} \sum_{t=1}^{T} \left[ r_i(t) - \mu_i \right]\left[ r_j(t) - \mu_j \right]$$

Note that when $i = j$, the above equation reduces to the variance of security $j$. Thus, we can measure the variance or investment risk of a portfolio by

$$V = \sum_{i=1}^{N} \sum_{j=1}^{N} \sigma_{ij}^2 x_i x_j = x^T Q x$$

where $Q_{(N \times N)} = [\sigma_{ij}^2]$ is the variance-covariance matrix of the $N$ securities.

   A rational investor may be interested in obtaining a certain average return from the portfolio at a minimum risk. This can be formulated as

$$\text{Minimize} \quad Z = x^T Q x$$

$$\text{Subject to} \quad \sum_{j=1}^{N} x_j \leqslant C$$

$$x_j \geqslant 0$$

$$\mu^T x \geqslant R$$

where $R$ is the minimum average return desired from the portfolio. Additional policy constraints of the type discussed in model I could be added to the above model. In this *risk-averse* model, the constraints of the problem are linear but the objective is a quadratic function.

### *Example 11.5   Constrained Regression Problem*

The classical regression problem involves determining the coefficients of a regression equation that gives a relationship between a dependent variable and a set of independent variables. The basic least squares regression problem may be stated as

$$\text{Minimize} \quad e^T I e$$

$$\text{Subject to} \quad Y = X\beta + e$$

where $\beta$ is an $m \times 1$ vector of regression coefficients to be estimated, $e$ is an $n \times 1$ vector of error variables, $Y$ is an $n \times 1$ vector of observations on the dependent variable, and $X$ is an $n \times m$ matrix of observations on the independent variables. It is clear that the classical regression problem is a quadratic program.

In many practical problems a number of additional constraints may be imposed based on prior information about the problem. These may involve restrictions on the regression coefficients $\beta_j$ ($j = 1, \ldots, m$) such as $\beta_j \geqslant 0$ and $\Sigma \beta_j = 1$. One may also decide to weight the error terms $e_i$'s unevenly. In other words, the objective function may be written as $\Sigma_i w_i e_i^2$. Thus a QP approach to regression problems allows more flexibility in the model. For more applications of quadratic programming, the reader is referred to McCarl, Moskowitz, and Furtan [7].

### 11.2.2   Kuhn–Tucker Conditions

Comparing the quadratic program [Eqs. (11.1)–(11.3)] to the general nonlinear program given in Chapter 5 [Sect. 5.5, Eqs. (5.10)–(5.12)], we get

$$f(x) = cx + x^T Q x$$

$$g(x) = x \geqslant 0$$

$$h(x) = Ax - b = 0$$

Using Eqs. (5.13)–(5.17), the associated Kuhn-Tucker conditions to the quadratic program can be written as

$$c + x^T(Q + Q^T) - u - vA = 0 \qquad (11.4)$$

$$Ax = b \qquad x \geq 0 \qquad (11.5)$$

$$ux = 0 \qquad (11.6)$$

$$u \geq 0 \qquad v \text{ unrestricted in sign} \qquad (11.7)$$

We shall illustrate the Kuhn-Tucker conditions with an example.

### Example 11.6

$$\text{Minimize} \quad f(x) = -6x_1 + 2x_1^2 - 2x_1x_2 + 2x_2^2$$

$$\text{Subject to} \quad x_1 + x_2 = 2$$

$$x_1, x_2 \geq 0$$

For this problem,

$$c = (-6, 0) \qquad Q = \begin{pmatrix} 2 & -1 \\ -1 & 2 \end{pmatrix} \qquad A = (1, 1) \qquad b = (2)$$

The Kuhn-Tucker conditions are given by

$$(-6, 0) + (x_1, x_2)\begin{pmatrix} 4 & -2 \\ -2 & 4 \end{pmatrix} - (u_1, u_2) - v_1(1, 1) = 0$$

or

$$-6 + 4x_1 - 2x_2 - u_1 - v_1 = 0$$

$$0 - 2x_1 + 4x_2 - u_2 - v_1 = 0$$

$$x_1 + x_2 = 2$$

$$x_1, x_2 \geq 0$$

$$u_1 x_1 = 0$$

$$u_2 x_2 = 0$$

$$u_1, u_2 \geq 0 \qquad v_1 \text{ unrestricted in sign}$$

Since the constraints of the quadratic program are linear, the constraint qualification is always satisfied and the Kuhn-Tucker necessity theorem applies. (See Theorem 5.2 of Section 5.6.) Hence, the Kuhn-Tucker conditions are necessary for the optimality of the quadratic program.

Similarly, when the matrix of quadratic form $Q$ is positive definite or positive semidefinite, the objective function is a convex function and the Kuhn-Tucker conditions become sufficient for the optimality of the quadratic program. Hence, when $Q$ is positive definite or positive semidefinite, it is sufficient to solve the Kuhn-Tucker conditions to find an optimal solution to the quadratic program. Note that Eqs. (11.4) and (11.5) represent a system of simultaneous equations that can be solved as a linear program. Hence, one of the earliest procedures for solving a quadratic program developed by Wolfe [8] solves Eqs. (11.4), (11.5), and (11.7) as a linear program using the Phase I simplex method by creating artificial variables if necessary (see Chapter 4). Equation (11.6) is implicitly handled in the simplex algorithm using a "restricted basis entry rule" (see Chapter 8, Section 8.2). This would prohibit a nonbasic variable $u_j$ from entering the basis if the corresponding $x_j$ is in the basis at a positive value. Unfortunately, Wolfe's method fails to converge when the matrix $Q$ is positive semidefinite (instead of positive definite).

A more efficient and simple method of solving the Kuhn-Tucker conditions (11.4) through (11.7) is the *complementary pivot method* developed by Lemke [9]. In a recent experimental study, Ravindran and Lee [10] have shown that this method is computationally more attractive than most of the methods available for solving QP problems when $Q$ is positive semidefinite. The complementary pivot method will be discussed in the next section as a general procedure for solving a special class of problems known as *complementary problems*, which includes linear and quadratic programs.


## 11.3  COMPLEMENTARY PIVOT PROBLEMS

Consider the general problem of finding a nonnegative solution to a system of equations of the following form:

Find vectors $w$ and $z$ such that

$$w = Mz + q \tag{11.8}$$

$$w \geqslant 0 \quad z \geqslant 0 \tag{11.9}$$

$$w'z = 0 \tag{11.10}$$

where $M$ is an $n \times n$ square matrix and $w, z, q$ are $n$-dimensional column vectors.

The above problem is known as a *complementary problem*. Note that there is no objective function to minimize or maximize in this formulation. Condition

(11.8) represents a system of simultaneous linear equations; condition (11.9) requires that the solution to condition (11.8) be nonnegative; condition (11.10) implies $w_i z_i = 0$ for all $i = 1, 2, \ldots, n$, since $w_i$, $z_i \geq 0$. Thus we have a single nonlinear constraint.

### Example 11.7

Consider a problem with

$$M = \begin{pmatrix} 1 & 2 & 3 \\ 4 & 5 & 6 \\ 6 & 7 & 8 \end{pmatrix}$$

and

$$q = \begin{pmatrix} 2 \\ -5 \\ -3 \end{pmatrix}$$

The *complementary problem* is given by

Find $\qquad w_1, w_2, w_3, z_1, z_2, z_3$

Such that $\qquad w_1 = z_1 + 2z_2 + 3z_3 + 2$

$$w_2 = 4z_1 + 5z_2 + 6z_3 - 5$$

$$w_3 = 6z_1 + 7z_2 + 8z_3 - 3$$

$$w_1, w_2, w_3, z_1, z_2, z_3 \geq 0$$

$$w_1 z_1 + w_2 z_2 + w_3 z_3 = 0$$

Consider a convex QP problem of the form:

$$\text{Minimize} \quad f(x) = cx + x^T Q x$$

$$\text{Subject to} \quad Ax \geq b$$

$$x \geq 0$$

where $Q$ is an $n \times n$ matrix. Assume $Q$ is symmetric and is positive definite or positive semidefinite.

In matrix notation, the Kuhn-Tucker optimality conditions to the above convex quadratic program can be written as follows.

Find vectors $x$, $y$, $u$, $s$ such that

$$u = 2Qx - A^Ty + c^T$$

$$s = Ax - b$$

$$x, y, u, s \geqslant 0$$

$$u^Tx + s^Ty = 0$$

Comparing the above system of equations to the complementary problem, we note that

$$w = \begin{pmatrix} u \\ s \end{pmatrix} \quad z = \begin{pmatrix} x \\ y \end{pmatrix} \quad M = \begin{pmatrix} 2Q & -A^T \\ A & 0 \end{pmatrix} \quad q = \begin{pmatrix} c^T \\ -b \end{pmatrix}$$

Thus an optimal solution to the convex quadratic program may be obtained by solving the equivalent complementary problem shown above. It should be again noted that the matrix $M$ is positive semidefinite, since $Q$ is positive definite or positive semidefinite.

The reduction of a linear program as a complementary problem can be easily achieved by setting the matrix $Q$ to zero. Hence, complementary pivot algorithms can be used to solve linear programs as well.

### Example 11.8

Consider a convex QP problem:

$$\text{Minimize} \quad f(x) = -6x_1 + 2x_1^2 - 2x_1x_2 + 2x_2^2$$

$$\text{Subject to} \quad -x_1 - x_2 \geqslant -2$$

$$x_1, x_2 \geqslant 0$$

For this problem,

$$A = (-1, -1) \quad b = (-2) \quad c^T = \begin{pmatrix} -6 \\ 0 \end{pmatrix}$$

and

$$Q = \begin{bmatrix} 2 & -1 \\ -1 & 2 \end{bmatrix}$$

The equivalent complementary problem is given by

$$M_{(3 \times 3)} = \begin{pmatrix} Q + Q^T & -A^T \\ A & 0 \end{pmatrix} = \begin{pmatrix} 4 & -2 & 1 \\ -2 & 4 & 1 \\ -1 & -1 & 0 \end{pmatrix}$$

and

$$q = \left( \begin{array}{c} c^T \\ -b \end{array} \right) = \left( \begin{array}{c} -6 \\ 0 \\ 2 \end{array} \right)$$

The values of $z_1$ and $z_2$ correspond to the optimal values of $x_1$ and $x_2$, since the matrix of the quadratic form (Q) is positive definite.

### Complementary Pivot Method

Consider the complementary problem: Find vectors $w$ and $z$ such that

$$w = Mz + q$$

$$w, z \geqslant 0 \qquad w'z = 0$$

### *Definitions*

A nonnegative solution $(w, z)$ to the system of the equation $w = Mz + q$ is called a *feasible solution* to the complementary problem.

A feasible solution $(w, z)$ to the complementary problem that also satisfies the complementarity condition $w^T z = 0$ is called a *complementary solution*.

The condition $w^T z = 0$ is equivalent to $w_i z_i = 0$ for all $i$. The variables $w_i$ and $z_i$ for each $i$ is called a *complementary pair* of variables. Note that if the elements of the vector $q$ are nonnegative, then there exists an obvious complementary solution given by $w = q$, $z = 0$. Hence the complementary problem is nontrivial only when at least one of the elements of $q$ is negative. This means that the initial basic solution given by $w = q$, $z = 0$ is *infeasible* to the complementary problem even though it satisfies the complementary condition $w^T z = 0$.

The complementary pivot method (Lemke [9]) starts with the infeasible basic solution given by $w = q$; $z = 0$. In order to make the solution nonnegative, an artificial variable $z_0$ is added at a sufficiently positive value to each of the equations in the $w - Mz = q$ system, so that the right-hand-side constants $(q_i + z_0)$ become nonnegative. The value of $z_0$ will be the absolute value of the most negative $q_i$. We now have a basic solution given by

$$w_i = q_i + z_0 \qquad z_i = 0 \qquad \text{for all } i = 1,\ldots, n$$

and

$$z_0 = -\min(q_i)$$

Note that, as previously discussed, even though this solution is nonnegative, satisfies the constraints, and is complementary ($w_i z_i = 0$), it is not feasible, because of the presence of the artificial variable $z_0$ at a positive value. We shall call such a solution an *almost complementary solution.*

The first step in the complementary pivot method is to find an almost complementary solution by augmenting the original system of equations ($w = \mathbf{M}z + q$) by an artificial variable $z_0$ as follows:

$$w - \mathbf{M}z - ez_0 = q$$

$$w, z, z_0 \geqslant 0$$

$$w^T z = 0$$

where

$$e_{(n \times 1)} = (1, 1, \ldots, 1)'$$

Thus, the initial tableau becomes

| Basis | $w_1$ | $\cdots$ | $w_s$ | $\cdots$ | $w_n$ | $z_1$ | $\cdots$ | $z_s$ | $\cdots$ | $z_n$ | $z_0$ | $q$ |
|---|---|---|---|---|---|---|---|---|---|---|---|---|
| $w_1$ | 1 | | | | | $-m_{11}$ | | $-m_{1s}$ | | $-m_{1n}$ | $-1$ | $q_1$ |
| $w_s$ | | | 1 | | | $-m_{s1}$ | | $m_{ss}$ | | $-m_{sn}$ | $-1$ | $q_s$ |
| $w_n$ | | | | | 1 | $-m_{n1}$ | | $-m_{ns}$ | | $-m_{nn}$ | $-1$ | $q_n$ |

where the $m_{ij}$'s are the elements of the **M** matrix.

**Step 1.** To determine the initial almost complementary solution, the variable $z_0$ is brought into the basis, replacing the basis variable with the most negative value. (Let $q_s = \min q_i < 0$.) This implies that $z_0$ replaces $w_s$ from the basis. Performing the necessary pivot operation yields Tableau 1.

where

$$q'_s = -q_s \qquad q'_i = q_i - q_s \qquad \text{for all } i \neq s \tag{11.11}$$

$$m'_{sj} = \frac{-m_{sj}}{-1} = m_{sj} \qquad \text{for all } j = 1, \ldots, n \tag{11.12}$$

$$m'_{ij} = -m_{ij} + m_{sj} \qquad \text{for all } j = 1, \ldots, n \quad \text{and } i \neq s \tag{11.13}$$

**Tableau 1**

| Basis | $w_1$ | $\cdots$ | $w_s$ | $\cdots$ | $w_n$ | $z_1$ | $\cdots$ | $z_s$ | $\cdots$ | $z_n$ | $z_0$ | $q$ |
|---|---|---|---|---|---|---|---|---|---|---|---|---|
| $w_1$ | 1 | | $-1$ | | 0 | $m'_{11}$ | | $m'_{1s}$ | | $m'_{1n}$ | 0 | $q'_1$ |
| $z_0$ | 0 | | $-1$ | | 0 | $m'_{s1}$ | | $m'_{ss}$ | | $m'_{sn}$ | 1 | $q'_s$ |
| $w_n$ | 0 | | $-1$ | | 1 | $m'_{n1}$ | | $m'_{ns}$ | | $m'_{nn}$ | 0 | $q'_n$ |

*Note*:

1. $q'_i \geqslant 0; \quad i = 1, \ldots, n$
2. The basic solution $w_1 = q'_1, \ldots, w_{s-1} = q'_{s-1}, z_0 = q'_s, w_{s+1} = q'_{s+1}, \ldots, w_n = q'_n$, and all other variables initially zero is an almost complementary solution.
3. The almost complementary solution becomes a complementary solution as soon as the value of $z_0$ is reduced to zero.

In essence, the complementary pivot algorithm proceeds to find a sequence of almost complementary solutions (tableaus) until $z_0$ becomes zero. To do this, the basis changes must be done in such a way that the following conditions are met:

(a) The complementarity between the variables must be maintained (i.e., $w_i z_i = 0$ for all $i = 1, \ldots, n$).

(b) The basic solution remains nonnegative (i.e., the right-hand-side constants must be nonnegative in all tableaus).

**Step 2.** In order to satisfy condition (a), we observe that the variables $w_s$ and $z_s$ are both out of the basis in Tableau 1. As long as either one of them is made basic, the complementarity between $w$ and $z$ variables will still be maintained. Since $w_s$ just came out of the basis, the choice is naturally to bring $z_s$ into the basis. Thus we have a simple rule for selecting the nonbasic variable to enter the basis in the next tableau. It is always the *complement of the basic variable that just left the basis in the last tableau*. This is called the *complementary rule*.

After selecting the variable to enter the basis, we have to determine the basic variable to leave. This is done by applying the *minimum ratio test* similar to the one used in the **LP** simplex method so that condition (b) is satisfied. Therefore, to determine the variable to leave the basis, the following ratios are formed:

$$\frac{q'_i}{m'_{is}} \qquad \text{for those } i = 1, \ldots, n \text{ for which } m'_{is} > 0$$

Let

$$\frac{q'_k}{m'_{ks}} = \min_{m'_{is} > 0} \left( \frac{q'_i}{m'_{is}} \right) \qquad (11.14)$$

This implies that the basic variable $w_k$ leaves the basis to be replaced by $z_s$. We now obtain the new tableau by performing the pivot operation with $m'_{ks}$ as the pivot element.

**Step 3.** Since $w_k$ left the basis, the variable $z_k$ is brought into the basis by the complementary rule, and the basis changes are continued as before.

### Termination of the Algorithm

1.  The minimum ratio is obtained in row $s$, and $z_0$ leaves the basis. The resulting basic solution after performing the pivot operation is the complementary solution.
2.  The minimum ratio test fails, since all the coefficients in the pivot column are nonpositive. This implies that *no* solution to the complementary problem exists. In this case, we say that the complementary problem has a *ray solution*.

### Remarks

1.  It has been shown that the complementary pivot method always terminates with a complementary solution in a finite number of steps whenever (a) all the elements of **M** are positive or (b) **M** has positive principal determinants (includes the case where **M** is positive definite).
2.  The most important application of complementary pivot theory is in solving convex QP problems. We have seen that under these cases, **M** is a positive semidefinite matrix. It has been proved that whenever **M** is positive semidefinite, the algorithm will terminate with a complementary solution if one exists for that problem. In other words, termination 2 implies that the given linear program or quadratic program has no optimal solution.

For a proof of the above remarks, the reader is referred to Cottle and Dantzig [11].

Let us illustrate the complementary pivot method for solving a convex quadratic program using Example 11.8.

$$\text{Minimize} \quad f(x) = -6x_1 + 2x_1^2 - 2x_1x_2 + 2x_2^2$$

$$\text{Subject to} \quad -x_1 - x_2 \geqslant -2$$

$$x_1, x_2 \geqslant 0$$

The equivalent complementary problem is

$$\underset{(3 \times 3)}{\mathbf{M}} = \begin{pmatrix} 4 & -2 & 1 \\ -2 & 4 & 1 \\ -1 & -1 & 0 \end{pmatrix} \quad \text{and} \quad q = \begin{pmatrix} -6 \\ 0 \\ 2 \end{pmatrix}$$

Since all the elements of $q$ are not nonnegative, an artificial variable $z_0$ is added to every equation. The initial tableau is given as Tableau 1.

**Tableau 1**

| Basis | $w_1$ | $w_2$ | $w_3$ | $z_1$ | $z_2$ | $z_3$ | $z_0$ | $q$ |
|---|---|---|---|---|---|---|---|---|
| $w_1$ | 1 | 0 | 0 | $-4$ | 2 | $-1$ | $\boxed{-1}$ | $-6$ |
| $w_2$ | 0 | 1 | 0 | 2 | $-4$ | $-1$ | $-1$ | 0 |
| $w_3$ | 0 | 0 | 1 | 1 | 1 | 0 | $-1$ | 2 |

The initial basic solution is $w_1 = -6$, $w_2 = 0$, $w_3 = 2$, $z_1 = z_2 = z_3 = z_0 = 0$. An almost complementary solution is obtained by replacing $w_1$ by $z_0$ and using Eqs. (11.11)–(11.13) as shown in Tableau 2. The almost complementary solution is given by $z_0 = 6$, $w_2 = 6$, $w_3 = 8$, $z_1 = z_2 = z_3 = w_1 = 0$. Since the complementary pair $(w_1, z_1)$ is out of the basis, either $w_1$ or $z_1$ can be made a basic variable without affecting the complementarity between all pairs of variables ($w_i z_i = 0$). Since $w_1$ just left the basis, we bring $z_1$ into the basis. Applying the minimum ratio test [Eq. (11.14)], we obtain the ratios as $(\frac{6}{4}, \frac{6}{6}, \frac{8}{5})$. This implies that $z_1$ replaces $w_2$ in the basis. Tableau 3 gives the new almost complementary solution after the pivot operation. By applying the complementary rule, $z_2$ is selected as the next basic variable ($w_2$ just left the basis). The minimum ratio test determines $w_3$ as the basic variable to leave. The next almost complementary solution after the pivot operation is shown in Tableau 4.

By the complementary rule, $z_3$ becomes the next basic variable. Application of the minimum ratio test results in the replacement of $z_0$ from the basis. This implies that the next tableau will correspond to a complementary solution as shown in Tableau 5.

**Tableau 2**

| Basis | $w_1$ | $w_2$ | $w_3$ | $z_1$ | $z_2$ | $z_3$ | $z_0$ | $q$ |
|---|---|---|---|---|---|---|---|---|
| $z_0$ | $-1$ | 0 | 0 | 4 | $-2$ | 1 | 1 | 6 |
| $w_2$ | $-1$ | 1 | 0 | $\boxed{6}$ | $-6$ | 0 | 0 | 6 |
| $w_3$ | $-1$ | 0 | 1 | 5 | $-1$ | 1 | 0 | 8 |

**Tableau 3**

| Basis | $w_1$ | $w_2$ | $w_3$ | $z_1$ | $z_2$ | $z_3$ | $z_0$ | $q$ |
|---|---|---|---|---|---|---|---|---|
| $z_0$ | $-\frac{1}{3}$ | $-\frac{2}{3}$ | 0 | 0 | 2 | 1 | 1 | 2 |
| $z_1$ | $-\frac{1}{6}$ | $\frac{1}{6}$ | 0 | 1 | $-1$ | 0 | 0 | 1 |
| $w_3$ | $-\frac{1}{6}$ | $-\frac{5}{6}$ | 1 | 0 | $\boxed{4}$ | 1 | 0 | 3 |

**Tableau 4**

| Basis | $w_1$ | $w_2$ | $w_3$ | $z_1$ | $z_2$ | $z_3$ | $z_0$ | $q$ |
|---|---|---|---|---|---|---|---|---|
| $z_0$ | $-\frac{1}{4}$ | $-\frac{1}{4}$ | $-\frac{1}{2}$ | 0 | 0 | $\boxed{\tfrac{1}{2}}$ | 1 | $\frac{1}{2}$ |
| $z_1$ | $-\frac{5}{24}$ | $-\frac{1}{24}$ | $\frac{1}{4}$ | 1 | 0 | $\frac{1}{4}$ | 0 | $\frac{7}{4}$ |
| $z_2$ | $-\frac{1}{24}$ | $-\frac{5}{24}$ | $\frac{1}{4}$ | 0 | 1 | $\frac{1}{4}$ | 0 | $\frac{3}{4}$ |

**Tableau 5**

| Basis | $w_1$ | $w_2$ | $w_3$ | $z_1$ | $z_2$ | $z_3$ | $z_0$ | $q$ |
|---|---|---|---|---|---|---|---|---|
| $z_3$ | $-\frac{1}{2}$ | $-\frac{1}{2}$ | $-1$ | 0 | 0 | 1 | 2 | 1 |
| $z_1$ | $-\frac{1}{12}$ | $-\frac{1}{12}$ | $\frac{1}{2}$ | 1 | 0 | 0 | $-\frac{1}{2}$ | $\frac{3}{2}$ |
| $z_2$ | $-\frac{1}{12}$ | $-\frac{1}{12}$ | $\frac{1}{2}$ | 0 | 1 | 0 | $-\frac{1}{2}$ | $\frac{1}{2}$ |

The complementary solution is given by $z_1 = \frac{3}{2}$, $z_2 = \frac{1}{2}$, $z_3 = 1$, $w_1 = w_2 = w_3 = 0$. Hence the optimal solution to the given quadratic program becomes

$$x_1^* = \tfrac{3}{2} \qquad x_2^* = \tfrac{1}{2} \qquad f(x^*) = -\tfrac{11}{2}$$

A computer program implementing the complementary pivot method is given by Ravindran [12]. Experimental studies with this code by Ravindran [13, 14] and Ravindran and Lee [10] have shown the superiority of the complementary pivot method for solving linear programs and convex quadratic programs, respectively. In the QP study [10], the convex simplex method (see Sect. 9.2.1) was a close second to the complementary pivot method for solving the convex quadratic programs. For the nonconvex quadratic programs, the convex simplex method should be preferred, since it guarantees convergence to a local optimum. On the other hand, for nonconvex problems the complementary pivot method may terminate with a *ray solution* (see termination 2) and fail to identify even a local optimum.

## 11.4  GEOMETRIC PROGRAMMING

Geometric programming is concerned with constrained optimization problems involving a special class of polynomial functions as the objective and constraints of the problem. A number of engineering design problems frequently fall into the geometric programming framework. Since the initial work of Zener in 1961, geometric programming has undergone considerable theoretical development, has experienced a proliferation of proposals for numerical solu-

tion algorithms, and has enjoyed considerable practical engineering application. In this section we review the main results of geometric programming
theory and discuss the available procedures for solving geometric programs.
For a more detailed treatment of geometric programming and its applications,
the reader is referred to the texts by Beightler and Phillips [15] and Duffin,
Peterson, and Zener [16].

### 11.4.1  Geometric Programming With Posynomial Functions

#### Definition

A *posynomial* function $f(x)$ with $N$ variables and $T$ terms is defined as

$$f(x) = \sum_{t=1}^{T} c_t p_t(x) \tag{11.15}$$

where $c_t$ are positive scalars, $x = (x_1, \ldots, x_N)^T$ is the vector of design variables, and the functions $p_t(x)$ are defined as

$$p_t(x) = \prod_{n=1}^{N} x_n^{a_{nt}} \tag{11.16}$$

with $a_{nt}$ being any real numbers—positive, negative, or zero. A posynomial
differs from a polynomial because the exponents $a_{nt}$ do not have to be positive
integers but the coefficients $c_t$ are restricted to be positive.

The function $f(x)$ may be considered as the cost of equipment that is
designed and the terms $c_t p_t(x)$ as the costs of individual components that
make up the equipment. An example of a posynomial involving three terms
and two design variables is

$$f(x) = c_1 x_1^{-3} x_2^{-2} + c_2 x_1^3 x_2 + c_3 x_1^{-3} x_2^3$$

where

$$c_1, c_2, c_3 > 0$$

The posynomial function, given by Eqs. (11.15) and (11.16), is in general a
nonconvex function. However, an interesting property of the posynomial is
that with the transformation of variables,

$$x_n = \exp(z_n) \qquad \text{for } n = 1, 2, \ldots, N \tag{11.17}$$

it can be transformed to a convex function because $f(x)$ reduces to

$$f(z) = \sum_{t=1}^{T} c_t \exp\left( \sum_{n=1}^{N} a_{nt} z_n \right) \tag{11.18}$$

Note that $f(z)$ is a sum of convex functions and hence is convex. This feature has been used to great advantage in the optimization of posynomial functions recently.

The earlier approaches to the optimization of posynomials used the *arithmetic-geometric mean* inequality in algebra, which states that the arithmetic mean is greater than or equal to the geometric mean. The use of this inequality gave rise to the name *geometric programming*.

The arithmetic-geometric mean inequality states that for any positive numbers $v_1, v_2, \ldots, v_T$ and a set of positive weights $w_1, w_2, \ldots, w_T > 0$ and $\Sigma w_t = 1$,

$$\sum_{t=1}^{T} v_t w_t \geq \prod_{t=1}^{T} v_t^{w_t} \tag{11.19}$$

with equality holding only when $v_1 = v_2 = \cdots = v_T$. Setting $v_t w_t = u_t > 0$, Eq. (11.19) reduces to

$$\sum_{t=1}^{T} u_t \geq \prod_{t=1}^{T} \left( \frac{u_t}{w_t} \right)^{w_t} \tag{11.20}$$

for all $u_t, w_t > 0$, and $\Sigma w_t = 1$. Inequality (11.20) holds as an equation if and only if $w_t = u_t / \Sigma u_t$.

Let

$$u_t = c_t p_t(x) = c_t \prod_{n=1}^{N} x_n^{a_{nt}}$$

Then, the posynomial $f(x)$ given by Eq. (11.15) becomes

$$f(x) = \sum_t c_t p_t(x) = \sum_t u_t$$

Using inequality (11.20), we get

$$\sum_t u_t \geq \prod_{t=1}^{T} \left( \frac{c_t \prod_{n=1}^{N} x_n^{a_{nt}}}{w_t} \right)^{w_t}$$

$$= \left[ \prod_{t=1}^{T} \left( \frac{c_t}{w_t} \right)^{w_t} \right] \left[ \prod_{n=1}^{N} x_n^{\sum_{t=1}^{T} a_{nt} w_t} \right] \tag{11.21}$$

In inequality (11.20), the values of the weights $w_t$ are arbitrary as long as they are positive and sum up to 1. Suppose we choose their values such that

$$\sum_{t=1}^{T} a_{nt} w_t = 0 \qquad \text{for } n = 1, 2, \ldots, N$$

then inequality (11.21) reduces to

$$f(x) = \sum_t u_t \geq \prod_{t=1}^{T} \left( \frac{c_t}{w_t} \right)^{w_t} \qquad \text{for all } w_t > 0$$

$$\sum_t w_t = 1$$

$$\sum_t a_{nt} w_t = 0 \qquad \text{for } n = 1, 2, \ldots, N$$

Since the equality in (11.21) is possible, we get

$$\min_{x_n > 0} f(x) = \min \sum_t u_t = \max \prod_t \left( \frac{c_t}{w_t} \right)^{w_t} \qquad (11.22)$$

where $w_t$ must satisfy

$$\sum_t w_t = 1$$

$$\sum_t a_{nt} w_t = 0 \qquad n = 1, \ldots, N$$

$$w_t > 0$$

Thus the minimization of a posynomial function is changed to the maximization of a nonlinear function [Eq. (11.22)] subject to linear constraints. Note that since equality is possible in (11.20) if and only if $w_t = u_t / \sum u_t$, it is possible to develop a relationship between the primal and dual variables. Specifically, the optimal values of $x_n$ (denoted by $x_n^*$) and $w_t$ (denoted by $w_t^*$) must satisfy the relation

$$w_t^* = \frac{c_t p_t(x^*)}{f(x^*)} = \frac{c_t \prod_{n=1}^{N} (x_n^*)^{a_{nt}}}{f(x^*)} \qquad (11.23)$$

Based on constructions such as these, Duffin et al. [16] developed a formal duality theory of geometric programming and a dual approach to solving the primal geometric program.

**The Dual Problem.** Let the primal geometric program be stated as

$$\text{Minimize} \quad f(x) = \sum_{t=1}^{T} c_t \prod_{n=1}^{N} x_n^{a_{nt}}$$

$$\text{Subject to} \quad x_n > 0$$

The dual geometric program is defined as

$$\text{(Dual) Maximize} \quad d(w) = \prod_{t=1}^{T} \left(\frac{c_t}{w_t}\right)^{w_t} \tag{11.24}$$

$$\text{Subject to} \quad \sum_{t=1}^{T} w_t = 1 \tag{11.25}$$

$$\sum_{t=1}^{T} a_{nt} w_t = 0 \quad \text{for } n = 1,\ldots, N \tag{11.26}$$

$$w_t \geq 0$$

Using the arithmetic-geometric mean inequality and Eqs. (11.22) and (11.23), we can establish the following relationships between the primal and the dual problems:

1. A sufficient condition for the existence of the optimum of the primal geometric program is that the dual problem has an optimal solution $w_t^*$.
2. At the respective optima,
    (a) $\min f(x) = \max d(w) = y^*$
    (b) The primal and dual optimal solutions are related by the following set of log-linear equations:

$$\sum_{n=1}^{N} a_{nt} \ln[x_n^*] = \ln\left[\frac{y^* w_t^*}{c_t}\right] \quad \text{for } t = 1,\ldots, T \tag{11.27}$$

   (c) Any feasible solution to the dual provides a lower bound on the optimal value of the geometric program:

$$y^* = \min f(x) \geq \prod_{t=1}^{T} \left[\frac{c_t}{w_t}\right]^{w_t} \quad \text{for any feasible set of } w_t \tag{11.28}$$

Since the weights are between 0 and 1, maximizing $d(w)$ is equivalent to maximizing $\log d(w)$, and the logarithmic form of the dual function is given by:

$$\text{Maximize} \quad z(w) = \ln \prod_{t=1}^{T} \left(\frac{c_t}{w_t}\right)^{w_t}$$

$$= -\sum_{t=1}^{T} w_t \ln\left(\frac{w_t}{c_t}\right)$$

Note that the logarithmic dual function is separable and each term $w_t \ln(w_t/c_t)$ is a convex function. Hence the transformed dual is a concave programming problem with linear constraints.

Let us illustrate the dual method of solving a geometric program with an example.

***Example 11.9***    *(Phillips, Ravindran, and Solberg [3])*

$$\text{Minimize} \quad f(x) = 60x_1^{-3}x_2^{-2} + 50x_1^3x_2 + 20x_1^{-3}x_2^3$$

$$\text{Subject to} \quad x_1, x_2, x_3 > 0$$

The dual problem is given by

$$\text{Maximize} \quad d(w) = \left(\frac{60}{w_1}\right)^{w_1}\left(\frac{50}{w_2}\right)^{w_2}\left(\frac{20}{w_3}\right)^{w_3}$$

$$\text{Subject to} \quad w_1 + w_2 + w_3 = 1$$

$$-3w_1 + 3w_2 - 3w_3 = 0$$

$$-2w_1 + w_2 + 3w_3 = 0$$

$$w_1, w_2, w_3 \geqslant 0$$

With three equations and three unknowns, the dual constraints have a unique feasible solution, which consequently must be the minimum point. Solving the dual constraints, we thus obtain the dual optimal solution as

$$w_1^* = 0.4 \qquad w_2^* = 0.5 \qquad w_3^* = 0.1$$

The maximum value of the dual objective is $d(w^*) = 125.8 = f(x^*)$. Using Eq. (11.23), the optimal values of $x_n$ can be recovered as follows:

$$60x_1^{-3}x_2^{-2} = 0.4(125.8) = 50.32$$

$$50x_1^3x_2 = 0.5(125.8) = 62.9$$

$$20x_1^{-3}x_2^3 = 0.1(125.8) = 12.58$$

Solution to the above system of equations gives $x_1^* = 1.12$ and $x_2^* = 0.944$.

**Degrees of Difficulty.** In Example 11.9, the dual constraints had a unique solution, since the number of dual variables equaled the number of dual constraints. Equivalently, the number of terms $T$ in the primal geometric program was equal to the number of variables $N$ plus one. In situations where $T$ is greater than $N + 1$, the constraints of the dual will possess an infinite number of feasible solutions, and an optimization problem has to be solved to determine the optimal values of $w_t$.

The difference between $T$, the number of dual variables, and $N + 1$, the number of dual constraints, is called the *degree of difficulty*. The larger this number, the higher the effective dimensionality of the dual problem. Since the logarithmic form of the dual objective is a separable concave function with linear constraints, any of the efficient methods discussed in Chapters 8 and 9 (e.g., separable programming, convex simplex method, successive linear approximation algorithms) could be used to solve the dual problem.

**Treatment of Inequality Constraints.**   Geometric programming techniques can also be used to optimize problems with posynomial constraints. We shall briefly discuss the modifications necessary to handle constraints. A complete discussion and derivation of these results are available in Beightler and Phillips [15].

Consider a constrained posynomial geometric program as given below:

$$\text{(Primal) Minimize} \quad f_0(x) = \sum_{t=1}^{T_0} c_{0t} \prod_{n=1}^{N} x_n^{a_{0nt}} \tag{11.29}$$

$$\text{subject to} \quad f_m(x) = \sum_{t=1}^{T_m} c_{mt} \prod_{n=1}^{N} x_n^{a_{mnt}} \leqslant 1 \qquad \text{for } m = 1, 2, \ldots, M$$

$$x_n > 0 \qquad \text{for all } n \tag{11.30}$$

Note that the posynomial $f_0(x)$ containing $T_0$ terms represents the objective function, while the posynomials $f_m(x)$ for $m = 1, 2, \ldots, M$ containing $T_m$ terms represent $M$ inequality constraints. By the definition of posynomial, all the coefficients $c_{mt}$ for $m = 0, 1, \ldots, M$ and $t = 1, 2, \ldots, T_m$ are positive. Once again, the primal problem has $N$ design variables $x_1, x_2, \ldots, x_n$. As in the unconstrained case, a dual problem can be defined, and, as before, the primal problem can be solved by solving an equivalent linearly constrained dual problem.

Let $\delta_{0t}$, for $t = 1, 2, \ldots, T_0$ be the weights associated with the $t$th term in the objective function, where

$$\delta_{0t} = \frac{c_{0t} \prod_{n=1}^{N} x_n^{a_{0nt}}}{f_0(x)} \qquad \text{for } t = 1, 2, \ldots, T_0 \tag{11.31}$$

Define $\lambda_m$ for $m = 1, 2, \ldots, M$ to be the Lagrange multipliers associated with constraint $m$ and $\delta_{mt}$ for $m = 1, 2, \ldots, M$ and $t = 1, 2, \ldots, T_m$ the dual variables associated with the $t$th term in constraint $m$, where

$$\lambda_m = \sum_{t=1}^{T_m} \delta_{mt}$$

and

$$\frac{\delta_{mt}}{\lambda_m} = c_{mt} \prod_{n=1}^{N} x_n^{a_{mnt}} \qquad \text{for } m = 1, 2, \ldots, M; \, t = 1, 2, \ldots, T_m \qquad (11.32)$$

Using these variables, we can now formally state the dual geometric program for the constrained posynomial case as follows:

$$\text{(Dual) Maximize} \quad Z(\delta) = \prod_{m=1}^{M} \prod_{t=1}^{T_m} \left[ \frac{c_{mt}\lambda_m}{\delta_{mt}} \right]^{\delta_{mt}} \qquad (11.33)$$

$$\text{Subject to} \quad \sum_{t=1}^{T_0} \delta_{0t} = 1 \qquad (11.34)$$

$$\sum_{m=0}^{M} \sum_{t=1}^{T_m} a_{mnt}\delta_{mt} = 0 \quad \text{for } n = 1, 2, \ldots, N \qquad (11.35)$$

$$\lambda_m = \sum_{t=1}^{T_m} \delta_{mt} \qquad \text{for } m = 1, 2, \ldots, M \qquad (11.36)$$

$$\delta_{mt} \geqslant 0 \qquad \text{for } m = 0, 1, \ldots, M; \, t = 1, 2, \ldots, T_m$$

$$\lambda_m \geqslant 0 \qquad \text{for } m = 1, 2, \ldots, M$$

with the assumption that $\lambda_0 \equiv 1$ and the convention that

$$\lim_{\delta_{mt} \to 0} \left( \frac{c_{mt}\lambda_m}{\delta_{mt}} \right)^{\delta_{mt}} - 1$$

Note that the number of independent dual variables equals

$$T = \sum_{m=0}^{M} T_m$$

which is the total number of terms in the objective function and constraints. Also, the number of independent dual constraints equals $N + 1$ [since Eq. (11.36) can be eliminated by replacing $\lambda_m$ by $\Sigma\delta_{mt}$]. Hence, the dual problem has zero degrees of difficulty if $T = N + 1$. When $T > N + 1$, a true optimization problem exists for the dual.

The following results relating the optimal solutions of the primal [Eqs. (11.29) and (11.30)] and the dual [Eqs. (11.33)–(11.36)] have been proved by Duffin et al. [16]:

1. At their respective optimal, $x^*$ and $\delta^*$, the optimal values of the objective functions are equal; that is,

$$f_0(x^*) = Z(\delta^*)$$

2. If a primal constraint is inactive at the optimum, then all the dual variables associated with that constraint will be zero at the optimum. In other words, if $f_m(x^*) < 1$, then $\delta^*_{mt} \equiv 0$ for all $t = 1, 2, \ldots, T_m$. [*Note.* This result is obvious, since $f_m(x^*) < 1$ implies that the Lagrange multiplier $\lambda^*_m = 0$, and using Eq. (11.36) and the fact that $\delta^*_{mt} \geqslant 0$, it follows that $\delta_{mt} \equiv 0$ for all $t$.]

3. The primal and dual solutions ($x^*$ and $\delta^*$) are related by the following log-linear equations, which are defined for those $t$ with $\delta^*_{mt} > 0$:

$$\sum_{n=1}^{N} a_{0nt} \ln(x_n^*) = \ln\left[\frac{\delta^*_{0t} f_0(x^*)}{c_{0t}}\right]$$

$$\text{for } t = 1, 2, \ldots, T_0; \ \delta^*_{0t} > 0$$

$$\sum_{n=1}^{N} a_{mnt} \ln(x_n^*) = \ln\left[\frac{\delta^*_{mt}}{\lambda^*_m c_{mt}}\right]$$

$$\text{for } t = 1, 2, \ldots, T_m; \ m = 1, 2, \ldots, M; \ \delta^*_{mt} > 0$$

*Note:* The above equations are obtained by taking the natural logarithms of Eqs. (11.31) and (11.32).

4. The logarithm of $Z(\delta)$ is a concave function that is continuously differentiable over the positive orthant. Hence, the dual problem with the logarithmic objective function is a linearly constrained concave programming problem.

### *Example 11.10  Metal Cutting*

One of the engineering applications of mathematical programming is the problem of determining the optimal machining parameters in metal cutting. A detailed discussion of different optimization models in metal cutting with illustrations is given in a survey paper by Philipson and Ravindran [17]. Here we shall elaborate on the use of geometric programming for a machining problem in which a single cutting tool turns a diameter in one pass.

The decision variables in this machining problem are the cutting speed $v$ and the feed per revolution $f$. Increasing the speed and feed reduces the actual machining time, and hence the machining cost, but it has an adverse effect on the life of the cutting tool and results in a higher tooling cost. In addition, the optimal values of $v$ and $f$ will also depend on labor and overhead costs of nonproductive time and tool-changing time. Hence, minimization of the total cost per component is the criterion most often used in selecting the optimal machining parameters, feed and speed.

The cost per component $c$, for a part produced in one pass, is given by Armarego and Brown [18]:

$c$ = (cost of nonproductive time/component) + (machining time cost)
  + (cost of tool-changing time/component) + (tool cost/component)

Material costs are not considered. The third and fourth terms may be stated more specifically as:

Cost of tool-changing time per component

$$= \frac{\text{cost rate} \times \text{tool-changing time}}{\text{number of parts produced between tool changes}}$$

and

tool cost per component $= \dfrac{\text{tool cost per cutting edge}}{\text{number of parts produced between tool changes}}$

The cost equation can be expressed mathematically as:

$$c = xT_L + xT_c + xT_d\left(\frac{T_{ac}}{T}\right) + y\left(\frac{T_{ac}}{T}\right)(\text{dollars}) \qquad (11.37)$$

where  $x$ = labor plus overhead cost rate, \$
  $T_L$ = nonproductive time (loading, unloading, and inspection time), min
  $T_c$ = machining time, including approach time, min
  $T_{ac}$ = actual cutting time (approximately equal to $T_c$), min
  $T$ = tool life, min [given by Eq. (11.39)]
  $T_d$ = tool-changing time, min
  $y$ = tool cost per cutting edge, \$
  $T/T_{ac}$ = number of parts produced between tool changes

The equation for machining time $T_c$ is

$$T_c = \frac{l}{\lambda v f} \quad (\text{min}) \qquad (11.38)$$

where  $l$ = distance traveled by the tool in making a turning pass, in.
  $\lambda = 12/\pi D$, where $D$ is the mean workpiece diameter, in.
  $v$ = cutting speed, surface feet/min
  $f$ = feed, in./rev

It has been found [18] that tool life, cutting speed, and feed are related as follows:

$$T = \frac{A}{v^{1/n} f^{1/n_1}} \quad (\min) \tag{11.39}$$

where $A$, $n$, and $n_1$ are constants.

Assuming that $T_{ac} \simeq T_c$, and inserting Eqs. (11.38) and (11.39) into Eq. (11.37), we obtain

$$c = xT_L + \frac{xl}{\lambda v f} + \left( xT_d \frac{l}{\lambda A} + \frac{yl}{\lambda A} \right) v^{(1/n)-1} f^{(1/n_1)-1} (\$) \tag{11.40}$$

The constraints imposed on $v$ and $f$ by the machine tool and by process and part requirements are:

(i) *Maximum and minimum available cutting speed*

$$v_{\min} \leqslant v \leqslant v_{\max}$$

(ii) *Maximum and minimum available feed*

$$f_{\min} \leqslant f \leqslant f_{\max}$$

(iii) *Maximum allowable cutting force* ($F_{t,\max}$). This constraint is necessary to limit tool-work deflections and their effect upon the accuracy of the turned part. Armarego and Brown [18] give the following expression for the tangential cutting force $F_t$:

$$F_t = c_t f^\alpha d_c^\gamma \tag{11.41}$$

where $c_t$, $\alpha$, and $\gamma$ are constants, and $d_c$ is the depth of cut, which is held constant at a given value. This constraint on the cutting force results in the following feed constraint:

$$f \leqslant \left[ \frac{F_{t\max}}{c_t d_c^\gamma} \right]^{1/\alpha}$$

(iv) *Maximum available horsepower*. The horsepower consumed in cutting can be determined from the equation

$$HP = \frac{F_t v}{33,000}$$

where $F_t$ is obtained from Eq. (11.41). If $P_{\max}$ is the maximum horsepower available at the spindle, then

$$v f^\alpha \leqslant \frac{P_{\max}(33,000)}{c_t d_c^\gamma}$$

For a given $P_{max}$, $c_t$, $\gamma$, and $d_c$, the right-hand side of the above inequality will be a constant.

(v) *Stable cutting region.* Certain combinations of $v$ and $f$ values are likely to cause chatter vibration, adhesion, and built-up edge formation. To avoid this problem, the following stable cutting region constraint is used:

$$v^\delta f \geqslant \beta$$

where $\delta$ and $\beta$ are given constants.

As an illustration [17], consider the case where a single diameter is to be turned in one pass using the feed rate and cutting speed that will minimize costs. The bar is 2.75 in. in diameter by 12.00 in. long. The turned bar is 2.25 in. in diameter by 10.00 in. long. In the cutting speed calculations, a mean diameter of 2.50 in. will be used. The lathe has a 15-hp motor and a maximum speed capability of 1500 rpm. The minimum speed available is 75 rpm. The cost rate, tool costs, idle time, tool-changing time, and tool life parameters are given below:

| | |
|---|---|
| $x = \$0.15/\text{min}$ | $T_L = 2.00 \text{ min}$ |
| $l = 10.00 \text{ in.}$ | $D = 2.50 \text{ in.}$ |
| $\lambda = 1.528$ | $T_d = 1.00 \text{ min}$ |
| $y = \$0.50$ | $A = 113{,}420$ |
| $n = 0.30$ | $n_1 = 0.45$ |
| $d_c = 0.25 \text{ in.}$ | |
| $N_{max} = 1500 \text{ rpm}$ | $N_{min} = 75 \text{ rpm}$ |
| $F_{t,max} = 1583.0 \text{ lb.}$ | $c_t = 344.7$ |
| $\alpha = 0.78$ | $\gamma = 0.9$ |
| $\delta = 2.0$ | |
| $\beta = 380000$ | machine drive efficiency $= 0.8$ |

machine hp $= 15.0$

When the fixed values are inserted into Eq. (11.37), the cost function becomes

$$\text{Minimize} \quad c = 0.30 + \frac{982.0}{vf} + 8.1 \times 10^{-9} v^{2.333} f^{1.222}$$

[*Note*: For ease of calculations, $f$ is expressed in thousandths of an inch per revolution rather than in inches per revolution.]

The constraints on $v$ and $f$ are given by

$$v \leqslant 982.0$$

$$v \geqslant 49.1$$

$$f \leqslant 35.0 \qquad \text{(cutting force)}$$

$$f \geqslant 1.0$$

$$vf^{0.78} \leqslant 4000.0 \qquad \text{(horsepower)}$$

$$v^{2.0}f \geqslant 380,000.0 \qquad \text{(stable cutting region)}$$

$$v, f \geqslant 0$$

To solve the above problem by geometric programming, we would write it in the following posynomial form:

$$\text{Minimize} \qquad f_0 = \frac{982.0}{vf} + 8.1 \times 10^{-9} v^{2.333} f^{1.222}$$

$$\text{Subject to} \qquad \frac{v}{982.0} \leqslant 1$$

$$\frac{49.1}{v} \leqslant 1$$

$$\frac{f}{35.0} \leqslant 1$$

$$\frac{1.0}{f} \leqslant 1$$

$$\frac{vf^{0.78}}{4000.0} \leqslant 1$$

$$\frac{380,000}{v^{2.0}f} \leqslant 1$$

$$(v, f) > 0$$

Note that the problem has eight terms and two variables. Hence, it will have five degrees of difficulty. Also, each constraint has only one term. Hence, Eq. (11.36) reduces to $\lambda_m = \delta_{m1}$ for $m = 1, 2, \ldots, 6$. The dual of the above posynomial problem is given by

$$\text{Maximize} \quad Z(\delta) = \left[\frac{982}{\delta_{01}}\right]^{\delta_{01}} \left[\frac{8.1 \times 10^{-9}}{\delta_{02}}\right]^{\delta_{02}} \left[\frac{1}{982}\right]^{\delta_{11}}$$

$$[49.1]^{\delta_{21}} \left[\frac{1}{35.0}\right]^{\delta_{31}} [1]^{\delta_{41}} \left[\frac{1}{4000}\right]^{\delta_{51}} [380,000]^{\delta_{61}}$$

$$\text{Subject to} \quad \delta_{01} + \delta_{02} = 1$$

$$-\delta_{01} + 2.333\delta_{02} + \delta_{11} - \delta_{21} + \delta_{51} \qquad - 2\delta_{61} = 0$$

$$-\delta_{01} + 1.222\delta_{02} + \delta_{31} - \delta_{41} + 0.788\delta_{51} - \delta_{61} = 0$$

$$\delta_{mt} \geqslant 0$$

Using the logarithmic form of the dual objective function, the problem is a concave programming problem with linear constraints. Hence, any of the methods discussed in Chapters 8 and 9 may be used to determine the dual optimal solution, which is given by

$$\delta_{01}^* = 0.7 \qquad \delta_{02}^* = 0.3 \qquad \delta_{31}^* = 0.3334$$

$$\delta_{11}^* = \delta_{21}^* = \delta_{41}^* = \delta_{51}^* = \delta_{61}^* = 0$$

and

$$\max z(\delta) = z^* = 0.2613$$

Hence, $f_0^* = z^* = 0.2613$ and $c^* = 0.3 + 0.2613 = 0.5613$. To determine the optimal values of the primal variables $v^*$ and $f^*$, we use Eqs. (11.31) and (11.32):

$$\frac{982}{v^*f^*} = (0.7)(0.2613) = 0.18291$$

$$8.1 \times 10^{-9}(v^*)^{2.333}(f^*)^{1.222} = (0.3)(.2613) = .07839$$

$$\frac{f^*}{35.0} = \frac{\delta_{31}^*}{\lambda_1^*} = \frac{0.3334}{0.3334} = 1$$

Solving the above equations, we get

$$f^* = 35.0 \quad \text{and} \quad v^* = 153.4$$

The reader is referred to Phillips and Beightler [19] for other geometric programming models in tool engineering.

**Computational Problems with the Dual Approach.** Even though the dual problem can sometimes be easier to solve, its direct maximization does present some potential computational problems.

1. The gradient of $\ln[z(\delta)]$ is not defined when any dual variable $\delta_{mt} = 0$.
2. If $\delta_{mt}^* = 0$ for some $t$, $1 \leqslant t \leqslant T_m$ and $m$, $1 \leqslant m \leqslant M$, then all the dual variables $\delta_{mt}^*$ associated with constraint $m$ must equal zero.
3. It is possible that the system of log-linear equations that must be solved to determine the optimal primal variables may lead to inaccurate solutions (if $\delta^*$ is not determined very accurately) or that its rank may be less than $N$.

Dembo [20, 21] has pointed out ways to overcome some of these computational problems. The problem of nondifferentiability when $\delta_{mt} = 0$ is usually handled by setting arbitrarily small lower bounds ($\delta_{mt} \geqslant \varepsilon$).

**The Convexified Primal Problem.** The posynomial primal geometric program given by Eqs. (11.29) and (11.30) is in general a nonconvex programming problem that, because of the nonlinearities of the constraints, can be expected to severely tax conventional NLP codes. However, despite the apparent difficulty in solving the primal problem, there are structural features of the posynomial functions that can be exploited to facilitate direct primal solutions.

We showed earlier [see Eqs. (11.17) and (11.18)] that a posynomial can be transformed to a convex function by a transformation of variables:

$$x_n = \exp(z_n) \qquad \text{for } n = 1, 2, \ldots, N$$

Using this transformation, the primal problem given by Eqs. (11.29) and (11.30) can be transformed to a convex programming problem as given below:

$$\text{Minimize} \quad f_0(x) = \sum_{t=1}^{T_0} c_{0t} \exp\left( \sum_{n=1}^{N} a_{0nt} z_n \right)$$

$$\text{Subject to} \quad f_m(x) = \sum_{t=1}^{T_m} c_{mt} \exp\left( \sum_{n=1}^{N} a_{mnt} z_n \right) \leqslant 1 \qquad \text{for } m = 1, 2, \ldots, M$$

$$z_n, \text{ unrestricted in sign}$$

$$n = 1, \ldots, N$$

Note that the objective function and the constraints are convex functions and the problem is basically a convex programming problem. This feature can be used to great advantage in computations, since the Kuhn-Tucker conditions are sufficient to determine the optimal solution. Also, it permits application of any of the convex programming algorithms discussed in Chapters 6, 8, and 9, such as the generalized reduced gradient methods, penalty function methods, cutting plane algorithms, and successive linear approximation methods.

In addition to the general-purpose NLP methods that can be used to solve the convexified primal formulation, there are several special-purpose geometric programming (GP) computer codes (see Dembo [21] for a review). Some of them solve the convexified primal by Kelley's cutting plane method [22] or solve the Kuhn-Tucker conditions of the convexified primal by the Newton-Raphson method [23]. Other special-purpose codes solve the dual formulation by a modified convex simplex algorithm [24] or as a series of quadratic programs [25].

**Computational Results (Posynomial Case).** A major computational study to investigate the performances or different GP algorithms has recently been completed by Fattler et al. [26]. On the basis of statistical tests, the convex primal formulation was shown to be intrinsically easiest to solve for the general problems. A general-purpose generalized reduced gradient code [27] applied to the convex primal turned out to be highly competitive with the reputedly best specialized GP codes currently available. It could be shown that the effectiveness of the highly regarded specialized codes GGP [22] and GPKTC [23] are largely due to the fact that these codes solve the convex primal formulation. Further details on the study and its results are given in Chapter 12.

### 11.4.2 Generalized Geometric Programming

Duffin, Peterson, and Zener's [16] pioneering work in posynomial geometric programming, now called "prototype GP's," involved an objective function and upper-bounded inequality constraints that are posynomial functions, that is,

$$f(x) = \sum_t c_t \prod_{n=1}^{N} x_n^{a_{nt}}$$

where all $c_t > 0$ and $a_{nt}$'s are arbitrary real numbers. Soon after this development, Passy and Wilde [28] reported extensions to geometric programming that allowed the coefficients $c_t$ to be negative and accommodated both upper- and lower-bounded inequality constraints. We briefly describe the generalized GP (GGP) case here. For a complete discussion, the reader is referred to the text by Beightler and Phillips [15].

### The Primal Problem

A *signomial* (generalized posynomial) function is defined as

$$f(x) = \sum_{t=1}^{T} \sigma_t c_t \prod_{n=1}^{N} x_n^{a_{nt}} \tag{11.42}$$

where $\sigma_t = \pm 1$ (called signum functions), and $c_t > 0$. Using Eq. (11.42), we can formulate the generalized geometric program (GGP) as

$$\text{(Primal) Minimize} \quad f_0(x) = \sum_{t=1}^{T_0} \sigma_{0t} c_{0t} \prod_{n=1}^{N} x_n^{a_{0nt}} \tag{11.43}$$

$$\text{Subject to} \quad f_m(x) = \sum_{t=1}^{T_m} \sigma_{mt} c_{mt} \prod_{n=1}^{N} x_n^{a_{mnt}} \leqslant \sigma_m$$

$$\text{for } m = 1,\ldots, M \tag{11.44}$$

Note that the values of the signum functions $\sigma_{mt}$ for $m = 0, 1,\ldots, M$ and $\sigma_m$ for $m = 1, 2,\ldots, M$ are known constants from the problem formulation. The above GGP problem is in general a nonconvex nonlinear program that may possess multiple local minima.

### The Dual Problem

As shown by Rijckaert [29], the GGP problem has associated with it a dual problem. Define the dual variables for each term in the objective function and constraints as $\delta_{mt}$ for $m = 0, 1, 2,\ldots, M$ and $t = 1, 2,\ldots, T_m$. Let $\lambda_m$ for $m = 1, 2,\ldots, M$ be the Lagrange multiplier associated with constraint $m$. Then the dual problem may be stated as:

$$\text{Dual (Optimize)} \quad z(\delta) = \left[ \sum_{m=0}^{M} \prod_{t=1}^{T_m} \left( \frac{c_{mt}\lambda_m}{\delta_{mt}} \right)^{\sigma_{mt}\delta_{mt}} \right] \tag{11.45}$$

$$\text{Subject to} \quad \sum_{t=1}^{T_0} \sigma_{0t}\delta_{0t} = 1 \tag{11.46}$$

$$\sum_{m=0}^{M} \sum_{t=1}^{T_m} \sigma_{mt} a_{mnt}\delta_{mt} = 0 \quad \text{for } n = 1, 2,\ldots, N \tag{11.47}$$

$$\lambda_m = \sigma_m \sum_{t=1}^{T_m} \sigma_{mt}\delta_{mt} \quad \text{for } m = 1, 2,\ldots, M \tag{11.48}$$

$$\delta_{mt} \geqslant 0 \quad \text{for } m = 0, 1,\ldots, M \text{ and } t = 1,\ldots, T_m$$

$$\lambda_m \geqslant 0 \quad \text{for } m = 1,\ldots, M$$

Once again, we will assume that $\lambda_0 \equiv 1$ and

$$\lim_{\delta_{mt} \to 0} \left( \frac{c_{mt}\lambda_m}{\delta_{mt}} \right)^{\sigma_{mt}\delta_{mt}} = 1$$

It can be shown that for every local optimal solution $x^*$ where $f(x)$ achieves a local minimum, there exists a stationary point $\delta^*$ to the dual such that $f(x^*) = z(\delta^*)$. Once the dual variables $\delta_{mt}^*$ are known, the corresponding values of the primal variables $x_n^*$ are determined using the following relations or their log-linear forms:

$$c_{0t} \prod_{n=1}^{N} (x_n^*)^{a_{0nt}} = \delta_{0t} y^* \qquad \text{for } t = 1, \ldots, T_0 \tag{11.49}$$

$$c_{mt} \prod_{n=1}^{N} (x_n^*)^{a_{mnt}} = \frac{\delta_{mt}}{\lambda_m} \qquad \text{for } t = 1, 2, \ldots, T_m \quad \text{and} \quad m = 1, 2, \ldots, M \tag{11.50}$$

The degree of difficulty of the GGP problem is given by $T - (N + 1)$, where $T = \sum_{m=0}^{M} T_m$.

Although the above relationships between the primal and the dual variables are valid at the corresponding stationary points, as in the posynomial GP case, the bounding relationship between the primal and dual objectives no longer holds in the generalized GP case. Instead, the maximization of the dual must be replaced by a search for dual stationary points [29]. This feature disallows direct maximization, and thus GGP dual solution requires numerical solution of the Lagrangian conditions applied to the dual. As in the posynomial case, this approach must be used with great care because of difficulties presented by vanishing dual variables [21].

### Example 11.11 [30]

Consider the problem of designing the natural gas pipeline transmission system illustrated in Figure 11.7. It is required to pump 100 MMSCF/day of natural gas through a distance of 600 mi. The compressor stations are to be placed at equal distances. The design variables are the diameter of the pipe ($D$ in.),

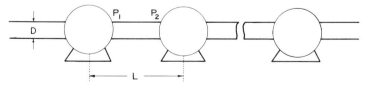

**Figure 11.7.** Gas pipeline transmission system.

compressor discharge and inlet pressures ($P_1$ and $P_2$ psia), and length between stations ($L$ mi). The optimum design should be such that the total cost of the pipeline facility, which consists of the fixed cost of the pipeline and operating costs of the compressor stations, is minimized.

Assume the following cost data (hp = compressor horsepower):

Initial compressor cost = \$50,000 + (\$290)(hp)
Pipeline cost = \$312/ton
Surveying, leasing, and engineering cost = \$10,000/mi
Installation of pipeline = \$820/mi per inch of pipe diameter
Labor and supervision at each compressor station = \$[1200 + 1.0 (hp)]/month
Maintenance and utilities = \$(0.30)(hp)/month
Internal charge for fuel = \$0.34/MSCF
Fuel consumption = 12 SCF/(hp · hr)
Depreciation rate = 15% per year

The fixed cost of the pipeline facility is the sum of the costs of the pipe and the compressors. These are prorated to annual costs using a 15% depreciation rate. The total annual cost (fixed + operating) can be computed in terms of the design variables $D$, $P_1$, $P_2$, and $L$.

## Cost of Compressor Stations

1.  Installed cost of the compressors:

$$\text{Number of compressors} = \frac{600}{L}$$

$$\text{Volumetric flow rate} = 100 \times 10^6 \text{ SCF/day}$$

$$= \frac{100 \times 10^6}{359} = 2.785 \times 10^5 \text{ moles/day}$$

Horsepower of each compressor can be obtained using the adiabatic compression work for an ideal gas, which is given by

$$W = \left(\frac{\gamma}{\gamma - 1}\right) \frac{RT_0}{\eta} [r^{\gamma/(\gamma-1)} - 1] \qquad (11.51)$$

where   $W$ = work in BTU/lb · mol
$r = P_1/P_2$   (design variable)
$R$ = gas constant = 1.987 BTU/(lb · mol · °R)

$$T_0 = \text{gas temperature} = 492°\text{R}$$
$$\gamma = \text{ratio of specific heats} = 1.28$$
$$\eta = \text{efficiency of compressor} = 0.75$$

From Eq. (11.51), the horsepower of each compressor is given by

$$\text{hp} = (2.7168 \times 10^4)[r^{0.219} - 1]$$

Hence, the annual fixed cost of all the installed compressors is

$$\$(0.15)\left(\frac{600}{L}\right)[50,000 + (290)(2.7168)(10^4)(r^{0.219} - 1)]$$

$$= \$\frac{4.5}{L}10^6 + \frac{7.08}{L}(10^8)(r^{0.219} - 1)$$

2. Annual fuel cost of all the compressors:

$$\$\left(\frac{0.34}{1000}\right)(12)(24)(365)\left(\frac{600}{L}\right)(2.7168)(10^4)(r^{0.219} - 1)$$

$$= \frac{5.82}{L}(10^8)(r^{0.219} - 1)$$

3. Annual maintenance and utilities costs of all the compressors:

$$\$(0.30)(2.7168)(10^4)(r^{0.219} - 1)(12)\frac{600}{L} = \frac{0.58}{L}(10^8)(r^{0.219} - 1)$$

4. Annual labor and supervision costs:

$$\$\left(\frac{600}{L}\right)[1200(12) + (1.0)(12)(2.7168)(10^4)(r^{0.219} - 1)]$$

$$= \frac{8.64}{L}(10^6) + \frac{1.96}{L}(10^8)(r^{0.219} - 1)$$

**Fixed Costs of Pipeline**

5. Installation cost $= \$820(600)D(0.15)$
$$= \$73,800D$$

6. Cost of surveying, leasing, and engineering:

$$\$10,000(600)(0.15) = \$900,000$$

7. Pipeline cost: The cost of the pipeline is a function of its weight, which

is given by

$$\text{Weight (tons/mi)} = \frac{\pi}{144}(D+t)t\left(\frac{5280 \text{ ft}}{\text{mi}}\right)\left(\frac{480 \text{ lb}}{\text{ft}^3}\right)\left(\frac{1 \text{ ton}}{2000 \text{ lb}}\right)$$

$$= 27.646(D+t)t$$

where $t$ is the thickness of the pipe in inches. The thickness can be obtained by the *hoop stress* formula:

$$t = \frac{DP_1}{2(S - 0.6P_1)} \simeq \frac{DP_1}{2S}$$

where $S$ is the tensile stress, 25,000 psia. Thus,

$$\text{Weight (tons/mi)} = (27.646)D^2\left(1 + \frac{P_1}{50,000}\right)\left(\frac{P_1}{50,000}\right)$$

$$\simeq (27.646)D^2\left(\frac{P_1}{50,000}\right)$$

$$\text{Annual cost of pipe} = \$(0.15)(600)(315)(27.646)\frac{D^2P_1}{50,000}$$

$$= \$15.68D^2P_1$$

The total annual cost of the pipeline $C(D, P_1, L, r)$, including the operation of compressors, is given by adding cost items (1) through (7). Thus,

$$C(D, P_1, L, r) = 15.68D^2P_1 + 9 \times 10^5 + 7.38 \times 10^4 D$$

$$+ \frac{13.14 \times 10^6}{L} + \frac{15.44 \times 10^8}{L}(r^{0.219} - 1) \quad (11.52)$$

The only constraint the variables must satisfy is the design relation involving the pressure drop in the pipeline given by

$$Q = 3.39\left[\frac{(P_1^2 - P_2^2)D^5}{fL}\right]^{1/2} \quad (11.53)$$

where $Q$ = volumetric flow, SCF/hr
       $f$ = friction factor = $0.008D^{-1/3}$

The above optimization problem involves four design variables, $D$, $P_1$, $L$, and $r$, and one equality constraint given by (11.53). The problem dimension can be reduced by eliminating one of the variables using Eq. (11.53). Since the pipeline is required to deliver 100 million ft³ of gas per day (4.17 × 10⁶ SCF/hr), Eq. (11.53) may be written as

$$P_1 = r \left[ \frac{1.21 \times 10^{10} L D^{-5.33}}{r^2 - 1} \right]^{1/2} \tag{11.54}$$

Substituting Eq. (11.54) into the cost expression given by Eq. (11.52), we get

$$C(D, L, r) = 17.22 \times 10^5 L^{1/2} r D^{-2/3} (r^2 - 1)^{1/2}$$

$$+ 9 \times 10^5 + 7.38 \times 10^4 D + \frac{13.14}{L} \times 10^6$$

$$+ \frac{15.44}{L} \times 10^8 (r^{0.219} - 1)$$

Defining $x_1 = L$, $x_2 = r$, and $x_3 = D$, and dropping the constant term, the objective function reduces to

$$\text{Minimize} \quad f(x) = 17.22 \times 10^5 x_1^{1/2} x_2 x_3^{-2/3} (x_2^2 - 1)^{-1/2}$$

$$+ 7.38 \times 10^4 x_3 + 13.14 \times 10^6 x_1^{-1}$$

$$+ 15.44 \times 10^8 x_1^{-1} (x_2^{0.219} - 1) \tag{11.55}$$

For the design to be feasible, we need $x_1 > 0$, $x_2$ (compression ratio) $> 1$, and $x_3 > 0$. The objective function given by Eq. (11.55) is not a signomial, due to the presence of the term $(x_2^2 - 1)^{-1/2}$. However, the problem can be reduced to a generalized geometric program by introducing a new variable $x_4$ and a new constraint

$$x_4 \leqslant x_2^2 - 1$$

Thus, the equivalent GGP problem is

$$\text{Minimize} \quad f(x) = 17.22 \times 10^5 x_1^{1/2} x_2 x_3^{-2/3} x_4^{1/2} + 7.38 \times 10^4 x_3$$

$$+ 15.44 \times 10^8 x_1^{-1} x_2^{0.219} - 1530.86 \times 10^6 x_1^{-1}$$

$$\text{Subject to} \quad x_4 x_2^{-2} + x_2^{-2} \leqslant 1$$

$$x_1, x_2, x_3, x_4 > 0$$

Let us solve the GGP problem using the dual formulation [refer to Eqs. (11.45)–(11.48)].

**Dual**

$$\text{Optimize} \quad z(\delta) = \left[\left[\frac{17.22 \times 10^5}{\delta_{01}}\right]^{\delta_{01}}\left[\frac{7.38 \times 10^4}{\delta_{02}}\right]^{\delta_{02}}\left[\frac{15.44 \times 10^8}{\delta_{03}}\right]^{\delta_{03}}\right.$$

$$\left[\frac{1530.86 \times 10^6}{\delta_{04}}\right]^{-\delta_{04}}\left[\frac{\lambda_1}{\delta_{11}}\right]^{\delta_{11}}\left[\frac{\lambda_1}{\delta_{12}}\right]^{\delta_{12}}\right]$$

$$\begin{aligned}
\text{Subject to} \qquad \delta_{01} + \delta_{02} &\quad + \delta_{03} - \delta_{04} &&= 1 \\
0.5\delta_{01} &\quad - \delta_{03} + \delta_{04} &&= 0 \\
\delta_{01} &\quad + 0.219\delta_{03} &\quad - 2\delta_{11} - 2\delta_{12} &= 0 \\
-0.678\delta_{01} + \delta_{02} & &&= 0 \\
-0.5\delta_{01} & &\quad + \delta_{11} &= 0 \\
& &\quad \lambda_1 - \delta_{11} - \delta_{12} &= 0
\end{aligned}$$

$$\delta_{ij} \geq 0$$

$$\lambda_1 \geq 0$$

Note that the dual problem has $T = 6$ terms and $N = 4$ variables. The degree of difficulty $= T - (N + 1) = 1$. Hence an optimization problem exists for the dual. Since the primal problem is a signomial, the dual problem may be nonconvex and possess multiple stationary points. It can be verified that the following stationary point solution to the dual corresponds to the primal optimum:

$$\delta_{01}^* = 0.4552 \qquad \delta_{02}^* = 0.3094 \qquad \delta_{03}^* = 5.152 \qquad \delta_{04}^* = 4.92$$

$$\delta_{11}^* = 0.2276 \qquad \delta_{12}^* = 0.7519 \qquad \lambda_1^* = 0.9795$$

$$z(w^*) = 591.6 \times 10^4 = f(x^*)$$

Using Eqs. (11.49) and (11.50), we obtain the optimal values of the design variable:

$$D = 24.8 \text{ in.} \qquad L = 52.6 \text{ mi} \qquad r = 1.87 \qquad P_1 = 284 \text{ psia}$$

The annual cost of the pipeline system $C(D^*, L^*, r^*) = \$6.816$ million.

**Equivalent GGP Formulations.** In addition to the primal and dual GGP forms discussed, there exist other equivalent GGP problem structures. A structurally more revealing but not necessarily computationally more advantageous form of the primal [Eqs. (11.42)–(11.44)] can be obtained by rewriting each signomial function as the difference of two posynomials:

$$f_m(x) = P_m(x) - Q_m(x)$$

where

$$P_m(x) = \sum_{t \in M^+} c_{mt} \prod_{n=1}^{N} x_n^{a_{mnt}}$$

$$Q_m(x) = \sum_{t \in M^-} c_{mt} \prod_{n=1}^{N} x_n^{a_{mnt}}$$

and $M^+$ and $M^-$ are the subsets of signomial $m$ whose signum functions $\sigma_{mt}$ are positive and negative, respectively.

As shown by Avriel and Williams [31], the GGP primal can then be written in the *complementary or quotient form*,

$$\text{Minimize} \quad x_0$$

$$\text{Subject to} \quad q_m(x) \leqslant 1 \qquad m = 1, \ldots, M + 1$$

$$x_0, x > 0$$

where each function $q_m(x)$ is a quotient of posynomials,

$$q_m(x) = \frac{P_m(x)}{Q_m(x) + 1} \qquad m = 1, \ldots, M$$

and

$$q_{M+1} = \frac{P_0(x)}{x_0 + Q_0(x)}$$

Note that the variable $x_0$ is simply a device used to transform the signomial objective to a constraint.

Duffin and Peterson [32] have demonstrated that any nonlinear algebraic program whose objective and constraints are real-valued functions formed by addition, subtraction, or exponentiation to real powers can be transformed to a signomial geometric program. This is accomplished primarily by introducing new variables together with their constraints so as to decompose all functions into sums or differences of individual terms. Moreover, Duffin and Peterson have also shown that since each signomial can be written as the difference of two posynomial functions, each signomial constraint can be replaced by two posynomial constraints one of which is a lower-bounded constraint. Specifically, if an artificial variable $y_m$ is introduced, each constraint

$$P_m(x) - Q_m(x) \leqslant 1$$

can be replaced

$$P_m(x) \leqslant y_m \leqslant 1 + Q_m(x)$$

or

$$y_m^{-1} P_m(x) \leqslant 1 \quad \text{and} \quad (1 + Q_m(x)) y_m^{-1} \geqslant 1$$

Therefore, at the expense of increasing the number of variables and constraints, the signomial program can be converted to a *reversed* geometric program, a problem in which all functions are posynomials but some are involved in upper-bounded or normal constraints while others are involved in lower-bounded or *reversed* constraints.

Thus, the *reversed GP* problem is defined as

$$\text{Minimize} \quad h_0(x)$$

$$\text{Subject to} \quad h_k(x) \leqslant 1 \quad k = 1, \ldots, K$$

$$h_k(x) \geqslant 1 \quad k = K + 1, \ldots, L$$

$$x > 0$$

where all $h_k(x)$ for $k = 0, 1, \ldots, L$ are posynomials.

### Exponential Primal Form

As in the posynomial GP case, each signomial can be recast to a sum of exponentials using the transformation

$$x_n = \exp(z_n)$$

Thus, the signomial

$$f_m(x) = \sum_{t=1}^{T_m} \sigma_{mt} c_{mt} \prod_{n=1}^{N} x_n^{a_{mnt}}$$

can be replaced by

$$f_m(z) = \sum_t \sigma_{mt} c_{mt} \exp\left(\sum_n a_{mnt} z_n\right)$$

where the original variables $x_n$ are constrained to be positive, the $z_n$ are unrestricted in sign. In the posynomial case, these exponential functions are convex functions, and use of this form of the primal in computation proves to be much preferable to direct primal solution [26]. In the signomial case, the transformed functions are in general nonconvex; hence some of the computational advantages may well be diminished. However, application of this transformation to the reversed primal results in a problem in which all functions

$$h_k(z) = \sum_t c_{kt} \exp\left(\sum_n a_{knt} z_n\right)$$

are convex but the feasible region is the intersection of a convex set generated by the inequalities

$$h_k(z) \leqslant 1 \qquad k = 1, \ldots, K$$

and a *reverse* convex set generated by the inequalities

$$h_k(z) \geqslant 1 \qquad k = K + 1, \ldots, L$$

A reverse convex set is simply the complement of a convex set. The exponential form of the reversed primal thus clearly reveals the underlying structure of GGP problems and clarifies the reason for the possible occurrence of multiple local minima.

**Solution Approaches to Generalized Geometric Programming.** The solution methods proposed for generalized GP problems have been of two types: sequential methods employing a series of approximating problems, and direct approaches to one of the equivalent GGP forms. Available fragmentary evidence indicates that the sequential methods are superior to direct approaches.

*Sequential Minimization*

The quotient and reversed GP forms of the primal suggest that if the denominators and reversed constraints, respectively, could be replaced by approximating single-term posynomials, then the resulting approximating problems would reduce to prototype GP's. Such approximations can readily be obtained via the condensation device proposed by Avriel and Williams [31] and Duffin [33].

Given a posynomial $P(x) = \sum_{t=1}^{T} u_t(x)$ and a set of nonnegative normalized parameters $\alpha_t$, $t = 1,\ldots, T$, then from the inequality between the arithmetic and geometric means it follows that

$$P(x) = \sum_t u_t(x) \geqslant \prod_t \left(\frac{u_t}{\alpha_t}\right)^{\alpha_t} \equiv \tilde{P}(x, \alpha) \qquad (11.56)$$

Thus, a multiterm posynomial $P(x)$ is approximated by a single-term posynomial $\tilde{P}(x, \alpha)$. Using this construction, Avriel and Williams [31] have proposed replacing the constraints of a GGP problem in quotient form,

$$q_m(x) = \frac{P_m(x)}{Q_m(x) + 1} \leqslant 1$$

with the approximation

$$\tilde{q}_m(x) = P_m(x)\left[\tilde{Q}(x, \alpha)\right]^{-1} \leqslant 1$$

Similarly, Duffin [33] proposed replacing the reversed constraint

$$h_k(x) \geqslant 1$$

with the posynomial approximation

$$\left[\tilde{h}_k(x, \alpha)\right]^{-1} \leqslant 1$$

Note that since inequality (11.56) is an equality if and only if $\alpha_t = u_t/\sum u_t$, the parameters $\alpha_t$ of the approximating functions in both the quotient and reversed GGP cases are updated by setting

$$\alpha_t^k = \frac{u_t(x^{(k-1)})}{P(x^{(k-1)})}$$

where $x^{(k-1)}$ is the solution of the $(k-1)$st approximating problem. Thus a series of approximating problems is generated and solved until the difference $\alpha_t^{(k)} - \alpha_t^{(k-1)}$ becomes sufficiently small for all $t$. It can be shown [31] that

1.  Any feasible point of an approximating problem will also be a feasible point of the GGP.
2.  The sequence of approximating problem solutions will converge to a local minimum of the GGP under mild assumptions.

These alternative primal approximation schemes reduce the solution of GGP problems to the solution of a series of prototypes (posynomial) GP problems but leave open the choice of which prototype formulation to solve and what algorithm to use for solution. In principle, any approach suitable for prototype GP problems can be employed in conjunction with condensation constructions, including direct minimization of the primal approximating problem.

### Direct Approaches

These can be of two types, direct minimization of one of the primal forms (original primal, exponential form of primal, quotient form of GGP, reversed GGP) or solution of the Kuhn-Tucker conditions corresponding to one of the GGP formulations by specialized Newton-Raphson method. However, no generally available software seems to have been produced as yet that will locate the global minimum of GGP problems.

For a complete discussion of the different solution approaches to GGP problems, refer to the text by Beightler and Phillips [15]. A good bibliography on generalized geometric programming is available in Rijckaert [29].

**Computational Results for Generalized Geometric Programming.** As a followup to the study on prototype GP's [26], Sin and Reklaitis [34, 35] have also completed a recent study on solving GGP problems. In that study, five codes (four specialized GP codes—GGP [22], GPKTC [23], QUADGP [25], and DAP [36]—as well as a general NLP code based on the generalized reduced gradient method [27] were used to solve four GGP problem formulations.

A total of 25 test problems, again representing both engineering applications and artificially constructed problems, were solved. Each test series involved solution of each test problem from up to 20 randomly generated starting points. On the basis of statistical tests, the preferred solution approach was shown to involve use of the quotient form of the signomial functions, condensation of the denominators of the quotients using Duffin's geometric mean construction, and solution of the condensed subproblems in their convexified subproblem form. The code GGP that employed this strategy appeared to be most effective. Direct GRG solution of the exponential form of the primal turned out to be next best. The use of condensation and the quotient representation of signomial functions thus appears to be the most computationally significant development arising from generalized GP research. As in the prototype GP study, a correlation analysis revealed that the number of primal variables and multiterm constraints were strongly exponentially correlated. Depending upon the primal form used, the solution time is strongly correlated with the number of negative terms or the number of reversed constraints. It appears that the number of multiterm constraints is more significant than the division between posynomial and signomial constraints.

For reports on other early studies on computational performances of GGP algorithms, the reader is referred to Dembo [21] and Sarma et al. [37].

### 11.4.3  Engineering Applications

In conclusion, we mention a number of engineering problems that have been successfully solved by GP techniques. A partial list of such successful applications includes gas transmission compressor designs, cofferdam design, optimal design of welded beams, design of industrial waste treatment plants, optimization of alkylation processes, investment for batch processes, inventory control, and optimal design of pressure vessels. For a good description of the various GP applications in engineering design, the reader is referred to the text by Beightler and Phillips [15, chaps. 4 and 11]. Rijckaert [38] also discusses a number of engineering applications of geometric programming.

## 11.5  SUMMARY

In this chapter we studied optimization problems with special structures and developed algorithms for their solution. We began with integer programming problems that required some or all of the design variables to be integer valued. For the case of linear integer programs (problems with linear constraints and linear objective functions), branch-and-bound methods were developed for their solution. They involve solution of a series of linear programs that employ "continuous relaxation" of the original mixed integer problem.

Optimization problems with quadratic objective function and linear constraints were discussed next. Known as quadratic programming (QP) problems, these could be solved as special cases of complementary problems using Lemke's complementary pivot method. The complementary problem basically represented the Kuhn-Tucker optimality conditions of the quadratic programs. Lemke's algorithm is an iterative procedure for finding a solution that satisfies the Kuhn-Tucker conditions.

Next, we discussed an important class of optimization problems known as geometric programming. These involved problems whose objective function can be expressed as the sum of generalized polynomial functions, and the design variables were restricted to be positive. We saw that an efficient way of solving the geometric programs was to transform them into equivalent convex NLP problems. The transformed problems can then be solved efficiently by the generalized reduced gradient (GRG) algorithm.

## REFERENCES

1.  Watters, L., "Reduction of Integer Polynomial Programming Problems to Zero-One Linear Programming Problems," *Oper. Res.*, **15**(6), 1171–1174 (1967).
2.  Glover, F., and E. Woolsey, "Converting 0–1 Polynomial Programming Problem to a 0–1 Linear Program," *Oper. Res.*, **22**, 180–182 (1974).

3. Phillips, D. T., A. Ravindran, and J. J. Solberg, *Operations Research: Principles and Practice*, Wiley, New York, 1976.

4. Plane, D. R., and C. McMillan, Jr., *Discrete Optimization*, Prentice-Hall, Englewood Cliffs, NJ, 1971.

5. *IBM General Information Manual*, "An Introduction to Modeling Using Mixed Integer Programming," Amsterdam, 1972.

6. Gupta, Omprakash K., and A. Ravindran, "Nonlinear Integer Programming and Discrete Optimization," in *Progress in Engineering Optimization* (R. W. Mayne and K. M. Ragsdell, Eds.,), ASME, New York, 1981, pp. 27–32.

7. McCarl, B. A., H. Moskowitz, and H. Furtan, "Quadratic Programming Applications," *Omega*, **5**(1), 43–55 (1977).

8. Wolfe, P., "The Simplex Method of Quadratic Programming," *Econometrica*, **27**, 382–398 (1959).

9. Lemke, C. E., "Bimatrix Equilibrium Points and Mathematical Programming," *Manage. Sci.*, **11**, 681–689 (1965).

10. Ravindran, A., and H. Lee, "Computer Experiments on Quadratic Programming Algorithms," *Eur. J. Oper. Res.*, **8**(2), 166–174 (1981).

11. Cottle, R. W., and G. B. Dantzig, "Complementary Pivot Theory and Mathematical Programming," *J. Linear Algebra Appl.*, **1**, 105–125 (1968).

12. Ravindran, A., "A Computer Routine for Quadratic and Linear Programming Problems," *Commun. ACM*, **15**(9), 818–820 (1972).

13. Ravindran, A., "A Comparison of Primal Simplex and Complementary Pivot Methods for Linear Programming," *Naval Res. Logist. Quart.*, **20**(1), 95–100 (1973).

14. Ravindran, A., "Computational Aspects of Lemke's Algorithm Applied to Linear Programs," *OPSEARCH*, **7**(4), 241–262 (1970).

15. Beightler, C. S., and D. T. Phillips, *Applied Geometric Programming*, Wiley, New York, 1976.

16. Duffin, R. J., E. L. Peterson, and C. Zener, *Geometric Programming*, Wiley, New York, 1967.

17. Philipson, R. H., and A. Ravindran, "Application of Mathematical Programming to Metal Cutting," *Math. Programming Study*, **19**, 116–134 (1979).

18. Armarego, E. J. A., and R. H. Brown, *The Machining of Metals*, Prentice-Hall, Englewood Cliffs, NJ, 1969.

19. Phillips, D. T., and C. S. Beightler, "Optimization in Tool Engineering Using Geometric Programming," *Amer. Inst. Ind. Engs. Trans.*, **2**, 355–360 (1970).

20. Dembo, R. S., "Dual to Primal Conversion in Geometric Programming," *J. Optimization Theory Appl.*, **26**, 355–360 (1978).

21. Dembo, R. S., "The Current State of the Art of Algorithms and Computer Software for Geometric Programming," *J. Optimization Theory Appl.*, **26**, 149–183 (1978).

22. Dembo, R. S., "GGP: A Computer Program for the Solution of Generalized Geometric Programming Problems," McMaster University, Hamilton, Ontario, Canada, *Users Manual*, 1975.

23. Rijckaert, M. J., and X. M. Martens, "GPKTC: A Generalized Geometric Programming Code," Rep. CE-RM-7601, Katholieke Universiteit, Leuven, Belgium, 1976.

24. Beck, P. A., and J. G. Ecker, "A Modified Concave Simplex Algorithm for Geometric Programming," *J. Optimization Theory Appl.*, **15**, 184–202 (1975).

25. Bradley, J., "The Development of Polynomial Programming Algorithms with Applications," Ph.D. thesis, Dept. Computer Sci., Dublin University, Sept. 1975.

26. Fattler, J. E., Y. T. Sin, R. R. Root, K. M. Ragsdell, and G. V. Reklaitis, "On the Computational Utility of Posynomial Geometric Programming Solution Methods," *Math. Prog.*, **22**, 163 (1982).

27. Gabriele, G. A., and K. M. Ragsdell, "OPT: A Nonlinear Programming Code in FORTRAN IV," *The Modern Design Series*, 1, Purdue Research Foundation, West Lafayette, 1976.

28. Passy, U., and D. J. Wilde, "Generalized Polynomial Optimization," *SIAM J. Appl. Math.*, **15**, 1344–1356 (1967).

29. Rijckaert, M. J., "Bibliographical Note on Geometric Programming," *J. Optimization Theory Appl.*, **26**, 325–337 (1978).

30. Sherwood, T. K., *A Course in Process Design*, MIT Press, Cambridge, MA, 1963.

31. Avriel, M., and A. C. Williams, "Complementary Geometric Programming," *SIAM J. Appl. Math.*, **19**, 125–141 (1970).

32. Duffin, R. J., and E. L. Peterson, "Geometric Programming with Signomials," *J. Optimization Theory Appl.*, **11**, 3–35 (1973).

33. Duffin, R. J., "Linearizing Geometric Programs," *SIAM Rev.*, **12**, 211–217 (1970).

34. Sin, Y. T., and G. V. Reklaitis, "On the Computational Utility of Generalized Geometric Programming Solution Methods: Part I—Review of Test Procedure Design," *Progress in Engineering Optimization—1981* (R. W. Mayne and K. M. Ragsdell, Eds.), ASME, New York, Sept. 1981, pp. 7–14.

35. Sin, Y. T., and G. V. Reklaitis, "Part II—Results and Interpretations," in *Progress in Engineering Optimization—1981* (R. W. Mayne and K. M. Ragsdell, Eds.), ASME, New York, Sept. 1981, pp. 15–26.

36. Reklaitis, G. V., and D. J. Wilde, "Geometric Programming Via a Primal Auxiliary Problem," *AIIE Trans.*, **6**, 308–317 (1974).

37. Sarma, P. V. L. N., X. M. Martens, G. V. Reklaitis, and M. J. Rijckaert, "A Comparison of Computational Strategies for Geometric Programs," *J. Optimization Theory Appl.*, **26**, 185–204 (1978).

38. Rijckaert, M. J., "Engineering Applications of Geometric Programming," in *Optimization and Design* (M. Avriel, M. J. Rijckaert and D. J. Wilde, Eds.), Prentice-Hall, Englewood Cliffs, NJ, 1974.

## PROBLEMS

**11.1.** Explain under what conditions rounding off continuous solutions is a good strategy for obtaining integral values. Cite some situations where such a strategy will not work and explain why.

**11.2.** Explain the concepts of "branching" and "bounding" used in the branch-and-bound algorithm.

**11.3.** What is the meaning of "fathoming" a node? Under what conditions can a node be fathomed in the branch-and-bound algorithm?

**11.4.** Suppose we have an integer program in which a variable $X_1$ only takes on a specified set of discrete values. Explain how to modify the branch and bound method to handle this variation.

**11.5.** Under what assumptions are the Kuhn–Tucker conditions (a) necessary and (b) sufficient for the optimality of a quadratic programming problem.

**11.6.** Cite conditions under which a complementary problem always has a complementary solution for all $q$.

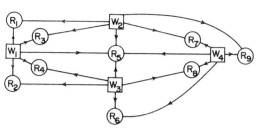

**Figure 11.8.**

**11.7.** Is Lemke's algorithm guaranteed to solve a quadratic programming problem always? Why or why not?

**11.8.** What is a "ray solution" and what is its significance in the complementary pivot method?

**11.9.** What is the difference between a "posynomial" and a "polynomial"?

**11.10.** What is a "prototype GP"? What approach is generally efficient to solve a "prototype GP"?

**11.11.** What is a Generalized GP problem? What are the different approaches to solving a Generalized GP? Which one would you recommend and why?

**11.12.** A firm has four possible sites for locating its warehouses. The cost of locating a warehouse at site $i$ is $\$K$. There are nine retail outlets, each of which must be supplied by at least one warehouse. It is not possible for any one site to supply all the retail outlets as shown in Figure 11.8. The problem is to determine the location of the warehouses such that the total cost is minimized. Formulate this as an integer program.

**11.13.** A company manufactures three products: A, B, and C. Each unit of product A requires 1 hr of engineering service, 10 hr of direct labor, and 3 lb of material. To produce one unit of product B requires 2 hr of engineering, 4 hr of direct labor, and 2 lb of material. Each unit of product C requires 1 hr of engineering, 5 hr of direct labor, and 1 lb of material. There are 100 hr of engineering, 700 hr of labor, and 400 lb of materials available. The cost of production is a nonlinear function of the quantity produced as shown below:

| Product A | | Product B | | Product C | |
|---|---|---|---|---|---|
| Production | Unit Cost | Production | Unit Cost | Production | Unit Cost |
| units | $ | units | $ | units | $ |
| 0–40 | 10 | 0–50 | 6 | 0–100 | 5 |
| 40–100 | 9 | 50–100 | 4 | over 100 | 4 |
| 100–150 | 8 | over 100 | 3 | | |
| over 150 | 7 | | | | |

Formulate a mixed integer program to determine the minimum-cost production schedule.

**11.14.** Explain how the following conditions can be represented as linear constraints by the use of binary (0-1) integer variables:

(a) Either $x_1 + x_2 \leqslant 2$ or $2x_1 + 3x_2 \geqslant 8$.

(b) Variable $x_3$ can assume values 0, 5, 9, and 12 only.

(c) If $x_4 \leqslant 4$, then $x_5 \geqslant 6$. Otherwise $x_5 \leqslant 3$.

(d) At least two out of the following four constraints must be satisfied:

$$x_6 + x_1 \leqslant 2$$

$$x_6 \leqslant 1$$

$$x_7 \leqslant 5$$

$$x_6 + x_1 \geqslant 3$$

**11.15.** Convert the following nonlinear integer program to a linear integer program:

$$\text{Minimize} \quad Z = x_1^2 - x_1 x_2 + x_2$$

$$\text{Subject to} \quad x_1^2 + x_1 x_2 \leqslant 8$$

$$x_1 \leqslant 2$$

$$x_2 \leqslant 7$$

$$x_1, x_2 \geqslant 0 \text{ and integer}$$

(*Hint:* Replace $x_1$ by $2^0 \delta_1 + 2^1 \delta_2$, and $x_2$ by $2^0 \delta_3 + 2^1 \delta_4 + 2^2 \delta_5$, where $\delta_i \in (0, 1)$ for $i = 1, 2, \ldots, 5$.)

**11.16.** Solve the following pure integer program by the branch-and-bound algorithm:

$$\text{Maximize} \quad Z = 21x_1 + 11x_2$$

$$\text{Subject to} \quad 7x_1 + 4x_2 + x_3 = 13$$

$$x_1, x_2, x_3 \text{ nonnegative integers}$$

**11.17.** Solve the following mixed integer program by the branch-and-bound algorithm:

$$\text{Minimize} \quad Z = 10x_1 + 9x_2$$

$$\text{Subject to} \quad x_1 \leqslant 8$$

$$x_2 \leqslant 10$$

$$5x_1 + 3x_2 \geqslant 45$$

$$x_1 \geqslant 0 \qquad x_2 \geqslant 0$$

$$x_2 \text{ is an integer}$$

**11.18.** A scientist has observed a certain quantity $Q$ as a function of a variable $t$. She has good reasons to believe that there is a physical law relating $t$ and $Q$ that takes the form:

$$Q(t) = a \sin t + b \tan t + c \qquad (11.57)$$

She wants to have the "best" possible idea of the value of the coefficients $a, b, c$ using the results of her $n$ experiments $(t_1, Q_1; t_2, Q_2; t_3, Q_3; \ldots; t_n, Q_n)$. Taking into account that her experiments are not perfect and also perhaps that Eq. (11.57) is not rigorous, she does not expect to find a perfect fit. So she defines an error term

$$e_i = Q_i - Q(t_i)$$

and has two different ideas of what "best" may mean:
(1) Her first idea is to minimize

$$Z_1 = \sum_{i=1}^{n} |e_i|$$

(2) Her second idea is to minimize

$$Z_2 = \sum_{i=1}^{n} e_i^2$$

She then discovers that for physical reasons the coefficients $a, b, c$ must be nonnegative. Taking these new constraints into consideration, she minimizes again:
(3)

$$Z_3 = \sum_{i=1}^{n} |e_i|$$

(4)

$$Z_4 = \sum_{i=1}^{n} e_i^2$$

Show, with detailed justification, that she will have to solve, respectively,

(a) A linear program in case (1).

(b) A system of linear equations in case (2). [*Hint*: First show that in case (2), the problem reduces to minimizing an unconstrained convex function.]

(c) A linear program in case (3).

(d) A problem of the form

$$\omega = Mz + q \qquad \omega \geq 0 \qquad z \geq 0 \qquad \omega^T z = 0 \text{ in case (4)}.$$

Specify what $M$, $z$, $q$, $\omega$ are. (*Hint*: Show first that the problem in this case reduces to a convex quadratic program.)

*Note*: Give all specifications you can on each problem: number of variables, number of constraints or size, etc.

**11.19.** A monopolist estimates the following price vs. demand relationships for four of his products:

$$
\begin{aligned}
\text{Product 1:} \quad & x_1 = -1.5143p_1 + 2671 \\
\text{Product 2:} \quad & x_2 = -0.0203p_2 + 135 \\
\text{Product 3:} \quad & x_3 = -0.0136p_3 + 0.0015p_4 + 103 \\
\text{Product 4:} \quad & x_4 = 0.0016p_3 - 0.0027p_4 + 19
\end{aligned}
$$

where $x_j$ and $p_j$ are the unknown demand and price for product $j$. The function to be maximized is $\sum_{j=1}^{4} p_j x_j$. The cost of production is fixed and hence can be discarded. Using the price-demand relationships, the objective function can be written as a quadratic function in either $p_j$ or $x_j$. The products use two raw materials for production, and they give rise to the following capacity constraints:

$$0.026x_1 + 0.8x_2 + 0.306x_3 + 0.245x_4 \leq 121$$

$$0.086x_1 + 0.02x_2 + 0.297x_3 + 0.371x_4 \leq 250$$

$$x_1, \ldots, x_4 \geq$$

$$p_1, \ldots, p_4 \geq 0$$

Formulate a quadratic program to determine the optimal production levels that will maximize the total revenue $\Sigma p_j x_j$, and solve.

**11.20.** (a) Write the equivalent Kuhn-Tucker problem for the following quadratic program:

$$\text{Minimize} \quad Z = x_1 - x_2 + x_1^2 + x_2^2 - x_1 x_2$$

$$\text{Subject to} \quad x_1 + x_2 = 1$$

$$x_1, x_2 \geqslant 0$$

(b) Show whether or not $x_1 = \frac{1}{6}$, $x_2 = \frac{5}{6}$ in an optimal solution using the Kuhn-Tucker conditions obtained in (a).

**11.21.** The electrical resistance network shown in Figure 11.9 is required to accommodate a current flow of 10 A from junction 1 to node 4. Given the resistance of each of the arcs, the problem is to determine the equilibrium distribution of current flows that will minimize the total power loss in the network.

(a) Formulate the problem as a quadratic program.

(b) Write down the Kuhn-Tucker conditions.

(c) Using (b) and the fact that the optimal current flows have to be positive, determine the equilibrium distribution of current flows in the network.

**11.22.** Consider the convex region shown in Figure 11.10, where $P_1$, $P_2$, $P_3$, and $P_4$ are the extreme points of the convex region. Let $P_0 = (-2, -1)$ be a specified point. The problem is to determine the point in the convex region "closest" to point $P_0$.

(a) Formulate the above problem as a quadratic program. (*Hint:* Any point $x$ in the convex region can be expressed as a convex combination

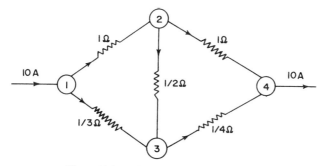

**Figure 11.9.**  Electrical resistance network.

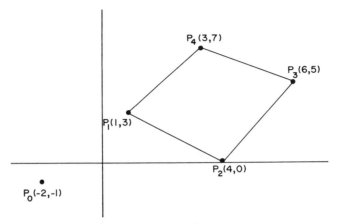

**Figure 11.10.**

of its extreme points $P_1$, $P_2$, $P_3$, and $P_4$, and the objective is to minimize $\|x - P_0\|^2$.)

(b) Solve, using the complementary pivot method.

**11.23.** (a) If $Q$ is positive semidefinite, prove that

$$\mathbf{M} = \begin{pmatrix} \mathbf{Q} & -\mathbf{A}^T \\ \mathbf{A} & 0 \end{pmatrix}$$

is also positive semidefinite for any $\mathbf{A}$ matrix.

(b) If $\mathbf{Q}$ is positive definite, what more can you say about the matrix $\mathbf{M}$?

**11.24.** Solve the following quadratic program by the complementary pivot algorithm.

$$\text{Minimize} \quad Z = -x_1 \quad - 2x_2 + \tfrac{1}{2}x_1^2 + \tfrac{1}{2}x_2^2$$

$$\text{Subject to} \quad x_1 + x_2 \leqslant 4$$

$$x_1, x_2 \geqslant 0$$

Explain in two *or* three lines why the algorithm is guaranteed to find a solution to this problem.

**11.25.** Solve the following complementary problem by Lemke's algorithm:

$$\mathbf{M} = \begin{pmatrix} 1 & 2 & 3 \\ 4 & 5 & 6 \\ 6 & 7 & 8 \end{pmatrix} \qquad q = \begin{pmatrix} 2 \\ -5 \\ -3 \end{pmatrix}$$

**11.26.** Write down the equivalent complementary problem for the following linear programs. (In each case, state clearly what $\mathbf{M}$, $q$, $w$, and $z$ are.)

(a) Maximize $Z = 2x_1 + 3x_2 + x_3$

Subject to $x_1 - x_2 + x_3 = 2$

$x_1, x_2, x_3 \geqslant 0$

(b) Minimize $Z = 2x_1 - 3x_2 + 6x_3$

Subject to $3x_1 - 4x_2 - 6x_3 \leqslant 2$

$2x_1 + x_2 + 2x_3 \geqslant 11$

$x_1 + 3x_2 - 2x_3 \leqslant 5$

$x_1, x_2, x_3 \geqslant 0$

**11.27.** Prove or disprove that the complementary problem $w = \mathbf{M}z + q$, $w$, $z \geqslant 0$, $w^T z = 0$, where

$$\mathbf{M} = \begin{pmatrix} 2 & 1 & 0 \\ 1 & 1 & 0 \\ 0 & 0 & 1 \end{pmatrix}$$

has a complementary solution for any $q$. (Give due references to any theorems/results that are applicable to your answer.)

**11.28.** Solve for the dual variables associated with the following geometric program:

$$\text{Minimize} \quad f(x) = (x_1 x_2)^{-1} + x_1^{1/2} + x_2^{3/4}$$

$$x_1, x_2 > 0$$

From the dual solution, determine the optimal solution to the given geometric program.

**11.29.** Solve the following posynomial GP problems:

(a) Minimize $f(x) = \dfrac{1}{x_1 x_2 x_3}$

Subject to $2x_1 + x_2 + 3x_3 \leqslant 1$

$x_1, x_2, x_3 > 0$

(b) Minimize $\quad f(x) = 5x_1^2 x_2^{-1} x_3 + 10 x_1^{-3} x_2^2 x_3^{-2}$

Subject to $\quad 0.357 x_1^{-1} x_3 + 0.625 x_1^{-1} x_3^{-1} \leqslant 1$

$\qquad x_1, x_2, x_3 > 0$

**11.30.** Solve the following signomial geometric program by the dual method:

$$\text{Maximize} \quad f = 5x_1^2 - x_2^2 x_3^4$$

$$\text{Subject to} \quad -5x_1^2 x_2^{-2} + 3x_2^{-1} x_3 \geqslant 2$$

$$x_1, x_2, x_3 > 0$$

**11.31.** Consider the welded beam problem discussed in Chapter 1. Solve it as a GP problem by:
(a) The dual approach
(b) The direct primal approach
(c) The transformed primal approach
Use appropriate NLP software for the solution.

**11.32.** Consider the "multistage compressor optimization" problem discussed in Chapter 8 (Example 8.1). Solve it as a geometric programming problem.

# Chapter 12

# Comparison of Constrained Optimization Methods

Elsewhere in this text we examine methods from primarily two perspectives. First, some methods are included because of their historical importance (or as tools to demonstrate fundamental concepts) such as Cauchy's method. Second, we include numerous methods that we feel are of practical importance in a modern computational environment. As we discuss these methods, we include, to the extent possible, remarks that delimit their advantages and disadvantages. It is, of course, nearly impossible to be complete in this effort, and in addition we have avoided extensive discussion of rate and region of convergence.

In this chapter we survey the major comparative studies up to this time and give selected results from a few of these experiments. The word "experiment" is important, since painfully little of what is currently known of the performance of these methods on "real" problems has come from purely theoretical considerations. Finally, a representative list of codes is given, along with information concerning availability.

## 12.1 A COMPARISON PHILOSOPHY

An important issue associated with any comparison is the selection of measures of merit. Even in the most qualitative setting, one must formulate merit measures that can be used as the basis of comparison. Presentation of a new algorithm or associated analysis in the literature will include performance indicators such as number of function and constraint evaluations and, less frequently, computing time. Unfortunately, the test problem set is typically limited, and comparative data for competing algorithms are seldom given. Most performance data in the literature are given in an effort to quantify algorithm speed or *efficiency*. The ability of a particular algorithm or software implementation to solve a wide variety of problems (that is, to be *robust*) is seldom specifically considered. This is truly regrettable, since most users consider robustness and *ease of use* to be at least equally as important as efficiency. It is important to recognize that the comparison of optimization algorithms and software is a multicriteria decision problem that has qualitative as well as quantitative aspects. Crowder, Dembo, and Mulvey [1] list eight measures of merit in their survey of comparative results: (1) CPU time, (2) terminal point accuracy, (3) number of iterations, (4) robustness, (5) number of function evaluations, (6) user friendliness, (7) storage requirements, and (8) basic operation count. We see that 1, 3, 5, and 8 are closely related and are indicators of efficiency.

Consider now the difference between algorithm and software testing. Results from both activities are of genuine interest. Recognize the difference between an *algorithm* and a computer program implementation of that algorithm, or *software*. An algorithm can be written down step by step and comparisons conducted at this level. For instance, we might make operation counts (that is, number of function and gradient calculations, or additions, subtractions, multiplications, and divisions required per iteration) or perform other calculations related to the basic logic of the algorithm as a basis of comparison. In order for these algorithms to be useful, we must translate them into machine-readable form, that is, we must create software. A number of details and decisions are required to effect this translation that are difficult if not impossible to describe adequately at the algorithm level. This, coupled with the desire to monitor actual performance on *representative* problems, is a strong motivation for software testing. The decision to test software brings with it a number of complicating factors, since machine-dependent parameters become a part of the experiment. Eason [2] has recently demonstrated experimentally the influence of several machine-dependent factors. He says that "the relative performance of optimization codes does depend on the computer and compiler used for testing, and this dependence is evident regardless of the performance measure." Even when the tests are performed with a single machine/compiler combination, several factors may influence the relative performance of codes. For instance, machine accuracy (word length), termination accuracy level, choice of test problems, and performance measure(s) employed may very well influence single-machine tests.

The difficulties associated with testing should be kept in mind during attempts to relate the performance of one algorithm or code implementation to another, especially when someone else's experimental results are employed. On the other hand, with the current state of development of the art and science of engineering optimization, imperfect empirical data to support decisions related to code selection seem superior to none at all. With this in mind, we survey in this chapter several recent experiments that may be useful in matching code and problem characteristics and therefore aid in code selection for practical applications.

## 12.2 A BRIEF HISTORY OF COMPARATIVE EXPERIMENTS

Since digital computers became widely available (approximately 1960 in the United States), there have been a growing number of reports in the literature of "new" algorithms, often complete with supporting data suggesting superiority of the new method. In addition, a number of survey articles (see, for instance, Box [3], Brooks [4], Leon [5], and Fletcher [6]) have attempted to compare methods within classes and from class to class. A common characteristic of these surveys is the mention of a relatively small number of test problems and methods. The situation changed dramatically with the appearance of Colville's study [7] in 1968. Colville sent eight problems having from 3 to 16 design variables, 0 to 14 constraints, and a standard timing routine to the developers of 30 codes. Each participant was invited to submit his "best effort" on each problem and the time required to execute the standard timing routine on his machine. The general characteristics of the Colville problems are listed in Table 12.1. Note that only two problems contain equality constraints and one contains no constraints at all.

The 30 codes tested represented quite a wide cross section of the available algorithms at that time. Colville placed the codes in one of five categories: search, small step gradient, large step gradient, second derivative, and miscellaneous methods. Included in the code set are four GRG codes, several methods based on the successive LP concept of Griffith and Stewart [8], a number of penalty function methods with a variety of unconstrained minimization strategies (including SUMT [9]), and a number of direct-search methods such as the complex method of Box [3]. Some codes required analytical expressions of the objective and constraint gradients, others employed numerical approximations to these gradients, while others did not employ gradient information at all.

Colville collected, organized the results, and concluded that the large step gradient methods were the most robust and efficient as a class. Unfortunately,

**Table 12.1 Colville Problem Set[a]**

| Number | Name and/or Source | $N$ | $J$ | $K$ | NC | NB |
|--------|-------------------|-----|-----|-----|-----|-----|
| 1 | Shell | 5 | 10 | 0 | 10 | 5 |
| 2 | Shell | 15 | 5 | 0 | 5 | 10 |
| 3 | Mylander/Research Analysis Corp. | 5 | 6 | 0 | 6 | 10 |
| 4 | Wood/Westinghouse | 4 | 0 | 0 | 0 | 8 |
| 5 | Efroymson/Esso | 6 | 4 | 0 | 4 | 0 |
| 6 | Huard/Electricite de France | 6 | 0 | 4 | 4 | 12 |
| 7 | Gauthier/IBM France | 16 | 0 | 8 | 8 | 32 |
| 8 | Colville/IBM | 3 | 14 | 0 | 14 | 6 |

[a]$N$ = number of variables, $J$ = number of inequality constraints, $K$ = number of equality constraints, NC = total number of constraints = $J + K$, NB = total number of bounds on variables.

the Colville study contains at last three flaws. Eason [10] has shown that Colville's timing routine does not adequately remove the effect of compiler and computing machine selection on code performance. Accordingly, the data collected at one site are not comparable to data collected at another. Furthermore, each participant was allowed to attack each problem as many times as he felt necessary to optimize the performance of his code. Thus another investigator (and certainly a general user) could not reasonably expect to produce similar results with the same code in the absence of the special insight that only its originator would possess. Finally, and possibly most important, no two participants reported solutions to the same accuracy.

In 1974, Eason and Fenton [11] reported on a comparative study involving 20 codes and 13 problems. Eason did his testing on one machine and included in his test-set the Colville problems plus several problems from mechanical design as shown in Table 12.2. In addition, Eason tested primarily penalty-type methods. The major contribution of Eason's study is the introduction of error curves as given in Figure 12.1. These data are obtained by storing intermediate iterations along the path to termination for each code. In this way a level of error is chosen as a basis of comparison after the experiments are complete. It is difficult to overestimate the value of this idea, since it makes it feasible to compare fairly the performance of codes from initial to terminal points. In effect, Eason removed the major flaws from the Colville experimental procedure. The major shortcomings of the Eason study are the lack of difficulty in the problem set and the failure to include the most powerful methods of the day. Many other comparisons have been reported, but the Colville and Eason studies mark the beginning of scientific comparison of nonlinear optimization methods.

#### Table 12.2  Eason Problem Set

| Number | Name and/or Source | $N$ | $J$ | $K$ | NC | NB |
|--------|--------------------|-----|-----|-----|-----|-----|
| 1 | Colville #1 | 5 | 10 | 0 | 10 | 5 |
| 2 | Post office parcel problem | 3 | 2 | 0 | 2 | 6 |
| 3 | Colville #3 | 5 | 6 | 0 | 6 | 10 |
| 4 | Colville #4 | 4 | 0 | 0 | 0 | 8 |
| 5 | Rosenbrock | 2 | 0 | 0 | 0 | 4 |
| 6 | Colville #6 | 6 | 0 | 4 | 4 | 12 |
| 7 | Journal bearing/Beightler | 2 | 1 | 0 | 1 | 4 |
| 8 | Flywheel/Siddall | 3 | 2 | 0 | 2 | 6 |
| 9 | Chemical reactor/Siddall | 3 | 9 | 0 | 9 | 4 |
| 10 | Gear train/Mischke | 2 | 0 | 0 | 0 | 4 |
| 11 | Cam design/Mischke | 2 | 2 | 0 | 2 | 4 |
| 12 | Mechanism synthesis/Eason | 4 | 0 | 0 | 0 | 8 |
| 13 | Gear train/Eason | 5 | 4 | 0 | 4 | 3 |

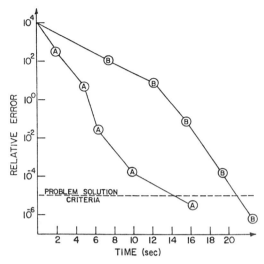

**Figure 12.1.** Error curves.

## 12.3 THE SANDGREN STUDY

In 1977 Eric Sandgren completed a major study [12, 13] of the leading nonlinear programming (NLP) methods of the day. We shall here examine the experimental procedure employed and the major results reported. The major steps in the study are:

1. Assembly of codes and problems.
2. Qualification of codes via a preliminary test set of 14 problems.
3. Application of 24 qualified codes to a full 35-problem test set.
4. Elimination of problems on which fewer than 5 codes were successful.
5. Compilation and tabulation of final results (24 codes on 23 problems).
6. Preparation of individual and composite utility curves.

Sandgren used the Colville and Eason problems as well as a set of geometric programming (GP) problems given by R. S. Dembo [14] (see Table 12.3). These problems provide a valuable link with previous comparisons, since they are well known and widely used. The GP problems are especially useful because of the availability of the powerful GP theory, which can be used to characterize the solution. In addition, this class of problems has a reputation of being difficult to solve using general-purpose NLP methods. The Sandgren test problems consist of the 13 Eason problems, Problems 2, 7, and 8 of Colville's study, the 8 Dembo problems, the welded beam problem, and 6 industrial design applications (all numbered in the order given). A complete description of these problems along with Fortran listings are given in reference 12. The problems vary in size and structure: the number of variables range from 2 to 48; the number of constraints range from 0 to 19; and the number of variable bounds range from 3 to 72. Unfortunately, only six problems contain equality constraints. Problems 1–8, 10–12, and 14–16 (a total of 14) were used in the preliminary tests of step 2, above. Problems 9, 13, 21, 22, and 28–30 were eliminated in step 4, because few codes made progress on them. Accordingly, the final test problem set in step 5 contains 23 problems.

The codes tested by Sandgren are given and briefly described in Table 12.4. Included in the tests were four generalized reduced gradient (GRG) codes, a method-of-multipliers (MOM) code, two methods based on extensions of linear programming, and a large variety of penalty methods. BIAS [15] is a method of multipliers code developed by Root and Ragsdell; whereas SEEK1, SEEK3, APPROX, SIMPLX, DAVID, and MEMGRD [16] were contributed by Siddall. GRGDFP [17] is a generalized reduced gradient code developed jointly by LaFrance, Hamilton, Ragsdell, and Whirlpool Corporation. RALP [18] is a variation of the Griffith-Stewart [8] method developed by Schuldt at Honeywell. RALP uses a Newton-type iteration to satisfy the equality constraints at each stage. GRG [19] is a GRG code developed by Lasdon and coworkers. OPT [20] is also a GRG code developed by Gabriele and Ragsdell. GREG [21] is a GRG code (in fact, the first to be generally available) developed by Abadie. COMPUTE II [22] is a penalty function package contributed by Gulf Oil Corporation. The package employs an exterior penalty formulation, several unconstrained search methods, and automatic constraint scaling. Mayne contributed the EXPEN and IPENAL package [23]. The package contains both exterior and interior penalty forms plus a variety of unconstrained search methods. SUMT IV [24] is the well-known penalty package developed by Fiacco, McCormick, Mylander, and others. Various unconstrained search methods are provided, as well as a choice of penalty parameter updating schemes in conjunction with a standard interior formula-

Table 12.3   Dembo Problem Set

| Number | Name and/or Source | $N$ | $J$ | $K$ | NC | NB |
|--------|--------------------|-----|-----|-----|-----|-----|
| 1[a] | Gibbs free energy | 12 | 3 | 0 | 3 | 24 |
| 2 | Colville #3 | 5 | 6 | 0 | 6 | 10 |
| 3 | Alkylation Process/Bracken and McCormick | 7 | 14 | 0 | 14 | 14 |
| 4[b] | Reactor Design/Rijckaert | 8 | 4 | 0 | 4 | 16 |
| 5 | Heat Exchanger/Avriel | 8 | 6 | 0 | 6 | 16 |
| 6 | Membrane Separation/Dembo | 13 | 13 | 0 | 13 | 26 |
| 7 | Membrane Separation/Dembo | 16 | 19 | 0 | 19 | 32 |
| 8 | Beck and Ecker | 7 | 4 | 0 | 4 | 14 |

[a]Dembo considered an unscaled and a scaled version of this problem.
[b]Dembo considered two versions of this problem.

tion. E04HAF [25] is a penalty method employing extrapolation techniques developed by Lootsma and contributed by Lill. The final code, COMET [26], is a moving truncations penalty method contributed by Himmelblau. All codes were converted to single precision, since the tests were conducted on a CDC-6500 with a 60-bit word. Any code that required analytical gradient information was altered to accept numerical gradient approximations using forward differences. Finally, all print instructions were removed from the basic iteration loop of the algorithms so that accurate measurements could be made of calculation time.

### 12.3.1 Preliminary and Final Results

A preliminary test problem set consisting of the Colville plus Eason problems, except Eason problems 9 and 13, was used to qualify codes. This was done to avoid the possibility of wasted effort (and cost) associated with applying all available codes to the full problem set. Partial results are given in Table 12.5. Codes that solved less than half of these 14 problems were eliminated from further study. It should be noted that in this phase any code that required more than three times the average time to solve a given problem was considered to have failed on that problem. Accordingly, codes 2, 4–7, 17, 18, 23–25, and 30 were eliminated at this preliminary stage. The codes employed in the final tests (that is, on the full problem set) are given in Table 12.6. Note that the final test set contains one MOM, four GRG, one Griffith-Stewart, one moving truncation, and 17 penalty type codes, for a total of 24. These 24 codes were applied to the full 35-problem test set.

Sandgren chose an error function that indicates degree of constraint violation and objective value:

$$\varepsilon_t = \varepsilon_f + \sum_{j=1}^{J} \langle g_j(x) \rangle_- + \sum_{k=1}^{K} |h_k(x)| \tag{12.1}$$

where

$$\varepsilon_f = \frac{|f(x) - f(x^*)|}{|f(x^*)|} \qquad \text{for } f(x^*) \neq 0 \tag{12.2}$$

and

$$\varepsilon_f = |f(x)| \qquad \text{for } f(x^*) = 0 \tag{12.3}$$

Recall the bracket operator notation, that is,

$$\langle \alpha \rangle_- = \begin{cases} 0 & \text{if } \alpha \geq 0 \\ -\alpha & \text{if } \alpha < 0 \end{cases} \tag{12.4}$$

**Table 12.4  Sandgren Test Codes**

| Code Number | Name | Source | Class | Unconstrained Search Method |
|---|---|---|---|---|
| 1 | BIAS | Root and Ragsdell | Multiplier method | Variable metric (DFP) |
| 2 | SEEK1 | Siddall | Interior penalty | Pattern + random check |
| 3 | SEEK3 | Siddall | Interior penalty | Pattern |
| 4 | APPROX | Siddall | Griffith-Stewart | None |
| 5 | SIMPLX | Siddall | Interior penalty | Simplex direct search |
| 6 | DAVID | Siddall | Interior penalty | Memory gradient |
| 8 | GRGDFP | Whirlpool/Purdue | Reduced gradient | Variable metric (DFP) |
| 9 | RALP | Schuldt/Honeywell | Griffith-Stewart | None |
| 10 | GRG | Lasdon and Waren | Reduced gradient | Variable metric (BFS) |
| 11 | OPT | Gabriele and Ragsdell | Reduced gradient | Conjugate gradient (FR) |
| 12 | GREG | Abadie | Reduced gradient | Conjugate gradient (FR) |
| 13 | COMPUTE II (0) | Gulf Oil | Exterior penalty | Hooke-Jeeves |
| 14 | COMPUTE II (1) | Gulf Oil | Exterior penalty | Conjugate gradient (FR) |
| 15 | COMPUTE II (2) | Gulf Oil | Exterior penalty | Variable metric (DFP) |
| 16 | COMPUTE II (3) | Gulf Oil | Exterior penalty | Simplex/pattern |

| 17 | EXPEN (1) | Afimiwala and Mayne | Exterior penalty | Pattern |
| 18 | EXPEN (2) | Afimiwala and Mayne | Exterior penalty | Steepest descent |
| 19 | EXPEN (3) | Afimiwala and Mayne | Exterior penalty | Conjugate direction |
| 20 | EXPEN (4) | Afimiwala and Mayne | Exterior penalty | Conjugate gradient (FR) |
| 21 | EXPEN (5) | Afimiwala and Mayne | Exterior penalty | Variable metric (DFP) |
| 22 | EXPEN (6) | Afimiwala and Mayne | Exterior penalty | Pattern |
| 23 | IPENAL (1) | Afimiwala and Mayne | Interior penalty | Univariate |
| 24 | IPENAL (2) | Afimiwala and Mayne | Interior penalty | Steepest descent |
| 25 | IPENAL (3) | Afimiwala and Mayne | Interior penalty | Conjugate direction |
| 26 | IPENAL (4) | Afimiwala and Mayne | Interior penalty | Conjugate gradient (FR) |
| 27 | IPENAL (5) | Afimiwala and Mayne | Interior penalty | Variable metric (DFP) |
| 28 | SUMT IV (1) | Mylander | Interior penalty | Newton + |
| 29 | SUMT IV (2) | Mylander | Interior penalty | Newton + + |
| 30 | SUMT IV (3) | Mylander | Interior penalty | Steepest descent |
| 31 | SUMT IV (4) | Mylander | Interior penalty | Variable metric (DFP) |
| 32 | E04HAF (0) | Lill and Lootsma | Mixed penalty | Conjugate direction |
| 33 | E04HAF (1) | Lill and Lootsma | Mixed penalty | Variable metric (BFS) |
| 34 | E04HAF (2) | Lill and Lootsma | Mixed penalty | Newton |
| 35 | COMET | Staha and Himmelblau | Moving truncations | Variable metric (BFS) |

Table 12.5  Preliminary Test Results

| Code | Name | Number of Problems Solved | Code | Name | Number of Problems Solved |
|---|---|---|---|---|---|
| 1 | BIAS | 12 | 19 | EXPEN (3) | 11 |
| 2 | SEEK1 | 5 | 20 | EXPEN (4) | 10 |
| 3 | SEEK3 | 8 | 21 | EXPEN (5) | 12 |
| 4 | APPROX | 6 | 22 | EXPEN (6) | 7 |
| 5 | SIMPLX | 6 | 23 | IPENAL (1) | 4 |
| 6 | DAVID | 5 | 24 | IPENAL (2) | 4 |
| 7 | MEMGRD | 2 | 25 | IPENAL (3) | 3 |
| 8 | GRGDFP | 14 | 26 | IPENAL (4) | 10 |
| 9 | RALP | 9 | 27 | IPENAL (5) | 11 |
| 10 | GRG | 14 | 28 | SUMT IV (1) | 8 |
| 11 | OPT | 13 | 29 | SUMT IV (2) | 9 |
| 12 | GREG | 14 | 30 | SUMT IV (3) | 2 |
| 13 | COMPUTE II (0) | 12 | 31 | SUMT IV (4) | 11 |
| 14 | COMPUTE II (1) | 8 | 32 | E04HAF (0) | 12 |
| 15 | COMPUTE II (2) | 10 | 33 | E04HAF (1) | 11 |
| 16 | COMPUTE II (3) | 10 | 34 | E04HAF (2) | 8 |
| 17 | EXPEN (1) | 4 | 35 | COMET | 11 |
| 18 | EXPEN (2) | 5 | | | |

Table 12.6  Final Test Codes

| Code | Name | Code | Name |
|---|---|---|---|
| 1 | BIAS | 20 | EXPEN (4) |
| 3 | SEEK3 | 21 | EXPEN (5) |
| 8 | GRGDFP | 22 | EXPEN (6) |
| 9 | RALP | 26 | IPENAL (4) |
| 10 | GRG | 27 | IPENAL (5) |
| 11 | OPT | 28 | SUMT IV (1) |
| 12 | GREG | 29 | SUMT IV (2) |
| 13 | COMPUTE II (0) | 31 | SUMT IV (4) |
| 14 | COMPUTE II (1) | 32 | E04HAF (0) |
| 15 | COMPUTE II (2) | 33 | E04HAF (1) |
| 16 | COMPUTE II (3) | 34 | E04HAF (2) |
| 19 | EXPEN (3) | 35 | COMET |

542

In other words, only the violated constraints appear in (12.1). Plots of $\varepsilon_t$ for each code on each problem were prepared such as those in Figures 12.1 and 12.2. After Eason, this allowed the comparison of codes at exactly the same error level. One can learn something of the character of an algorithm by examining these curves. For instance, the point on a given curve (that is, corresponding to a particular code) of lowest error value indicates the ability of the algorithm to sharply define a solution. In addition, the shape of the curve is an indication of the algorithm's rate of convergence.

Since the ability to solve a large number of problems in a reasonable amount of time was Sandgren's comparison criterion, he ranked the codes on the basis of the number of problems solved within a series of specified limits on relative solution time. The limits are based on a fraction of the average time for all codes on each problem. Furthermore, each solution time for a problem was normalized by dividing by the average solution time on that problem. This produces a low normalized solution time for a code with a relatively fast solution time, and high normalized solution time for a code with a relatively slow solution time. This normalization process allows direct comparison of results on problems of different levels of difficulty. Examine the performance of the final test codes at an $\varepsilon_t$ level of $10^{-4}$ as given in Table 12.7. From the table it is easy to identify the fast codes, since they have the large numbers in the first column labeled .25. The first column contains the number of problems solved by each code in 25% of the average solution time, and so on for the other columns. At this time, seven problems were excluded from the final rated set (problems 9, 13, 21, 22, and 28–30), because fewer than five codes were successful. In addition, only 14 of the codes in the final tests solved half or more of the rated test problems.

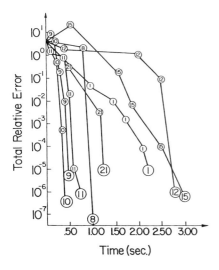

**Figure 12.2.** Code performance on welded beam problem.

Table 12.7   Number of Problems Solved at $\varepsilon_t = 10^{-4}$

| Code Number | Name and/or Source | Number of Problems Solved | | | | | |
|---|---|---|---|---|---|---|---|
| | | $.25^a$ | .50 | .75 | 1.00 | 1.50 | 2.50 |
| 1 | BIAS | 0 | 9 | 14 | 17 | 19 | 20 |
| 3 | SEEK3 | 0 | 1 | 2 | 3 | 6 | 10 |
| 8 | GRGDFP | 10 | 15 | 17 | 17 | 18 | 18 |
| 9 | RALP | 12 | 13 | 13 | 14 | 14 | 16 |
| 10 | GRG | 15 | 17 | 19 | 19 | 19 | 19 |
| 11 | OPT | 16 | 21 | 21 | 21 | 21 | 21 |
| 12 | GREG | 14 | 18 | 20 | 23 | 23 | 23 |
| 13 | COMPUTE II (0) | 2 | 7 | 9 | 11 | 13 | 15 |
| 14 | COMPUTE II (1) | 2 | 4 | 6 | 6 | 8 | 9 |
| 15 | COMPUTE II (2) | 4 | 9 | 11 | 15 | 15 | 15 |
| 16 | COMPUTE II (3) | 1 | 5 | 7 | 8 | 9 | 11 |
| 19 | EXPEN (3) | 0 | 0 | 2 | 3 | 7 | 11 |
| 20 | EXPEN (4) | 1 | 4 | 10 | 10 | 11 | 11 |
| 21 | EXPEN (5) | 7 | 13 | 14 | 16 | 16 | 16 |
| 22 | EXPEN (6) | 2 | 4 | 7 | 8 | 8 | 9 |
| 26 | IPENAL (4) | 0 | 1 | 4 | 6 | 9 | 10 |
| 27 | IPENAL (5) | 2 | 7 | 11 | 14 | 14 | 17 |
| 28 | SUMT IV (1) | 0 | 0 | 0 | 0 | 3 | 9 |
| 29 | SUMT IV (2) | 0 | 0 | 0 | 1 | 2 | 7 |
| 31 | SUMT IV (4) | 0 | 1 | 3 | 5 | 9 | 13 |
| 32 | E04HAF (0) | 0 | 0 | 0 | 4 | 8 | 15 |
| 33 | E04HAF (1) | 0 | 2 | 4 | 7 | 13 | 15 |
| 34 | E04HAF (2) | 0 | 1 | 2 | 4 | 8 | 10 |
| 35 | COMET | 0 | 2 | 5 | 7 | 9 | 15 |

$^a$Times the average solution time.

Rather than a simple ranking, Sandgren gave the performance of the final code set on the final (rated) problem set in graphical form. He normalized the results with regard to problem difficulty by using fraction of average time as the abscissa. Number of problems solved (robustness) is the ranking criterion and is therefore the ordinate. *Average time* is the average solution time to a given termination error for all successful codes on a given problem. Figure 12.3 gives a graphical representation of the performance of the 14 codes that solved half or more of the rated test problems. All codes in this final category exhibited superior speed and robustness. The shape of an individual utility curve gives insight to a code's character. The perfect code would solve all 23 problems in no time, which is obviously impossible. On the other hand, codes that have a utility curve with a steep slope and a high final ordinate are both fast and robust. In Figure 12.4 the effect of the termination criterion is given. We see that all codes have difficulty solving the problems when the termina-

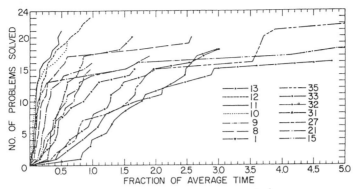

**Figure 12.3.** Algorithm utility ($\varepsilon_t = 10^{-4}$).

tion criterion becomes very small. Since several code implementations of the same (or similar) algorithm were tested, one wonders if *algorithm class* trends can be seen in the performance data. The three algorithm classes considered are generalized reduced gradient, exterior penalty, and interior penalty. The average performance of the best three codes in each class is shown in Figure 12.5. Clearly, the generalized reduced gradient codes are the most robust and fastest tested. Sandgren has shown [27] that the trends in Figure 12.5 are indeed statistically significant.

Sandgren also tested a hybrid code consisting of a method of multipliers (BIAS) to find the neighborhood of the solution, and a GRG code (OPT) to sharply refine the solution. He found this hybrid method to be the fastest and most robust when compared to the other codes tested. Finally, Sandgren used the Rosen and Suzuki [28] approach to generate NLPs with quadratic objective

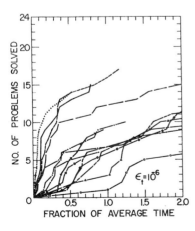

**Figure 12.4.** Algorithm utility.

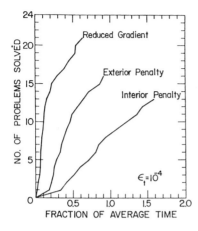

**Figure 12.5.** Average algorithm performance.

and constraint functions. In this way he conveniently generated problems with known characteristics, including the solution. He studied the effect of increase in number of design variables, inequality constraints, equality constraints, and nonlinearity in the objective and constraint functions on solution time for a MOM code (BIAS), a repetitive LP code (RALP), two GRG codes (GRG and OPT), two exterior penalty methods [COMPUTE II(2) and EXPEN(5)], and an interior penalty code [SUMT IV(4)]. He discovered a rather uniform increase in the solution time as a function of $N$. RALP displayed the least sensitivity (in solution time) to number of inequality constraints, with a 59% increase in time corresponding to a 100% increase in number of inequality constraints. The GRG codes were significantly less sensitive to an increase in equality constraints, with OPT actually recording a *decrease* in solution time with an increase in number of equality constraints from one to three! Mayne's EXPEN(5) was the least sensitive to increase in nonlinearity, with Lasdon's GRG a close second. The interested reader is referred to Sandgren's thesis [12] for additional details.

## 12.4   THE SCHITTKOWSKI STUDY

In 1980 Schittkowski published [29] the results of a very extensive study of NLP codes. His experiments included 20 codes (actually more, considering minor modifications of the same code) and 180 randomly generated test problems with predetermined character and multiple starting points. There is great similarity between the Schittkowski and Sandgren studies, with several differences. The most important difference is the inclusion of *successive quadratic programming* (SQP) methods in the Schittkowski study. The codes he tested are listed and briefly described in Table 12.8. Six of these codes were

## Table 12.8 Schittkowski Test Codes

| Code Number | Code Name | Code Source | Class | Unconstrained Search Method |
|---|---|---|---|---|
| 1 | OPRQP | Biggs [30] | SQP | Variable metric (BFS) |
| 2 | XROP | Biggs [30] | SQP | Variable metric (BFS/Powell) |
| 3 | VF02AD | Powell [31] | SQP | Variable metric (BFS/Powell) |
| 4 | GRGA | Abadie [32] | GRG | Conjugate gradient (FR) |
| 5 | OPT | Gabriele and Ragsdell [33] | GRG | Conjugate gradient (FR) |
| 6 | GRG2(1) | Lasdon et al. [34] | GRG | Variable metric (BFS) |
| 7 | GRG2(2) | Lasdon et al. [34] | GRG | Variable metric (BFS) |
| 8 | VF01A | Fletcher [35] | MOM | Quasi-Newton/Fletcher |
| 9 | LPNLP | Pierre and Lowe [36] | MOM | Variable metric (DFP) |
| 10 | SALQDR | Gill and Murray [37] | MOM | Variable metric (BFS) |
| 11 | SALQDF | Gill and Murray [37] | MOM | Variable metric (BFS) |
| 12 | SALMNF | Gill and Murray [37] | MOM | Modified Newton |
| 13 | CONMIN | Haarhoff and Buys [38] | MOM | Variable metric (DFP) |
| 14 | BIAS(1) | Root and | MOM | Variable metric (DFP) |
| 15 | BIAS(2) | Ragsdell [39] | MOM | Variable metric (DFP) |
| 16 | FUNMIN | Kraft [40] | MOM | Variable metric (BFS) |
| 17 | GAPFPR | Himmelblau [41] | MOM | Quasi-Newton/Fletcher |
| 18 | GAPFQL | Newell and Himmelblau [41] | MOM | Quasi-Newton/Fletcher |
| 19 | ACDPAC | Best and Bowler [42] | MOM | Conjugate direction |
| 20 | FMIN(1) | Lootsma and Kraft [43] | Mixed penalty | Variable metric(BFS) |
| 21 | FMIN(2) | Lootsma and Kraft [43] | Mixed penalty | Variable metric (BFS) |
| 22 | FMIN (3) | Lootsma and Kraft [43] | Mixed penalty | Modified Newton |
| 23 | NLP | Rufer [44] | Exterior penalty | Quasi-Newton/Fletcher |
| 24 | SUMT | Mylander | Interior penalty | Variable metric (DFP) |
| 25 | DFP | Indusi [45] | Exterior penalty | Variable metric (DFP) |
| 26 | FCDPAK | Best [46] | SQP/ Robinson | Conjugate direction |

also tested by Sandgren, but Schittkowski's test set obviously contains the most powerful methods available to date. He tested 4 penalty, 11 MOM, 3 GRG, and 4 SQP codes. Space will not allow detailed examination of Schittkowski's experimental design or extensive results. On the other hand, we here attempt to highlight the important aspects, and refer the interested reader to ref. [29].

Codes were evaluated using the following criteria: (1) efficiency, (2) reliability, (3) global convergence; ability to solve (4) degenerate problems, (5) ill-conditioned problems, and (6) indefinite problems; sensitivity to (7) slight problem variation and (8) position of the starting point; and (9) ease of use.

Schittkowski applied each code to all of the 180 randomly generated test problems, collecting the data much in the same manner as Sandgren. Each of the nine evaluation criteria contain several subcriteria; for instance, efficiency has five subcriteria. He then chose weights using Saaty's priority theory [47] as outlined by Lootsma [48] in order to calculate a single average efficiency

Table 12.9    Schittkowski's Code Ranking

| Code Number | Code Name | Class | Rank |
|---|---|---|---|
| 3 | VF02AD | SQP | 1 |
| 4 | GRGA | GRG | 2 |
| 7 | GRG2(2) | GRG | 3 |
| 26 | FCDPAK | SQP/Robinson | 4 |
| 6 | GRG2(1) | GRG | 5 |
| 1 | OPRQP | SQP | 6 |
| 2 | XROP | SQP | 7 |
| 19 | ACDPAC | MOM | 8 |
| 14 | BIAS(1) | MOM | 9 |
| 5 | OPT | GRG | 10 |
| 17 | GAPFPR | MOM | 11 |
| 8 | VF01A | MOM | 12 |
| 11 | SALQDF | MOM | 13 |
| 12 | SALMNF | MOM | 14 |
| 23 | NLP | Exterior penalty | 15 |
| 9 | LPNLP | MOM | 16 |
| 10 | SALQDR | MOM | 17 |
| 15 | BIAS(2) | MOM | 18 |
| 21 | FMIN(2) | Mixed penalty | 19 |
| 22 | FMIN(3) | Mixed penalty | 20 |
| 18 | GAPFQL | MOM | 21 |
| 20 | FMIN(1) | Mixed penalty | 22 |
| 16 | FUNMIN | MOM | 23 |
| 24 | SUMT | Interior penalty | 24 |
| 13 | CONMIN | MOM | 25 |
| 25 | DFP | Exterior penalty | 26 |

**Table 12.10    Performance Measure Priorities**

| | |
|---|---|
| Efficiency | .32 |
| Reliability | .23 |
| Global convergence | .08 |
| Ability to solve degenerate problems | .05 |
| Ability to solve ill-conditioned problems | .05 |
| Ability to solve indefinite problems | .03 |
| Sensitivity to slight problem variation | .03 |
| Sensitivity to the position of the starting point | .07 |
| Ease of use | .14 |
| | 1.00 |

measure, and similarly for the other criteria. Using the same priority theory, he chose weights for each of the nine criteria to form an overall performance measure. This produces the ranking given in Table 12.9. The weights used for each criterion are listed in Table 12.10. These priorities (weights) are a subjective matter, and a particular user might find the various criteria to be of much different relative importance and therefore would choose different weights and produce a correspondingly different ranking.

Based on his very extensive tests, Schittkowski draws some interesting algorithm class conclusions and gives recommendations for software development. He concludes that with regard to efficiency (speed) the newer methods exhibit significant improvement; and he ranks the algorithms classes more or less as (1) SQP, (2) GRG, (3) MOM, and (4) other penalty methods. In addition, his tests strongly support the not widely understood fact that code reliability is more a reflection of *coding* than *algorithm*. He also noted that the performance of GRG codes such as GRGA and OPT was relatively independent of problem condition. In fact, in some cases the performance of these codes actually improved as the test problems became increasingly ill-conditioned. This is probably due to the strict constraint-following strategy employed. This also agrees with our own experience, which suggests that GRG-based codes are relatively insensitive to problem condition and are therefore *not* generally in need of special problem-scaling procedures.

## 12.5    THE FATTLER GEOMETRIC PROGRAMMING STUDY

In 1974 Reklaitis and Ragsdell [49–51] began a study of the relative merits of GP formulation strategies and solution techniques. They considered methods that address prototype posynomial GP problems (see Section 11.4). These problems are in general nonconvex because of the nonlinearities of the constraints, and can be expected to severely tax conventional NLP codes. However, despite the apparent difficulty of primal problems, there are structural features of the generalized posynomial functions that can be exploited to facilitate direct primal solutions. In addition, various transformations can be employed to give equivalent formulations. The additional formulations considered in this study are the convexified primal, the transformed primal, the dual, and the transformed dual, as described in Chapter 11. Other comparative studies of prototype GP solution approaches have been reported in the literature. Two of these, Rijckaert and Martens [52] and Dembo [53], primarily focused on generalized GP's but did include prototype problems in their test sets. The study by Dinkel et al. [54] was restricted to the examination of alternative cutting plane methods used for the solution of the convex primal.

Ten codes or code variants were used in this study. The first four are general-purpose NLP codes included because of favorable performance in the Sandgren study. These codes are listed as 1 (BIAS), 9 (RALP), 11 (OPT), and 27 [IPENAL(5)] in Table 12.4. In addition, Root [55] prepared a special version of the method of multipliers code, BIAS-SV, which exploits the convexified GP primal formulation for use in the study. BIAS-SV employs Newton's method for the unconstrained iterations and uses second derivatives in the line searches. This was conveniently accomplished due to the availability of first and second derivatives in the convexified primal form. The remaining five codes are special-purpose GP codes. GPKTC [56], a code developed by Martens and Rijckaert, solves the convexified primal using a Kuhn-Tucker optimality condition iteration; GGP [57], a code developed by Dembo, uses Kelley's cutting plane method and also addresses the convexified primal; MCS [58] is a GP specialization of the convex simplex method of Zangwill as modified by Beck and Ecker; QUADGP [59] is a GP code developed by Bradley, which solves the transformed dual as a series of quadratic programs; and finally DAP [60] is an adaptation of the differential algorithm of Beightler and Wilde given by Reklaitis.

Forty-two prototype GP problems were selected from those available in the literature, with about half representing engineering applications. Many of the problems had been used previously by Dembo and by Beck and Ecker; complete source references are given by Fattler et al. [50]. A review of the problem set reveals the following general characteristics: the number of primal variables range from 2 to 30, primal terms from 8 to 197, constraints from 1 to 73, and exponent matrix density from 3.3 to 83%.

This experiment was conducted along the lines of the Sandgren study with the following exceptions: multiple, randomly generated starting points were used; a computer program was written that essentially automated the data-collection process; formulation effects were explored; and the significance of the results was statistically verified. Primal and dual starting points for each problem were generated randomly by sampling points on a hypersphere whose center is the solution. Only feasible points were used. Typically, two radii were used and 10 points retained for each radius for each problem. The obvious difficulty associated with multiple starting points is the increase in computational effort. In this study, some 10,000 test runs were made, of which 6000 produced useful results. A Lagrangian-type error function is used for purposes of comparison:

$$\varepsilon_L = \left| \frac{g_0 - g_0^*}{g_0^*} + \sum_k \lambda_k^* \frac{|g_k - g_k^*|}{g_0^*} \right| \tag{12.5}$$

The starred quantities are the known optimal values of the problem functions and Lagrange multipliers. The sum includes only those constraints that are tight at the solution. Values of this error function were computed for the intermediate iteration points recorded during each run. The intermediate solution times (which excluded all I/O time required for recording test data) and error function values were fitted to polynomials, which were used as interpolating functions to determine solution times at specified function levels. Mean solution times and standard deviations were tabulated for each problem code or problem-formulation combination at relative error levels of $10^{-2}, 10^{-3}, 10^{-4}$, for all successful runs and at termination for all runs. Each code was run with a fixed set of program parameters, which were determined in preliminary tests and usually were very close to the values suggested by the various code authors. Since adjustment of these values was not allowed during the formal tests, some codes did not achieve a $10^{-4}$ error level on some problems.

The results from this study are simply too voluminous to adequately describe here, but some brief remarks can be given along with general conclusions. First it is important to note the effect of starting point selection on solution time. As can be seen from Table 12.11, the variation in solution times obtained for problem 13 using OPT started from 10 different points can be considerable. For instance, for primal solution the range of times is from 1.776 to 5.72 s, with a mean of 3.65 and standard deviation of 1.42. Similarly, for convex primal solution the range is from 0.83 to 1.1 s, with mean of 0.878 and standard deviation of 0.165. Note that for a given starting point the ratio of primal to convex primal solution times changes substantially. For instance, it is 1.8 for starting point 5 and 8.3 for starting point 8. Yet these computation times are obtained using the same code with identical termination parameters. A similar, although less pronounced, variation can be noted when the OPT convexified primal solution times are compared to those obtained with the GGP program. The solution time ratio ranges from 0.53 for problem 8 to 1.0 for problem 10. These results indicate that code ranking based on performance with a single starting point, as is commonly done in the literature, is a very questionable procedure. If at all possible, a sufficiently large number of points must be used so that reliable means and standard deviations can be computed. Comparisons must then be made via statistical tests such as those of Student. A summary of the number of problems solved and runs attempted with each code-formulation pair is given in Table 12.12. As can be seen, not all problems were attempted with some codes because it was sometimes impossible but more often, unnecessary due to well-established trends. The general NLP codes RALP and IPENAL(5) seem particularly prone to failure, as was the code OPT when used on the transformed primal formulation. Most surprising was the erratic performance of GPKTC, which at times produced extremely fast solutions but in other cases failed completely. Similarly surprising is the high number of failures of the special version of BIAS, especially in view of the high reliability of the regular version. Both solved the convexified form of the primal; the former used a modified Newton algorithm with analytic derivatives, the latter a DFP algorithm with numerical derivatives. The most reliable performance seems to have been achieved by the general NLP codes OPT and BIAS when applied to the convexified primal formulation. The next best performance was attained by the specialized codes GGP and DAP. The results were aggregated into the following classes:

1. Comparison of solution times of various algorithms for a given GP problem formulation.
2. Cross-comparison of solution times for the various formulations all solved using the same algorithm.
3. Cross-comparisons of the most successful algorithms found for each GP formulation type.
4. Examination of how solution time varies with problem characteristic dimensions for each of the various formulations.

Finally, statistical tests were administered to discern the significance of the variation in code performance. Assuming that the solution times $x$ and $y$ of two codes for any given problem are normally distributed variables each with their own variance $\sigma_x^2$ and $\sigma_y^2$, then code solution time comparison is equivalent

**Table 12.11    The Effect of Starting Points on Solution Times**

| Starting Point ($R = .5$) | OPT Primal, s | OPT Primal / OPT Convex Primal | OPT Convexified Primal, s | OPT Convex Primal / GGP | GGP, s |
|---|---|---|---|---|---|
| 1 | 2.991 | 2.7 | 1.094 | .87 | 1.253 |
| 2 | 4.224 | 6.2 | .684 | .58 | 1.174 |
| 3 | 2.161 | 2.8 | .770 | .58 | 1.335 |
| 4 | 3.323 | 4.4 | .748 | .58 | 1.296 |
| 5 | 1.776 | 1.8 | .995 | .87 | 1.141 |
| 6 | 2.115 | 2.6 | .824 | .63 | 1.305 |
| 7 | 4.309 | 4.1 | 1.045 | .68 | 1.535 |
| 8 | 5.673 | 8.3 | .683 | .53 | 1.298 |
| 9 | 5.722 | 6.8 | .837 | .65 | 1.320 |
| 10 | 4.214 | 3.8 | 1.100 | 1.00 | 1.104 |

**Table 12.12    Number of Problems Attempted and Solved in Fattler Study**

| Code | Problems Attempted | Runs Attempted | Runs Failed | % Unsuccessful Attempts |
|------|------|------|------|------|
| OPT-P | 40 | 399 | 27 | 6.77 |
| OPT-CP | 41 | 616 | 1 | 0.16 |
| OPT-TP | 25 | 452 | 124 | 27.43 |
| GGP | 41 | 598 | 24 | 4.01 |
| GPKTC | 39 | 589 | 240 | 40.75 |
| MAYNE | 31 | 379 | 61 | 16.09 |
| RALP-P | 34 | 446 | 146 | 32.74 |
| RALP-CP | 37 | 552 | 115 | 20.83 |
| BIAS-P | 39 | 260 | 1 | 0.38 |
| BIAS-CP | 40 | 457 | 0 | 0.0 |
| BIAS-SV | 29 | 166 | 21 | 12.65 |
| MCS | 26 | 412 | 60 | 14.56 |
| DAP | 40 | 406 | 14 | 3.45 |
| QUAD-GP | 34 | 149 | 13 | 8.72 |

to the problem of testing whether the true mean solution times $M_x$ and $M_y$ of the two codes for the given problem are equal. This is the Behrens-Fisher problem of statistics [61]. Appropriate expressions were used to test the difference in the means of the various solution times. For several problems, the means were not significantly different. These calculations were completed at the $10^{-2}$, $10^{-3}$, and $10^{-4}$ relative error levels. The results were used to determine the number of problems for which each code achieved the best or second best solution time. A 90% significance level was required before means were considered to be different; otherwise they were considered to be the same. The results of this ranking process are given in Table 12.13

**Table 12.13    Fattler Code Ranking**

| Relative Error Ranking | $10^{-2}$ | | $10^{-3}$ | | $10^{-4}$ | |
|------|------|------|------|------|------|------|
| | 1st | 2nd | 1st | 2nd | 1st | 2nd |
| OPT-CP | 13 | 14 | 18 | 8 | 14 | 7 |
| DAP | 16 | 1 | 8 | 5 | 4 | 2 |
| GGP | 6 | 9 | 7 | 7 | 3 | 8 |
| QUADGP | 10 | 5 | 5 | 7 | 5 | 6 |
| MCS | 3 | 6 | 7 | 4 | 13 | 1 |
| OPT-P | 0 | 4 | 0 | 4 | 0 | 3 |

The major conclusions of this study are:

1. The convex primal is inherently the most advantageous formulation.
2. A general-purpose GRG code applied to the convex primal will dominate even the reputedly best specialized GP codes currently available.
3. The differences between the primal and convex primal formulations lie mainly in scaling and function evaluation time.
4. Transformed primal solution approaches are not likely to lead to more efficient GP solution than the convex primal.
5. The dual approaches are likely to be competitive only for tightly constrained problems of a small degree of difficulty.
6. Posynomial GP problem difficulty as measured in solution time is best correlated to an exponential of the number of variables in the formulation being solved and is proportional to the total number of multiterm primal constraints.

Sin and Reklaitis [62, 63] report on an extension of this study to include generalized GP problems; that is, the function terms may have + or − signs. The tests and results follow the posynomial study closely, except that the specialized code GGP due to Dembo performed slightly better than the GRG code, OPT.

## 12.6   CODE AVAILABILITY

Giving advice on optimization software is somewhat like marriage counseling in that it helps to know a great deal about the particular application before presuming to advise. Others have made similar efforts from which we have profited much (see, for instance, the excellent decision tree given by Fletcher [6, ps. 72 and 73]). In fact, to some extent this entire text addresses this single issue; that is, knowing the character of my problem or problem class, what algorithm or code should I use? Throwing caution to the wind, we here give recommendations that some will find useful when considering available optimization software.

Anyone seriously considering a software development effort or acquisition of an optimization package should first carefully read the paper by Waren and Lasdon [64], the dissertation by Sandgren [12], and the report by Schittkowski [29]. Next, write for additional information describing the technical aspects of the algorithm, the author's expectation of efficiency and robustness, and the author's estimate of range of applicability for one or more of the following recommended software packages:

### *State of the Art NLP Software*

1. **GRGA**: GRG/Fletcher-Reeves
   Contact:    J. Abadie
                 University of Paris VI
                 Institut de Programmation
                 4, Place Jussieu
                 Paris, France

2. **GRG2**: GRG/BFS
   Contact:    L. S. Lasdon
                 General Business
                 University of Texas
                 Austin, TX 78712

3. **OPT**: GRG/Fletcher-Reeves
   Contact:    K. M. Ragsdell
                 Design Optimization Laboratory
                 Aerospace and Mechanical Engineering
                 The University of Arizona
                 Tucson, AZ 85721

4. **MINOS**/Augmented: Augmented Lagrangian/GRG for linearly constrained subproblems.
   Contact:    M. Saunders
                 Systems Optimization Laboratory
                 Operations Research
                 Stanford University
                 Stanford, CA 94305

5. **COMPUTE II**: Penalty package with various penalty forms and
   unconstrained methods.
   Contact:    A. B. King
                 Gulf Research and Development
                 P. O. Box 2038
                 Pittsburgh, PA 15230

6. **BIAS**: MOM/DFP/automatic scaling
   Contact:    K. M. Ragsdell
                 Design Optimization Laboratory
                 Aerospace and Mechanical Engineering
                 The University of Arizona
                 Tucson, AZ 85721

7. **OPRQP**: SQP/BFS
   Contact:    M. C. Bartholomew-Biggs
                 Numerical Optimization Centre

> Hatfield Polytechnic
> 19 St. Albans Road
> Hatfield, Herts, UK

8.  **VF02AD**: SQP/BFS +
    Contact:  M. J. D. Powell
    Computer Science and Systems Div.
    A.E.R.E.
    Harwell, Oxfordshire, UK

9.  **VMCON1**: Similar to VF02AD
    Contact:  L. C. Ranzini
    National Energy Software Center
    Argonne National Laboratory
    9700 S. Cass Avenue
    Argonne, IL 60439


*For Education Use*


10.  **OPTIVAR**: A library of very-easy-to-use NLP methods; includes most modern methods
     Contact:  J. N. Siddall
     Faculty of Engineering
     McMaster University
     Hamilton, Ontario, Canada

11.  (No name): A group of interior and exterior penalty methods with various unconstrained techniques
     Contact:  R. W. Mayne
     Mechanical Engineering
     SUNY at Buffalo
     Buffalo, NY

12.  **OPTLIB**: A library of very-easy-to-use NLP methods
     Contact:  K. M. Ragsdell
     Design Optimization Laboratory
     Aerospace and Mechanical Engineering
     The University of Arizona
     Tucson, AZ 85721


At this writing most would call OPRQP and VF02AD experimental codes, but certainly more generally useful versions will become available in the near future. Hopefully, all of these codes are doomed to obsolescence, and much more powerful algorithms and codes will be available in the future.

## 12.7  SUMMARY

In this chapter we have discussed the need for empirical data to discern the relative merits of the various modern NLP methods. In order to learn something of the nature of the various algorithms, we test software and create a noise in the data that is both troublesome and pragmatic—troublesome in that it is sometimes impossible to point to specific algorithm characteristics as causes of particular numerical results, and pragmatic in that all algorithms must be coded in machine-readable form to be of value on modern engineering problems.

We have given a brief history of testing and briefly examined the Sandgren, Schittkowski, and Fattler studies. Finally, we have combined our understanding of currently available comparative data and our own prejudice to recommend codes worthy of additional investigation.

## REFERENCES

1. Crowder, H. P., R. S. Dembo, and J. M. Mulvey, "Reporting Computational Experiments in Mathematical Programming," *Math. Programming*, **15**, 316–329 (1978).

2. Eason, E. D., "Evidence of Fundamental Difficulties in Nonlinear Optimization Code Comparisons," *Proc. U.S. Bur. Stands/Math. Programming Soc. Conf. Testing Validating Algorithms Software*, Boulder, CO, Jan. 1981.

3. Box, M. J., "A New Method of Constrained Optimization and a Comparison with Other Methods," *Computer J.*, **8**(1), 42–51 (1965).

4. Brooks, S. H., "A Comparison of Maximum Seeking Methods," *J. Oper. Res.*, **7**(4), 430–437 (1959).

5. Leon, A., "A Comparison Among Eight Known Optimizing Procedures," in *Recent Advances in Optimization Techniques* (A. Lavi and W. Vogl, Eds.), Wiley, New York, 1966, pp. 23–42.

6. Fletcher, R., "Mathematical-Programming Methods—A Critical Review," in *Optimum Structural Design: Theory and Applications* (R. H. Gallagher and O. C. Zienkiewicz, Eds.), Wiley, New York, 1973, pp. 51–77.

7. Colville, A. R., "A Comparative Study of Nonlinear Programming Codes," Tech. Rep. No. 320-2949, IBM New York Scientific Center, June 1968.

8. Griffith, F. E., and R. A. Stewart, "A Nonlinear Programming Technique for Optimization of Continuous Processing Systems," *Manage. Sci.*, **7**, 379–392 (1961).

9. Mylander, W. C., R. L. Holmes, and G. P. McCormick, *A Guide to SUMT-Version*, Vol. 4, Research Analysis Corporation, RAC-P-63, 1964.

10. Eason, E. D., "Validity of Colville's Time Standardization for Comparing Optimization Codes," *ASME Des. Eng. Tech. Conf.*, paper no. 77-DET-116, Chicago, Sept. 1977.

11. Eason, E. D., and R. G. Fenton, "A Comparison of Numerical Optimization Methods for Engineering Design," *ASME J. Eng. Ind. Ser. B*, **96**(1), 196–200 (Feb. 1974).

12. Sandgren, E., "The Utility of Nonlinear Programming Algorithms," Ph.D. dissertation, Purdue University, University Microfilm, 300 North Zeeb Road, Ann Arbor, MI, document no. 7813115, Dec. 1977.

13. Sandgren, E., and K. M. Ragsdell, "The Utility of Nonlinear Programming Algorithms: A Comparative Study—Parts 1 and 2," *ASME J. Mech. Des.*, **102**(3), 540–551 (July 1980).

14. Dembo, R. S., "A Set of Geometric Programming Test Problems and Their Solutions," Dept. Management Sci. University of Waterloo, Ontario, Canada, Working Paper 87, Dec. 1974.

15. Root, R. R., and K. M. Ragsdell, "BIAS: A Nonlinear Programming Code in Fortran IV—Users Manual," Purdue Research Foundation, West Lafayette, IN, Sept. 1978.

16. Siddall, J. N., "OPTI-SEP: Designers Optimization Subroutines," McMaster University, Canada, 1971.

17. LaFrance, L. J., "User's Manual for GRGDFP, An Optimization Program," Herrick Lab. Rep. No. 45, Purdue University, Apr. 1974.

18. Schuldt, S., "RALP User Instructions," Honeywell Corporation, Bloomington, MN, personal communication.

19. Lasdon, L. S., A. D. Waren, M. W. Ratner, and A. Jain, "GRG User's Guide," Cleveland State University *Tech. Memo.* CIS-75-02, Nov. 1975.

20. Gabriele, G. A., and K. M. Ragsdell, "OPT: A Nonlinear Programming Code in Fortran IV—Users Manual," Purdue Research Foundation, West Lafayette, IN, Jan. 1976.

21. Guigou, J., "Presentation et utilisation du code, GREG," Department traitement de l'information et etudes mathematiques," Electricite de France, Paris, France, 1971.

22. Gulf Oil Corporation, "Compute II. A General-Purpose Optimizer for Continuous, Nonlinear Models—Description and User's Manual," Gulf Computer Sciences Inc. Houston, TX, 1972.

23. Afimiwala, K. A., "Program Package for Design Optimization," Department of Mechanical Engineering, State University of New York at Buffalo, 1974.

24. Mylander, W. C., R. L. Holmes, and G. P. McCormick, "A Guide to SUMT—Version 4," Research Analysis Corporation, RAC-P-63, 1974.

25. Lill, S., "E04HAF User's Guide," The University of Liverpool Computer Laboratory, Numerical Algorithms Group Document no. 737, 1974.

26. Staha, R. L., "Documentation for Program Comet. A Constrained Optimization Code," The University of Texas, Austin, Apr. 1973.

27. Sandgren, E., "A Statistical Review of the Sandgren-Ragsdell Comparative Study," *Mathematical Programming Society COAL Meeting Proceedings, Boulder, CO*, pp. 72–90, Jan. 5–6 1981.

28. Rosen, J. B., and S. Suzuki, "Construction of Nonlinear Programming Test Problems," *ACM Commun.* **8**(2), Feb. 1965. Schittkowski, K., "Nonlinear Programming Codes: Information Tests, Performance," Lecture Notes in Economics and Mathematical Systems, Vol. 183, Springer-Verlag, New York, 1980.

29. Schittkowski, K., "Nonlinear Programming Codes: Information, Tests, Performance," Lecture Notes in Economics and Mathematical Systems, Vol. 183, Springer-Verlag, New York, 1980.

30. Biggs, M. C., "Constrained Minimization Using Recursive Equality Quadratic Programming," in *Numerical Methods for Nonlinear Optimization*, (F. A. Lootsma, Ed.), Academic Press, London, 1971.

31. Powell, M. J. D., "A Fast Algorithm for Nonlinearly Constrained Optimization Calculations," in *Proceedings 1977 Dundee Conference on Numerical Analysis*, Lecture Notes in Mathematics, Springer-Verlag, Berlin, 1978.

32. Abadie, J., and J. Carpenter, "The Generalization of the Wolfe Reduced Gradient Method to the Case of Nonlinear Constraints," in *Optimization* (R. Fletcher, Ed.), Academic Press, New York, 1969.

33. Gabriele, G. A., and K. M. Ragsdell, "The Generalized Reduced Gradient Method: A Reliable Tool for Optimal Design," ASME *J. Eng. Ind.*, **99**(2), 394–400 (May 1977).

34. Lasdon, L. S., A. D. Waren, A. Jain, and M. Ratner, "Design and Testing of a Generalized Reduced Gradient Code for Nonlinear Programming," ACM *Trans. Math. Software*, **4**(1), 34–50 (1978).

35. Fletcher, R., "An Ideal Penalty Function for Constrained Optimization," in *Nonlinear Programming*, Vol. 2, (O. L. Mangasarian, R. R. Meyer, S. M. Robinson, Eds.), Academic Press, New York, 1975.

36. Pierre, D. A., and M. J. Lowe, *Mathematical Programming via Augmented Lagrangians: An Introduction with Computer Programs*, Addison-Wesley, Reading, MA, 1975.

37. Gill, P. E., and W. Murray (Eds.), *Numerical Methods for Constrained Optimization*, Academic Press, New York, 1974.

38. Hearhoff, P. C., and J. D. Buys, "A New Method for the Optimization of a Nonlinear Function Subject to Nonlinear Constraints," *Computer J.*, **13**, 1978–184 (1970).

39. Root, R. R., and K. M. Ragsdell, "Computational Enhancements to the Method of Multipliers," ASME *J. Mech. Des.*, **102**(3), 517–523 (July 1980).

40. Kraft, D., *Nichtlineare Programming—Grundlagen, Verfohren, Beispiele, Forschingsbericht*, DFVLR, Oberpfoffenhofen, West Germany, 1977.

41. Newell, J. S., and D. M. Himmelblau, "A New Method for Nonlinear Constrained Optimization," *AICHE J.*, **21**(3), 479–486 (1975).

42. Best, M. J., and K. Ritter, "An Accelerated Conjugate Direction Method to Solve Linearly Constrained Minimization Problems," *J. Computer Syst. Sci.*, **11**(3), 295–322 (1975).

43. Lootsma, F. A., "The Algol 60 Procedure MINIFUN for Solving Nonlinear Optimization Problems," in *Design and Implementation of Optimization Software* (H. J. Greenberg, Ed.), Sijthoff and Noordhoff, Alphen aan den Rijn, The Netherlands, 1978.

44. Rufer, D., "User's Guide for NLP—A Subroutine Package to Solve Nonlinear Optimization Problems," Rep. No. 78-07, Fachgruppe fur Automatik, Eidgenossische Technische Hochschule, Zurich, Switzerland, 1978.

45. Indusi, J. P., "A Computer Algorithm for Constrained Minimization," in *Minimization Algorithms* (G. P. Szego, Ed.), Academic Press, London, 1972.

46. Best, M. J., "A Feasible Conjugate Direction Method to Solve Linearly Constrained Optimization Problems," *JOTA*, **16**(1–2), 25–38 (1975).

47. Saaty, T. L., "A Scaling Method for Priorities in Hierarchical Structures," *J. Math. Psych.*, **15**, 234–281 (1977).

48. Lootsma, F. A., "Ranking of Nonlinear Optimization Codes According to Efficiency and Robustness," in *Konstruktive Methoden der finiten nichtlinearen Optimierung* (Collatz L., G. Meinardus, and W. Wetterling, Eds.) Birkhauser, Basel, Switzerland, 1980, pp. 157–158.

49. Sarma, P. V. L. N., X. M. Martens, G. V. Reklaitis, and M. J. Rijckaert, "A Comparison of Computational Strategies for Geometric Programs," *JOTA*, **26**(2), 185 203 (1978).

50. Fattler, J. E., Y. T. Sin, R. R. Root, K. M. Ragsdell, and G. V. Reklaitis, "On the Computational Utility of Posynomial Geometric Programming Solution Methods," *Math. Programming*, **22** 163–201 Mar. 1982.

51. Ragsdell, K. M., "The Evaluation of Optimization Software for Engineering Design," *Proceedings of the U.S. Bureau of Standards/Mathematical Programming Society Conference on Testing and Validating Algorithms and Software, Boulder, CO*. Springer-Verlag, Lecture Notes in Economics and Mathematical Systems, Vol. 199, pp. 358–379, 1982.

52. Rijckaert, M. J., and X. M. Martens, "A Comparison of Generalized GP Algorithms," Rep. CE-RM-7503, Katholieke Universiteit, Leuven, Belgium, 1975.

53. Dembo, R. S., "The Current State-of-the-Art of Algorithms and Computer Software for Geometric Programming," *JOTA*, **26**, 149 (1978).

54. Dinkel, J. J., W. H. Elliot, and G. A. Kochenberger, "Computational Aspects of Cutting Plane Algorithms for Geometric Programming Problems," *Math. Prog.*, **13**, 200 (1977).

55. Root, R. R., "An Investigation of the Method of Multipliers," Ph.D. dissertation, Purdue University, Dec. 1977.

56. Rijckaert, M. J., and X. M. Martens, "GPKTC: A Generalized Geometric Programming Code," Rep. CE-RM-7601, Katholieke Universiteit, Leuven, Belgium, 1976.

57. Dembo, R. S., "GGP: A Computer Program for the Solution of Generalized Geometric Programming Problems," McMaster University, Hamilton, Ontario, Canada, 1975.

58. Beck, P. A., and J. G. Ecker, "A Modified Concave Simplex Algorithm for Geometric Programming," *JOTA*, **15**, 184–202 (1975).

59. Bradley, J., "The Development of Polynomial Programming Algorithms with Applications," Ph.D. thesis, Dept. of Computer Science, Dublin University, Sept. 1975.

60. Reklaitis, G. V., and D. J. Wilde, "A Differential Algorithm for Posynomial Programs," *Dechema Monogr. Ban* **67**, 324–350 (1970).

61. Hoel, P. G., *Introduction to Mathematical Statistics*, John Wiley, New York, 1962, p. 279.

62. Sin, Y. T., and G. V. Reklaitis, "On the Computational Utility of Generalized Geometric Programming Solution Methods: Review and Test Procedure," *Progress in Engineering Optimization—1981* (R. W. Mayne, and K. M. Ragsdell, Eds.), ASME, New York, Sept. 1981, pp. 7–14.

63. Sin, Y. T., and G. V. Reklaitis, "On the Computational Utility of Generalized Geometric Programming Solution Methods: Results and Interpretation," *Progress in Engineering Optimization—1981* (R. W. Mayne, and K. M. Ragsdell, Eds.), ASME, New York, pp. 15–26, Sept. 1981.

64. Waren, A. D., and L. S. Lasdon, "The Status of Nonlinear Programming Software," *Oper. Res.*, **27**, 431–457 (1979).

# Chapter 13

# Strategies for Optimization Studies

In the preceding chapters we have concentrated on the analysis of contemporary optimization algorithms in order to understand their underlying logic as well as to develop an appreciation of their strengths and weaknesses. On the basis of this study we are now sufficiently knowledgeable to select the algorithm most appropriate for a given engineering problem and to verify whether or not a point generated by the selected algorithm is in fact a solution. While this knowledge is a necessary prerequisite for the successful execution of an optimization study, it is by no means sufficient. Many additional steps and considerations are involved in conducting such a study. It is necessary to define the optimization problem itself; to prepare it for solution; to select a suitable optimization algorithm; to choose or prepare an efficient computer implementation of this algorithm; to execute a campaign of computer runs involving possible readjustments of problem and algorithm parameters; and, finally, having obtained a reliable solution, to interpret this solution in terms of the real system as well as to implement it.

Given that in most engineering applications the optimization algorithm will be available as a "canned" program obtained from the company's program library or purchased/leased from software vendors, the most expensive and time-consuming parts of an optimization study will be problem formulation, preparation for solution, and preliminary troubleshooting runs. Unfortunately, these tasks involve skills that are not easily categorized, analyzed, and taught. Rather, they are acquired through practice and the careful analysis of past successes and failures.

In this chapter we present some guidelines and suggest via examples some alternative approaches to problem formulation, preparation, and troubleshooting. Our discussion is by no means intended as an exhaustive treatment of these topics. It simply represents a synthesis of the experiences of the authors and those of other practitioners of the art of engineering optimization that could be gleaned through informal discussions or from casual remarks scattered in the applications literature.

## 13.1 MODEL FORMULATION

As noted in Chapter 1, a problem suitable for the application of optimization methods consists of a performance measure, a set of independent variables, and a set of equality and inequality constraints that constitutes the *model* of the system under consideration. The definition and formulation of a model that represents the real system is the most important step in conducting an optimization study, because it will dictate whether the solution obtained will be physically meaningful and realizable.

The process of optimization via a model can be considered as an alternate route that allows the optimum of the real system to be found without experimenting directly with the real system. As shown in Figure 13.1, the "direct" route is replaced by a path that involves formulation of a model, optimization of the model, and translation of the model results into physically meaningful and applicable terms. Clearly, the indirect route to system optimization inherently involves the use of a representation of the real system that is less than perfect. In formulating this approximation or model, it is necessary to select the significant aspects of the system that must be included and those of secondary importance that can be ignored; to identify the assumptions that can be made; and to choose the form in which the model will be cast, the level of model detail, and the manner in which the model will be generated. These decisions, which together constitute the model-formulation task, are inherently characterized by a certain amount of arbitrariness. Equally competent engineers viewing the same real system are quite likely to develop models that are quite different in form. Yet none of these various models of the same system, no matter how detailed and complex, can be anointed as the "true" model. Instead, the various models can at best be ranked on the basis of how adequately they approximate the behavior of the real system over the operating range of interest. Thus models cannot be evaluated on the basis of structure or form but only on the accuracy of their predictions. A model involving nonlinear functions is to be preferred to a model involving linear functions only if it represents the real system more accurately.

Moreover, the accuracy itself can often not be assessed precisely, and hence the selection between models can involve a considerable degree of subjectivity. For instance, one model may be more accurate than a competitive formulation over one subregion but less accurate over another subregion. In this case the

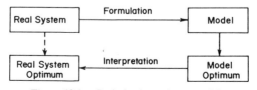

**Figure 13.1.** Optimization using a model.

engineer may well choose a model with overall lower accuracy because it is more accurate in a critical subregion. In other cases, it may, in fact, not even be possible to assess in advance the accuracy of the complete model but only that of some of its constituent equations.

The key points to note are that the ties between a model and the system that it is intended to represent are at best ties of plausible association. It is part of the nature of models as simplifications of reality that there are no absolute criteria by means of which competitive models can be ranked. There are always factors involved that require subjective judgment and a "feel" for the behavior of the real system. Thus, there always is room for reasonable and equally competent engineers to disagree about the details of alternative model formulations. As a consequence, it is essential that the model builder be thoroughly familiar with the system to be modeled, that he understand the engineering principles that underlie the components of the model, and, in the case of design optimization, that he has mastered all of the computations that are necessary to obtain a feasible design.

### 13.1.1   Levels of Modeling

Since model development is an activity that requires the concentrated involvement of competent engineers thoroughly familiar with the system under study, it usually is the most expensive part of an optimization study. Since model development costs increase dramatically with the level of detail of the model, it is necessary to use considerable judgment in selecting the level that is appropriate to the goals of the study and consistent with the quality of the information available about the system. It is clearly undesirable to develop a detailed dynamic model of a plant operation if an overall input-output model will suffice. Furthermore, it is useless to develop a complex model when the data available to evaluate the model coefficients are sparse and unreliable. On the other hand, since it is the model that will be optimized and not the real system, it is pointless to carry out an optimization study with a simplistic model that will not adequately approximate the true system optimum.

The ideal sought after in model development is sometimes called the "principle of optimum sloppiness"; the model should be only as detailed as necessary for the goals of the study for which it is constructed. Achievement of this goal, however, is always difficult. While it is possible through a history of successes and failures to develop an a priori feel for the appropriate level of model complexity for specific types of systems, usually it is difficult to generalize this experience to other applications. The only foolproof way to the development of a model with optimum level of sloppiness is to proceed in steps with model development and optimization, starting with a very simple model and concluding with the version in which the refinements to the value of the optimum attained approach the accuracy of the data incorporated in the model. This, however, is a painstaking and methodical process whose time requirements sometimes exceed the short times that the demands of a competitive business environment allow for producing results. A common recourse to meet such deadline pressures or to avoid the labor of stepwise model development is to tailor the model to the optimization technique that is currently in favor, or with which the engineer is most familiar at the time, or that was used in a previous study. For instance, the temptation often arises to formulate all problems as linear programs or geometric programs simply because the engineer feels comfortable with this type of problem structure.

To be sure, it sometimes is necessary to force a problem into a particular model form. For instance, in modeling problems involving integer variables, it is essential that the problem be forced into a tractable form such as that of a network if it is to be successfully solved. The important point is to know the impact of forcing a problem to a particular model form on the quality of the solution. Also, the capabilities and limitations of available optimization methods must be kept in mind when developing the model. It is clearly inappropriate to expect to be able to solve an NLP problem as large in dimensionality as can be accommodated by commercial LP software. However, at the modeling stage such concerns should be secondary rather than dominant considerations.

To illustrate the different levels of models that might be developed for a given system, we will consider at a qualitative level a study to optimize a multiproduct plant.

### Example 13.1    Optimization of Polymer Plant

**Problem Definition.**    A multiproduct polymer plant consists of several serial stages of parallel processors with in-process storage between production stages, as shown in Figure 13.2. The first stage contains several nonidentical polymerization reactor trains; the second stage consists of several parallel extrusion lines of different capacity; the third stage is a blending stage, consisting of five lines, each containing one or more blending units. The final processing stage is the packaging section, which provides several packaging alternatives: bulk loading, packaging into boxes, and packaging into sacks. The plant produces a

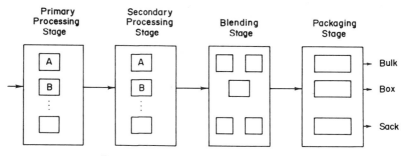

Figure 13.2.    Conceptual diagram of process.

number of different types of polymer grades and types and consequently is not operated on a continuous basis. To allow for product changeovers, each processing stage has associated with it a limited amount of in-process storage. Moreover, the blending stage has associated with it several large-capacity storage vessels, which are required to assure adequate inventory for preparing product blends.

A decision is made to "optimize" the process. It is suspected that since the current product slate is substantially different from the product slate projected when the plant and its operating procedures were designed, the plant profitability could be substantially improved.

**Preliminary Studies: Linear Models.** As a first step in the study, a simple linear input-output model is developed to investigate what the preferred product slate might be. The linear model is based on the average production rates and processing times required by each product in each processing stage, uses average "off-specification" production rates (those not meeting product specifications), and assumes current levels of feed availability, current market prices, and unrestricted demand. The linear model is solved using a commercial LP code, and on the basis of several case runs it is unequivocally concluded that some products in the current slate ought to be dropped and that the production of others ought to be increased.

Upon further study it is revealed that as a result of increasing the production rate of some of the products and deletion of others, the production of intermediate off-specification material will be increased. This material is normally recycled internally. The increased internal recycling is likely to cause bottlenecks at the extrusion stage. The initial linear model is consequently expanded to include the key internal material flows as shown in Figure 13.3, as well as to incorporate explicit constraints on the maximum and minimum allowable intermediate processor rates. The revised model, considerably increased in dimensionality, remains a linear model suitable for optimization by

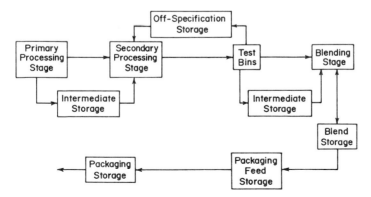

**Figure 13.3.** Process with explicit off-specification recycle.

LP methods. The case studies conducted using the expanded model indicate that the capacities of the extrusion and blending stages do, in fact, severely limit the initially projected optimum product slate. It appears that some limited expansion of capacity would be desirable.

**Nonlinear Models.** To determine the optimum investment in additional equipment, the model is revised to incorporate nonlinear cost terms for the investment and operating costs of the extra units as a function of their throughput. The extra production capacity is easily incorporated into the linear constraint set. However, as a result of the cost terms, the objective function becomes nonlinear. The resulting linearly constrained NLP problem is solved using available software. On the basis of such optimization, an appropriate capital expenditure is proposed for plant upgrading. At this stage it might be noted by team members that the selected additional capacity was all based on average production rates and the assumption that the same product sequence was used as is currently being followed. It is suggested that alternative sequences and production run sizes ought to be considered. This introduces new considerations such as batch sizes, run lengths, product sequencing, and changeover policies into the plant model.

To incorporate these elements to even a limited degree, the model must be modified to include 0-1 variables, which will be used to reflect the sequencing part of the problem. Computational considerations now come into play: available mixed integer programming codes cannot accommodate nonlinearities. Consequently, each nonlinear cost term is approximated by a fixed cost and a linear cost term. The resulting mixed integer linear program is solved, and the solution might indicate that moderate changes in the production sequence together with subdivision into shorter product runs are attractive. At this point it may be appropriate to include time periods in the planning model as well as to incorporate the regional warehouses and the distribution network in the study. In this case, the MIP model would be further expanded to accommodate these additional elements.

**Simulation Models.** Alternatively, questions could arise of the significance of unit shutdown and start-up transients as well as off-specification production during these periods on the plant economics. Furthermore, in-process and final product inventory levels might also need to be given closer scrutiny. If transients and inventory changes do become an important concern, then a simulation model could be developed to represent the system dynamics. In this case optimization would be carried out either by case studies or by direct search over a few variables using a direct-search optimization algorithm.

Finally, if it appears that fluctuations in the production rates or down-times or changeover times of some or all of the processing stages are significant, then the simulation model may need to be refined further to include Monte Carlo sampling of these fluctuating variables. In this case it will be necessary to collect statistical data on the observed fluctuations, to identify the appropriate distribution functions, and to insert those stochastic features into the model. For a discussion of such a model for the above plant, the reader is invited to consult reference 1. With this type of model, direct optimization will normally be very costly in terms of computer time, and optimization will best be performed by case studies involving a few parameter changes.

In summary, depending upon the depth and direction of the study, the model used in plant optimization will range from a moderate-sized linear model to a large-scale linear program, to a large-scale mixed integer program, to a nonlinear program, to a deterministic simulation model, and finally to Monte Carlo simulation. Clearly it would be wasteful to begin the study with the most complex model if the simplest model could serve to indicate that the potential for improvement is small. Similarly, it would be meaningless to develop and to use the detailed simulation model if no data were available to evaluate the stochastic model parameters.

### 13.1.2   Types of Models

From the discussion of Example 13.1, it is evident that in addition to selecting the level of detail at which the system is to be modeled, the engineer also has to choose the type of model that will be employed. Three basic model types are commonly used in optimization studies:

1.  Phenomenologically based equation-oriented models
2.  Response surface models
3.  Procedure- or simulation-oriented models

In the first case, the model is composed of the basic material-balance and energy-balance equations, engineering design relations, and physical properties equations, all assembled as explicit sets of equations or inequalities. To be amenable to the methods discussed in this book, the equations must be real-valued functions that can be evaluated for selected values of the independent variables. In principle, the equations could involve integral or differential operators. However, in practice it is generally best to eliminate such terms by quadrature formulas or approximations so as to result in purely algebraic model functions. Most of the examples discussed in previous chapters were equation-oriented algebraic models. Because equation-oriented models describe the system behavior at the level of basic engineering principles, they normally are valid over wider ranges of system conditions than the next category of models, response surface models.

In a response surface model, the entire system or its component parts are represented by approximating equations derived by fitting coefficients to selected equation forms using directly or indirectly measured systems response data. Typical approximating equations are the linear correlation function and the quadratic function with or without cross-correlation terms. Such models are used if the system response is poorly understood or too complex to allow detailed modeling from basic engineering principles. Because they are in essence correlations, response surface models are normally valid only over limited ranges of the system variables. They do, however, have the advantage of simplified structure.

Finally, in the third case, the basic equations describing the behavior of the system are grouped into separate modules or subroutines that represent a particular piece of equipment or the collection of activities associated with a change in the state of the system. Each of these models or subroutines is usually a self-contained entity that may involve internal numerical procedures; equation solving, integration, or logical branching procedures. Simulation or procedure-oriented models are used when evaluation of the equations is complex, involving implicity determined variables; when the selection of a logical block of the calculation procedure or appropriate equations is subject to the state of the system; and when the model must involve stochastic elements that require Monte Carlo sampling techniques. Models of this type are generally much more complex than the previous two types and usually are much more demanding of computer resources.

In most cases the choice of model type will be dictated by the level of information available about the system, the level of understanding of the phenomena occurring within the system, and the inherent complexity of the system itself. If a choice is available, then equation-oriented models are to be preferred, because they are most conveniently treated by conventional NLP techniques. Although procedure-oriented models can sometimes be more efficient, because they allow specialized treatment of recurring blocks of equations, they do make it inconvenient to access values of intermediate dependent variables, to fix their values, or to impose bounds on them, as well as to interchange the roles of dependent and independent variables. We illustrate the basic differences of the three types of models with the following example.

***Example 13.2   Flash Separator Model***

**System Description.**   An equilibrium flash separator is a device in which a multicomponent mixture is reduced in pressure and cooled, the resulting mixture is permitted to settle into liquid and vapor phases, and the two phases are separated. As shown in Figure 13.4, such a system might consists of a valve, a heat exchanger, and a tank with an inlet nozzle and two outlet nozzles. Typically, a wire mesh pad or demister is included to trap liquid droplets entrained with the exiting vapor. Given a feed mixture with specified component mole fractions $z_i$ (moles of component $i$ per mole of mixture), and feed rate $F$ (moles/hr), the key design parameter of the system is the ratio of vapor to feed flows $V/F$ obtained for various values of the tank temperature $T$ and pressure $P$. Suppose that a model describing the relationship between these variables is desired for a feed of fixed composition.

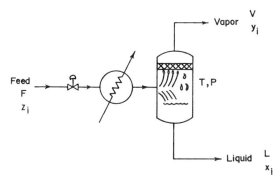

**Figure 13.4.** Equilibrium flash separator.

**Response Surface Model.** In the absence of other information about the system, one way of describing the relationship between the ratio $V/F$ and the tank $T$ and $P$ is to generate a response surface model for this feed mixture over a suitable range of $T$ and $P$ combinations. This requires that experiments be conducted over the selected range of $T$, $P$ values and that these experimental results be used to determine the coefficients of a correlating function of appropriate form. For instance, if a quadratic function of $T$ and $P$ is assumed, then the response surface will take the general form

$$\frac{V}{F} = c_1 + (b_1, b_2)\binom{T}{P} + (T, P)^T \begin{pmatrix} a_{11} & a_{21} \\ a_{12} & a_{22} \end{pmatrix}\binom{T}{P} \qquad (13.1)$$

Since $a_{21} = a_{12}$, the response surface model will involve six constants and hence will require at least six experiments to allow the determination of these constants. Clearly, the response surface will be valid only over the range

$$T^{(L)} \leqslant T \leqslant T^{(U)}$$

$$P^{(L)} \leqslant P \leqslant P^{(U)}$$

covered by the experiments. Moreover, even when restricted to that range, it will be subject to a prediction error, because the actual dependence of $V/F$ on $T$ and $P$ is not likely to be exactly quadratic.

**Equation-Oriented Model.** It is known that under suitably ideal conditions, the mole fractions of vapor and liquid phases in equilibrium with each other can be related via Raoult's law [2]. That is,

$$\frac{y_i}{x_i} = \frac{p_i(T)}{P} \qquad (13.2)$$

where $p_i(T)$ is the partial pressure of component $i$ at temperature $T$. The dependence of $p_i$ on $T$ can be approximated using a semiempirical equation of the form

$$\ln p_i = A_i - \frac{B_i}{T - C_i} \tag{13.3}$$

where $A_i$, $B_i$, $C_i$ are constants empirically determined for each chemical component $i$.

An equation-oriented model relating $V$, $L$, $T$, and $P$ can be constructed by using Eqs. (13.2) and (13.3) as well as the principle of conservation of mass. Thus, since the input of each component $i$ must be equal its output rate from the system, we have

$$Fz_i = Vy_i + Lx_i \tag{13.4}$$

Moreover, since $y_i$ and $x_i$ are mole fractions, they must sum to one; that is,

$$\sum_i x_i = 1 \tag{13.5}$$

$$\sum_i y_i = 1 \tag{13.6}$$

Combining (13.2) and (13.3), we also have

$$\frac{y_i}{x_i} = \frac{1}{P} \exp\left[ A_i - \frac{B_i}{T - C_i} \right] \tag{13.7}$$

If the feed has three components, then Eqs. (13.4)–(13.7) will be a system of eight equations describing the relationship between $V, L, T, P$ and the six additional variables $x_i$, $y_i$. For known $F$, $z_i$ and specified $T$, $P$, the nonlinear equation-oriented model will serve to determine the values of the eight variables $V$, $L$, $x_i$, and $y_i$. For purposes of optimization, Eqs. (13.4)–(13.7) would be treated as equality constraints. Moreover, these would probably be supplemented with the inequality constraints

$$T^{(L)} \leqslant T \leqslant T^{(U)}$$

$$P^{(L)} \leqslant P \leqslant P^{(U)}$$

$$V, L, x_i, y_i \text{ all} \geqslant 0$$

**Procedure-Oriented Model.**  The system of Eqs. (13.4)–(13.7) does not admit direct algebraic solution for $V/F$. However, it is possible to reduce it to a single nonlinear equation in the ratio $V/F$ and the variables $T$, $P$. For fixed values of $T$ and $P$, this equation can then be solved using a root-finding method to determine $V/F$. Specifically, it can be shown (see ref. 3) that by suitable algebraic rearrangement the equation will take the form

$$\sum_i \left[ \frac{z_i \{1 - P\exp[B_i/(T - C_i) - A_i]\}}{1 - (V/F)\{1 - P\exp[B_i/(T - C_i) - A_i]\}} \right] = 0 \qquad (13.8)$$

Since $A_i$, $B_i$, $C_i$, and $z_i$ are constants, this equation involves only $T$, $P$, and $V/F$.

The numerical solution of Eq. (13.8) for given values of $T$ and $P$ can be implemented as a self-contained subroutine or procedure that we will call FLASH. Using this routine, our model of the relationship between $V/F$, $T$, $P$ reduces to

$$\frac{V}{F} = \text{FLASH}(T, P) \qquad (13.9)$$

Therefore, rather than treating Eqs. (13.4)–(13.7) as equality constraints and variables $V$, $L$, $x_i$, $y_i$, $T$, $P$ all as independent variables, the ratio $V/F$ is simply a dependent variable whose value is fixed by those of the independent variables $T$, $P$. In contrast to the response-surface model, Eq. (13.1), the procedure-oriented model will require the internal execution of a root-finding calculation to solve Eq. (13.8) each time relation (13.9) is evaluated. Note also that if for purposes of optimization it is appropriate to impose bounds on $V/F$, then these bounds must be imposed on the procedure FLASH.

An obvious advantage of the use of procedure FLASH is that if the system in question involves several flash units, then by suitable transfer of the parameters $z_i$, $A_i$, $B_i$, $C_i$ appropriate to each flash unit, the same procedure can be used to model each unit. An obvious disadvantage is that the procedure has a fixed calculation structure. Consequently, if in subsequent versions of the system model, it proves desirable to fix $V/F$ and, say, to treat the temperature $T$ as variable, then FLASH must be discarded and a new procedure written to take its place. By contrast, in the equation-oriented model, no real difficulty is posed in fixing the value of $V/F$ and letting $T$ be a variable: the model equations need not be altered.

## 13.2 PROBLEM IMPLEMENTATION

Regardless of what level and type of model is selected, the engineer has the further choices of the manner in which the model will be implemented or assembled, the devices that will be used to prepare the resulting optimization problem for solution, and the computational strategies that will be used in problem execution. These choices each have a considerable impact on the overall solution efficiency as well as on the preparation time that will be required and are therefore important components of an optimization study.

### 13.2.1  Model Assembly

In general, a model for an optimization study can be assembled manually and then coded into suitable problem function and derivative routines, or it can be generated using computer aids of various levels of sophistication. In the case of the equation-oriented model, the simplest form of which is a linear program, the linear equations and inequalities can be written and assembled by hand and the coefficients coded into an array suitable for processing by the LP code. Alternatively, a *matrix generator* can be used to automatically assemble the coefficients for certain classes of constraints. For instance, constraints of the form $\Sigma x_i = 1$ could easily be generated by merely defining the indices of the variables in the sum and the numerical value of the right-hand-side constant. The matrix generator would then generate the appropriate array entries. All commercial LP codes require that input data be in a standard MPS form in which each array entry is identified by its row, column, and numerical value, with each such triplet constituting a data record. Manual generation of such data files is very tedious, and hence some form of matrix generator is commonly used.

In applications involving interconnected subsystems of various types of regular structures, it is usually efficient to use *equation generators*. An equation generator is written for each type of subsystem so that the assembly of the model merely involves identifying the subsystems that are present and their interconnections. For example, Bethlehem Steel Corporation has recently reported on an equation generator for their LP model that optimizes plant operations [4]. Similarly, Westerberg and coworkers [5] as well as Hutchison and coworkers [6] have developed equation generators that are designed to prepare chemical process flowsheet models for case evaluation or direct optimization. The former generates nonlinear models and the latter linear material- and energy-balance models. Use of nonlinear equation generators is particularly attractive in processing applications, because such applications involve recurring components—pumps, compressors, heat exchangers, and so on—whose representation in terms of equations can be developed in general form. Because of their complexity and high development costs, equation generators are certainly not justifiable for one-time optimization studies. However, if a number of studies are contemplated, each of which could employ the generator, then development is well worthwhile. Use of a generator allows model standardization, provides convenient documentation, and minimizes model-coding errors as well as errors of omission.

In the case of response-surface models, the system or its components can be employed directly to generate data from which approximating equations relating the dependent and independent variables can be derived. Typically, the data will be gathered manually and fed to a regression program to determine the constants of the approximating equation. This can, of course, be automated so that data logging, fitting, and optimization are all performed without manual intervention. Often more complex models of system components are used to automatically generate response-surface models for subsequent optimization. For instance, Pollack and Lieder [7] report on the use of simulations of the fluidized catalytic cracker, hydrocracker, and catalytic reformer reactors of a refinery to generate linear response-surface models, which are incorporated into an overall refinery LP model. The generalized automatic generation of linear as well as quadratic response-surface models from detailed nonlinear models of process flowsheets is further considered by Hughes and coworkers [8, 9] as well as McLane et al. [10]. The automated generation of response surface models from Monte Carlo simulations and their use to optimize simulated system performance is reviewed by Biles and Swain [11]. Unquestionably, automated generation and updating are preferred for all applications that will require repeated solutions.

Finally, simulation or procedure-oriented models can be assembled by coding them *ab initio* in their entirety or by making use of simulation packages or libraries. Using such systems, a model can be assembled using the building block concept: Individual modules that represent the separate system components and are provided by the simulation package are linked together by the user either through data statements or through coding. Use of such systems is continually increasing in the processing industries [12]. Specialized packages are available for the simulation of networks involving stochastic elements [13] as well as systems involving discrete or combined discrete-continuous dynamics (see ref. 14, chapter XIII-11 for a good review). In general, simulation or procedure-oriented models tend to be complex and hence require considerable CPU time for evaluation. Use in optimization studies is thus normally limited to a modest number of independent variables.

In summary, the bulk of engineering applications tend to involve either manually prepared equation-oriented models or specially prepared stand-alone, procedure-oriented models. Typically, automated generation of equation-oriented models is employed only with linear or mixed integer programming models, although their use is receiving increased attention in process industry applications. Response-surface models are most frequently used in conjunction with complex simulation models to avoid direct optimization with the simulation models.

### 13.2.2  Preparation for Solution

Once the model is formulated and its manner of assembly is selected, the problem must be prepared for the application of an appropriate optimization algorithm. Problem preparation basically involves three types of activities:

1. Modifications to avoid potential numerical difficulties
2. Model rearrangement to enhance solution efficiency
3. Analysis to identify likely problem solution features

**Eliminating Numerical Difficulties.** Numerical difficulties in problem solution that result in premature termination or an execution abort can normally be traced to four common causes: poor scaling, inconsistencies between function and derivative routines, nondifferentiable model elements, and inadequate safeguards for function arguments. In most cases these situations can be recognized from a careful review of the model and can be eliminated by simple problem modifications.

### Scaling

Scaling refers to the relative magnitudes of the quantities that appear in the model to be optimized. *Variable* scaling refers to the relative magnitudes of the variables in the problem. *Constraint* scaling refers to the relative magnitudes or sensitivities of the individual constraints. Ideally, all variables in the model should be defined in such a way that their likely values fall in the range 0.1–10. This will ensure that search-direction vectors and quasi-Newton update vectors will take on reasonable values. If some variable values at a likely estimate of the solution fall outside this range, it is a simple matter to replace those variables by new variables multiplied by suitable constants so that the new variables are well scaled.

Similarly, it is straightforward to evaluate the constraints at the solution estimate and to examine their sensitivity with respect to changes in the variable values. As a further check, one can examine the elements of the matrix of constraint gradients at the estimated solution. Generally, it is best if all constraints exhibit a similar sensitivity to changes in the variable values and if the gradients of the constraints take on the same range of values. This will ensure that violations of constraints will be equally weighted and that matrix operations involving the constraint Jacobian will not give rise to numerical inaccuracies. One remedy is to multiply the constraint functions by suitable scalars so that typical constraint values fall in the range 0.1–10 and/or that constraint gradient components take on values in the range 0.1–10. The following simple example illustrates the effects of variable and constraint scaling.

### Example 13.3  Problem Scaling

(a) Consider the scaling of the constraint set

$$h_1(x) = x_1 + x_2 - 2 = 0$$

$$h_2(x) = 10^6 x_1 - 0.9 \times 10^6 x_2 - 10^5 = 0$$

at the point (1.1, 1.0).

(b) Consider the scaling of the constraint set

$$h_1(x) = x_1 + 10^6 x_2 - 2 \times 10^6 = 0$$

$$h_2(x) = x_1 - 0.9 \times 10^6 x_2 - 10^5 = 0$$

at the point $(10^6, 0.9)$.

**Solution.** (a) At the given point, the function values are

$$h_1(1.1, 1.0) = 0.1 \qquad h_2(1.1, 1.0) = 10^5$$

The given point really is reasonably close to the solution of the equation set $x^* = (1, 1)$. But, because of the poor constraint scaling, the constraint sensitivities are quite disparate. Note also that the matrix of constraint coefficients

$$\begin{pmatrix} 1 & 1 \\ 10^6 & -.9 \times 10^6 \end{pmatrix}$$

has a determinant of $-1.9 \times 10^6$. The situation is easily corrected by merely dividing $h_2(x)$ by $10^6$. The result is

$$h_1(x) = x_1 + x_2 - 2 = 0$$

$$h_2'(x) = \frac{h_2(x)}{10^6} = x_1 - 0.9x_2 - 0.1 = 0$$

Observe that $h_2'(1.1, 1.0) = 0.1$ and the Jacobian becomes

$$\begin{pmatrix} 1 & 1 \\ 1 & -0.9 \end{pmatrix}$$

with determinant $-1.9$.

(b) At the given point, the function values are

$$h_1(10^6, 0.9) = -10^5 \qquad h_2(10^6, 0.9) = 0.9 \times 10^5$$

and the constraint Jacobian is

$$\begin{pmatrix} 1 & 10^6 \\ 1 & -0.9 \times 10^6 \end{pmatrix}$$

with determinant $-1.9 \times 10^6$. Again the given point is reasonably close to the solution point $x^* = (10^6, 1.0)$. Suppose variable $x_1$ is scaled by introducing the new variable $y$ defined as $y = 10^{-6}x_1$. In terms of the new variable, the equation set becomes

$$h_1(y, x_2) = 10^6 y + 10^6 x_2 - 2 \times 10^6 = 0$$

$$h_2(y, x_2) = 10^6 y - 9 \times 10^5 x_2 - 10^5 = 0$$

If both equations are further scaled by dividing them by $10^6$, we obtain

$$y + x_2 - 2 = 0$$

$$y - 0.9x_2 - 0.1 = 0$$

At the given point $(1, 0.9)$,

$$h_1' = -0.1 \qquad h_2' = 0.09$$

From the example it is apparent that constraint and variable scaling can be viewed in terms of the row and column scaling of the associated Jacobian matrix. Moreover, scaling may have to be performed in several successive passes, since, as was the case in Example 13.3b, scaling of a variable can impose the need to scale one or more of the constraints. A problem-scaling procedure that uses successive passes to arrive at a selection of scale factors was investigated by Root and Ragsdell [15].

The objective of the scaling procedure is to determine a row scaling vector $r$ and a column scaling vector $c$ such that the elements

$$J_{ij}' = r_i J_{ij} c_j \tag{13.10}$$

where $J_{ij}$ is the $(i, j)$th element of the matrix, whose rows are the constraint gradient vectors, lie within ratios $\varepsilon_1$ and $\varepsilon_2$ of the magnitude of the "average" element of $\mathbf{J}$. The "average" element $z$ is defined by finding the element of $\mathbf{J}$ of largest absolute value, identifying the corresponding row $i^*$ and column $j^*$, and calculating the average absolute value of all nonzero elements of $\mathbf{J}$ excluding those of row $i^*$ or column $j^*$. Thus, if NZ is the number of nonzero elements of $J$ excluding those in row $i^*$ or column $j^*$, then

$$z = \frac{1}{NZ} \sum_{i \neq i^*} \sum_{j \neq j^*} |J_{ij}|$$

In terms of this notation, the goal of scaling is thus to find $r$ and $c$ such that

$$\varepsilon_2 z \leqslant |J_{ij}'| \leqslant \varepsilon_1 z$$

for some reasonable values of $\varepsilon_1$ and $\varepsilon_2$, say 10 and 0.1, respectively. The procedure outlined below does not guarantee attainment of the above condition for all $i$ and $j$, but nonetheless tends to approach it.

## Recursive Scaling Method

At a given point $x^0$, calculate $\mathbf{J}$ and $z$. For given values of $\varepsilon_1$ and $\varepsilon_2$, execute the following steps:

**Step 1.** For all columns $j$ not previously scaled, calculate

$$c_j = \frac{n_j z}{\sum\limits_i |J_{ij}|}$$

where $n_j$ is the number of nonzero elements in column $j$ of $\mathbf{J}$. If $\varepsilon_2 \leqslant c_j \leqslant \varepsilon_1$, reset $c_j = 1$. Otherwise, retain $c_j$ as scale factor, update the column of $J$ using Eq. (13.10), and recalculate $z$.

**Step 2.** If at least one column $j$ was scaled in step 1, continue. Otherwise, stop.

**Step 3.** For all rows $i$ not previously scaled, calculate

$$r_i = \frac{m_i z}{\sum\limits_j |J_{ij}|}$$

where $m_i$ is the number of nonzero elements in row $i$ of $J$. If $\varepsilon_2 \leqslant r_i \leqslant \varepsilon_1$, reset $r_i = 1$. Otherwise, retain $r_i$ as scale factor, update the row of $J$ using Eq. (13.10), and recalculate $z$.

**Step 4.** If at least one row $i$ was scaled in step 3, go to step 1. Otherwise stop.

Note that in this procedure each row and column is scaled only once. In numerical tests performed using the MOM code BIAS [15], it was shown that a specialized form of the above scaling procedure can significantly increase solution efficiency for problems solved with BIAS. These results are confirmed by other investigators, who employed scaling procedures in conjunction with penalty function methods [16]. On the other hand, Lasdon and Beck [17] investigated a more elaborate scaling method in the context of GRG algorithms and observed rather mixed results: enhanced performance in some cases, but deteriorated performance in other cases. These results support the widely held belief that GRG and projection-based methods are less sensitive to scaling then penalty or multiplier methods. Lasdon and Beck [17] do suggest that simple scaling procedures such as those illustrated in Example 13.3 should be employed even with GRG methods, but that more elaborate automatic scaling methods remain a topic for further research. In the experience of the authors, simple scaling should be a regular component of problem preparation for optimization using any and all optimization methods.

*Inconsistencies*

Inconsistencies between model function and derivative values can be insidious because they can catastrophically mislead the search algorithm. The simplest way to check the consistency of function and gradient values is to compute gradients by function value differences and compare these to the analytical gradients generated by the derivative routines. To avoid such difficulties, many practitioners always compute gradients by numerical differencing of function values. However, generally speaking, the use of analytical gradients is computationally more efficient, particularly if care is exercised in saving values of repeatedly recurring variable groupings.

*Nondifferentiability*

Nondifferentiable model elements arise most commonly in two forms: conditional statements that cause different expressions or model segments to be executed depending upon the values of selected variables or functions, and max/min operators. Such nondifferentiable elements cause difficulties principally when gradients are evaluated numerically in the vicinity of the switching point of the conditional statement or max/min operator. If the differencing parameter is $\delta$, then at $x^0 + \delta$ one expression may be evaluated while at $x^0$ another may have been used. The gradient thus becomes unreliable. As noted in Chapter 9 (Section 9.5), max/min operators can be eliminated by replacing them with multiple inequalities. Thus,

$$g(x) \equiv \max_j \{ g_j(x); \quad j = 1, \ldots, J \}$$

can be replaced by

$$g(x) \equiv y \geqslant g_j(x) \qquad j = 1, \ldots, J$$

However, if a model involves extensive use of conditional expressions, then it is best to use an algorithm that does not rely on gradient values.

*Invalid Arguments*

Finally, one frequently overlooked but very common cause of computer aborts of optimization runs is failure to insert safeguards against invalid function arguments. In engineering models, terms of the form $[g(x)]^b$ or $\log[(g(x)]$, where $g(x)$ is some expression in $x$, occur quite frequently. If during the course of the optimization $g(x)$ is allowed to take on negative or zero values, then a program abort will take place. To avoid this problem, constraints of the form $g(x) \geqslant \varepsilon > 0$, where $\varepsilon$ is some small tolerance value, should be added to the problem formulation. Some investigators also go so far as to suggest that all occurrences of division by variables should be eliminated from model formulations so as to avoid the possible encounters of poles of the functions or their derivatives [18].

**Enhancing Solution Efficiency.** As noted in our discussion of NLP algorithms, the difficulty of nonlinear optimization problems increases in an exponential fashion with the number of variables as well as the number of equality and inequality constraints. As a rule, linear constraints present less difficulty than nonlinear constraints, and inequality constraints less difficulty than equality constraints. Consequently, there is some incentive during the problem preparation phase to attempt to rearrange the model to reduce the number of constraints, particularly the nonlinear ones, and to reduce the number of variables. These objectives can be achieved through the use of function and variable transformations, the elimination of redundant constraints, and the use of equation-sequencing techniques.

### Function Transformation

In the present context, a function transformation is simply any algebraic rearrangement of the function or concatenation of a given function with another. In order to enhance the solution efficiency of a problem, transformations are normally chosen so as to convert nonlinear constraints to linear constraints or to convert an equality to an inequality. For instance, by cross multiplication of all terms it is obviously possible to convert a nonlinear constraint of the form

$$y^{-1}g(x) = b$$

where $g(x)$ is linear in $x$, to a linear constraint

$$g(x) - by = 0$$

Similarly, nonlinear product terms of the form

$$\prod_n x_n^{a_{ni}} \leqslant c$$

can, by taking logarithms of both sides of the inequality, be converted to a constraint that is linear in the logarithms of the variables,

$$\sum_n a_{nt}\ln x_n \leqslant \ln c$$

Of course, this transformation is only applicable if all $x_n$ and the constant $c$ are constrained to be positive.

Another popular constraint-transformation device is to replace an equality constraint by two inequalities of opposite direction. Thus, $h_k(x) = 0$ is replaced by

$$h_k(x) \geqslant 0 \quad \text{and} \quad h_k(x) \leqslant 0$$

This device is only useful if, from the structure of $h_k$ and the objective function, it is obvious that one or the other of the inequalities will be limiting at the optimum, so that only one inequality need be retained in the problem formulation. If only one inequality can be used, then the feasible region will have been artificially expanded, thus making it easier to generate a feasible starting pont as well as to approach the constrained solution. However, if both inequalities must be retained, then no real computational advantage will have been achieved. Any point that is strictly interior to one inequality will be infeasible with respect to the other. Moreover, at any feasible point, both inequalities must be satisfied as equations, and hence the result of the transformation is to actually increase the number of equality constraints. To avoid such difficulties, the above inequalities are sometimes relaxed by introducing a suitable tolerance $\varepsilon > 0$,

$$h_k(x) \leqslant \varepsilon \qquad h_k(x) \geqslant -\varepsilon$$

The tolerance parameter serves to open up the feasible region to allow excursion from the surface $h_k(x) = 0$. At the same time, however, the solution obtained will be in error by an amount proportional to the tolerance parameter.

### Example 13.4   Constraint Transformation

Consider the problem

$$\text{Minimize} \quad f(x) = x_1^{-1}x_3^{-1}x_4$$

$$\text{Subject to} \quad h_1(x) = x_1^2 + x_2^2 - 1 = 0$$

$$h_2(x) = x_2^{-1}x_3 - 2 = 0$$

$$g_1(x) = 3 - x_4^{-2}x_5^{-1/3} \geqslant 0$$

$$g_2(x) = 4 - x_5 \geqslant 0$$

$$x_i \geqslant 0 \qquad i = 1,\ldots, 5$$

Observe first of all that $h_2(x)$ can readily be rewritten as a linear constraint:

$$h_2'(x) = x_3 - 2x_2 = 0$$

Next, by taking logarithms of both sides of the inequality,

$$x_4^{-2} x_5^{-1/3} \leqslant 3$$

Constraint $g_1$ can be converted to

$$g_1'(x) = \ln 3 + 2 \ln x_4 + \tfrac{1}{3} \ln x_5 \geqslant 0$$

Finally, equality $h_1(x)$ can be written as two inequalities:

$$g_3(x) = x_1^2 + x_2^2 - 1 \geqslant 0$$

$$g_4(x) = x_1^2 + x_2^2 - 1 \leqslant 0$$

Note, however, that $f(x)$ will be minimized by making $x_1$ and $x_3$ (hence, from constraint $h_2'$, also $x_2$) as large as possible. Therefore, the solution will be limited by the second inequality, and the first can be deleted.

Thus, with the variable definitions

$$y_1 = \ln x_4 \qquad y_2 = \ln x_5$$

the optimization problem takes the form

$$\text{Minimize} \qquad f' = x_1^{-1} x_3^{-1} \exp(y_1)$$

$$\text{Subject to} \quad h_2'(x) = x_3 - 2x_2 = 0$$

$$g_1'(x) = \ln 3 + 2 y_1 + \tfrac{1}{3} y_2 \geqslant 0$$

$$g_2'(x) = \ln 4 - y_2 \geqslant 0$$

$$g_4(x) = 1 - x_1^2 - x_2^2 \geqslant 0$$

$$x_1, x_2, x_3 \geqslant 0$$

where $y_1$ and $y_2$ are unrestricted in sign.

Of course, since minimizing $f'$ is equivalent to minimizing $\ln f'$, we can also replace $f'$ by the separable function

$$\ln f' = -\ln x_1 - \ln x_3 + y_1$$

The problem of minimizing $\ln f'$ subject to the above transformed constraints is a convex programming problem with only one nonlinear inequality ($g_4$) and three linear constraints. The original problem is nonconvex with three nonlinear constraints. Undoubtedly, the transformed problem will be much easier to solve than the original.

## Variable Transformation

A second device for reformulating a problem to enhance its solution efficiency is to use variable transformation. As illustrated in part in Example 13.4, variable transformations can serve to eliminate or to simplify constraints. For instance, the transformation

$$x_i = y_i^2$$

can be used to replace a variable $x_i$ and its nonnegativity condition $x_i \geqslant 0$. Similarly, a constraint of the form

$$\sum_i x_i = 1 \qquad \text{with } x_i \geqslant 0$$

can be eliminated by defining

$$y_i = \frac{x_i}{\sum x_i}$$

and expressing all remaining occurrences of $x_i$ by $y_i$. The use of various clever transformations of this type is discussed in more detail by Box [19]. However, as pointed out by Sisser [20], variable transformations must be used with some care because of the following complications they can induce:

1. The occurrence of extraneous local optima
2. The degradation of convexity
3. The inhibition of convergence

The following examples from Sisser [20] illustrate some of these adverse effects of poorly chosen variable transformations.

### Example 13.5   Variable Transformation Difficulties

(a) Consider the problem,

$$\text{Minimize} \quad f(x) = (x_1 + 1)^2 + (x_2 - 4)^2$$

$$x_1, x_2 \geqslant 0$$

under the transformation $x_i = (y_i^2 - 1)^2$.

(b) Consider the problem

$$\text{Minimize} \quad f(x) = 3x_1^2 - 2x_1x_2 + 3x_2^2$$

$$\text{Subject to} \quad x_1, x_2 \geqslant 0$$

under the transformation $x_i = y_i^2$.

**Solution.**   (a) The result of the substitution is

$$F(y) = \left(y_1^4 - 2y_1^2 + 2\right)^2 + \left(y_2^4 - 2y_2^2 - 3\right)^2$$

Since   $\nabla F = [2(y_1^4 - 2y_1^2 + 2)(4y_1^3 - 4y_1), 2(y_2^4 - 2y_2^2 - 3)(4y_2^3 - 4y_2)]$

and   $\dfrac{\partial^2 F}{\partial y_1^2} = 32y_1^2\left(y_1^2 - 1\right)^2 + 8\left(y_1^4 - 2y_1^2 + 2\right)\left(3y_1^2 - 1\right)$

$\dfrac{\partial^2 F}{\partial y_2^2} = 32y_2^2\left(y_2^2 - 1\right)^2 + 8\left(y_2^4 - 2y_2^2 - 3\right)\left(3y_2^2 - 1\right)$

it is easy to show that $F(y)$ has six local minima. Four of these, $y = (\pm 1, \pm\sqrt{3})$, correspond to the minimum of $f(x)$, namely $x^* = (0, 4)$; but two, $y = (\pm 1, 0)$, do not correspond to local minima of $f(x)$. The transformation has clearly introduced extraneous solutions.

(b) The substituted objective function is

$$F(y) = 3y_1^4 - 2y_1^2 y_2^2 + 3y_2^4$$

The second derivative of $F$ is given by

$$\nabla^2 F = 4 \begin{pmatrix} (9y_1^2 - y_2^2) & (-2y_1 y_2) \\ (-2y_1 y_2) & (9y_2^2 - y_1^2) \end{pmatrix}$$

which is not positive definite. For instance, along the ray $y_2 = 3y_1$, the determinant of the first principal minor is zero, while that of the second principal minor is $-4y_1^2$. On the other hand, since the matrix of second derivatives of $f$ is constant and positive definite,

$$\nabla^2 f = \begin{pmatrix} 6 & -2 \\ -2 & 6 \end{pmatrix}$$

the original function is convex. Thus the transformation has destroyed a desirable problem feature.

Sisser [20] gives conditions that the variable transformations and the transformed problem must satisfy to ensure that the above complications are avoided. The interested reader is invited to consult reference 20 for the details. From a practical point of view, the most important difficulty to be avoided is the introduction of spurious solutions. The use of monotonic one-to-one transformations will normally achieve that aim.

### Redundancy Elimination

Another means of enhancing solution efficiency is to delete redundant constraints from the problem formulation, since this obviously reduces the total number of problem constraints. By a redundant constraint we mean a constraint that is not required to define the boundaries of the feasible region. For instance, the inequalities

$$x_1 \leqslant a \quad \text{and} \quad x_1 \leqslant b \qquad \text{where } b > a > 0$$

are both legitimate constraints. However, since the latter lies outside of the former, it is not required to define the boundary of the feasible region in $x_1$ and can be deleted. While redundant bounds can, of course, be very easily recognized, there are no practical procedures available that can be used to identify redundant constraints in general. Procedures are available in the linearly constrained case [21]; however, the computations involved prove worthwhile only if the optimization problem must be solved more than once, for example, with several different objective functions or with a nonlinear objective function.

In spite of the lack of efficient general procedures that can guarantee the identification of all redundant constraints, it is nonetheless often possible to recognize cases of redundancy merely by inspection. Recall, for instance, that in the design problem discussed in Section 9.5 we noted by inspection that inequalities (9.53)

$$\frac{1.2B_1}{R_2} \geqslant \frac{1.2B_1}{R_3}$$

$$\frac{1.5B_2}{R_2} \geqslant \frac{1.5B_2}{R_3}$$

$$\frac{1.1B_3}{R_2} \geqslant \frac{1.1B_3}{R_3}$$

could be collapsed to the single inequality $R_3 \geqslant R_2$. Similarly, we observed that constraints (9.52)

$$T_1 \geqslant \frac{1.2B_1}{R_1} \quad \text{and} \quad T_1 \geqslant \frac{1.2B_1}{R_2}$$

were subsumed by constraints (9.56)

$$T_1 \geqslant \frac{1.2B_1}{R_1} + 3 + \frac{1.2B_1}{R_2}$$

since all variables were required to be nonnegative. Identification and elimination of such clearly redundant constraints should always be considered in the course of preparing a problem for solution.

### Equation Sequencing

Finally, as discussed in Chapter 7, the dimensionality and number of equality constraints of a problem can be substantially reduced by explicitly or implicitly solving some of the equality constraints and using them to eliminate variables. In general, this amounts to selecting a set of independent variables and sequencing the constraints to allow the constraints to be solved for the dependent variables with as little iteration as possible. While in fortuitous situations all equalities can be eliminated without any iterative solution, normally some equations will remain that cannot be solved directly for one or more of the dependent variables. In this case, two approaches are possible: The constraints that cannot be sequenced further are solved iteratively for their dependent variables each time the problem functions must be evaluated, or these remaining equations are treated explicitly as equality constraints and the remaining dependent variables as independent variables. In the former case, the iterative calculations must be carried out with sufficiently tight tolerances so as not to cause errors in the search method logic. Specifically, the tolerance on equations solved iteratively as part of the problem function calculations must be tighter than the tolerance used by the search logic in satisfying explicit equality constraints. This is inherently less efficient than applying the same tolerance to all equality constraints directly as part of the normal search logic. Consequently, the strategy in which the equations that cannot be further sequenced are treated as explicit equality constraints is usually to be preferred. As noted in Section 7.1, in all cases, bounds or inequality constraints imposed on the dependent variables must be retained in the problem formulation by expressing them in terms of the chosen set of independent variables.

**Analysis of Problem Features.** The last stage of activities before proceeding with the actual problem optimization consists of an analysis of the final problem formulation to identify, if possible, some of the structural features that will have an impact on the solution process. These features are convexity, boundedness of the feasible region, uniqueness of the solution, and existence of a feasible point. Although the concepts themselves appear abstract and largely of theoretical interest, they in fact do have important practical significance, and hence some modest attempt at identifying these features is well worthwhile.

### Convexity

The concept of convexity has been discussed amply in previous sections, and its consequences elaborated both in terms of optimality criteria and in terms of the increased reliability of the search logic. While it can be cumbersome to prove problem convexity, it generally is rather straightforward to identify formulation elements that will make the problem nonconvex. First and foremost, it is easy to check whether the problem involves nonlinear equality constraints. If at least one is present, the problem is nonconvex. If none are present, the nonlinear inequality constraints should be tested for convexity. Two types of tests can be used: one based upon the sign of the Hessian and the other that decomposes the function as a sum of convex functions. In either case, it is advisable to begin the process of testing with the simplest nonlinear inequality and to proceed to inequalities of increasing complexity. This ordering is based on the pessimistic, but also not unrealistic, supposition that the majority of engineering problems are nonconvex and can be identified as such even from the simpler inequalities (if not from the equalities). Finally, only after the constraint set is found to be convex is it worthwhile to test for the convexity of the objective function. If at the end of the analysis the problem is found to be convex, then we first of all have assurance that any point satisfying the Kuhn-Tucker conditions will be the global minimum. Second, we will be able to select from a much broader range of algorithms for solution, and we can, in general, expect that the convergence behavior of general-purpose methods will be better than in the nonconvex case. On the other hand, if the problem is identified as nonconvex, then we must prepare to face the very real issue of multiple local solutions.

### Boundedness

In either the convex or the nonconvex case, there is one further important characteristic that must be checked: the boundedness of the problem. In the present context, we say that a problem is *bounded* if the set of feasible solutions with objective function values less than that of some nominal (initial feasible) solution can be enclosed in a finite hypercube. In engineering practice, optimization studies always involve a balancing of factors and costs to achieve finite, optimum variable values. Before any design variable can become arbitrarily large or small, some technical or economic factors will prevent it from doing so. In carrying out the optimization, however, we use a model, an approximation to the real system. The model can quite easily give rise to unbounded solutions because the modeler either intentionally or unwittingly deleted some elements or limitations of the real system. Clearly, the fact that the model is based on a real system is no assurance that mathematical pathologies will not be encountered. Cases of model unboundedness can often be identified by means of a straightforward examination of the model and can always be prevented by imposing reasonable upper and lower bounds on all variables. However, in keeping with our dictum to avoid redundant constraints, it is appropriate to first check whether these additional bounds are required before automatically imposing them.

Although no guaranteed test of boundedness is known to the authors, two simple checks can be constructed that will usually identify the obvious cases of unboundedness. The first test, which we call the *variable test*, simply involves a one-at-a-time check of each variable to establish whether its value will be bounded above and below when all remaining variables are assigned fixed values (say, those of some starting point). This test can often be carried out by inspection, without actually performing any evaluation of terms, and can normally be confined to the inequality constraints.

The second test, called the *partial derivative* test, can be employed when function gradients are available or can be estimated. From the Kuhn-Tucker conditions, stated in terms of inequality constraints only,

$$\nabla f = \sum_j u_j \nabla g_j \quad \text{and} \quad u_j \geqslant 0$$

it is obvious that in order to be able to express the partial derivative $\partial f/\partial x_i$ of the objective function with respect to variable $x_i$ as a convex combination of inequality constraint gradients $\partial g_j/\partial x_i$, it is necessary that there exist at least one constraint, say $l$, with a negative partial derivative $\partial g_l/\partial x_i$ and at least one, say $m$, with a positive value, $\partial g_m/\partial x_i > 0$. This statement, of course, assumes that the optimum value of $x_i$ will not be unconstrained ($\partial f/\partial x_i = 0$) or lie on a constraint surface, but rather that it will lie at the intersection of two or more constraints. This is a reasonable and conservative assumption. The boundedness test based on this observation merely consists of checking whether for each variable $x_i$ there exists a constraint with a positive partial derivative and one with a negative partial derivative. If this is not the case for some variable, then the imposition of a bound is prudent. If no negative partial exists, an upper bound should be added; if no positive partial exists, a lower bound should be added. For purposes of this test, we again restrict our attention to the inequality constraints and, if many are present in the formulation, initially to the ones with simplest structure. We illustrate these simple tests with the following straightforward example.

### Example 13.6 Application of Boundedness Tests

Consider the problem

$$\text{Minimize} \quad f(x) = x_1^{-1} + x_2^{-2} + x_3^{-3}$$

$$\text{Subject to} \quad g_1(x) = x_1 + x_2 + x_3 - 1 \geqslant 0$$

$$g_2(x) = 2 - x_1^2 - x_2^2 \geqslant 0$$

$$g_3(x) = x_1 x_2 x_3 - 2 \geqslant 0$$

$$\text{and all } x_i \geqslant 0$$

Let us first apply the *variable test*.

When $x_2$ and $x_3$ are fixed with values $a$ and $b$, respectively, the constraints reduce as follows:

$g_1$ is equivalent to $x_1 \geqslant 1 - a - b$          (a lower bound)

$g_2$ is equivalent to $x_1 \leqslant +(2 - a^2)^{1/2}$    (an upper bound)

and

$$x_1 \geqslant 0$$

When $x_1$ and $x_3$ are similarly fixed, the constraints become

$g_1$ is equivalent to $x_2 \geqslant 1 - a - b$          (a lower bound)

$g_2$ is equivalent to $x_2 \leqslant (2 - a^2)^{1/2}$    (an upper bound)

and, of course,

$$x_2 \geqslant 0$$

Finally, when $x_1$ and $x_2$ are fixed at $a$ and $b$, respectively, we obtain

$g_1$ is equivalent to $x_3 \geqslant 1 - a - b$    (a lower bound);

$g_3$ is equivalent to $x_3 \geqslant \dfrac{2}{ab}$          (another lower bound);

$g_2$ imposes no bound on $x_3$    and $x_3 \geqslant 0$

Thus, $x_1$ and $x_2$ appear to be upper and lower bounded, while $x_3$ is only lower bounded. Prudence would suggest the addition of a reasonable upper bound on $x_3$, say $x_3 \leqslant 25$. In fact, without this bound, any search method applied to this problem would tend toward the limiting solution, $x^* = (\sqrt{2}, \sqrt{2}, +\infty)$.

A similar conclusion would have been found by the partial derivative test. The nonnegativity conditions (as well as $g_1$ and $g_3$) will yield positive partials for all $x_i$. But constraint $g_2$ has negative partial derivatives for only $x_1$ and $x_2$ $[\nabla g_2 = (-2x_1, -2x_2, 0)]$. To compensate for the missing negative partial with respect to $x_3$, we should impose an upper bound on $x_3$.

If we consider the Kuhn-Tucker gradient condition for this problem,

$$\begin{pmatrix} -x_1^{-2} \\ -x_2^{-2} \\ -x_3^{-2} \end{pmatrix} = u_1 \begin{pmatrix} 1 \\ 1 \\ 1 \end{pmatrix} + u_2 \begin{pmatrix} -2x_1 \\ -2x_2 \\ 0 \end{pmatrix} + u_3 \begin{pmatrix} x_2 x_3 \\ x_1 x_3 \\ x_1 x_2 \end{pmatrix} + (u_4, u_5, u_6) \begin{pmatrix} 1 & 0 & 0 \\ 0 & 1 & 0 \\ 0 & 0 & 1 \end{pmatrix}$$

where the identity matrix corresponds to the nonnegativity conditions on the variables, it is obvious that for no finite, positive values of $x_i$ can there be a set of nonnegative multipliers $u_i$ that will satisfy the above conditions. The imposition of the upper bound on $x_3$ adds the column $(0, 0, -1)^T$ and rectifies the situation.

Note that if an equality constraint involving $x_3$ had been included in the above formulation, then we would have ignored it for purposes of our tests. An equality may or may not have served to transfer bounds imposed on the other variables to $x_3$. For instance, $h_1 = x_3 - x_1^2 - x_2^2 = 0$ would serve to transfer an upper bound on $x_3$, since, from $g_2$,

$$x_3 = x_1^2 + x_2^2 \leqslant 2$$

However, $h_2 = x_3 + \log(x_1 - x_2) = 0$ would not. Disregard of equality constraints is thus a conservative policy but one that simplifies the testing.

The simple boundedness tests may be too awkward to apply in very complex situations such as those involving simulation models. The decision to use them, as always, rests with the user. The key point is that the possibility of model unboundedness should always be considered. Optimization software is not as yet distinguished by its diagnostic capabilities. In our experience, if often takes many wasted runs before a user is convinced that run aborts are caused by model unboundedness rather than by a coding error in the vendor software or the user-supplied routines.

### Solution Uniqueness

A third important consideration in problem analysis is recognition of the possibility of *nonunique* solutions or *multiple local minima*. In general, a convex problem may have only a single optimum objective function value but may take on that value at a family of points. Thus, although convexity guarantees a global optimum, it does not guarantee a unique solution. On the other hand, in order for a problem to possess more than one local minimum it must be nonconvex, but nonconvexity is not enough to guarantee the occurrence of multiple local minima. Evidently, the convexity test discussed earlier in this section does not provide all the answers. Additional analysis is desirable to determine whether the problem will have a nonunique solution and whether it is likely to have multiple local solutions.

Normally, the occurrence of a nonunique solution is an artifact of the model rather than a natural feature of the real system. The most common cause for nonuniqueness is the occurrence throughout the model of a fixed combination of two or more variables. For instance, each time either $x$ or $y$ appears in the problem functions, they are in the fixed combination $xy$. In principle, the product $xy$ could be replaced by $z = xy$ and the problem reduced in dimensionality by 1. However, this type of nonuniqueness generally arises because some constraints or relations involving $x$ or $y$ separately were omitted in the model formulation. Consequently, the appropriate remedy, having recognized the situation, is to impose the omitted constraints.

### Example 13.7  Nonunique Solution

Consider the problem

$$\text{Minimize} \quad f_2(x) = x_1^{-1} x_2$$

$$\text{Subject to} \quad g_1(x) = x_2 - x_1 \geqslant 0$$

$$x_1, x_2 \geqslant 0$$

Suppose we construct the Kuhn-Tucker gradient conditions for the problem. Since $\Delta f = (-x_2 x_1^{-2}, x_1^{-1})$ and $\nabla g = (-1, 1)$, we must have

$$\begin{pmatrix} -x_2 x_1^{-2} \\ x_1^{-1} \end{pmatrix} = u_1 \begin{pmatrix} -1 \\ 1 \end{pmatrix} + (u_2, u_3) \begin{pmatrix} 1 & 0 \\ 0 & 1 \end{pmatrix}$$

At a constrained minimum, $g_1$ must be tight; therefore, $x_2 = x_1$, and the above condition reduces to

$$\begin{pmatrix} -x_1^{-1} \\ x_1^{-1} \end{pmatrix} = u \begin{pmatrix} -1 \\ 1 \end{pmatrix} + (u_2, u_3) \begin{pmatrix} 1 & 0 \\ 0 & 1 \end{pmatrix}$$

Note that any positive value of $x_1$ will satisfy the Kuhn-Tucker conditions with $u_1 = x_1^{-1}$ and $u_2 = u_3 = 0$. The minimum value of the objective function is $f(x) = 1$, and that value will be attained at all points such that $x_1^{-1} x_2 = 1$, that is, $x_2 = x_1$.

The occurrence of $x_1$ and $x_2$ in fixed combination becomes more obvious if we rewrite $g_1$ as a quotient:

$$g_1' = x_1^{-1} x_2 - 1 \geqslant 0$$

because then we see that these variables occur in a fixed combination in both $f$ and $g_1'$.

In any practical case it is highly unlikely that the designer will be indifferent to the choice of $x_1 = x_2$. Typically, some features of the real system will have been deleted from the model, perhaps, for instance, the condition $x_1 + x_2 = 2$. This additional constraint in which $x_1$ and $x_2$ occur separately will force the unique solution $x_1 = x_2 = 1$.

Unfortunately, there are no convenient, guaranteed procedures for establishing the existence of nonunique solutions. In the case of both prototype and generalized geometric programming problems, it is sufficient to check the rank of the matrix of variable exponents [22, 23]. If the rank is less than $N$, the number of primal variables, then nonuniqueness is assured. In such a case, it is also easy to construct a variable grouping that can be removed from the formulation [22]. This is a powerful test but does require that the problem be recast to GP form. By way of illustration, Rijckaert and Martens [24] used the exponent matrix test to show that a well-known process optimization benchmark problem in use for some 15 years had a nonunique solution. Thus, by reformulating the problem to a geometric program, it was possible to explain the number of apparently different solutions with similar objective function values that had been reported by various researchers. Since the effort involved in recasting a problem as a GP and performing a rank test is nontrivial, it can in many cases be avoided by merely inspecting the formulation to verify whether certain variable groupings tend to occur throughout the problem functions. Only if grouping is suspected is it prudent to proceed with the GP constructions. Again, the key point is to recognize that nonuniqueness is an artifact of models that can happen in practice. If it could not be identified prior to solution, it should be suspected to be occurring if multiple optimization runs from different starting points yield termination points that have similar objective function values but considerably different values of some of the variables.

As noted earlier, although multiple local minima can occur whenever a nonconvex problem is solved, nonconvexity itself does not guarantee multiple local minima. In fact, it is generally difficult to prove that multiple local minima do exist except by actually identifying these points. Because the threat of local solutions is ever-present in engineering applications, while their actual existence is difficult to establish, there is a tendency to ignore this issue in application studies. In point of fact, the issue is real and must always be addressed unless it can be shown that local optima *cannot* exist.

The problem need not be particularly complex to admit multiple local solutions. Recall that in Example 8.7 the existence of one quadratic equality constraint and two quadratic inequalities in an otherwise linear problem was sufficient to produce three distinct Kuhn-Tucker points. As shown in the following example, drawn from reference 25, multiple local minima can arise in even small problems.

### Example 13.8 Multiple Minima

Consider the simplified network with two heat exchangers shown in Figure 13.5. In this system, one stream with flowrate $F_1$ is to be heated from 100°F to 300°F by exchanging heat with two hot streams with flowrates $F_2$ and $F_3$, available at 600 and 900°F, respectively. Assume that all three flows are equal to $10^4$ lb/hr; all stream heat capacities are equal to 1 Btu/(lb · °F); both

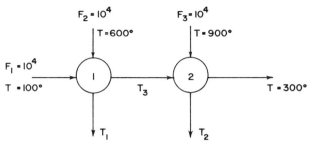

**Figure 13.5.** Two-exchanger network, Example 13.8.

overall heat transfer coefficients $U_1$ and $U_2$ are equal to 200 Btu/(hr · ft$^2$ · °F); and exchanger costs are given by

$$\text{Cost} = 35(\text{area, ft}^2)^{0.6}$$

The objective is to select the unknown temperatures $T_1$, $T_2$, and $T_3$ so that the combined cost of the two exchangers is minimized.

**Formulation.** An elementary model of a heat exchanger consists of an energy balance on each stream relating the heat transferred $Q$ to the stream temperature difference and a design equation relating the heat transfer area $A$ to the heat transferred and the temperature differences between the streams. The system model will, therefore, consist of the following equations:

**Exchanger 1**
Energy balances:   $Q_1 = 10^4(1)(T_3 - 100)$
                   $Q_1 = 10^4(1)(600 - T_1)$
Design equation:   $Q_1 = U_1 A_1 \text{LMTD}_1 = 200 A_1 \text{LMTD}_1$

where
$$\text{LMTD}_1 = \frac{(600 - T_3) + (T_1 - 100)}{2}$$

**Exchanger 2**
Energy balances:   $Q_2 = 10^4(1)(300 - T_3)$
                   $Q_2 = 10^4(1)(900 - T_2)$
Design equation:   $Q_2 = U_2 A_2 \text{LMTD}_2 = 200 A_2 \text{LMTD}_2$

where
$$\text{LMTD}_2 = \frac{(900 - 300) + (T_2 - T_3)}{2}$$

Note that in this case the usual log mean driving force LMTD is replaced by the arithmetic mean driving force, since the flows and heat capacities of the contacting streams are equal. Because of the simple structure of the equations as well as to compress the problem for purposes of the subsequent analysis, we will select $T_3$ as independent design variable and reduce the problem as follows. First, the exchanger 1 energy balances can be used to relate $T_1$ to $T_3$, and then the result can be used in the design equation to express the area of exchanger 1 in terms of $T_3$. Thus,

$$A_1 = \frac{50(T_3 - 100)}{600 - T_3}$$

Similarly, using the exchanger 2 energy balances to express $T_2$ as a function of $T_3$, the exchanger 2 design equation can be used to solve for $A_2$. As a result, the area of the second exchanger is given by

$$A_2 = 50\frac{300 - T_3}{600}$$

Since the objective function is

$$35A_1^{0.6} + 35A_2^{0.6}$$

the problem becomes

$$\text{Minimize} \quad f(T_3) = 35\left[\frac{50(T_3 - 100)}{600 - T_3}\right]^{0.6} + 35\left[\frac{50(300 - T_3)}{600}\right]^{0.6}$$

$$\text{Subject to} \quad 100 \leqslant T_3 \leqslant 300$$

**Analysis.** At $T_3 = 100°$, $f(T_3) = 189.3$, while at $T_3 = 300°$, $f(T_3) = 286.9$. The plot of $f(T_3)$ over the range $100° \leqslant T_3 \leqslant 300°$ is shown in Figure 13.6. Clearly, the function has two local minima, one at $T_3 = 100°$ and one at $T_3 = 300°$, and one maximum, at about $T_3 = 280°$. Presumably, any search

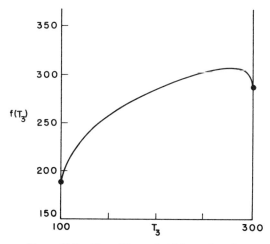

**Figure 13.6.** Plot of Example 13.8 cost function.

method initiated with a temperature estimate $T_3 > 280°$ will converge to the higher local minimum; while initiation with an estimate $T_3 < 280°$ will converge to the lower local minimum, which is the global minimum. Note from the plot that $f(T_3)$ is a *concave* function.

In addition to illustrating a case of multiple local solutions, Example 13.8 also offers a clue that can be used in identifying situations involving local solutions. Note that the objective function contains variables raised to a power less than 1. Expressions of the form $x^b$, where $b < 1$, are *concave* functions in $x$. Such expressions occur quite frequently in process optimization, because the capital cost of process equipment is typically correlated in terms of power functions $ax^b$, where $x$ is some measure of equipment capacity [26]. The exponent $b$ of such correlations is typically less than 1, because the cost of equipment per unit of capacity decreases as the capacity of the equipment increases. This is another instance of the phenomenon of economy of scale. Since such cost terms introduce concave elements into the objective function, we can reasonably expect that problems involving such terms will give rise to multiple local solutions. The reader should note, however, that this is a rule of thumb, not a guarantee. Recall that in Section 9.5.3 we considered a design optimization problem that had as objective function the total capital cost of the plant equipment. In that problem, all the exponents were less than 1; however, it can be shown that the constraint structure is such that only one local, hence global, minimum exists [27].

In summary, before proceeding with the problem solution, it is expedient to analyze the problem for the possibilities of nonunique solutions and multiple local optima. To check for the former situation, the formulation ought to be reviewed to identify possible occurrence of fixed-variable combinations. To check for the latter, the problem is first tested for nonconvexity. If the problem is nonconvex, then the presence of power-law objective function terms should be taken as a special indication of the likely occurrence of local solutions.

*Feasibility*

As the final element of problem presolution analysis, the formulation should be tested for *feasibility*. While the preceding stages of problem preparation and analysis may have resulted in a numerically stable, bounded, and nonredundant formulation, it is always possible that, as a result of poor data or calculation errors, the problem constraints simply exclude all possible solutions. Thus, whether or not the optimization algorithm selected for use requires a feasible starting point, it is good practice to devote some effort to generate a feasible starting point. Obviously, if no feasible starting point can be obtained, there is little point in proceeding with optimization. Instead, the model must be inspected again and validated in a piecemeal fashion until the sources of error are identified. If a feasible point can be generated, and if the variable values at the generated point (or points) appear reasonable, then one can proceed with problem solution with a fair degree of confidence that the optimization runs will be productive.

There are three procedures in common use for generating feasible starting points: random sampling, minimization of unconstrained penalty-type functions, and sequential constraint minimization. As discussed in Section 7.1.2, generation of random starting points simply involves the use of Eq. (7.1) and a pseudo-random-number generation routine to obtain trial points that are then tested for feasibility. Infeasible points are rejected, and the sampling is repeated until a feasible point it found. Unfortunately, even for moderate-sized nonlinear inequality-constrained problems (say, 15 variables and 10 inequalities), excessive computation time may be required (for instance, in excess of 500 CPU seconds on a CDC 6500 [28]. Consequently, for higher dimensionality and tightly constrained problems, the two other strategies are often preferable.

A very common way of generating feasible starting points is *direct minimization of the constraint infeasibilities*. Under this strategy the phase I procedure consists of solving an unconstrained minimization problem whose objective function is an exterior penalty function. The starting point is thus obtained as the solution of the problem

$$\text{Minimize} \quad f(x) \equiv \sum_k [h_k(x)]^2 + \sum_j \{\min[0, g_j(x)]\}^2$$

Clearly a feasible point is one that will result in an objective function value of zero. Hence, the unconstrained minimization will be terminated when $f(x)$ becomes sufficiently small. Generally, the minimization can be simplified if the problem is posed in equality-constraint-free form.

### Example 13.9   Starting Point Generation, Infeasibility Minimization

Consider the problem of Example 7.2. Generation of a feasible starting point could be accomplished by solving the unconstrained problem

$$\text{Minimize} \quad (x_1 + x_2 + x_3 - 1)^2 + (x_1^2 x_3 + x_2 x_3^2 + x_2^{-1} x_1 - 2)^2$$

$$+ [\min(0, x_1)]^2 + \{\min[0, (\tfrac{1}{2} - x_1)]\}^2 + [\min(0, x_3)]^2$$

$$+ \{\min[0, (\tfrac{1}{2} - x_3)]\}^2$$

or, alternatively, upon substituting $x_1$ by $1 - x_2 - x_3$, the problem

$$\text{Minimize} \quad [(1 - x_2 - x_3)^2 x_3 + x_2 x_3^2 + x_2^{-1}(1 - x_2 - x_3) - 2]^2$$

$$+ [\min(0, 1 - x_2 - x_3)]^2 + [\min(0, x_2 + x_3 - \tfrac{1}{2})]^2$$

$$+ [\min(0, x_3)]^2 + [\min(0, \tfrac{1}{2} - x_3)]^2$$

In either case, one of the methods discussed in Chapter 3 could be used to obtain a solution. In general, either no solution, one solution, or an infinite number of solutions can be found. In the present case, the point (0.39606, 0.20788, 0.39606) with an objective value of $9 \times 10^{-10}$ is a solution, but an infinite number of other feasible points also exist. At this point $h_1(x) = 0.0$, $h_2(x) = 3 \times 10^{-3}$, and the variable bounds are all satisfied.

Some optimization algorithms require that the starting point $x^0$ not only be feasible but also lie strictly within the inequality constraints, that is, $g_j(x^0) > 0$ for all $j$. In such cases the penalty function can be modified to include a tolerance parameter $\varepsilon > 0$, as follows:

$$\text{Minimize} \quad \sum_k [h_k(x)]^2 + \sum_j \{\min[0, g_j(x) - \varepsilon]\}^2$$

The effect of the tolerance parameter is to force the generation of a solution $x^0$ such that $g_j(x^0) = \varepsilon > 0$. However, for sufficiently large values of $\varepsilon$ there may be no feasible solution satisfying $g_j(x) = \varepsilon$. Thus, it is sometimes necessary to repeat the minimization with a reduced value of $\varepsilon$ if for the given $\varepsilon$ the objective value cannot be reduced sufficiently close to zero.

The minimization of the sum of squares of the constraint infeasibilities is the most frequently used phase I construction. However, the resulting objective function is quite nonlinear, and consequently, for tightly constrained highly nonlinear problems, it can be difficult to achieve an objective function value sufficiently close to zero.

A more complicated but usually more reliable alternative to direct minimization of the constraint infeasibilities is a *sequential minimization procedure* in which each constraint is used as the objective function of a constrained minimization problem. We state the procedure for the inequality constrained problem.

### Sequential Minimization Procedure

Given the constraint set

$$g_j(x) \geqslant 0 \quad j = 1, \ldots, J$$

$$x_i^{(L)} \leqslant x_i \leqslant x_i^{(U)} \quad i = 1, \ldots, N$$

and a tolerance parameter $\varepsilon > 0$.

**Step 1.** Solve the problem

$$\text{Minimize} \quad -g_1(x)$$

$$\text{Subject to} \quad x^{(L)} \leqslant x \leqslant x^{(U)}$$

terminating at the point $x^{(1)}$ such that $g_1(x^{(1)}) \geqslant \varepsilon$.

For $j = 2, \ldots, J$, solve the series of problems as follows:

**Step 2.**

$$\text{Minimize} \quad -g_j(x)$$

$$\text{Subject to} \quad g_m(x) \geq 0 \quad m = 1, \ldots, J - 1$$

$$x^{(L)} \leq x \leq x^{(U)}$$

with starting point $x^{(j-1)}$, terminating at the point $x^{(j)}$ such that $g_j(x^{(j)}) \geq \varepsilon$.

**Step 3.** If $g_j(x) \geq \varepsilon$ cannot be achieved for some $j$, reduce $\varepsilon$ and repeat the subproblem solution. If $\varepsilon$ becomes less than some specified zero tolerance, terminate the procedure.

Note that each subproblem involves the solution of an inequality-constrained optimization problem with known *feasible* starting point. Thus the method that will be used to solve the original problem can be used to solve each of the subproblems. Once, the $J$th subproblem is satisfactorily solved, a feasible point will have been obtained to the original constraint set. If $\varepsilon$ is selected to be too large, a subproblem solution yielding $g_j(x) \geq \varepsilon$ may not exist, and hence $\varepsilon$ must be reduced. To avoid an infinite computational loop, once $\varepsilon$ is reduced to a value sufficiently close to zero, the procedure must be terminated, with the conclusion that no feasible point may exist.

*Example 13.10   Starting Point Generation, Sequential Procedure*

Consider the feasible region

$$g_1(x) = x_1 + x_2 - 1 \geq 0$$

$$g_2(x) = x_2 - x_1^2 \geq 0$$

$$g_3(x) = x_1 - x_2^2 \geq 0$$

$$0 \leq (x_1, x_2) \leq 2$$

and suppose the sequential procedure is used to generate a feasible point with $\varepsilon = 0.1$. The first subproblem will be

$$\text{Minimize} \quad -g_1(x) = 1 - x_1 - x_2$$

$$\text{Subject to} \quad 0 \leq x_1 \leq 2$$

$$0 \leq x_2 \leq 2$$

and solution can be terminated at any point for which $g_1(x) \geq 0.1$. Obviously, the optimum point $x^{(1)} = (2, 2)$ will satisfy this requirement.

The second subproblem will be

$$\text{Minimize} \quad -g_2(x) = x_1^2 - x_2$$

$$\text{Subject to} \quad g_1(x) = x_1 + x_2 - 1 \geq 0$$

$$0 \leq (x_1, x_2) \leq 2$$

The starting point $x^{(1)} = (2, 2)$ is clearly feasible, and the suboptimal point $x^{(2)} = (0, 1)$ will clearly satisfy the requirement $g_2(x^{(2)}) \geq 0.1$, as will the subproblem optimum $(0, 2)$.

The third subproblem is

$$\text{Minimize} \quad -g_3(x) = x_2^2 - x_1$$

$$\text{Subject to} \quad g_2(x) \geq 0$$

$$g_1(x) \geq 0$$

$$0 \leq (x_1, x_2) \leq 2$$

Again, the starting point $x^{(2)} = (0, 1)$ is clearly feasible. The suboptimal point $x^{(3)} = (1/2, 1/2)$ satisfies $g_3(x^{(3)}) \geq 0.1$, as does the optimum $(0.618, 0.382)$ (see Figure 13.7).

Observe that in each case the subproblem solution iterations would normally have been stopped well short of the subproblem optimum. Hence the computational effort required per subproblem is relatively small.

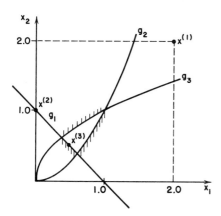

**Figure 13.7.** Feasible region for Example 13.10.

Equality constraints can be accommodated in the sequential approach only if the constraint-elimination device is used to express the subproblems in terms of the selected independent variables.

This completes our discussion of model preparation and analysis activities. In the next section, we briefly review some of the solution execution strategies that might be considered in carrying out the actual optimization.

### 13.2.3  Execution Strategies

Depending upon the type of model, the properties of the formulation, and its structural features, a number of different strategies can be used to actually carry out the optimization. While direct optimization using an appropriate NLP method is always possible, in many cases it is fruitful to consider alternatives such as sequential optimization using a series of subproblems or two-level approaches in which intermediate approximate models are employed. In addition, if multiple local minima are to be expected, then some strategy must be adopted to identify the global minimum. In this section we will very briefly review these strategies and their consequences.

**Strategies for Equation-Oriented Models.**  Equation-oriented models as well as models based on response surfaces are generally solved either directly or using a sequential approach. In the direct approach, the only real choice lies in whether the problem possesses sufficient structure to be amenable to some of the specialized optimization methods of Chapter 11, or whether general-purpose NLP algorithms must be used. Normally, specialized techniques will, by definition, be preferable to general-purpose methods and should always be investigated if repeated solution of the same type of problem will be required. However, it may take more time and effort to set up a problem for solution as a specially structured problem than to solve it as a general NLP. For instance, if a problem can be posed as a geometric programming problem, then it ought to be solved as such. However, if the problem is to be solved only once and then put aside, then it may be more efficient of the engineer's time to simply solve it as a general NLP.

Sequential approaches are those in which the problem is solved by optimizing a series of restricted subproblems. The basic idea is to attack a problem whose direct solution may be difficult by dividing the problem variables into those whose presence make the solution difficult and those whose determination would be relatively easy if the difficult variable values were fixed. The "easy" variable and "difficult" variable subproblems are solved separately with some suitable coordinating calculations to link the two. For such a strategy to be effective, the subproblems must necessarily be easier to solve than the original problem, and the subproblem coordinating calculations must be such as to assure overall convergence. A detailed discussion of such strategies is certainly beyond the scope of this text; instead, the reader is directed to reference 29. We have encountered several instances of this general type of solution approach in the previous chapters. Recall, for instance, the recursive approach to solving the pooling problem as a series of linear programs (Example 8.7); the solution of generalized geometric programs via condensation and a series of prototype geometric programs (Sect. 11.4); and the branch-and-bound solution of integer linear programs using LP subproblems (Section 11.1.3). Sequential approaches must be tailored to the specific problem structure, and hence are worthwhile if repeated solution of a class of problems with similar structure is projected.

**Strategies for Simulation Models.** The optimization of simulation or procedure-oriented models can be executed either directly or by using various *two-level strategies*. In the direct approach, the entire simulation or procedure acts as a function routine that calculates performance index and constraint values and is accessed by a suitable optimization method. If the simulation or procedure output variables are in principle continuously differentiable with respect to the input variables, then any of the gradient-based constrained or unconstrained optimization algorithms can be employed. Otherwise, direct-search methods such as the complex method or a random search method ought to be used. As an example of the former type, Sargent [30] formulates the design calculations of a multicomponent staged distillation column as a procedure-oriented model and performs the design optimization using a gradient projection method. Gradients with respect to the independent optimization variables are calculated by numerical differencing. A similar strategy is used by Ravindran and Gupta [31] to optimize a procedure-oriented model of a transportation network using a GRG algorithm. By contrast, Reklaitis and Woods [32] optimize a nondifferentiable simulation model of a nonideal multistage compressor system by using the Complex method. Direct-search optimization of simulation models using either the Complex or random sampling methods is, without doubt, the most frequently used approach.

Direct approaches are generally suitable for studies involving no more than five optimization variables and relatively simple inequality constraints on the optimization variables. The simulation or procedure-oriented model must be structured to allow the model to be called as a subroutine by the optimizer. Moreover, if the optimization routine requires separate calls to evaluate the objective function and the constraints, then the model may have to be executed both times.

There are three very common situations that arise in the direct optimization of simulation models and can cause difficulties in execution and thus merit reiteration. These are:

1.  Implicit constraints on dependent (internal) variables
2.  Implied constraints that underly the model itself
3.  Numerical procedures used within the simulation

It frequently is the case that a simulation model will be constructed assuming that certain dependent variables will be within a physically meaningful range or will satisfy some reasonable engineering conditions. Unless provisions are made in advance, the values of the independent variables selected by the optimization method may force violation of these implicit constraints and thus may either cause a program abort or lead to meaningless results. The safest way to treat such constraints is to define them as explicit constraints that will be implicit nonlinear functions of the independent variables. This will, in general, require that the simulation be executed to evaluate the constraints.

A similar situation and solution will be the case with implied model constraints. Often a simulation model is valid only over certain ranges of key dependent variables. Unless these ranges are treated as explicit constraints, the optimizer may force a solution that is well outside the range of validity of the model. This can be particularly insidious if an internal variable is forced outside its valid range and values of this variable are not monitored during the execution of the study. The obvious solution, again, is to define these implied constraints as explicit constraints.

Finally, as mentioned earlier, if the simulation model involves internal numerical procedures, such as integration, equation solving, or the use of quadrature approximations, then care must be exercised to ensure that the tolerances used in the numerical methods are tighter than those used by he optimizer in its numerical procedures or checks. Typical optimizer numerical procedures might be differencing to obtain gradient values, Newton iterations to ensure constraint feasibility, and the usual feasibility and optimality tests. Internal numerical calculations that are conducted with looser tolerances usually will cause the optimization algorithm to be misled and hence can cause premature termination or failure to converge properly.

The occurrence of the above three situations in a simulation model and the solutions outlined to resolve the associated potential difficulties can make the resulting optimization problem very difficult for direct optimization because of excessive computer time requirements. If this is the case, then it becomes increasingly appropriate to use various forms of *two-level strategies*. In the two-level approach, the simulation or procedure-oriented model is used to generate a response-surface model expressed in terms of the independent optimization variables, and the response-surface model is optimized using a suitable optimization method. The process is repeated using the response surface updated at the previously obtained optimum point until differences between successive solutions become appropriately small. Two-level strategies differ primarily in the type of approximating function used, in the level of detail at which the response surfaces are generated, and in the optimization techniques employed.

Most commonly, a quadratic approximating function is used, and the function coefficients are calculated from the results of a suitable pattern of model runs positioned about the current best estimate of the optimum [33] (a star or full factorial design might, for example, be used). Since the quadratic

**Table 13.1    Nominal Daily Production Requirements
(All Units in Total Gallons)**

| Product Type | Total | Pints | Quarts | Half-Gallons | Gallons |
|---|---|---|---|---|---|
| 3.2% Milk | 20,000 | 2000 | 5000 | 10,000 | 3000 |
| 3% Milk | 5,000 | 500 | 1000 | 3,000 | 500 |
| 2% Milk | 10,000 | 1000 | 2000 | 5,000 | 2000 |
| Skim milk | 15,000 | 1000 | 2000 | 10,000 | 2000 |
| Orange juice | 20,000 | 3000 | 5000 | 10,000 | 2000 |
| Fruit drink | 5,000 | 100 | 400 | 500 | 4000 |

function is well behaved, any gradient-based method can be used for the
optimization. The following example describes the optimization of a simulation
model using a two-level strategy.

### Example 13.11    Optimization via a Two-Level Approach

**Problem Description.**    The fluid-processing facility shown in Figure 13.8 consists of two parallel lines in which fluid products are pasteurized, followed by
four parallel, equal-size intermediate storage vessels and four parallel, nonidentical packaging lines. In the four packaging lines the pasteurized fluids are
packaged into pint, gallon, half-gallon, and gallon containers, respectively. The

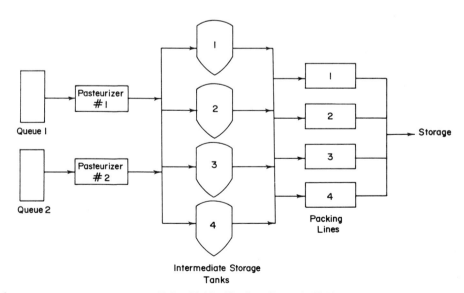

**Figure 13.8.**    Fluid milk plant, Example 13.11.

plant has a 16-h operating cycle during which a projected slate of eight products must be processed and packaged into the four container sizes. Given the design daily production requirement shown in Table 13.1, the production rates of the pasteurizer and packaging lines and the storage tank volume are to be selected such as to minimize the capital cost of the equipment.

**Model Formulation.** As described in reference 34, the system can be represented using a SLAM simulation model [35]. In this model, the products are divided among the production queues associated with each of the pasteurizers based on processing considerations. The queue of products associated with each processor is ordered according to largest volume product first. The events that occur during the 16-h processing day are simulated over time using a combined discrete-continuous representation. The continuous portions of the system are represented in terms of state variables. Typically the state variables represent the quantity of current product remaining to be processed on each semicontinuous unit or the current level in each storage tank. Changes in the state variables are all described by first-order difference equations (state equations), since the processing units are assumed to either operate at constant rate or be shut down. Important decision points in the time history of the system are identified in terms of specified values of the associated state variables. These decision points or events and their consequences are summarized in Table 13.2.

### Table 13.2  Simulation State Events

| Event Index | Event Definition and Consequent Action |
|---|---|
| 1 | A storage tank has become full. Shut off valve inputting material from pasteurizer. |
| 2 | Positive threshold value in a tank has been crossed (3/4 full). This allows the packagers to begin removing product. |
| 3 | A storage tank has become empty. Shut off valve removing material. Initiate search for tank with same material |
| 4 | A packager has completed its current product. Initiate search of storage tanks for its next product. |
| 5 | A pasteurizer has completed its current product. Remove next product from queue and initiate search for empty tank or tank with same product. |
| 6 | Check for an empty tank. |
| 7 | Check if there is an idle packager for acceptance of product. |
| 8 | Error detection event. |
| 9 | Negative threshold value in a storage tank has been crossed (1/4 empty). This allows the pasteurizers to begin refilling the tank. |

The key assumptions built into the model are:

1. No product mixing is allowed in the storage tanks.
2. Once a product begins running on a pasteurizer or packager it continues running until the production quota for the product is met.
3. Once a packager finishes processing a product, it selects the next product to process on a first come, first served basis.
4. A tank can be refilled once it reaches one-fourth of its volume and can commence to be emptied once it is three-fourths full.

The typical CPU time required to execute one 16-hr production simulation was 16 sec on a CDC 6500.

**Optimization Procedure.**  A two-level strategy is used to optimize the design of the system modeled via the simulation. At a given estimate of the design variables, a quadratic response surface is fitted to the simulation results. Since the problem involves seven design variables—the two pasteurizer rates, the storage tank volume, and the four packager rates—a factorial design using 36 simulation runs is required to evaluate all the coefficients. The response surface thus gives:

$$\text{Production time required} = q \, (\text{design variables})$$

The approximate optimization problem thus becomes

$$\text{Minimize} \quad \text{Cost of equipment} = f \, (\text{design variables})$$

$$\text{Subject to} \quad q \, (\text{design variables}) \leqslant 16 \text{ hr}$$

and

$$\text{Upper and lower bounds on design variables.}$$

The approximate problem is solved using a GRG algorithm. Typically, 20 sec (CDC 6500) is required for a subproblem optimization. At the resulting solution, a new response surface is generated and the process continued until successive values of the optimum cost are sufficiently close. Typical results are shown in Table 13.3. For the sake of brevity, the details of the nonlinear cost function and the numerical coefficients of the seven-variable response function are not given here. The interested reader is directed to reference 34 for the details. It should be noted that since one simulation run requires, on the average, 16 CPU seconds, optimization with respect to seven design variables is quite time-consuming. The five iteration cycles shown above consumed some 3000 sec or 180 simulation runs. On balance, the use of a response surface is nonetheless effective, because with 180 runs a trial optimization using the complex search made only very modest progress.

**Table 13.3  Results of Optimization of the Fluid Milk Process**

| Iteration Number | Pasteurizing Units, gal/hr | | 5 Storage Tanks, gal | Packaging Machines, gal/hr | | | |
|---|---|---|---|---|---|---|---|
| | 1 | 2 | | 1 | 2 | 3 | 4 |
| 1 | 6000 | 6000 | 3000 | 1000 | 25000 | 35000 | 1500 |
| 2 | 4000 | 3000 | 15000 | 500 | 1900 | 2900 | 1200 |
| 3 | 5000 | 3763 | 1000 | 500 | 1300 | 2600 | 1000 |
| 4 | 5500 | 400 | 800 | 500 | 1116 | 2450 | 900 |
| 5 | 5400 | 4094 | 800 | 499 | 1033 | 2400 | 900 |

| Simulated Time, hr | System Cost, $ |
|---|---|
| 12.12 | 897,490 |
| 17.04 | 743,314 |
| 18.05 | 672,045 |
| 16.26 | 644,172 |
| 16.74 | 639,855 |

Two-level approaches such as that illustrated in Example 13.11 can be used at varying levels of aggregation [10]. In Example 13.11, the entire system was represented in terms of a single-response-surface model. For more complex systems, it may be appropriate to generate a separate response surface for each major subsystem or even each separate component of the system. These individual response surfaces can then be combined into an approximate overall system model, which is optimized using a suitable NLP algorithm. At the resulting point the response surfaces are all updated, and the procedure is continued until convergence is attained as in the single-response-surface case. This approach has been used in the context of chemical processing systems and can be particularly effective if the process network involves recycles or loops.

It is worth noting that the two-level strategy was initially investigated as a means of accelerating the convergence of flowsheet calculations involving difficult recycles [10, 12], and only subsequently has its use been advocated in performing process optimization [8, 9]. Both linear and quadratic approximate models can be used, and the optimization is carried out most effectively using either GRG or CVM algorithms. Because typically each separate component model being approximated will involve only a few different independent optimization variables, this approach can be used successfully with large-scale optimization problems involving, say, up to 25 variables. However, if each separate response surface involves most of the optimization variables, then, because of the number of simulation runs required to generate the surfaces, the dimension of the optimization problem must be restricted to a much smaller number of variables [33].

The execution of this type of multiple-response-surface strategy should normally be automated to as high a degree as possible. Ideally the optimization program should

1. Control the execution of portions of the simulation model to generate data for the response-surface models.
2. Direct the computations required to evaluate the response-surface model coefficients.
3. Transmit the calculated coefficients to the optimization model.
4. Transfer control of the optimization to the selected optimization algorithm.
5. Recover the results of the optimization and check for convergence of the two-level procedure.

While full automation of the strategy is highly desirable, it clearly can be an expensive undertaking in terms of program development cost. It is therefore most appropriate for applications that will require repeated optimization studies.

**Strategies for Global Optimization.** The successful optimization of models that are likely to involve multiple local minima requires that some strategies be employed to provide assurance that the global minimum has been found. It is desirable to find the global minimum not only because it is the best possible solution to the problem but also because local minima can severely confound the interpretation of the results of studies investigating the effects of model parameters. If, with one set of, say, cost parameters the algorithm terminates at one local minimum, while with another set of parameter values a different local minimum results, then it becomes difficult to extract the true effect of the parameter changes on the solution, because much of the observed change will be due to merely having obtained a different local optimum. This difficulty is, of course, obviated if in all cases the global optimum is found.

Strategies for finding global optima are still very much the subject of intensive research. A good review of the state of the art is given by Archetti and Szego [36]. The available strategies can be divided into deterministic and stochastically based approaches, Both categories include both heuristic and theoretically based strategies. The simplest and most widely used approach involves multiple optimization runs, each initiated at a different starting point. Hence, it is sometimes called the *multistart strategy*. Under this strategy the starting points are selected using either a deterministic grid or random sampling. In the former case, the feasible region is subdivided into disjoint sectors and the optimization is performed over each sector separately. In random-sampling approaches, starting points are selected simply by sampling from a uniform distribution as discussed in Section 7.1.2. In both cases, the global minimum is defined to be the local minimum with the lowest objective function value among all local minima that can be identified. Both strategies are clearly heuristic. In the former case, the division into sectors is necessarily ad hoc: There is no assurance that each sector contains only one local minimum. In the latter case, there is no assurance that the starting points are adequately enough distributed to each lie within the neighborhood of one of the local minima. Certainly, the level of confidence in the fact that all local minima have been identified increases as the number of sectors or restart points is increased. However, that assurance is obtained at the price of a large number of optimization runs.

The only successful theoretically based global optimization approaches have been those addressed to specially structured problems. For instance, Westerberg and Shah [25] developed an algorithm suitable for separable, serially structured problems that employs bounds obtained using the Lagrangian function. Cabot and Francis [37], and others since, have reported procedures suitable for linearly constrained nonconvex problems. These strategies make use of the fact that all local minima must be corner points of the region and thus develop systematic enumeration procedures for finding the best extreme point. For a discussion of other approaches for structured problems, the reader should consult the paper by McCormick [38]. Except for the approaches designed for structured problems, there are no known procedures that will find the global minimum of a general problem in a finite number of problem optimization runs. At present, the only practical strategy for locating global minima is multistart with random sampling. This strategy does force the user to decide when sufficient restarts have been used to assure that the global solution has been identified. Nevertheless, despite the possibly high costs of the repeated runs, this procedure should be used with all problems that are suspected to possess multiple local minima.

## 13.3   SOLUTION EVALUATION

While it may seem that the optimization study is completed once the solution has been obtained, in fact the solution only serves as the basis of the most important parts of the study: solution validation and sensitivity analysis. In most studies, it is not the numerical solution itself that is most useful; rather it is the information about the state of the system in the vicinity of the solution that provides the key insights. The answers to questions such as What constraints are active at the solution? What cost terms dominate? What is the sensitivity of the solution to parameter changes? really are the key results of the entire study. The active constraints will indicate what system resources are limiting or what design elements limit the further improvement of the system. The cost-term magnitudes indicate which components of the system cost should be further refined. The sensitivity of the solution to model parameter changes will suggest what parameter estimates need to be further improved to clearly define the true solution. In this section we briefly review the considerations involved in validating the solution and exploring its sensitivity.

### 13.3.1    Solution Validation

Clearly, the first issue to be addressed in studying the result of an optimization run is that of the validity of the solution. In this context, a solution is valid if it is a realizable state of the system under consideration and if it is the optimum state of the system. A realizable state is simply one in which it is possible for the system to exist. Normally if the model is a reasonably good representation of the system, then it will contain the appropriate constraints and bounds that will assure that a solution is mathematically feasible if and only if it is physically feasible or realizable. However, all models have limitations, all correlations have ranges of validity, and all data have limited accuracy. Consequently, it is necessary to review the solution in the light of the model limitations to verify that the solution does not exceed these limitations. If it does, constraints and bounds must be added to ensure that these limitations are satisfied, and the optimization runs must be repeated.

Once it is established that the solution is realizable, then it is appropriate to verify whether it is reasonable to expect that it is an optimum solution for the system. The concern here is not in mathematically proving that the solution satisfies the necessary and sufficient optimality criteria, but in interpreting the solution and in being able to explain *why* it is the optimum. For a solution to be credible, it must be possible to explain the reasons for the obtained variable values on a qualitative, even intuitive, basis. Otherwise, the validity of the solution can be accepted only as an act of faith in mathematics and computers.

One approach toward gaining this insight is, as suggested by Geoffrion [39], to make use of simplified auxiliary models, which can elucidate the reasons for trends unobscured by mathematical complexity. The use of simplified auxiliary models to check trends is a device widely used in engineering and applied mathematics. For instance, in design case studies, Sherwood [40] makes considerable use of simplified, easily solved approximations to explain trends in more detailed optimal design calculations. The overall methodology recommended by Geoffrion [39] consists of the following steps:

1. Reduce the model detail until it can be solved by simple algebra.
2. From the auxiliary model, derive the general trend of the optimal solution as a function of key model parameters.
3. Calculate specific predictions from the auxiliary model, and test them by rerunning the full model.
4. If the numerical optimization results confirm the trends projected from the auxiliary model, then we have succeeded in explaining the key trade-off involved.

Normally, this process might be continued with several different auxiliary models, each of which reflects certain key factors or trade-offs in the full model or system. As a result of interpretive studies of this type, we not only will have verified the solution but also will have achieved some understanding of the physical meaning behind the calculated optimal solution. Thus, we will have begun to close the gap between the model optimum and the system optimum.

### 13.3.2  Sensitivity Analysis

The next stage in evaluating the solution is to determine the sensitivity of the solution to changes in the model parameters or assumptions. This is called *sensitivity analysis*. There are a number of reasons for performing a detailed sensitivity analysis:

1.  To find one or more parameters with respect to which the optimal solution is very sensitive. If such parameters exist, then it may be worthwhile to change the corresponding system features. For instance, if it is discovered that the optimal solution is very sensitive to labor availability (a constraint placed on the system), then it may be possible to increase labor availability by employing overtime labor and thus increase the overall system profitability.

2.  To extract information about additions or modifications to the system so as to improve the overall operation. Thus, one can obtain information as to the advisability of adding new production capacity or increasing intermediate storage.

3.  To clarify the effect on the system of variations in imprecisely known parameters. Some model parameters may be subject to considerable uncertainty. Sensitivity analysis can indicate whether it is worthwhile to expend resources to obtain better estimates of these parameter values. Alternatively, it may be that parameters that initially appeared to be critical, turn out to be unimportant and hence need not be further refined.

4.  To suggest the likely effects of variations in uncontrollable external parameters. Some system inputs, for example, product demands, may be outside of our control. Parameter sensitivity will give estimates of the effects of product demands on profits and hence allow the user to plan for a range of economic returns.

Because this type of information is so important in implementing a solution on the real system, a detailed sensitivity analysis is, in many cases, more valuable than the actual optimal solution itself.

Sensitivity information is normally extracted in two ways: through the Lagrange multiplier values and through parameter case study runs. The reader will recall that in the case of linear programming we can also easily obtain information about sensitivity with respect to the objective function coefficients without recalculating the optimum. However, that option is available only with LP models. As discussed in Chapter 5, the Lagrange multipliers (as well as the Kuhn-Tucker multipliers) are measures of the rate of change of the objective function with respect to a change in the right-hand side of the constraint.

Thus, given a constraint $h_k(x) = b_k$, the multiplier $v_k$ is equal to

$$v_k = \frac{\partial f}{\partial b_k}$$

Similarly, for an inequality constraint $g_j(x) \geqslant d_j$, the multiplier $u_j$ is equal to

$$u_j = \frac{\partial f}{\partial d_j}$$

As a first-order approximation, the change of the objective function value resulting from changes in the right-hand side of constraints will be given by

$$f(x) - f(x^*) = \sum_k \left(\frac{\partial f}{\partial b_k}\right) \Delta b_k + \sum_j \left(\frac{\partial f}{\partial d_j}\right) \Delta d_j = \sum_k v_k \Delta b_k + \sum_j u_j \Delta d_j$$

$$(13.33)$$

This estimate of the change in the optimal objective function value is likely to be quite good provided that the changes $\Delta d_j$, $\Delta b_k$ are small, that the same constraints remain tight at the optimum, and that only a few constraints are perturbed at a time.

In the next example we illustrate the use of Eq. (13.33) to estimate optimal objective function changes.

### Example 13.12    Use of Lagrange Multipliers

Consider the welded beam design problem of Example 1.2. With the parameter values specified in that example, the design optimization problem takes the form:

Minimize    $f(x) = 1.10471x_1^2 x_2 + 0.04811 x_3 x_4 (14 + x_2)$

Subject to    $g_1(x) = \dfrac{\tau_d}{F} - \left\{ \dfrac{1}{2x_1^2 x_2^2} + \dfrac{3(28 + x_2)}{x_1^2 x_2 \left[x_2^2 + 3(x_1 + x_3)^2\right]} \right.$

$$\left. + \dfrac{4.5(28 + x_2)^2 \left[x_2^2 + (x_1 + x_3)^2\right]}{x_1^2 x_2^2 \left(x_2^2 + 3(x_1 + x_3)^2\right)^2} \right\}^{1/2} \geqslant 0$$

where $F = 6000$ lb and $\tau_d = 13{,}600$ psi.

$$g_2(x) = x_3^2 x_4 - 12.8 \geqslant 0$$

$$g_3(x) = x_4 - x_1 \geqslant 0$$

$$g_4(x) = x_3 x_4^3 (1 - 0.02823 x_3) - 0.09267 \geqslant 0$$

$$g_5(x) = x_1 - 0.125 \geqslant 0$$

$$g_6(x) = x_3^3 x_4 - 8.7808 \geqslant 0$$

$$x_2 \text{ and } x_3 \geqslant 0$$

Suppose that the problem is solved using a penalty function method (SUMT-IV) to obtain the solution shown in Table 13.4. Estimate the effects on the objective function value of a change in $\tau_d$ from 13,600 to 13,240 psi.

**Solution.** From Table 13.4 it is evident that the first four constraints are tight at the optimum. In order to perform the sensitivity analysis, we first need to obtain values of the multipliers of the tight constraints if they are not provided automatically by the optimization routine. Thus, we must solve the system

$$\nabla f(x^*) = \Sigma u_j \nabla g_j(x^*)$$

for $u_j \geqslant 0$.

**Table 13.4  Solution of Welded Beam Problem**

| Quantity | Starting Point | Optimum Point |
|----------|---------------|---------------|
| $x_1$ | 2.0 | 0.2536 |
| $x_2$ | 4.0 | 7.141 |
| $x_3$ | 4.0 | 7.1044 |
| $x_4$ | 3.0 | 0.2536 |
| $f(x)$ | 28.0671 | 2.34027 |
| $g_1$ | 1.719225 | $2.327 \times 10^{-5}$ |
| $g_2$ | 35.2 | $1.863 \times 10^{-3}$ |
| $g_3$ | 1.0 | $7.341 \times 10^{-6}$ |
| $g_4$ | 95.7119 | $3.392 \times 10^{-6}$ |
| $g_5$ | 1.875 | 0.1280 |
| $g_6$ | 183.2192 | 82.169 |

In the present case, the system reduces to

$$
\begin{pmatrix}
9.222 & 0.24249 & 0.28518 & 0 \\
0 & 0 & 3.604 & 50.473 \\
-1 & 0 & 0 & 1.0 \\
0 & 0 & 0.00977 & 1.0961
\end{pmatrix}^T
\begin{pmatrix}
u_1 \\
u_2 \\
u_3 \\
u_4
\end{pmatrix}
=
\begin{pmatrix}
4.0018 \\
0.1577 \\
0.25797 \\
7.226
\end{pmatrix}
$$

In general this system need not be square. In fact, normally there will be more equations than unknown multipliers, and hence a linear least squares solution will have to be obtained. In the present case, however, the number of variables is exactly equal to the number of tight inequalities, and thus the Kuhn-Tucker conditions can be solved directly to yield

$$
u = (0.650553, 0.008193518, 1.997586, 4.3292586)
$$

Now, if $\tau_d$ is changed to 13,240, then the change in the constant associated with $g_1$ is from 2.2667 to 2.20667. Since the constraint is written

$$
g_1 = 2.2667 - q(x) \geqslant 0
$$

or

$$
-q(x) \geqslant -2.2667
$$

we have $d_1 = -2.2667$. Hence, $\Delta d_1 = (-2.20662) - (-2.2667)$ or $\Delta d_1 = 0.06$. From Eq. (13.33), the predicted change in the objective function corresponding to this change in the constraint will be

$$
\Delta f = u_1 \Delta d_1 = 0.650553(0.06) = 0.039033
$$

Thus, the new objective function value will be

$$
f(x) = 2.34027 + 0.039033 = 2.379303
$$

By way of comparison, when the problem is resolved using SUMT with the new value of $\tau_d$, the actual new objective function value is 2.38043. The prediction using the Lagrange multiplier is thus quite good. Note that the magnitudes of the multiplier values further indicate that the objective function value will be most sensitive with respect to the fourth constraint, the buckling constraint, and least sensitive with respect to the second constraint.

While the multipliers yield useful information about constraint sensitivity, they do not indicate sensitivity with respect to individual function parameters. Thus, it is expedient to execute a series of case studies in which selected model parameters are reassigned values and the problem is rerun.

## 13.4 SUMMARY

This chaper focused on the steps essential to the successful execution of an optimization study. The basic concepts involved in model formulation were reviewed, and the levels and types of optimization models (equation-oriented, response-surface, and procedure-oriented), as well as their relative advantages and disadvantages, were considered. Next, problem implementation details were summarized: how models can be implemented or assembled, what devices can be employed to prepare the problem for optimization, and what computational strategies should be used in executing the optimization itself. Problem preparation was found to involve three types of activities: modification to avoid numerical difficulties, model rearrangement to enhance solution efficiency, and problem analysis to establish essential problem features. The topics discussed under these categories include problem scaling, function and variable transformations, elimination of redundant constraints, and equation sequencing. Simple constructions useful in identifying key problem features such as convexity, boundedness, solution uniqueness, and problem feasibility were reviewed. Then optimization strategies for both equation-oriented and simulation models were considered, as were problems associated with identifying global optima. We found that while much research is under way on the latter issue, in general, the multistart strategy is presently the only practical recourse.

The chapter concluded with a consideration of issues involved in evaluating the solution provided by the optimizer. We observed that quantitative as well as qualitative validation and interpretation of a solution to establish its credibility are highly desirable. Finally, the concept of sensitivity analysis was discussed, and the use of Lagrange multipliers to estimate the sensitivity of the solution to parameter changes was reviewed.

We hope that this chapter has helped the reader realize that the execution of an optimization study cannot merely be reduced to the running of a canned program. It involves careful consideration of many aspects of the application problem itself, the model used to represent the application, and the algorithms employed to perform the number crunching.

## REFERENCES

1. Embury, M. C., G. V. Reklaitis, and J. M. Woods, "Scheduling and Simulation of a Staged Semi-Continuous Multi-Product Process," *Proceedings of Pacific Chemical Engineering Congress*, Denver, Colorado, Aug., 1977, Vol. 2.
2. Bett, K. E., J. S. Rawlinson, and G. Saville, *Thermodynamics for Chemical Engineers* The MIT Press, Cambridge, MA, 1975.
3. King, J. C., *Separation Processes*, McGraw-Hill, New York, 1971.
4. Scribner, B. F., and D. H. Berlau, "Optimal Planning for Product/Production Facility Studies (OPPLAN)," *Proc. Amer. Inst. Ind. Eng. Conf.*, San Francisco, pp. 383–387, May, 1979.

5.  Benjamin, D. R., M. H. Locke, and A. W. Westerberg, "ASCEND: Friendly Interactive Flowsheeting System," paper 25b, AIChE Summer Meeting, Detroit, August, 1981.

6.  Kilkas, A. C., and H. P. Hutchison, "Process Optimization Using Linear Models, *Computers & Chem. Eng.*, **4**, 39–48 (1980).

7.  Pollack, H. W., and W. D. Lieder, "Linking Process Simulators to a Refinery Linear Programming Model" in *Computer Applications to Chemical Engineering*, R. G. Squires and G. V. Reklaitis (eds), ACS Symposium Series Vol. 124, American Chemical Society, Washington, DC (1980).

8.  Parker, A. L., and R. R. Hughes, "Approximation Programming of Chemical Processes: Parts 1 and 2," *Computers Chem. Eng.* **5**, 123–131 (1981).

9.  Biegler, L. T., and R. R. Hughes, "Approximation Programming of Chemical Processes with Q/LAP", *Chem. Eng. Prog.*, **77**, 76–83 (1981).

10. McLane, M., M. K. Sood, and G. V. Reklaitis, "A Hierarchical Strategy for Large Scale Process Calculations," *Computers Chem. Eng.*, **3**, 383–394 (1979).

11. Biles, W. E., and J. J. Swain, *Optimization and Industrial Experimentation*, Wiley-Interscience, New York, 1980, Chapter 7.

12. Rosen, E. M., "Steady State Chemical Process Simulation: A State of the Art Review," in *Computer Applications to Chemical Engineering* (R. G. Squires and G. V. Reklaitis, Eds., ACS Symposium Series Vol. 124, American Chemical Society, Washington, D.C. (1980).

13. Pritsker, A. A. B., *Modeling and Analysis Using Q-GERT Networks*, Wiley-Halsted, New York, 1977.

14. Salvendy, G. (Ed), *Handbook of Industrial Engineering*, Wiley-Interscience, New York, 1981.

15. Root, R. R., and K. M. Ragsdell, "Computational Enhancements of the Method of Multipliers," *ASME J. Mech. Des.*, **102**, 517–523 (1980).

16. Keefer, D. L., and B. Gottfried, "Differential Constraint Scaling in Penalty Function Optimization," *AIIE Trans.*, **2**, 281–286 (1970).

17. Lasdon, L. S., and P. O. Beck, "Scaling Nonlinear Programs," *Oper. Res. Letters*, **1**, 6–9 (1981).

18. Berna, T. J., M. H. Locke, and A. W. Westerberg, "A New Approach to Optimization of Chemical Processes," *AIChE J.*, **26**, 37–43 (1980).

19. Box, M. J., "A Comparison of Several Current Optimization Methods and the use of Transformations in Constrained Problems," *Computer J.*, **9**, 67–77 (1966).

20. Sisser, F. S., "Elimination of Bounds in Optimization Problems by Transforming Variables," *Math. Prog.*, **20**, 110–121 (1981).

21. Luenberger, D. G., *Introduction to Linear and Nonlinear Programming* Addison-Wesley, Reading, MA, 1973, chap. 5.

22. Reklaitis, G. V., and D. J. Wilde, "A Differential Algorithm for Posynomial Programs," *DECHEMA Monogr.*, **67**, 324–350 (1970).

23. Reklaitis, G. V., and D. J. Wilde, "Geometric Programming via a Primal Auxiliary Problem," *AIIE Trans.*, **6**, 308–317 (1974).

24. Rijckaert, M. J., and X. M. Martens, "Analysis and Optimization of the Williams-Otto Process by Geometric Programming," *AIChE J.*, **20**, 742–750 (1974).

25. Westerberg, A. W., and J. V. Shah, "Assuring a Global Optimum by the Use of an Upper Bound on the Lower (Dual) Bound," *Computers Chem. Engr.*, **2**, 83–92 (1978).

26. Guthrie, K. M., *Process Plant Estimating, Evaluation, and Control*, Craftsman Book Company, Solana Beach, CA (1974).

27. Knopf, F. C., M. R. Okos, and G. V. Reklaitis, "Optimal Design of Batch/Semi-Continuous Processes," *Ind. Engr. Chem. Proc. Design Develop.*, **21** 79–86 (1982).

28. Fattler, J. E., G. V. Reklaitis, Y. T. Sin, R. R. Root, and K. M. Rasgsdell, "On the Computational Utility of Polynomial Geometric Programming Solution Methods," *Math. Programming*, **22**, 163–201 (1982).

29. Geoffrion, A. M., "Generalized Benders Decomposition," *JOTA*, **10**, 237–251 (1972).

30. Sargent, R. W. H., and K. Gannibandara, "Optimum Design of Plate Distillation Columns," in *Optimization in Action* (L. C. W. Dixon, Ed.), Academic Press, New York, 1976.

31. Gupta, A., and A. Ravindran, "Optimization of Transportation Networks," in *Stochastic Modeling of Large Scale Transportation Networks* (J. J. Solberg, Ed.), Final Rep., Automotive Transportation Center, Purdue University, April, 1981.

32. Reklaitis, G. V., and J. M. Woods, "Optimization of a Non-ideal Staged Compression Facility," *Proc. 1974 Purdue Compressor Technol. Conf.*, Purdue University, July 1974, pp. 415–422.

33. Biles, W. E., and J. J. Swain, *Optimization and Industrial Experimentation*, Wiley, New York, 1980, Chapter 7.

34. Knopf, F. C., G. V. Reklaitis, and M. R. Okos, "Combined Design and Scheduling of Regenerative Noncontinuous Processes," *Symp. Prod. Planning Multi-Product Plants*, AIChE Nat. Meet., Houston, TX, Apr. 1981.

35. Pritsker, A. A. B., and C. P. Pegden, *Introduction to Simulation and SLAM*, Systems Publ. Corp. and Halsted Press, New York, 1979.

36. Archetti, F., and G. P. Szego, "Global Optimization Algorithms," in *Nonlinear Optimization: Theory and Algorithms* (L. C. W. Dixon, E. Spedicato, and G. P. Szego, Eds.), Birkhauser, Boston, 1980.

37. Cabot, A. V., and R. L. Francis, "Solving Nonconvex Quadratic Minimization Problems by Ranking Extreme Points," *Oper. Res.*, **18**, 82–86 (1970).

38. McCormick, G. P., "Attempts to Calculate Global Solutions to Problems that may have Local Minima," in *Numerical Methods for Nonlinear Optimization* (F. A. Lootsma, Ed.), Academic Press, New York, 1972.

39. Geoffrion, A. M., "The Purpose of Mathematical Programming in Insight, not Numbers," *Interfaces*, **7**, 81–92 (1976).

40. Sherwood, T., *A Course in Process Design*, MIT Press, Cambridge, MA, 1963.

## PROBLEMS

**13.1.** Apply the scaling procedure to the two problems of Example 13.3, and compare the results to the scaling obtained in that example.

**13.2.** Determine row and column scale factors and the resulting scaled form of the Jacobian matrix

$$\begin{bmatrix} 2 & 10^2 & 10^{-4} \\ 10^4 & 10^7 & 0 \\ 10^{-3} & 10^5 & 1 \end{bmatrix}$$

**13.3.** Determine scale factors for the constraint set

$$g_1(x) = x_1 + 2x_2 - x_3 - 2 \geqslant 0$$

$$g_2(x) = 100x_1^2 + 0.02x_2^2 + 10x_3^2 \geqslant 0$$

$$g_3(x) = 10^5 x_1 + 10^3 x_2^2 - 4 \times 10^6 x_3^2 \geqslant 0$$

in the vicinity of the feasible point (0.01, 1, 0.01).

**13.4.** Consider the problem

$$\text{Minimize} \quad f(x) = x_1^2 + x_2^2$$

$$\text{Subject to} \quad g(x) = x_2 - \text{Max}[2x_1^{-1}, \tfrac{3}{2} - \tfrac{1}{2}(x_1 - 3)^2]$$

$$x_1 \quad \text{and} \quad x_2 \geq 0$$

(a) show that $\nabla g$ is discontinuous at the point (2, 1).

(b) Reformulate constraint $g(x)$ so that this problem is avoided.

**13.5.** A model for the optimization of a through circulation dryer can be formulated as follows [Chung, S. F. *Can. J. Chem. Eng.* **51**, 262

$$\text{Maximize} \quad f(x) = 0.033x_1 \left\{ \frac{0.036}{1 - \exp\left(-107.9x_2 x_1^{-0.41}\right)} \right.$$

$$+ 0.095 - \frac{9.27 \times 10^{-4} x_1^{0.41}}{x_2}$$

$$\left. \times \ln\left[ \frac{1 - \exp\left(-5.39x_2/x_1^{0.41}\right)}{1 - \exp\left(-107.9x_2/x_1^{0.41}\right)} \right] \right\}^{-1}$$

$$\text{Subject to} \quad g_{1(x)} = 2 \times 10^3 - 4.62 \times 10^{-6} x_1^{2.85} x_2 - 1.055x_1 \geq 0$$

$$g_2(x) = 175 - 9.84 \times 10^{-4} x_1^{1.85} x_2 \geq 0$$

$$g_3(x) = 2 - 109.6 \frac{x_2}{x_1^{0.41}} \left\{ \frac{0.036}{1 - \exp\left(-107.9x_2/x_1^{0.41}\right)} \right.$$

$$+ 0.095 - \frac{9.27 \times 10^{-4} x_1^{0.41}}{x_2} i$$

$$\left. \times \ln\left[ \frac{1 - \exp\left(-5.39x_2/x_1^{0.41}\right)}{1 - \exp\left(-107.9x_2/x_1^{0.41}\right)} \right] \right\}$$

where $f(x)$ is the drying rate as a function of the air mass flowrate $x_1$ and the bed thickness $x_2$. A feasible initial point is (800, 0.5). Suggest a reformulation of the problem that will reduce possible numerical difficulties and enhance efficiency of solution using gradient-based methods.

**13.6.** The following model has been proposed for the optimization of an alkylation process [Sauer, R. N., et al., *Hydro. Proc. Petrol. Refiner.* **43**, 84 (1964)].

Minimize     $f(x) = 0.063x_4x_7 - 5.04x_1 - 0.035x_2 - 10x_3 - 3.36x_5$

Subject to   $h_1(x) = \dfrac{x_4}{100x_1} - 112 - 13.167x_8 - 0.6667x_8^2 = 0$

$$h_2(x) = x_8 - \frac{x_5 + x_2}{x_1} = 0$$

$$h_3(x) = x_4 - x_1 - x_5 + 0.22x_4 = 0$$

$$h_4(x) = 10^3 x_3 - x_4 x_9 \left( \frac{x_6}{98 - x_6} \right) = 0$$

$$h_5(x) = x_9 - 35.82 - 0.222x_{10} = 0$$

$$h_6(x) = x_{10} + 133 - 3x_7 = 0$$

$$h_7(x) = x_7 - 86.35 - 1.098x_8 + 0.038x_8^2 + \left( \frac{1.3}{4} \right)(x_6 - 89) = 0$$

and the bounds

$$0 \leqslant x_1 \leqslant 2000$$
$$0 \leqslant x_2 \leqslant 16000$$
$$0 \leqslant x_3 \leqslant 120$$
$$0 \leqslant x_4 \leqslant 5000$$
$$0 \leqslant x_5 \leqslant 2000$$
$$85 \leqslant x_6 \leqslant 93$$
$$90 \leqslant x_7 \leqslant 95$$
$$3 \leqslant x_8 \leqslant 12$$
$$1.2 \leqslant x_9 \leqslant 4$$
$$145 \leqslant x_{10} \leqslant 162$$

(a) Generate a starting feasible solution to the problem.

(b) Obtain a reformulation of the problem suitable for the application of direct-search methods.

(c) Obtain a reformulation of the problem suitable for the application of LP-based methodology.

13.7. In the model defined in problem 13.6, constraints $h_1(x)$, $h_4(x)$, $h_5(x)$, and $h_7(x)$ are derived from correlations of experimental data, which naturally have some scatter.

(a) Reformulate these equality constraints to reflect the possible error in them.

(b) Use a suitable gradient-based method to resolve the problem for each of the possible relative error levels of 0, 1, 5, and 10% in the correlations.

(c) Explain the results obtained, and compare them to the predictions obtained from the multiplier values at the 0% error solution.

**13.8.** The cost minimization problem for a multimedia filter in a municipal water treatment plant can be posed as follows:

Minimize
$$f(x) = Q^{-1}(0.00656x_1 + 0.002189x_2 + 0.03284x_3 + 2345t^{-1})$$

Subject to
$$h_1(x) = Q^{-0.34}d_e^{-0.51}t^{-1.70}L^{0.53} - 0.007 = 0$$

$$h_2(x) = Q^{1.4}L^{0.8}d_e^{-1.1}H^{-1}t - 403.23 = 0$$

$$h_3(x) = 1.337x_1 L^{-1} + 0.5086x_2 L^{-1} + 0.4x_3 L^{-1} - d_e = 0$$

$$h_4(x) = H - H_t + 1.33 = 0$$

$$h_5(x) = L - x_1 - x_2 - x_3 = 0$$

$$g_1(x) = 45 - 1.41x_1 - 1.35x_2 - 1.1x_3 \geqslant 0$$

$$g_2(x) = L - d_e^3 H_t Q \geqslant 0$$

and the bounds
$$t \geqslant 40 \qquad H_t \leqslant 8 \qquad Q \geqslant 3.0$$

The variables are defined as follows:

$$L = \text{depth of filter bed}$$

$$Q = \text{filtration rate per unit area of filter}$$

$$t = \text{filtration time}$$

$$x_i = \text{depth of filter medium } i(i = 1, 2, 3)$$

$$d_e = \text{equivalent grain size of the multimedia filter}$$

$$H = \text{headloss change over filtration period}$$

$$H_t = \text{headloss at time } t$$

and all must naturally be nonnegative.

(a) Generate a starting feasible solution to the problem. On physical grounds it can be expected that $L \leqslant 50$, $Q \leqslant 10$, and $t \leqslant 500$.

(b) Suggest transformations and reformulations of the problem that would make it easier and more stable for solution.

**13.9.** Compare solution of problem 13.8 using a linearization-based code and a direct-search method ($f^* = 18.339$). Use the results of the former to conduct a sensitivity analysis at the optimum solution.

**13.10.** Consider the problem [Hesse, R., *Oper. Res.* **21**, 1267 (1973)]

Maximize $\quad f(x) = 25(x_1 - 2)^2 + (x_2 - 2)^2 + (x_3 - 1)^2 + (x_4 - 1)^2$

$$+ (x_5 - 1)^2 + (x_6 - 4)^2$$

Subject to $\quad g_1(x) = (x_3 - 3)^2 + x_4 - 4 \geqslant 0$

$$g_2(x) = (x_5 - 3)^2 + x_6 - 6 \geqslant 0$$

$$g_3(x) = 2 + x_1 - x_2 \geqslant 0$$

$$g_4(x) = 2 - x_1 + 3x_2 \geqslant 0$$

$$g_5(x) = 6 - x_1 - x_2 \geqslant 0$$

$$g_6(x) = x_1 + x_2 - 2 \geqslant 0$$

and the bounds

$$1 \leqslant x_3 \leqslant 5 \qquad x_4 \leqslant 6$$
$$1 \leqslant x_5 \leqslant 5 \qquad x_6 \leqslant 10$$

and all variables nonnegative.

(a) Show that the problem can have multiple local maxima (it has 18).

(b) Suggest how you might prove that the point $(5, 1, 5, 0, 5, 10)$ is the global maximum point.

**13.11.** Use a constrained optimization program of your choice to find at least three local maxima to problem 13.10. Prove that the points satisfy the necessary conditions for a local maximum.

**13.12.** A local minimum of the problem [Rosen J. B. and S. Suzuki, *Commun. ACM* **8**, 113 (1965)]

Minimize $\quad f(x) = x_1^2 + x_2^2 + 2x_3^2 + x_4^2 - 5x_1 - 5x_2 - 21x_3 + 7x_4$

Subject to $\quad g_1(x) = 8 - x_1^2 - x_2^2 - x_3^2 - x_4^2 - x_1 + x_2 - x_3 + x_4 \geqslant 0$

$$g_2(x) = 10 - x_1^2 - 2x_2^2 - x_3^2 - 2x_4^2 + x_1 + x_4 \geqslant 0$$

$$g_3(x) = 5 - 2x_1^2 - x_2^2 - x_3^2 - 2x_1 + x_2 + x_4 \geqslant 0$$

is $x^* = (0, 1, 2, -1)$, with $f^* = -44$.

(a) Is this the global minimum? Are other local minima possible?

(b) Perform a sensitivity analysis to estimate the changes in $f^*$ with 5% changes in the constraint constants 8, 10, and 5.

**13.13.** Consider the problem

$$\text{Maximize} \quad f(x) = x_1^2 + x_2^2 + x_3^2 + (x_3 - x_4)^2$$

$$\text{Subject to} \quad g_1(x) = 16 - 4(x_1 - \tfrac{1}{2})^2 - 2(x_2 - 0.2)^2$$

$$- x_3^2 - 0.1 x_1 x_2 - 0.2 x_2 x_3 \geqslant 0$$

$$g_2(x) = 2 x_1^2 + x_2^2 - 2 x_3^2 - 2 \geqslant 0$$

$$g_3(x) = x_2 + x_4 \geqslant 0$$

(a) Check what bounds may be appropriate in the problem formulation.

(b) Check whether the problem is likely to possess multiple local solutions.

(c) Generate a feasible starting point.

**13.14.** Solve problem 13.13 with the bounds

$$-2.3 \leqslant x_i \leqslant 2.7 \qquad i = 1, \ldots, 4$$

using a suitable constrained optimization code ($f^* - 11.68$; Luus, R. *AICHE J.* **20**, 608 (1974)].

**13.15.** Consider the problem

$$\text{Minimize} \quad f(x) = x_1 x_3^2 + x_2^{-1} x_3^{-1}$$

$$\text{Subject to} \quad g(x) = x_1 - x_2^3 x_3 - x_1 x_2 x_3 \geqslant 0$$

$$\text{all } x_i \geqslant 0$$

(a) Show that the problem has a nonunique solution. (*Hint:* Let $z_1 = x_1 x_3^2$ and $z_2 = x_2 x_3$.)

(b) Solve the resulting problem algebraically to show that a family of values of the $x_i$ will yield the same optimal objective function value.

**13.16.** Show that the problem

$$\text{Minimize} \quad f(x) = 2x_1 x_2 + x_1^{-2} x_2^{-2}$$

$$\text{Subject to} \quad g(x) = 1 - \tfrac{1}{2}x_1 - \tfrac{1}{3}x_2 \geqslant 0$$

$$x_1, x_2 \text{ both } \geqslant 0$$

attains its optimum value of $f(x^*) = 3$ at a family of points satisfying $x_1 x_2 = 1$ and $g(x) \geqslant 0$.

**13.17.** Suggest a reformulation of the problem

$$\text{Minimize} \quad f(x) = x_1^2 x_4^4$$

$$\text{Subject to} \quad h(x) = 0.25 - x_1^{3/4} x_2 = 0$$

$$g(x) = x_1 x_2^4 + x_1^{1/2} x_2^2 - 1 \geqslant 0$$

$$\text{all } x_i \geqslant 0$$

that would be easier to solve.

**13.18.** Generate an initial feasible solution to the problem

$$\text{Minimize} \quad f(x) = x_1 x_2 x_3 x_4$$

$$\text{Subject to} \quad g_1(x) = 4x_1^{1/2} + x_2^2 x_3 - 1 \geqslant 0$$

$$g_2(x) = 2 - \tfrac{1}{8}x_3^{-2} - x_{1\cdot\cdot 2} x_4^{-1} \geqslant 0$$

$$g_3(x) = 4 - 4x_1^2 x_2^2 - x_3^2 x_4^2 \geqslant 0$$

$$16 \geqslant x_i \geqslant 0$$

Use a random search method to generate 1000 trial points, and record the ratio of feasible to infeasible points generated.

# Chapter 14

# Engineering Case Studies

In this chapter we present three case studies involving the optimization of large-scale engineering systems. The first study concerns the optimal location of coal-blending plants within a region composed of a number of coal mines and coal users. The second involves the optimal preliminary design of a chemical plant for the production of ethylene oxide and ethylene glycol. The third considers the design of a compressed air storage system to balance peak demands on electric power plants. Each case study illustrates specific formulation and model-preparation strategies appropriate to that system. Because of space limitations, the model-formulation elements are given only in outline; however, the reader is directed to references that contain the full model details. In addition to discussing the selection of optimization methods, the case studies exemplify various stratagems for enhancing solution efficiency, for treating discrete or 0-1 variables, and for reducing the effective dimensionality of the optimization problem.

This chapter thus demonstrates the application of the techniques discussed in Chapter 13 and the algorithms developed in Chapters 1–12 to three real engineering examples. The chapter appropriately serves as a capstone to our study of engineering optimization.

## 14.1 OPTIMAL LOCATION OF COAL-BLENDING PLANTS BY MIXED INTEGER PROGRAMMING

In this case study we describe an application of mixed integer programming for solving a facility-location problem. The study conducted by Ravindran and Hanline [1] examined the feasibility of establishing centralized coal-blending plants in Indiana that could take coals of varying sulfur content and produce a coal product that would meet the state's coal consumer while meeting $SO_2$ emission standards. The case study will also illustrate the use of binary variables for handling fixed costs in the objective function.

### 14.1.1  Problem Description

Coal is one of the keys to making the United States independent of the uncertain supply of foreign oil. However, a major environmental problem in relying on coal is the presence of sulfur in coal as it is mined. Most midwestern and eastern coal reserves have high sulfur content [2]. Coal-fired furnaces are the major source of sulfur dioxide ($SO_2$) pollution, which is harmful to crops and animals. For this reason, strict $SO_2$ emission regulations have been set for each state by the U.S. Environmental Protection Agency (EPA) [3].

Currently coal users located in areas governed by the most restrictive $SO_2$ standards have limited options such as utilizing flue gas desulfurization techniques (scrubbers) or burning low-sulfur coal. According to the National Rural Electric Cooperative Association [4], "stack gas scrubbers," which the utilities are forced to install, produce 8–9 ft$^3$ of mudlike sludge for every ton of coal burned; sludge that must be disposed of, creating another environmental problem. Although the option to use low-sulfur coal has been attractive to some utilities and industries, the increased utilization of low-sulfur coal has resulted in higher coal prices for western coal and a depressed market for midwestern high-sulfur coal. The option studied here is the establishment of centralized coal-blending plants that could take coals of varying sulfur content (western, midwestern, and eastern coals) and produce a coal blend that would meet customers' needs while complying with environmental standards.

### 14.1.2  Model Formulation

Low-sulfur coal and high-sulfur coal are mined at different locations. By a mechanical crushing process these coals are blended to form environmentally acceptable mixtures for consumers. The questions to be studied are (1) How many blending plants should be established? (2) Where should these plants be located? (3) Which consumers and which mines should each blending plant expect to serve? (4) What are the optimal blending ratios of different coals that will meet EPA standards? A single-period (static) model is developed to answer these questions. The combinatorial nature of the site-selection problem introduces 0-1 binary variables, and the final formulation reduces to a mixed integer LP model.

**Decision Variables.** Indices $i$, $j$, and $k$ are used as subscripts to denote mine sources, blending plant sites, and consumer locations, respectively. Lowercase letters will be used to denote variables and capital letters for input parameters and constants. The decision variables are defined as follows:

$$v_{ij} = \text{volume (tons) shipped from source } i \text{ to blending plant } j$$

$$u_{jk} = \text{volume shipped from blending plant site } j \text{ to consumer } k$$

$w_{ij}$ = (0, 1) binary variable. If coal is shipped from source $i$ to site $j$, $w_{ij}$ = 1. Otherwise, $w_{ij}$ = 0.

$x_{jk}$ = (0, 1) binary variable. If coal is shipped from site $j$ to consumer $k$, $x_{jk}$ = 1. Otherwise, $x_{jk}$ = 0.

$y_j$ = (0, 1) binary variable. If site $j$ is used, $y_j$ = 1. Otherwise, $y_j$ = 0.

**System Objective.** The objective is to minimize the total system costs, which include

1. Costs of different coals at the mine source in dollars per ton.
2. Blending costs, which consist of a fixed cost of constructing the blending plant and a variable cost proportional to the volume of coal blended. A blending facility requires a large coal-storage area and material-handling equipments such as a conveyor system, weighing and analyzing equipment, and mixing units. The construction cost, along with other fixed costs, is discounted over the expected life of the facility on an annual basis.
3. Transportation costs, which are a function of origin, destination, and volume shipped. The pricing mechanism for rail-freight rates are very complex, with several different rates depending on the volume and on whether railroad cars are railroad owned, privately owned, or leased. Often, special rate structures are also used (e.g., unit train rates). In this study, we assume that the transportation costs consist of a fixed cost independent of the volume shipped and a variable cost proportional to the distance and volume shipped. The fixed transportation cost also allows for the expansion of the current transportation system or building new transportation modes. For example, fixed cost may include (a) acquisition of new rail cars instead of using those owned by the railroads (the freight rates are much lower for consumer-owned cars), (b) acquisition of new trucks, (c) building of slurry pipelines for coal transport. Fixed costs are converted to an annual basis by discounting over the expected life of the invested equipment. Thus, the system objective becomes:

$$\text{Minimize} \quad f(v, u, w, x, y) = \sum_i D_i \left( \sum_j v_{ij} \right) + \sum_j C_j y_j$$

$$+ \sum_j B_j \left( \sum_i v_{ij} \right) + \sum_i \sum_j (S_{ij} w_{ij} + T_{ij} v_{ij})$$

$$+ \sum_j \sum_k (F_{jk} x_{jk} + G_{jk} u_{jk}) \qquad (14.1)$$

where $D_i$ = mine-mouth cost of coal at source $i$ per unit volume, tons
$\quad S_{ij}$ = fixed cost of shipping from source $i$ to blending plant $j$
$\quad T_{ij}$ = cost of shipping a unit volume from source $i$ to plant $j$
$\quad C_j$ = annual fixed cost of blending in plant $j$
$\quad B_j$ = variable cost of blending in plant $j$ per unit volume
$\quad F_{jk}$ = fixed cost of shipping from blending plant $j$ to consumer $k$
$\quad G_{jk}$ = cost of shipping a unit volume from blending plant $j$ to consumer $k$

**System Requirements.** The following constraints describe the modeled system:

1. The volume of coal shipped from a mine source cannot exceed its available supply:

$$\sum_j v_{ij} \leqslant M_i \qquad \text{for all } i \qquad (14.2)$$

where $M_i$ = the supply available at source $i$.

2. The volume shipped to a consumer should meet the needs (demand) of that consumer:

$$\sum_j u_{jk} \geqslant N_k \qquad \text{for all } k \qquad (14.3)$$

Notice that this model does not allow for shortages. In the case of coal rationing, or as an alternative, coal shortages can be handled as follows:

$$N_k' \leqslant \sum_j u_{jk} \leqslant N_k$$

where $N_k'$ = the minimum acceptable demand and $N_k$ = the full demand. In this case, a shortage cost would become part of the cost function. Assuming a linear shortage cost, the additional shortage cost term would be

$$A_k \left( N_k - \sum_j u_{jk} \right)$$

where $A_k$ = shortage cost for consumer $k$ per unit volume.

3. The volume shipped from a blending plant cannot exceed the capacity of that blending plant:

$$\sum_k u_{jk} \leqslant P_j y_j \qquad \text{for all } j \qquad (14.4)$$

where $P_j$ = the capacity of blending plant $j$ (assume zero inventory of blended coal), and $y_j = 1$ if plant $j$ is operational.

4. The total volume of usable coal shipped from a blending plant does not exceed the volume shipped from the mines to the blending plant times a conversion factor:

$$E \sum_i v_{ij} \geqslant \sum_k u_{jk} \qquad \text{for all } j \qquad (14.5)$$

where $E$ = the efficiency of the blending process.

5. It is desirable that constraints assure that the sulfur content of the blended coal shipped to a consumer from a blending plant meets the consumer's specifications. For this study, we have assumed that the different coals used in the blend combine in a linear relationship. This assumption is made, since empirical data are unavailable to prove or disprove it.

If $H_i$ is the sulfur content (percent sulfur by weight) of coal from source $i$, then the net sulfur content of the blended coal at the blending plant $j$ is $\sum_i H_i v_{ij} / \sum_i v_{ij}$. If the blending plant $j$ supplies customer $k$ (i.e., if $x_{jk} = 1$), then we require $\sum_i H_i v_{ij} / \sum_i v_{ij} \leqslant L_k$, where $L_k$ is the maximum sulfur content of coal acceptable to customer $k$. This conditional constraint can be written as follows:

$$\sum_i H_i v_{ij} - L_k \sum_i v_{ij} \leqslant M(1 - x_{jk}) \qquad \text{for all } j \text{ and } k \qquad (14.6)$$

where $M$ is a large positive number. This constraint forces the blending plant to make a mixture that will meet the most stringent requirements of all the customers it serves. Note that when $x_{jk} = 1$, inequality (14.6) reduces to $\sum_i H_i v_{ij} / \sum v_{ij} \leqslant L_k$, that is, the required sulfur content of customer $k$ will be met by the net sulfur content of blended coal at site $j$. On the other hand, if customer $k$ is not supplied from site $j$, then inequality (14.6) becomes inactive, with $x_{jk} = 0$.

6. Miscellaneous constraints:

(a) The following constraints ensure that fixed transportation costs are included in the model.

$$u_{jk} - M x_{jk} \leqslant 0 \qquad \text{for all } j \text{ and } k \qquad (14.7)$$

$$v_{ij} - M w_{ij} \leqslant 0 \qquad \text{for all } i \text{ and } j \qquad (14.8)$$

where $M$ is a large positive number.

(b) $w_{ij}$, $x_{jk}$, $y_j$ are 0-1 integer variables.

(c) $v_{ij}$, $u_{jk}$ are nonnegative.

**Data.**  There are 17 coal mines in Indiana that account for 94% of the coal produced in the state [2]. Coal data were collected on all known utilities, industries, and state institutions (e.g., prisons, hospitals, universities, schools) that use over 1000 tons of coal annually [2, 5]. This included 33 utility plants and 46 industrial-institutional customers. However, because of computer size limitations, Indiana was divided into 15 coal-consuming regions. These regions were chosen based upon consumer demand patterns. Both the coke and the cement industries have special coal requirements and hence were not included in this study. Twelve possible blending plant sites were chosen based on discussions with the Indiana state energy officials [5]. Figure 14.1 illustrates the location of coal mines, major consumer locations, and blending plant sites used in the study.

The existing transportation system was assumed to be adequate for coal movement in Indiana, and fixed transportation costs were not included. The variable transportation costs were based upon average coal freight rates (cents per ton-mile) published in the *Coal Industry Manual* [2]. Using the coal freight rates and the distance between the origin and destination, the transportation

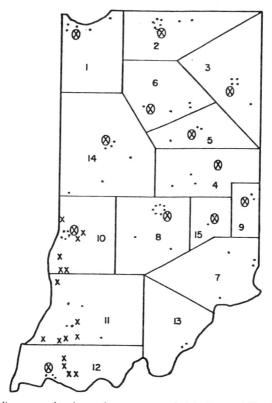

**Figure 14.1.**  Indiana map showing regions, source and sink sites, and blend plant locations.

cost coefficients were computed [1]. Railroad mileages were estimated from railroad data available [6].

To calculate the sulfur content of coal from various mines, the mean sulfur content of coal seams was found for each mine based on the *Geological Survey Report of the State of Indiana* [7]. An estimate was made of the acceptable sulfur level in the coal burned by each consumer. The method of estimation is based on the EPA standards as they relate to sulfur dioxide emissions and is described in reference 1.

As there were no blending plants commercially available, restrictions pertaining to blending plant capacities were ignored [Eq. (14.4)]. It was also assumed that any loss of coal due to the blending process was negligible. Hence, there was a conservation of mass through each blending plant, and $E$ was set to 1.0 in Eq. (14.5). Coal-blending plants were estimated to cost up to $12 million. The model was solved over a cost of $1.8–11.3 million. Life expectancy of a coal plant was assumed to be 30 years. The cost was discounted over 30 years at a discounting factor of 10%, thus giving the annual cost $C_j$. The operating costs $B_j$ (excluding coal costs) of blending plants were estimated at $0.25 per ton of coal blended. (These estimates were made by a large utility company that had contemplated building a blending facility.)

### 14.1.3 Results

An illustrative example is shown in which the annual fixed cost of a blending facility is $130,000. With an expected lifetime of 30 years and a discounting factor of 10%, this is equivalent to an investment of approximately $11.3 million. The final mixed integer program had 69 constraints, 406 variables, 12 binary variables, and 1767 nonzero elements. A computer code called MIPZI [8] based on Balas's implicit enumeration method [9] was used to solve the problem. It took 56.8 s of CPU time on the CDC 6500 computer at Purdue University.

The results show that six blending plant sites were selected for use. The six plant sites and their capacities are as follows:

| Site | Location | Volume Handled, tons coal | Consumer Areas Served |
|------|----------|---------------------------|------------------------|
| 1 | Gary | 4,895,000 | 1 |
| 2 | South Bend | 772,900 | 2 |
| 6 | Logansport | 3,529,000 | 6 |
| 8 | Indianapolis | 6,792,000 | 3, 7 |
| 10 | Terre Haute | 9,729,100 | 4, 8, 10, 14 |
| 11 | Evansville | 5,610,900 | 5, 9, 11, 12, 13, 15 |

An analysis of these results shows that all the coal produced in Indiana is used, and an additional 9 million tons is brought in. Blending plant 2 gets minimum volume, yet it is still economical to build the plant. EPA requirements are met or exceeded at all blending sites.

Since there were no commercially operated blending plants in existence, a sensitivity analysis with respect to the blending plant costs was done. The model was run with lower fixed costs for blending plants; however, the optimum number of sitings did not change. This is attributable to the huge volume of coal shipped. Blending plant costs were not very significant in comparison to transportation costs.

When the model was altered so that all 12 blending plant sites were to be used to supply the coal to the surrounding consumers, the annual costs increased from $45.1 million to $51.7 million, a 14.5% increase. This increase consists of $0.8 million for blending plant costs and $5.8 million for additional transportation costs. This emphasizes the importance of careful considerations and selection of blending plant sites.

## 14.2.   OPTIMIZATION OF AN ETHYLENE GLYCOL–ETHYLENE OXIDE PROCESS

In this section we outline the development and optimization of a simplified model of a chemical process. Space limitations preclude the complete elaboration of the process model and the presentation of constraint sequencing and reformulation mechanics. These details are given in references 10 and 11, including a listing of the final Fortran subroutines used in the optimization runs. The objective of the present discussion is to outline the form of the model and to highlight the tasks involved in arriving at a successful solution of the optimization problem.

### 14.2.1   Problem Description

A preliminary design optimization is to be performed for a plant to produce ethylene oxide by direct oxidation of ethylene with oxygen and coproduce ethylene glycol by hydration of a portion of the ethylene oxide. The key plant design parameters are to be selected so as to maximize net profit before taxes.

In the process, shown schematically in Figure 14.2, ethylene, oxygen, nitrogen, and recycle gases are mixed and fed to a tubular reactor packed with silver catalyst. Nitrogen is used as a diluent to keep the mixture outside the flammability limits. The oxidation reactions

$$C_2H_4 + \tfrac{1}{2}O_2 \rightarrow C_2H_4O$$

and

$$C_2H_4 + 3O_2 \rightarrow 2CO_2 + 2H_2O$$

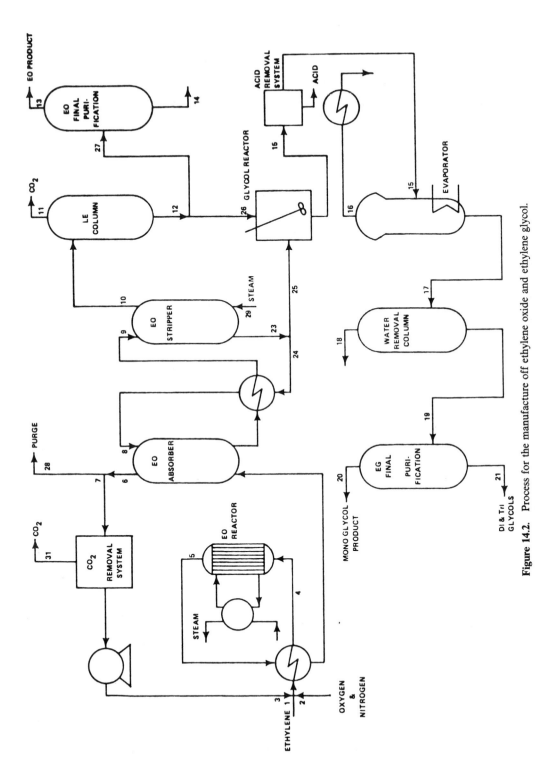

**Figure 14.2.** Process for the manufacture off ethylene oxide and ethylene glycol.

are highly exothermic, and temperature is controlled by transferring heat to boiling Dowtherm. The reactor effluent gases are fed to an absorber in which water is used to absorb $C_2H_4O$. Except for a small vent, the residual gases are sent to a carbon dioxide removal system and are then compressed to the operating pressure of the reactor. The solution from the ethylene oxide absorber is fed to a stripping column, where steam is used to remove the ethylene oxide. Most of the bottoms stream from the stripper is returned to the top of the absorber and the remainder is sent to the glycol reactor. The ethylene oxide–rich stream from the top of the stripper is compressed and condensed and sent to the light ends removal column in which residual carbon dioxide and other minor undesirable constituents are removed. The bottoms from this column now contains only ethylene oxide and water. A fraction of this stream is sent to the final purification column, where pure ethylene oxide is recovered as overhead product.

The streams from the stripper and the light ends column are mixed with makeup water and fed to a stirred tank reactor. Here ethylene oxide is hydrated to form mono, di, tri, and higher polyglycols in the presence of sulfuric acid. Monoglycol is the preferred product. The reactor effluent is passed through a sulfuric acid neutralization system and then fed to an evaporator. The bulk of the water is removed in the evaporator, and the remainder in a distillation column. The mixture is then fed to a final purification column, where monoglycol is recovered as overhead product.

### 14.2.2 Model Formulation

The model for this system consists basically of a set of equations and relations describing the individual processing units. In general, each such set will involve the following categories of equations: material balances, energy balances, physical property correlations, equipment design equations, and various types of inequality constraints. In this section we will only present selected examples of these types of equations. A more complete model formulation is given in references 10 and 11.

**Material-Balance Equations.**  The simplest of the model equations will usually be the material-balance equations. These are typically linear in the species flows and, in general, one balance equation can be written for each species processed in a unit. For example, in the case of the absorber, the balances on $CO_2$ and on ethylene oxide take the form

$$CO6 + CO9 = CO5$$

$$EO6 + EO9 = EO5 + EO8$$

where  $COi$ = flow of $CO_2$ in stream $i$, moles/hr

$EOi$ = flow of $C_2H_4O$ in stream $i$, moles/hr

**Energy-Balance Equations.** The energy-balance equations in chemical process applications typically exclude mechanical energy flows and therefore consider only enthalpy changes. In cases in which thermal properties such as heat capacities or latent heats can be assumed to be constant with temperature, the balances reduce to, at worst, quadratic equations in the temperatures and species flows. In other cases, constant properties may not be assumed, and then the balance equation will include additional nonlinear terms. For instance, in the ethylene oxide reactor, the isothermal heat load $Q$ will be given by

$$Q = \Delta H_0 + \sum_{\text{all } i} (\text{N5} - \text{N4})_i \int_{77°}^{T} C_{P_i} \, dt$$

where the first term corresponds to the enthalpy change due to reaction, and the second term corresponds to the difference in the sensible heat of the products and reactants. In particular,

$$\Delta H_0 = (\text{EO5} - \text{EO4})45{,}300 + (\text{CO5} - \text{CO4})284{,}000$$

Each $C_{P_i}$ is the gas heat capacity of species $i$ given by

$$C_{P_i} = \alpha_i + \beta_i T + \gamma_i T^2$$

where $\alpha_i$, $\beta_i$, and $\gamma_i$ are empirical constants, and $\text{N5}_i$, $\text{N4}_i$ are the molar flows of species $i$ in streams 5 and 4, respectively.

**Equilibrium Relations.** In addition ot the thermal properties, chemical process models also require vapor-liquid equilibrium distribution coefficients. These can be rather complex empirical expressions such as the following expression for the Henry's law constant for $CO_2$, $k_{CO_2}$ [12]:

$$\ln(k_{CO_2}) = \left(16.43 - \frac{2698}{T}\right) - \frac{P}{R}\left(\frac{203.14}{T} - \frac{9.7.56 \times 10^4}{T^2} - 0.00662R\right)$$

$$+ \left(\frac{P}{R}\right)^2\left(\frac{119.76}{T} + \frac{552.22}{T^2} - \frac{3.964 \times 10^7}{T^3} + \frac{9.518 \times 10^8}{T^4}\right)$$

where $T$, $P$, and $R$ are the temperature, pressure, and universal gas constant, respectively.

**Equipment Design Equations.** The equipment sizing or design equations range from algebraic equations to complex sets of differential equations. The former can either be expressions in which design variables can be explicitly evaluated given the other unit parameters or else involve implicit expressions that would require iterative solution for their evaluation.

For instance, the design of the *ethylene oxide absorber* can be modeled in a straightforward manner, assuming constant pressure and isothermal operation, and using the Kremser equation [13] to calculate the number of theoretical stages $N$. With ethylene oxide as the key species, this equation takes the form

$$N = \frac{\ln\left[\left(\frac{\text{YEO6} - m\text{XEO8}}{\text{YEO5} - m\text{XEO8}}\right)\left(1 - \frac{1}{A}\right) + \frac{1}{A}\right]}{\ln A} \tag{14.9}$$

where $m$ is the equilibrium constant for ethylene oxide in water (a function of temperature and pressure); $A = L/mG$, where $L/G$ is the ratio of liquid to gas molar flows in the absorber; and YEO$i$ and XEO$i$ are the vapor and liquid mole fractions of $C_2H_4O$ in stream $i$.

By contrast, the design equations for the *water-removal column*, although also algebraic, are more complex. The design equations for preliminary design of multicomponent distillation columns consist of the Fenske equations, the Underwood equations, and the Gilliland correlation [14].

In the present case, the column can be designed with water and monoglycol as the key components. Using the Fenske equation, the minimum number of theoretical separation stages, $N_m$, is calculated as follows:

$$N_m = \frac{\log[(\text{W18/EGl18}) \times (\text{EGl19/W19})]}{\log \bar{\alpha}_w} \tag{14.10}$$

where     $\bar{\alpha}_{w1} = (\alpha_{w1}^t \times \alpha_{w1}^b)^{1/2}$

$$\alpha_{w1}^t = \frac{p_w(T_t)}{p_{EGl}(T_t)}$$

$$\alpha_{w1}^b = \frac{p_w(T_b)}{p_{EGl}(T_b)}$$

$T_b$ = temperature at the bottom of the column

$T_t$ = temperature at the top of the column

$p_i(T)$ = partial pressure of species $i$ expressed as an empirical function of temperature $T$

and EGl$j$ and W$j$ are the monoglycol and water flows in stream $j$.

The temperatures at the top and bottom of the column are related to the compositions at the corresponding locations by assuming that stream 18 is at its dewpoint and stream 19 at its bubble-point temperature. This gives rise to two equations with implicit temperature dependence,

$$\frac{1}{P} \sum_i p_i(T_b) Xi17 = 1.0$$

$$P \sum_i \frac{Yi18}{p_i(T_t)} = 1.0$$

where $i$ = water, monoglycol, diglycol; and $P$ = total column pressure.

The minimum reflux ratio is calculated using the Underwood equation. Assuming saturated liquid feed, the equation

$$\sum_i \frac{\alpha_i \, Xi17}{\alpha_i - \theta} = 0 \quad \text{with } \alpha_{w1} \leqslant \theta \leqslant \alpha_{21} \tag{14.12}$$

fixes the value of the Underwood parameter $\theta$. The equation

$$R_m + 1 = \sum \frac{\alpha_{i1} \, Xi18}{\alpha_{i1} - \theta} \tag{14.13}$$

serves to define the minimum reflux ratio $R_m$. The actual reflux is usually selected as a design multiple of $R_m$ or can itself be selected by optimization. With $R$ known, the total condenser cooling duty and reboiler heating duty can be calculated from simple energy balances. Finally, with $N_m$, $R_m$, and $R$ determined, the actual number of stages can be established using Eduljee's analytical expression [15] for the Gilliland correlation.

$$\frac{N - N_m}{N + 1} = 0.75 \left[ 1 - \left( \frac{R - R_m}{R + 1} \right)^{0.5668} \right] \tag{14.14}$$

The design equations for the *tubular ethylene reactor* with external wall cooling will consist of a system of differential equations describing the changes in the species flows, stream enthalpy, and stream pressure as the feed mixture flows through the catalyst tubes. For instance, the production of $C_2H_4O$ will be given by the equation

$$\frac{dN_{EO}}{dw} = r_{EO}$$

where $w$ is the weight of catalyst; $r_{EO}$ is the rate equation for the ethylene oxide reaction given by reference 16,

$$r_{EO} = \frac{k_1 p_E p_{O_2}^{1/2} \cdot F}{1 + 24.55 p_E} \quad \frac{\text{lb} \cdot \text{mol of EO}}{\text{(hr)(lb of catalyst)}}$$

$$F = \frac{1}{1 + 30.63 p_{H_2O} + 7.68 \left( p_{CO_2} + p_{EO} \right)}$$

and $k_1$ is the temperature-dependent rate parameter

$$k_1 = 0.09428 \exp\left[ 13.41 \left( 1 - \frac{923}{T} \right) \right]$$

In these expressions, $p_E$, $p_{O_2}$, $p_{H_2O}$, $p_{EO}$ are partial pressures of ethylene, oxygen, water, and ethylene oxide, respectively. Similar rate expressions are used for the production of $CO_2$. The system of differential equations is not analytically integratable. As an approximation that will allow the reactor model to be more readily treated by NLP methodology, the numerical solutions of the differential equations generated over a range of anticipated operating conditions were fit with response surfaces of the form

$$w = A0\,YE4^{A1}\,YO4^{A2}\,X^{A3}P^{A4}(e^{-T})^{A5}\ln(2-X)^{A6} \qquad (14.15)$$

for the catalyst weight $w$ and an equation of the form

$$S = \frac{1}{1 + A7\,YO4^{A8}}$$

for the selectivity $S$. The parameters $A0$ through $A8$ are regression constants, $X$ is the overall conversion of ethylene in the reactor and $YE4$ in the mole fraction of ethylene and $YO4$ the mole fraction of oxygen in stream 4. The response surfaces inherently have only a limited range of validity and hence may need to be refitted if their limits are exceeded during use of the model.

**Inequality Constraints.**  In addition to the large number of nonlinear and linear equations of the type described above, the process model also has associated with it a number of inequalities. These can be grouped into three classes:

1. The explicit constraints that are an integral part of the model
2. The implicit constraints required to avoid numerical difficulties
3. Upper and lower bounds on key variables that are imposed to restrict the model to a given domain for a particular application.

The most important explicit constraints are the flammability concentration limits on the ethylene oxide reactor feed, the bounds on the Underwood parameter $\theta$ in Eq. (14.12), and the requirement that the saturated temperature of the stripping steam be above the stripper feed temperature to avoid excessive steam condensation.

The implicit constraints consist of the arguments of various logarithmic and power functions. In principle, all such arguments should be bounded to avoid possible arithmetic errors or undefined function values. However, in practice the majority of the resulting large number of inequality constraints can be deleted during initial model runs, since the argument values may be very far from the critical values. In the present case the following were found to be important:

1. The arguments of the logarithmic terms in the Kremser equation for the EO stripper and absorber [Eq. (14.9)]

2.  The argument of the Gilliland correlation in the case of the glycol
    purification column [Eq. (14.14)]

3.  The minimum reflux for the glycol-water column

4.  The temperature differences in the absorber liquid heat exchanger and
    in the evaporator

The two temperature-difference constraints are necessary to avoid calculation
of negative heat transfer rates and, thus, negative heat transfer areas.

The third type of inequality constraint is the set of upper and lower bounds
on various model variables, which will be specific for the given application.

**Objective Function.**   For the present study, the objective function consists of
the net profit before taxes expressed on an hourly basis. Net profit is sales
revenue minus total annual cost. The latter consists of utility and raw materials
costs plus an appropriate fraction of the total installed equipment cost. The
total installed equipment cost is the sum of the costs of each of the major items
of equipment. These individual equipment costs are estimated from cost-capac-
ity data for various equipment types reported by Guthrie [17] fitted to a power
function. For example, the installed cost of a tower of stainless steel–clad
construction with stainless steel bubble-cap trays is given by

$$2\left\{(674 + 4.63P)\left(D^{1.021}H^{0.965}(D + H)^{-0.175}\right)\right.$$

$$\left. + \left(88.42\,D^{1.454}H^{1.13}(D + H)^{-0.275}\right)\right\}$$

where $P$, $D$, and $H$ are the operating pressure, diameter, and height of the
tower, respectively.

The utility costs include the costs of steam for stripping, evaporation, and
reboiling; electricity for the recycle gas compressor; process water for the
glycol reactor; and sulfuric acid. The raw material costs consist of the costs of
ethylene, oxygen, and nitrogen. Particular values of these various parameters
are given in reference 10.

### 14.2.3   Problem Preparation

The complete equation-oriented flowsheet model consists of more than 100
equations and an even larger number of bounds and inequalities. Because of
the network structure of the underlying physical problem, the problem variables and equations can be sequenced for solution so that only a reduced set of
22 independent variables and associated constraints has to be explicitly considered.

Two types of equations resisted this type of sequencing: (1) the material-
and energy-balance equations required to close key recycle loops and (2)
subsets of the individual equipment-design equations that could not be solved

explicitly for one or more of the design variables. The former consisted of the material balances associated with the ethylene oxide loop and the energy balance associated with the absorber-stripper loop. The equipment equations that could not be sequenced are those associated with the glycol-water column, the bubble and dewpoint equations, as well as the instances of the Underwood equation (14.12).

The resulting NLP problem thus consists of 22 variables, 7 equality constraints, 12 inequality constraints, and a virtually complete set of upper and lower bounds. Note that by virtue of the nonlinear equality constraints, and because of the presence of numerous power-law terms in the objective function, the problem is nonconvex and highly likely to possess multiple local solutions.

### 14.2.4.  Discussion of Optimization Runs

The problem was implemented for optimization using a GRG code [18]. The initial runs with GRG using the original model were unsuccessful, often resulting in negative arguments for power and logarithmic functions and in premature termination because of scaling difficulties. These problems were remedied in part by imposing the class (2) inequality constraints discussed in Section 14.2.2. Three of the variables had to be scaled: the variables corresponding to the nitrogen feed rate (multiplied by $10^3$), the ratio of water to $C_2H_4O$ in the glycol reactor (multiplied by $10^2$), and the fraction of $C_2H_4O$ sent to the purification column (multiplied by $10^2$). With these changes, difficulties were still encountered with arithmetic errors and invalid arguments during the constraint-adjustment phases of the algorithm. Apparently, the modest constraint infeasibilities permitted by the algorithm during this phase were sufficient to cause abortion of the run. To surmount this type of difficulty, the model was modified by adding checks of function arguments and resetting these arguments to acceptable lower bounds if invalid arguments were encountered. Such resetting, in principle, introduces derivative discontinuities at the reset points; however, since the optimal solution was sufficiently far from the reset points, difficulties with discontinuities were not observed in the runs.

The modified model was successfully optimized using the GRG code from four different starting points. As shown in Table 14.1, a typical optimization run required less than 100 CPU time on a CDC 6500. In three out of four cases the objective function attained the same value, 2938.43, and in the fourth it differed only slightly. Hence, the objective value of 2938.43 is probably the desired global optimum.

Three variables—the fraction of ethylene oxide sent to the final purification column, the operating pressure in the ethylene oxide stripper, and the liquid/gas ratio in the ethylene oxide absorber—reached their lower bound. Four variables—conversion in the ethylene oxide reactor, the temperatures in the two reactors, and the fraction of water removed in the evaporator—reached their

Table 14.1 Optimization Results

| Variable | Starting Points | | | | Corresponding Optimal Points | | | |
|---|---|---|---|---|---|---|---|---|
| | 1 | 2 | 3 | 4 | 1 | 2 | 3 | 4 |
| X(1) | 4952.8 | 5892.785 | 12901.7 | 14237.1 | 3257.33 | 3257.33 | 3257.33 | 3257.33 |
| X(2) | 1.094 | 1.552 | 0.75 | 0.75 | 0.6667 | 0.6667 | 0.6667 | 0.6667 |
| X(3) | 240.0 | 219.99 | 220.0 | 220.04 | 242.0 | 242.0 | 242.0 | 242.0 |
| X(4) | 0.1938 | 0.1613 | 0.068 | 0.060 | 0.3 | 0.3 | 0.3 | 0.3 |
| X(5) | 2.4 | 4.4893 | 4.5 | 4.66 | 1.0362 | 1.0362 | 1.0362 | 0.0632 |
| X(6) | 34.0 | 38.997 | 39.0 | 35.0 | 30.0 | 30.0 | 30.0 | 30.0 |
| X(7) | 0.37 | 0.3737 | 0.35 | 0.34 | 0.385 | 0.385 | 0.385 | 0.385 |
| X(8) | 14.1 | 30.684 | 30.0 | 23.62 | 5.428 | 5.428 | 5.428 | 5.428 |
| X(9) | 14.0 | 14.0 | 14.0 | 14.0 | 12.72 | 12.72 | 12.72 | 12.72 |
| X(10) | 44.4 | 19.99 | 20.0 | 23.65 | 44.41 | 44.41 | 44.41 | 44.41 |
| X(11) | 2.1 | 2.8636 | 3.0 | 2.99 | 2.0 | 2.0 | 2.0 | 2.0 |
| X(12) | 2.42 | 1.888 | 2.0 | 1.92 | 2.4235 | 2.4235 | 2.4235 | 2.4235 |
| X(13) | 0.55 | 0.5 | 0.5 | 0.5 | 0.5 | 0.5 | 0.5 | 0.5 |
| X(14) | 44.4 | 49.61 | 50.0 | 51.06 | 44.41 | 44.41 | 44.41 | 44.41 |
| X(15) | 2410.0 | 2800.0 | 2800.0 | 2744.4 | 2753.58 | 2756.03 | 2756.03 | 2756.03 |
| X(16) | 74.0 | 66.0 | 66.0 | 66.0 | 75.0 | 75.0 | 75.0 | 75.0 |
| X(17) | 0.067 | 0.0575 | 0.05 | 0.0522 | 0.0173 | 0.0173 | 0.0173 | 0.0173 |
| X(18) | 0.89 | 0.9046 | 0.9 | 0.9 | 0.98 | 0.98 | 0.98 | 0.98 |
| X(19) | 159.0 | 175.798 | 177.0 | 177.3 | 160.62 | 176.4 | 176.41 | 176.41 |
| X(20) | 68.0 | 82.21 | 85.0 | 83.44 | 70.95 | 83.62 | 83.63 | 86.63 |
| X(21) | 1.36 | 0.138 | 0.14 | 0.1399 | 2.518 | 0.1429 | 0.1429 | 0.1429 |
| X(22) | .276 | 0.5 | 0.5 | 0.5263 | .2843 | 0.492 | 0.492 | 0.492 |
| Net profit | 2657.22 | 2100.7 | 1505.02 | 1653.38 | 2936.23 | 2938.43 | 2938.43 | 2938.43 |
| CPU time, s | — | — | — | — | 60.1 | 63.8 | 75.8 | 93.7 |

upper bounds. Fifteen variables are within their bounds. The constraints on the flammability limits and two temperature differences are binding at the optimum.

The temperatures in the reactors and the conversion of ethylene are expected to reach their upper bounds, because higher values of these variables favor higher production of ethylene oxide and ethylene glycol, which in turn increases the profit. The upper-bound value for the fraction of water removed in the evaporator suggests that at that upper-bound value it is still cheaper to remove water in the evaporator rather than in a distillation column. The flow split, $X(6)$, determines the fraction of ethylene oxide purified and sold as ethylene oxide product, the remainder being converted to ethylene glycol. Its optimum value reached the lower bound of 30%, which implies that it is more profitable to sell ethylene oxide as ethylene glycol than as ethylene oxide. This can be expected in view of the higher price of 19¢/lb used for the former compared to 15¢/lb used for ethylene oxide.

As described by Sarma and Reklaitis [10], further runs were carried out both to test the validity of the model and to study the effects of parameter changes on the optimal design. Such studies are important to support the credibility of the model as well as to indicate model elements that may require more detailed study using a rigorous process flowsheet simulator [11].

In summary, the lessons to be found in this case study are that model formulation and preparation for solution are difficult and time-consuming steps that require detailed knowledge of the system being studied. Even after careful preparation, it proved necessary to scale some of the variables and to add some inequality constraints to avoid arithmetic errors and invalid arguments. Since the GRG algorithm does allow infeasible excursions, it was necessary to add checks to the model subroutine to prohibit excursions that would cause fatal arithmetic errors. Finally, once the problem would successfully execute, it proved prudent to repeat solution with several different starting points to increase confidence in the solution accepted as optimal.

## 14.3  OPTIMAL DESIGN OF A COMPRESSED AIR ENERGY STORAGE SYSTEM

Finally in this section we use a case study involving the design of a compressed air energy storage system (CAES) to demonstrate formulation technique and several of the methods and concepts discussed in previous chapters. It is appropriate to devote considerable attention to problem formulation technique, since it remains to a large degree an art, which requires considerable practice to master. We demonstrate here a decomposition approach, involving the identification of coupling variables, suitable for application to large NLP problems. The work was carried out in two locations, with the support of the U.S. Department of Energy: by F. W. Ahrens and co-workers at Argonne

National Laboratory in Chicago, Illinois, and K. M. Ragsdell and students at Purdue in West Lafayette, Indiana. The employed decomposition approach very significantly enhanced the dual design team interaction, since each site had primary responsibility for a single subsystem optimization. Numerical results are given for a CAES plant employing the Media, Illinois Galesville aquifer formation. Additional modeling details and results are given in a number of related publications [19–21].

### 14.3.1  Problem Description

Compressed air energy storage is one of the technologies that is currently available to electric utilities to supply peak power using stored energy previously generated during periods of excess capacity [22]. A goal of this study was the determination of the economic viability of CAES peaking plants employing below-ground naturally occurring aquifers. The use of energy-storage systems can be economically advantageous to utilities, since they improve the utilization of high-efficiency base plants and permit a reduction in the amount of premium fuel required to generate peak power.

Consider the CAES power plant diagrammed in Figure 14.3, which is based on a split Brayton cycle and comprises four equipment groups: a reversible motor/generator, air-compression equipment, an air-storage reservoir (with associated piping), and power extraction equipment. The couplings on the motor/generator allow the plant operators to either use electrical power from the utility grid to compress air or use the stored compressed air and some premium fuel to generate power. The CAES system configuration shown in Figure 14.3 is typical, but many variations in equipment arrangement are possible [22]. In terms of its interaction with the other equipment groups, the turbine system can be characterized by its design inlet pressure ($p_{ti}$) and its mass flowrate per unit power output ($\dot{m}'$). The latter depends on the turbine inlet temperatures (premium fuel consumption) and equipment arrangement and design. Details of these relationships are given by Kim and Kartsounes [23]. Because of the requirements for storing large amounts of high-pressure compressed air (e.g., $10^7$–$10^8$ ft$^3$ at 50 atm for a typical 200-MW plant), it is known that underground air reservoirs are an economic necessity. The reservoir can be either a cavern (in hard rock or in a salt dome) or a porous rock layer (most commonly an aquifer), such as an edgewater aquifer.

The storage reservoir design requirements (capacity, pressure level, piping design, etc.) are interdependent with the selection (performance characteristics and operating conditions) of the above-ground compression and power-generation equipment, and the desired power output and operating cycle of the CAES plant. In turn, the decision on the power level and duty cycle is impacted by the economic and technical characteristics of the utility grid, by the cost of premium fuel, and so forth. Thus many technical and economic tradeoffs must be considered in specifying a CAES plant design.

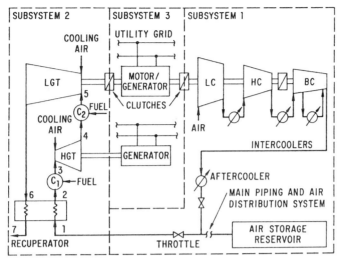

Figure 14.3.    CAES power plant.

## 14.3.2    Model Formulation

We have seen in Figure 14.3 the basic components of a CAES plant. Although the utility grid is not physically part of the CAES plant, the interaction should be considered, since the cost of base load power and the utility load cycle may have a strong influence on design cost of the CAES facility. Correspondingly, the CAES costs will influence the cost of power sold by the utility.

**Decomposition Strategy.**    We find it convenient, if not essential, to decompose the system into three groups or subsystems as seen in Figure 14.3. Subsystem 1 contains the air-storage reservoir, air-compression train, and main piping and air-distribution system. Subsystem 2 contains the power-generation train, and subsystem 3 contains the motor/generator and the utility grid. With the subsystems formed in this way, it is possible to choose coupling and internal variables so that the subsystems can be designed with a degree of independence from the other subsystems. The exact dependence is contained in the coupling variable relationships. For instance, we assume that subsystem 3 (the utility grid) affects the rest of the system through a variable $U_L$, the utility load cycle as shown in Figure 14.4. This single variable could, of course, represent many variables in the utility load cycle, but this is not pursued here, since our interest is primarily with subsystems 2 and 3. Finally, we suggest that the direct interactions (or coupling) between subsystems 1 and 2 are dependent on only two variables, the inlet pressure to the power-generation train $p_{ti}$ and the specific mass flowrate $\dot{m}'$. As the figure suggests, there is also the indirect influence of the utility load cycle. We quantify this effect by choosing the utility load cycle.

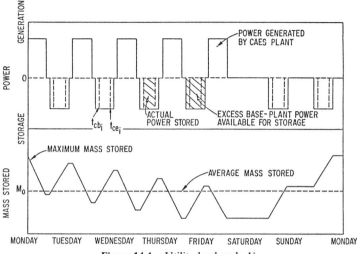

**Figure 14.4.** Utility load cycle $U_L$.

The criterion for optimal design is chosen to be the total normalized cost $C$ of the system (i.e., cost per unit of electricity *generated* by the CAES power plant). This total cost is the sum of the individual costs, which include fuel cost, maintenance, and charge rate on capital. The costs have to be minimized subject to various performance and technical constraints. The implication for CAES plant design is that for a given utility load cycle an optimization of subsystem 1 would provide the minimum subsystem operating cost $C_1^0$ and values for the corresponding subsystem design variables as a function of the coupling variables $p_{ti}$ and $\dot{m}'$. Similar optimization for subsystem 2 would yield $C_2^0$ (the minimum operating cost of subsystem 2) and its optimum design as a function of the coupling variables only. Finally, the sum of $C_1^0$ and $C_2^0$ can be minimized to determine the optimum values of the coupling variables, the minimum plant cost $C^*$, and the optimal plant design.

**Subsystem 1: Storage.** The aquifer (originally water-filled) should have the shape of an inverted saucer to prevent migration of the compressed air. The air bubble is formed by displacing the innate water; the compressed air is contained between the airtight caprock and a bottom layer of water. The compressor train included in this subsystem follows the recommendations of United Technologies Research Center [24]. A detailed discussion of the model employed is given by Ahluwalia et al. [25]. Here, we focus only on the conceptual formulation of the optimal design problem.

In the optimization of subsystem 1, the objective is to determine the combination of internal design variables that minimizes the subsystem operating cost for given values of the coupling variables $p_{ti}$, $\dot{m}'$, and $U_L$. The set of design variables can be classified into two subsets. The first subset includes variables that are restricted to take a limited number of discrete values. Engineering considerations require that the main piping diameter, the type of low-pressure compressor, and the reservoir well-bore diameters be restricted to discrete, commercially available designs. As the number of alternatives is limited, the simple method of incorporating these discrete variables into the optimization via an exhaustive search in all discrete dimensions is suitable. The remaining internal optimization variables of subsystem 1, which are assumed to be continuous, are the four geometric parameters of the reservoir design, $N_W$, $H$, $A_{act}$, and $d$ illustrated in Figure 14.5, and the energy storage process variables $t_{cb_i}$ and $t_{ce_i}$. The variables $t_{cb_i}$ represent the times during the weekly cycle when energy storage processes begin, and $t_{ce_i}$ are the ending times of these processes. The storage (charging) time variables are shown for a typical

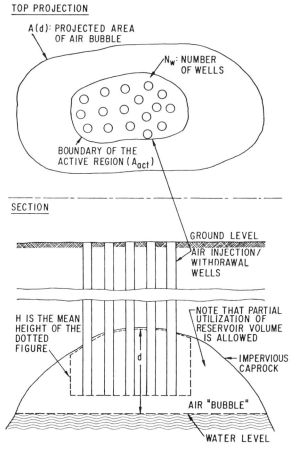

**Figure 14.5.** Aquifer reservoir geometry.

cycle in Figure 14.4. The operating cost, to be minimized, can be written as

$$C_1\left(N_W, H, A_{\text{act}}, d, t_{cb_i}, t_{ce_i}\right) = K_1(U_L)C_T + K_2(U_L)P_{c_i}\sum_i\left(t_{ce_i} - t_{cb_i}\right)$$

(14.16)

In the equation above, $K_1$ and $K_2$ are functions of the coupling variable $U_L$ but are treated as constants for the purpose of optimization. Similar notation is used to represent functions of other coupling variables and functions of the three discrete internal variables. The first term in Eq. (14.16) represents the operating cost due to the annual charge rate on capital, $C_T$, of subsystem 1, where $C_T$ is the sum of capital costs of the various components:

$$C_T\left(N_W, H, A_{\text{act}}, d, t_{cb_i}, t_{ce_i}\right) = \text{WC} + \text{LC} + \text{BC} + \text{CC} + K_3(\text{piping}).$$

(14.17)

$K_3$(piping) is the capital cost of the main piping and distribution system, which depends upon the piping design selected. The capital cost of wells is

$$\text{WC}(N_W, H, A_{\text{act}}) = N_W\left[K_{W_1} + K_{W_2}\{H - F(A_{\text{act}})\}\right]$$ (14.18)

with constants $K_{W_1}$, $K_{W_2}$, and $F(A_{\text{act}})$, a known function of $A_{\text{act}}$ determined from reservoir geometry. The term within curly brackets in Eq. (14.18) is the depth to which wells must be bored. The second term on the right in Eq. (14.17) is the cost of purchasing the land over the proposed reservoir,

$$\text{LC}(d) = K_l A(d)$$ (14.19)

where $A(d)$ is the land area over the air reservoir, a known geometric function of $d$.

In this simplified model, the capital cost of initially displacing water from the aquifer, or bubble development, is calculated in terms of energy required to compress the volume of air in the bubble, which is a function of $d$. Finally, the capital cost, CC, of the compressor train is expressed as

$$\text{CC}\left(N_w, H, A_{\text{act}}, d, t_{cb_i}, t_{ce_i}\right) = K_{cl_1} + K_{cl_2}M_c + K_{cb_1}\left[\dot{M}_c\left\{\frac{p_c}{K_{cl_3}} - 1\right\}\right]^{K_{cb_3}}$$

(14.20)

Here, $K_{cl_1}$, $K_{cl_2}$, and $K_{cl_3}$ are parametric constants determined by the choice of compressor train design. $\dot{M}_c$ is the air mass flowrate during the storage processes, chosen to be the same during all storage processes due to compressor performance considerations. The remaining unknown term in Eq. (14.20) is $p_c$, the discharge pressure required of the compressor train. This pressure can be calculated using the pressure-drop models given by Sharma et al. [19]. The second term in Eq. (14.16) is the subsystem operating cost incurred due to compressor power consumption, $p_c$, which is given by

$$P_c\left(N_w, H, A_{\text{act}}, d, t_{cb_i}, t_{ce_i}\right) = \left[K_{P_1} + K_{P_2}P_c + K_{P_3}P_c^2 + K_{P_4}P_c^3\right]\dot{M}_c$$

(14.21)

Engineering intuition, aquifer geology and geometry, and the utility load cycle suggest bounds and functional constraints on the design variables. These have been completely developed and explained by Ahrens et al. [20] and will not be repeated here. In summary, we place bounds on the storage process times as suggested in Figure 14.4, and upper and lower limits on the four physical variables $N_w$, $H$, $A_{\text{act}}$, and $d$ as defined in Figure 14.5. Functional constraints are imposed that require that all storage processes end after they begin, that the wells be placed close enough to ensure full utilization of reservoir volume, that the well bores not physically interfere one with another, that the "bubble" be no larger than the land purchased, that the wells not be drilled to a depth that would cause "coning," that the required compressor power be less than that which the utility is willing and/or able to supply at any time, and, finally, that the pressure requirement of subsystem 2 be met at all times by subsystem 1. In summary, there are 16 design variables with 32 bounds and 12 constraints, 4 of which are nonlinear.

**Subsystem 2: Generation.** Subsystem 2 of the CAES system is composed of the high- and low-pressure turbines, their combustors, and the recuperator, as indicated in Figure 14.3. The most interesting design tradeoffs for this subsystem are (1) larger, more effective recuperator vs. greater premium fuel consumption in the combustors, for preheating the air entering the turbines, and (2) advanced, high inlet-temperature turbines with high cost but high performance vs. conventional, lower temperature, lower cost turbines. An additional tradeoff, of secondary importance, is the pressure ratio split between the high-pressure turbine and the low-pressure turbine.

The performance model for subsystem 2 is based on a thermodynamic analysis of the components. The detailed equations are given by Kim and Kartsounes [23]. It should be mentioned, that the model includes the effect that as the turbine inlet temperatures exceed 1600°F it is necessary to use an increased function of the compressed air from storage for the turbine blades and other turbine components.

For the purpose of calculating subsystem 2 performance, the coupling variables $p_{ti}$ (the subsystem inlet pressure) and $\dot{m}'$ (the specific turbine system air flowrate, $\text{lb}_m/\text{kWh}$), and $P_{\text{gen}}$ (the total power output from the two turbines) are regarded as inputs. Because of this, it is not possible to independently specify both turbine inlet temperatures $T_3$ and $T_5$ if fixed, state-of-the-art turbine efficiencies are assumed. In the present mode, $T_5$ (low-pressure turbine inlet temperature) was considered a design variable and $T_3$, along with several intermediate variables, was subsequently determined during the iterative solution of the model equations. The other design variables of subsystem 2 are the recuperator effectiveness $\varepsilon$ and the low-pressure turbine pressure ratio $r_p$. The variable $r_p$ takes on two discrete values, 11 and 16, corresponding to the current practice of turbomachinery manufacturers.

In the optimization of subsystem 2, the objective is to find the combination of internal design variables that minimizes the subsystem operating cost for given values of the coupling variables. During a particular optimization process, the coupling variables $p_{ti}$, $m'$, and $U_L$ are fixed, so they will be omitted from the functional relationships that follow. The discrete variable $r_p$ is also omitted, since an optimization is performed separately for each of its values. The operating cost to be minimized can be written as

$$C(T_5, \varepsilon) = K_1(U_L)C_{\text{cap}} + K_F \dot{Q}' + K_{0m} \qquad (14.22)$$

The first term represents the operating cost due to the annual charge rate on the capital $C_{\text{cap}}$ of subsystem 2, where $C_{\text{cap}}$ is the sum of capital costs:

$$C_{\text{cap}}(T_5, \varepsilon) = C_{\text{LGT}} + C_{\text{HGT}} + C_R + C_{\text{BAL}} \qquad (14.23)$$

According to expression (14.23), the capital cost is the sum of $C_{\text{LGT}}$, the cost of the low-pressure turbine (including the increased expense of cooling air for high operating temperature); $C_{\text{HGT}}$, the cost of the high pressure turbine; $C_R$, the cost of the recuperator; and $C_{\text{BAL}}$, the cost of the balance of the plant. The second term in Eq. (14.22) is the cost of the premium fuel used in the combustors. $K_F$ is taken as $2.50/10^6$ Btu. The heat rate $\dot{Q}'$ is dependent on $T_3(T_5, \varepsilon)$, and $T_5$. The final term in Eq. (14.22) is the operating and maintenance cost of the plant. Finally, we place upper and lower bounds on the three design variables $\varepsilon$, the recuperator effectiveness, and $T_3$ and $T_5$, the turbine inlet temperatures.

### 14.3.3 Numerical Results

Using the transformation code BIAS [26] for subsystem 1, and the GRG code OPT [27] for subsystem 2, we have sought the plant design that minimizes the normalized operating cost of generation of 600 MW for 10 hr each weekday. We have employed the Media, Illinois, Galesville aquifer as the storage reservoir. Contour maps and material properties for this aquifer and other problem parameters are given by Sharma et al. [19], Katz and Lady [28], and

Ahluwalia et al. [25]. The subsystem 1 problem was solved for a number of combinations of $p_{ti}$ and $\dot{m}'$, using BIAS with automatic scaling.

Contours of constant minimized operating cost for subsystem 1 are shown in Figure 14.6. A very significant cost variation is evident. Figure 14.6 indicates that small $\dot{m}'$ values (i.e., low air flowrates) are favored. This is due primarily to the higher cost of the air-storage reservoir as the quantity of air stored is increased.

The optimum subsystem 1 designs corresponding to points in Figure 14.6 were also found to vary widely. Of particular interest is the number of wells required. It was found to vary from a low of 54 in the lower left (low-cost) region to values in the 200–500 range in the upper right region. Finally, it should be noted that the effects of the discrete variables (low-pressure compressor-compression ratio, well bore diameter, and main pipe diameter) have been studied and are reported by Ahrens [29].

Contours of constant minimized operating cost for subsystem 2 are presented in Figure 14.7 for a range of $p_{ti}$ and $\dot{m}'$ values. The minimum cost contour (22 mills/kWh) corresponds approximately to designs having the minimum allowed (1500°F) turbine inlet temperatures $T_3$ and $T_5$. These correspond to conventional designs proposed for CAES plants. The maximum cost contour (24.5 mills/kWh) shown is near to the constraint boundary representing the upper limit (2400°F) on turbine inlet temperatures. The

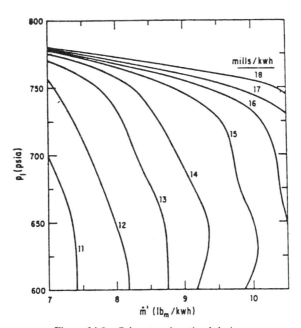

**Figure 14.6.** Subsystem 1 optimal designs.

results in Figure 14.7 are based on $r_p = 16$. It was found that use of $r_p = 11$ yielded similar, but slightly higher, cost results throughout the region explored. The optimum recuperator effectiveness $\varepsilon$ was found to vary from 0.52 to 0.77 for the ranges of coupling variables yielding solutions.

By the nature of the decomposition strategy employed in this study, the optimum CAES plant (the design that minimizes the power generation cost for the specified utility load cycle and aquifer site) may be easily found by superimposing the results from Figures 14.6 and 14.7. The resulting minimized cost contours are shown in Figure 14.8. Interestingly, even though the individual subsystem contours are open, their sum exhibits an overall optimum that is *within* the coupling variable domain considered. Figure 14.8 demonstrates that the power generation (operating) cost of the optimum CAES plant is slightly under 35.75 mills/kWh, and that the optimum values for the coupling variables are, approximately, $p_{ti} = 625$ psia and $\dot{m}' = 8.5$ lb$_m$/kWh. Knowing the optimum coupling variables, we can readily obtain the optimum values of other design variables from the separate subsystem 1 and 2 optimization results.

### 14.3.4  Discussion

The optimal design approach affords a significant opportunity for cost savings in the construction and operation of compressed-air energy-storage systems, as

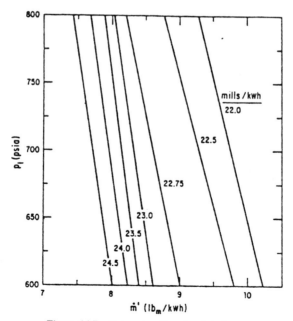

**Figure 14.7.**  Subsystem 2 optimal designs.

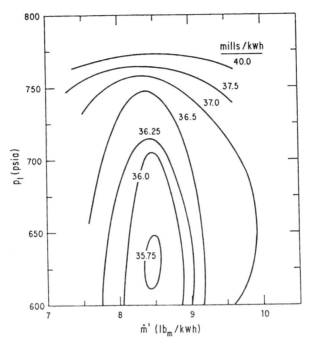

**Figure 14.8.**   CAES plant optimization results.

can be seen from the previously given results. On the other hand, the models necessary to adequately represent such a practical physical system can be quite complex. We have given what we feel to be the least complex system model that will produce a meaningful optimal design. Even with our simplified approach, the complete CAES system optimization (including subsystems 1 and 2) involves 20 design variables, 4 discrete design parameters, 8 linear constraints, upper and lower bounds on all design variables, and a nonlinear objective function. Furthermore, the model includes functions that require calculation of the modified Bessel functions of the first and second degree and first and second kind, and various spline approximations for empirical data.

The decomposition proved to be an effective means of attacking the problem, not only because of the savings in solution time but also because the subsystem optimization problems have utility within themselves. That is, these subgroup results provide insight that would be difficult at best to gather in any other way. Finally, the decomposition strategy employed here allows the use of an orderly modular approach of design. That is, we might envision a different storage system, (such as a hard rock cavern), which would produce a different subsystem 1 model. We could perform the subsystem 1 optimizations and synthesize the overall system results just as before.

## 14.4  SUMMARY

Three case studies in engineering optimization were developed in this chapter. The first resulted in a large-scale linear program that required the introduction of 0-1 variables to accommodate fixed costs in the objective function. Solution necessitated a branch-and-bound strategy with LP subproblems. The second case study led to a large nonlinear program whose size could be considerably reduced by explicit sequencing of equality constraints. Because of the presence of "hard" constraints whose violation would disallow model evaluation, a GRG solution approach proved effective. The third case study involved a large nonlinear program with some discrete variables and a decomposable structure. The decomposition separated the problem into three subproblems: two involving the optimization of each major subsystem for fixed values of selected coupling variables, and the third involving an overall optimization with respect to the coupling variables. Optimization with respect to the discrete variables could be carried out by exhaustive enumeration. The case studies illustrated quite clearly that the solution strategy must be carefully selected to exploit the characteristics of the system under study. Identification of these characteristics requires detailed knowledge of the engineering system as well as a mastery of optimization methodology. These case studies thus reaffirm that the optimization of engineering systems is of necessity the domain of engineers informed in optimization methods. We hope that this book has helped the reader to achieve this state of knowledge.

## REFERENCES

1. Ravindran A. and D. L. Hanline, "Optimal Location of Coal Blending Plants by Mixed Integer Programming," *Amer. Inst. Ind. Eng. Trans.*, **12**(2), 179–185 (1980).

2. *Key Stone Coal Industry Manual*, McGraw-Hill, New York, 1975.

3. *Compilation of Air Pollutant Emission Factors*, 2nd ed., U.S. Environmental Protection Agency, April 1973.

4. National Rural Electric Cooperative Association, *Rural Electrification*, Washington, DC, March 1978.

5. Indiana Energy Office, State Department of Commerce, Indianapolis, IN, private communication.

6. *Rand McNally Railroad Atlas of the United States*, Rand McNally & Co., Chicago, IL, 1971.

7. Wier, C. E., *Coal Resources of Indiana*, Indiana Geological Survey, Bloomington, IN, 1973.

8. *MIPZI—Documentation of a Zero-One Mixed Integer Programming Code*, Department of Agricultural Economics, Purdue University, West Lafayette, IN, Sept., 1973.

9. Balas, E., "An Additive Algorithm for Solving Linear Programs with Zero-One Variables," *Oper. Res.*, **13**, 517–547 (1965).

10. Sarma, P. V. L. N., and G. V. Reklaitis, "Optimization of a Complex Chemical Process Using an Equation Oriented Model," *Math. Programming Study*, **20**, 113–160, 1982.

11. Sarma, P. V. L. N., "Strategy of Modeling and Optimization in Chemical Process Design," Ph.D. dissertation, Purdue University, 1978.

12. Stewart, P. B., and P. Mungal, "Correlation Equation for Solubility of Carbon Dioxide in Water, Seawater, and Seawater Concentrates," *J. Chem. Eng. Data*, **2**, 170 (1971).

13. Smith, B. D., *Design of Equilibrium Stage Processes*, McGraw-Hill, New York, 1963.

14. Treybal, R. E., *Mass Transfer Operations*, McGraw-Hill, New York, 1968.

15. Eduljee, H. E., "Equations Replace Gilliland Plot," *Hydrocarbon Proc.*, **54**(9), 120 (1975).

16. Buntin, R. R., "A Kinetic Study of Oxidation of Ethylene on a Silver Catalyst," Ph.D. thesis, Purdue University, Aug. 1961.

17. Guthrie, K. M., "Data and Techniques for Preliminary Capital Cost Estimating," in *Modern Cost Engineering Techniques* (H. Popper, Ed.), McGraw-Hill, New York, 1970.

18. Lasdon, L. S., A. D. Warren, A. Jain, and M. W. Ratner, "Design and Testing of Generalized Reduced Gradient Code for Nonlinear Programming," *ACM Trans. Math. Software*, **4**, 34–50 (Mar. 1978).

19. Sharma, A., H. H. Chiu, F. W. Ahrens, R. K. Ahluwalia, and K. M. Ragsdell, "Design of Optimum Compressed Air Energy Storage Systems," *Energy*, **4**, 201–216 (1979).

20. Ahrens, F. W., A. Sharma, and K. M. Ragsdell, "Computer Aided Optimal Design of Compressed Air Energy Storage Systems," ASME *J. Mech. Des.*, **102**, 437–445 (1980).

21. Ahrens, F. W., A. Sharma, and K. M. Ragsdell, "Optimal Design of Compressed Air Energy Storage Systems," *Proceedings of the Sixth UMR-DNR Conference on Energy*, 431–441 (Oct. 1979).

22. Vosburgh, K. G., "Compressed Air Energy Storage," *AIAA J. Energy*, **2**(2), 106–112 (1978).

23. Kim, C. S., and G. T. Kartsounes, "A Parametric Study of Turbine Systems for Compressed Air Energy Storage Plants," Final Report FY1977, *Argonne Nat. Lab. Rep.* **ANL / ES-64**, 1978.

24. Davison, W. R., and R. D. Lessard, "Study of Selected Turbomachinery Components for Compressed Air Storage Systems," prepared by United Technologies Research Center for Argonne National Laboratory, **ANL / EES-TM-14**, 1977.

25. Ahluwalia, R. K., A. Sharma, and F. W. Ahrens, *Proc. Compressed Air Energy Storage Technology Symposium, Pacific Grove, CA*, (1979), pp. 295–325.

26. Root, R. R., and K. M. Ragsdell, "Computational Enhancements to the Method of Multipliers," ASME *J. Mech. Des.*, **102**(3), 517–523 (1980).

27. Gabriele, G. A., and K. M. Ragsdell, "Large Scale Nonlinear Programming Using the Generalized Reduced Gradient Method," *ASME J. Mech. Des.*, **102**(3), 566–573 (1980).

28. Katz, D. L., and E. R. Lady, *Compressed Air Storage for Electric Power Generation*, Ulrich's Books, Ann Arbor, MI, 1976.

29. Ahrens, F. W., "The Design Optimization of Aquifer Reservoir-Based Compressed Air Energy Storage Systems," *Proc. 1978 Mech. Magnetic Energy Storage Contractors' Rev. Meet., Luray, VA*, **CONF-781046**, 404–415 (1978).

# Appendix A

# Review of Linear Algebra

## A.1 SET THEORY

A *set* is a well-defined collection of things. By well defined we mean that given any object it is possible to determine whether or not it belongs to the set.

The set $S = \{x | x \geqslant 0\}$ defines the set of all nonnegative numbers. $x = 2$ is an element of the set $S$ and is written as $2 \in S$ (2 belongs to $S$).

The *union* of two sets $P$ and $Q$ defines another set $R$ such that $R = P \cup Q = \{x | x \in P \text{ or } x \in Q, \text{ or both}\}$.

The *intersection* of two sets, written $P \cap Q$, defines a set $R = \{x | x \in P \text{ and } x \in Q\}$.

$P$ is a *subset* of $Q$, written $P \subset Q$, if every element of $P$ is in $Q$.

*Disjoint* sets have no elements in common. If $P$ and $Q$ are disjoint sets, then $x \in P$ implies $x \notin Q$, and vice versa.

The *empty set*, denoted by $\varnothing$, is a set with no elements in it.

## A.2 VECTORS

A *vector* is an ordered set of real numbers. For instance, $a = (a_1, a_2, \ldots, a_n)$ is a vector of $n$ elements or components.

If $a = (a_1, a_2, \ldots, a_n)$, and $b = (b_1, b_2, \ldots, b_n)$, then

$$a + b = c = (a_1 + b_1, a_2 + b_2, \ldots, a_n + b_n)$$
$$a - b = d = (a_1 - b_1, a_2 - b_2, \ldots, a_n - b_n)$$

For any scalar $\alpha$ positive or negative,

$$\alpha a = (\alpha a_1, \alpha a_2, \ldots, \alpha a_n)$$

The vector $O = (0, 0, \ldots, 0)$ is called the *null* vector.

The inner product of two vectors, written $a \cdot b$ or simply $ab$, is a number given by

$$a_1 b_1 + a_2 b_2 + \cdots + a_n b_n$$

For example, if $a = (1, 2, 3)$ and $b = \begin{pmatrix} 4 \\ 5 \\ 6 \end{pmatrix}$, then $a \cdot b = 4 + 10 + 18 = 32$.

A set of vectors $a_1, a_2, \ldots, a_n$ is *linearly dependent* if there exist scalars $\alpha_1, \alpha_2, \ldots, \alpha_n$ not all zero, such that

$$\sum_{i=1}^{n} \alpha_i a_i = 0$$

In this case, at least one vector can be written as a *linear combination* of the others. For example,

$$a_1 = \lambda_2 a_2 + \lambda_3 a_3 + \cdots + \lambda_n a_n$$

If a set of vectors is not dependent, then it must be *independent*.

A *vector space* is the set of all $n$-component vectors. This is generally called the *euclidean n-space*.

A set of vectors is said to *span* a vector space $V$ if every vector in $V$ can be expressed as a linear combination of the vectors in that set.

A *basis* for the vector space $V$ is a set of linearly independent vectors that spans $V$.

## A.3 MATRICES

A *matrix* **A** of size $m \times n$ is a rectangular array (table) of numbers with $m$ rows and $n$ columns.

*Example A.3.1*

$$\underset{(2\times3)}{\mathbf{A}} = \begin{bmatrix} 1 & 2 & 3 \\ 4 & 5 & 6 \end{bmatrix}$$

is a matrix of two rows and three columns. The $(i, j)$th element of **A**, denoted by $a_{ij}$, is the element in the $i$th row and $j$th column of **A**. In Example A.3 1, $a_{12} = 2$ while $a_{23} = 6$. In general an $m \times n$ matrix is written as

$$\underset{(m\times n)}{\mathbf{A}} = [a_{ij}]$$

The elements $a_{ij}$ for $i = j$ are called the *diagonal* elements; while the $a_{ij}$ for $i \neq j$ are called the *off-diagonal* elements.

The elements of each column of a matrix define a vector called a *column vector*. Similarly, each row of a matrix defines a *row vector*. In Example A.3-1, the vectors

$$a_1 = \begin{pmatrix} 1 \\ 4 \end{pmatrix} \qquad a_2 = \begin{pmatrix} 2 \\ 5 \end{pmatrix} \qquad a_3 = \begin{pmatrix} 3 \\ 6 \end{pmatrix}$$

are the column vectors of the matrix $\mathbf{A}$, while the vectors $b_1 = (1, 2, 3)$ and $b_2 = (4, 5, 6)$ are the row vectors of $\mathbf{A}$.

Thus a vector may be treated as a special matrix with just one row or one column. A matrix with an equal number of rows and columns is called a *square matrix*.

The *transpose* of a matrix $\mathbf{A} = [a_{ij}]$, denoted by $\mathbf{A}'$ or $\mathbf{A}^T$ is a matrix obtained by interchanging the rows and columns of $\mathbf{A}$. In other words, $\mathbf{A}^T = [a'_{ij}]$ where $a'_{ij} = a_{ji}$. The transpose of $\mathbf{A}$ defined in Example A.3-1 is given by

$$\mathbf{A}^T_{(3\times2)} = \begin{bmatrix} 1 & 4 \\ 2 & 5 \\ 3 & 6 \end{bmatrix}$$

The matrix $\mathbf{A}$ is said to be *symmetric* if $\mathbf{A}^T = \mathbf{A}$. The *identity matrix*, denoted by $\mathbf{I}$, is a square matrix whose diagonal elements are all 1 and the off-diagonal elements are all zero.

A matrix whose elements are all zero is called a *null matrix*.

### A.3.1  Matrix Operations

The *sum* or *difference* of two matrices $\mathbf{A}$ and $\mathbf{B}$ is a matrix $\mathbf{C}$ (written $\mathbf{C} = \mathbf{A} \pm \mathbf{B}$) where the elements of $\mathbf{C}$ are given by

$$c_{ij} = a_{ij} \pm b_{ij}$$

For two matrices $\mathbf{A}$ and $\mathbf{B}$ the product $\mathbf{AB}$ is defined if and only if the number of columns of $\mathbf{A}$ is equal to the number of rows of $\mathbf{B}$. If $\mathbf{A}$ is an $m \times n$ matrix and $\mathbf{B}$ is an $n \times r$ matrix, then the product $\mathbf{AB} = \mathbf{C}$ is defined as a matrix whose size is $m \times r$. The $(i, j)$th element of $\mathbf{C}$ is given by

$$c_{ij} = \sum_{k=1}^{n} a_{ik}b_{kj}$$

**Example A.3.2**

$$\mathbf{A}_{(2\times3)} = \begin{bmatrix} 1 & 2 & 3 \\ 4 & 5 & 6 \end{bmatrix} \qquad \mathbf{B}_{(3\times2)} = \begin{bmatrix} 1 & 2 \\ 3 & 4 \\ 5 & 6 \end{bmatrix}$$

$$\mathbf{AB} = \mathbf{C}_{(2\times2)} = \begin{bmatrix} 22 & 28 \\ 49 & 64 \end{bmatrix}$$

**Example A.3.3**

Let

$$\mathbf{A} = \begin{bmatrix} 1 & 2 & 3 \\ 4 & 5 & 6 \end{bmatrix}, \qquad x = \begin{bmatrix} 2 \\ 3 \\ 4 \end{bmatrix} \quad \text{and} \quad y = (2, 3)$$

Then

$$\mathbf{A}x = \underset{(2\times1)}{b} = \begin{bmatrix} 20 \\ 47 \end{bmatrix}$$

while

$$y\mathbf{A} = \underset{(1\times3)}{d} = (14, 19, 24)$$

Note that $b$ is a column vector and $d$ is a row vector. For any scalar $\alpha$,

$$\alpha\mathbf{A} = [\alpha a_{ij}]$$

Matrix operations satisfy the following properties:

1.  $(\mathbf{A} + \mathbf{B}) + \mathbf{C} = \mathbf{A} + (\mathbf{B} + \mathbf{C})$
2.  $\mathbf{A} + \mathbf{B} = \mathbf{B} + \mathbf{A}$
3.  $(\mathbf{A} + \mathbf{B})\mathbf{C} = \mathbf{A}\mathbf{C} + \mathbf{B}\mathbf{C}$
4.  $(\mathbf{A}\mathbf{B})\mathbf{C} = \mathbf{A}(\mathbf{B}\mathbf{C})$
5.  $\mathbf{I}\mathbf{A} = \mathbf{A}\mathbf{I} = \mathbf{A}$
6.  $(\mathbf{A} + \mathbf{B})^T = \mathbf{A}^T + \mathbf{B}'$
7.  $(\mathbf{A}\mathbf{B})^T = \mathbf{B}'\mathbf{A}^T$

In general, $\mathbf{A}\mathbf{B} \neq \mathbf{B}\mathbf{A}$.

## A3.2   Determinant of a Square Matrix

The determinant of a square matrix $\mathbf{A}$, denoted by $|\mathbf{A}|$, is a number obtained by certain operations on the elements of $\mathbf{A}$. If $\mathbf{A}$ is a $2 \times 2$ matrix, then,

$$|\mathbf{A}| = \begin{vmatrix} a_{11} & a_{12} \\ a_{21} & a_{22} \end{vmatrix} = a_{11}a_{22} - a_{12}a_{21}$$

If $\mathbf{A}$ is an $n \times n$ matrix, then

$$|\mathbf{A}| = \sum_{i=1}^{n} a_{i1}(-1)^{i+1}|\mathbf{M}_{i1}|$$

where $\mathbf{M}_{i1}$ is a submatrix obtained by deleting row $i$ and column 1 of $\mathbf{A}$. For example, if

$$\underset{(3\times3)}{\mathbf{A}} = \begin{bmatrix} 1 & 2 & 3 \\ 4 & 5 & 6 \\ 7 & 8 & 9 \end{bmatrix}$$

then

$$|A| = 1\begin{vmatrix} 5 & 6 \\ 8 & 9 \end{vmatrix} - 4\begin{vmatrix} 2 & 3 \\ 8 & 9 \end{vmatrix} + 7\begin{vmatrix} 2 & 3 \\ 5 & 6 \end{vmatrix}$$

$$= (45 - 48) - 4(18 - 24) + 7(12 - 15) = 0$$

A matrix is said to be *singular* if its determinant is equal to zero. If $|A| \neq 0$, then **A** is called *nonsingular*.

### A.3.3   Inverse of a Matrix

For a nonsingular square matrix **A**, the inverse of **A**, denoted by $A^{-1}$, is a nonsingular square matrix such that

$$AA^{-1} = A^{-1}A = I \quad \text{(identity matrix)}$$

The inverse matrix $A^{-1}$ may be obtained by performing row operations on the original matrix **A**. The row operations consist of:

1.   Multiply or divide any row by a number.
2.   Multiply any row by a number and add it to another row.

To find the inverse of a matrix **A**, one starts by adjoining an identity matrix of similar size as [**A**]. By a sequence of row operations, **A** is reduced to **I**. This will reduce the original **I** matrix to $A^{-1}$, since

$$A^{-1}[AI] = [IA^{-1}]$$

*Example A.3.4*

$$A = \begin{bmatrix} 1 & 1 \\ 1 & -1 \end{bmatrix}$$

Since $|A| = -2$, **A** is nonsingular, and hence $A^{-1}$ exists. To compute $A^{-1}$, start with the matrix

$$(AI) = \begin{bmatrix} 1 & 1 & | & 1 & 0 \\ 1 & -1 & | & 0 & 1 \end{bmatrix}$$

Subtract row 1 from row 2:

$$\begin{bmatrix} 1 & 1 & | & 1 & 0 \\ 0 & -2 & | & -1 & 1 \end{bmatrix}$$

Divide row 2 by $-2$:

$$\begin{bmatrix} 1 & 1 & | & 1 & 0 \\ 0 & 1 & | & 1/2 & -1/2 \end{bmatrix}$$

Subtract row 2 from row 1:

$$\begin{bmatrix} 1 & 0 & \vdots & 1/2 & 1/2 \\ 0 & 1 & \vdots & 1/2 & -1/2 \end{bmatrix}$$

Thus,

$$\mathbf{A}^{-1} = \begin{bmatrix} 1/2 & 1/2 \\ 1/2 & -1/2 \end{bmatrix}$$

Verify that $\mathbf{AA}^{-1} = \mathbf{A}^{-1}\mathbf{A} = \mathbf{I}$.

### A.3.4  Condition of a Matrix

The condition of the matrix is a measure of the numerical difficulties likely to arise when the matrix is used in calculations. Consider, for example, the quadratic function

$$\mathbf{Q}(x) = x^T \mathbf{A}^{-1} x$$

$\mathbf{A}$ is said to be ill-conditioned if small changes in $x$ produce large changes in the value of $\mathbf{Q}$. Ill-conditioning is often associated with $\mathbf{A}$ being very nearly singular.

The *condition number* of a matrix as used in this book is given by

$$K(\mathbf{A}) = \left| \frac{\lambda_h}{\lambda_l} \right|$$

where $\lambda_h$ and $\lambda_l$ are the eigenvalues of greatest and smallest modulus respectively. A matrix with large condition number is ill-conditioned; a matrix with condition number close to 1 is well-conditioned.

### A.3.5  Sparse Matrix

A *sparse* matrix is one that has relatively few nonzero elements. For example, a $100 \times 100$ identity matrix will have $10^4$ elements but only 100 of these (the diagonal elements) are nonzero; hence, this matrix has only 1% nonzero elements. To avoid storage of and arithmetic operations with the zero elements, sparse matrices are usually stored in packed form, operations involving matrix elements are carried out involving only the nonzero elements, and the operations are sequenced to generate as few new nonzero elements as practical. This extra logic constitutes an overhead that must always be balanced against the savings obtained by sparse matrix methods.

One simple packed form of storage is to store the nonzero elements in a linked list supplemented by a column pointer vector. Thus array elements are stored in a long vector consisting of a string of triplets $(i, a, p)$. The entries of each triplet are associated with a specific nonzero element: $i$ is the row index of the element, $a$ is the value of the element, and $j$ is the pointer that gives the location of the triplet associated with the next nonzero element in that row of the matrix. The column pointer vector is simply an $n$-component vector that gives the location of the triplet associated with first nonzero row element of each column of the matrix. For further discussion of sparse matrices and methods, the reader is directed to R. P. Tewarson, *Sparse Matrices*, Academic Press, New York, 1973.

## A.4   QUADRATIC FORMS

A function of $n$ variables $f(x_1, x_2, \ldots, x_n)$ is called a *quadratic form* if

$$f(x_1, x_2, \ldots, x_n) = \sum_{i=1}^{n} \sum_{j=1}^{n} q_{ij} x_i x_j = x^T Q x$$

where $Q_{(n \times n)} = [q_{ij}]$ and $x^T = (x_1, x_2, \ldots, x_n)$. Without any loss of generality, $Q$ can always be assumed to be symmetric. Otherwise, $Q$ may be replaced by the symmetric matrix $(Q + Q')/2$ without changing the value of the quadratic form.

### Definitions

1.  A matrix $Q$ is *positive definite* if and only if the quadratic form $x^T Q x > 0$ for all $x \neq 0$. For example,

$$Q = \begin{bmatrix} 2 & -1 \\ -1 & 2 \end{bmatrix}$$

   is positive definite.

2.  A matrix $Q$ is *positive semidefinite* if and only if the quadratic form $x^T Q x \geq 0$ for all $x$ and there exists an $x \neq 0$ such that $x^T Q x = 0$. For example,

$$Q = \begin{bmatrix} 1 & -1 \\ -1 & 1 \end{bmatrix}$$

   is positive semidefinite.

3.  A matrix $Q$ is *negative definite* if and only if $-Q$ is positive definite. In other words, $Q$ is negative definite when $x^T Q x < 0$ for all $x \neq 0$. For example,

$$Q = \begin{bmatrix} -2 & 1 \\ 1 & -3 \end{bmatrix}$$

   is negative definite.

4.  A matrix $\mathbf{Q}$ is *negative semidefinite* if $-\mathbf{Q}$ is positive semidefinite. For example,

$$\mathbf{Q} = \begin{bmatrix} -1 & 1 \\ 1 & -1 \end{bmatrix}$$

is negative semidefinite.

5.  A matrix $\mathbf{Q}$ is *indefinite* if $x^T\mathbf{Q}x$ is positive for some $x$ and negative for some other $x$. For example,

$$\mathbf{Q} = \begin{bmatrix} 1 & -1 \\ 1 & -2 \end{bmatrix}$$

is indefinite.

### A.4.1  Principal Minor

If $\mathbf{Q}$ is an $n \times n$ matrix, then the *principal minor* of order $k$ is a submatrix of size $k \times k$ obtained by deleting any $n - k$ rows and their corresponding columns from the matrix $\mathbf{Q}$.

*Example A.4.1*

$$\mathbf{Q} = \begin{bmatrix} 1 & 2 & 3 \\ 4 & 5 & 6 \\ 7 & 8 & 9 \end{bmatrix}$$

Principal minors of order 1 are essentially the diagonal elements 1, 5, and 9. The principal minor of order 2 are the following ($2 \times 2$) matrices:

$$\begin{bmatrix} 1 & 2 \\ 4 & 5 \end{bmatrix} \quad \begin{bmatrix} 1 & 3 \\ 7 & 9 \end{bmatrix} \quad \text{and} \quad \begin{bmatrix} 5 & 6 \\ 8 & 9 \end{bmatrix}$$

The principal minor of order 3 is the matrix $\mathbf{Q}$ itself.

The determinant of a principal minor is called the *principal determinant*. For an $n \times n$ square matrix, there are $2^n - 1$ principal determinants in all.

The *leading principal minor* of order $k$ of an $n \times n$ matrix is obtained by deleting the last $n - k$ rows and their corresponding columns. In Example A.4.1, the leading principal minor of order 1 is 1 (delete the last two rows and columns). The leading principal minor of order 2 is $\begin{bmatrix} 1 & 2 \\ 4 & 5 \end{bmatrix}$, while that of order 3 is the matrix $\mathbf{Q}$ itself. The number of leading principal determinants of an $n \times n$ matrix is $n$.

There are some easier tests to determine whether a given matrix is positive definite, negative definite, positive semidefinite, negative semidefinite, or indefinite. *All these tests are valid only when the matrix is symmetric.* (If the matrix $\mathbf{Q}$ is not symmetric, change $\mathbf{Q}$ to $(\mathbf{Q} + \mathbf{Q}')/2$ and then apply the tests.)

*Tests for Positive Definite Matrices*

**1.** All diagonal elements must be positive.
**2.** All the leading principal determinants must be positive.

*Tests for Positive Semidefinite Matrices*

**1.** All diagonal elements are nonnegative.
**2.** All the principal determinants are nonnegative.

*Remarks*

**1.** Note the major difference in the tests for positive definite and positive semidefinite matrices. For the positive definite property, it is sufficient that the *leading* principal determinants are positive, while for positive semidefinite, we must test whether *all* the principal determinants are nonnegative.
**2.** To prove that a matrix is negative definite (negative semidefinite), test the negative of that matrix for positive definite (positive semidefinite).
**3.** A sufficient test for a matrix to be indefinite is that at least two of its diagonal elements are of the opposite sign.

### A.4.2  Completing the Square

Another approach to determine whether a symmetric matrix $\mathbf{Q}$ is positive definite, negative definite, or indefinite is to rewrite the quadratic form $x^T \mathbf{Q} x$ as a sum of perfect squares. The property of the quadratic form can then be determined from the coefficients of the squared terms.

*Example A.4.2*

Given the quadratic function

$$q(x) = ax_1^2 + 2bx_1x_2 + cx_2^2$$

collect the terms in $x_1$;

$$q(x) = \frac{1}{a}\left(a^2x_1^2 + 2abx_1x_2\right) + cx_2^2$$

Next, add and subtract an appropriate term in $x_2^2$ so that the term in parentheses becomes a perfect square:

$$q(x) = \frac{1}{a}\left(a^2x_1^2 + 2abx_1x_2 + b^2x_2^2 - b^2x_2^2\right) + cx_2^2$$

$$q(x) = \frac{1}{a}\left(ax_1 + bx_2\right)^2 - \frac{b^2}{a}x_2^2 + cx_2^2$$

Finally, grouping the coefficients of the $x_2^2$ term, we obtain

$$q(x) = \frac{1}{a}(ax_1 + bx_2)^2 + \frac{1}{a}(ac - b^2)x_2^2$$

Now, if $1/a > 0$ and $(ac - b^2) > 0$, then clearly $q(x)$ will be $> 0$ for any choice of $x \in R^N$. Similarly, if $1/a < 0$ and $(ac - b^2) > 0$, then $q(x)$ will be negative for all choices of $x$. For any other combination of coefficient signs, $q(x)$ will go through a sign change and hence will be indefinite.

Completing the square can be viewed as a change of variables that results in a perfect square. For instance, if in the above example we let

$$z_1 = ax_1 + bx_2$$

$$z_2 = x_2$$

or

$$x_1 = \frac{1}{a}(z_1 - bx_2) = \frac{1}{a}(z_1 - bz_2)$$

$$x_2 = z_2$$

then we obtain

$$q(x) = ax_1^2 + 2bx_1x_2 + cx_2^2$$

$$= a\left(\frac{z_1 - bz_2}{a}\right)^2 + 2b\left(\frac{z_1 - bz_2}{a}\right)z_2 + cz_2^2$$

$$= \frac{1}{a}\left(z_1^2 - 2bz_1z_2 + b^2z_2^2\right) + \frac{1}{a}\left(2bz_1z_2 - 2b^2z_2^2\right) + cz_2^2$$

$$q(x) = \frac{1}{a}z_1^2 + \frac{1}{a}(ac - b^2)z_2^2$$

It will be helpful to examine these operations in matrix form. Begin, once again, with the quadratic

$$q(x) = [x_1, x_2]\begin{bmatrix} a & b \\ b & c \end{bmatrix}\begin{Bmatrix} x_1 \\ x_2 \end{Bmatrix}$$

and introduce the change of variables:

$$\begin{Bmatrix} x_1 \\ x_2 \end{Bmatrix} = \begin{bmatrix} \dfrac{1}{a} & \dfrac{-b}{a} \\ 0 & 1 \end{bmatrix}\begin{Bmatrix} z_1 \\ z_2 \end{Bmatrix}$$

Accordingly, we obtain

$$q(z) = [z_1, z_2] \begin{bmatrix} \dfrac{1}{a} & 0 \\ \dfrac{-b}{a} & 1 \end{bmatrix} \begin{bmatrix} a & b \\ b & c \end{bmatrix} \begin{bmatrix} \dfrac{1}{a} & \dfrac{-b}{a} \\ 0 & 1 \end{bmatrix} \left\{ \begin{matrix} z_1 \\ z_2 \end{matrix} \right\}$$

$$= [z_1, z_2] \begin{bmatrix} \dfrac{1}{a} & 0 \\ \dfrac{-b}{a} & 1 \end{bmatrix} \begin{bmatrix} 1 & 0 \\ \dfrac{b}{a} & \left( c - \dfrac{b^2}{a} \right) \end{bmatrix} \left\{ \begin{matrix} z_1 \\ z_2 \end{matrix} \right\}$$

$$q(z) = [z_1, z_2] \begin{bmatrix} \dfrac{1}{a} & 0 \\ 0 & \dfrac{1}{a}(ac - b^2) \end{bmatrix} \left\{ \begin{matrix} z_1 \\ z_2 \end{matrix} \right\}$$

Clearly, the process of completing the square is equivalent to introducing a change of variable that will reduce a general symmetric matrix to *diagonal* form. Alternatively, we can view these operations as equivalent to finding the factors **D** and **P** of a symmetric matrix **Q**, where **D** is a diagonal such that

$$\mathbf{Q} = \mathbf{P}^T \mathbf{D} \mathbf{P}$$

Fortunately, this can easily be done using the *Gaussian elimination procedure* (see Appendix C) or *reduction to row echelon form*, with some slight modifications.

We illustrate the procedure by continuing with our example. Start with

$$\mathbf{Q}_1 = \begin{bmatrix} a & b \\ b & c \end{bmatrix}$$

and store $a$ as $d_1$. Following the standard reduction procedure, divide the first row by $a$, the pivot element, and subtract $b$ times the result from the second row:

$$\mathbf{Q}_2 = \begin{bmatrix} 1 & \dfrac{b}{a} \\ 0 & \left( c - \dfrac{b^2}{a} \right) \end{bmatrix}$$

Next, the second row is divided by its first nonzero element, the pivot, and that element is stored as $d_2$. In our simple example, we have arrived at the desired row eschelon form, and the process terminates. Furthermore, we observe that with the final pivot removed,

$$\mathbf{Q}_3 = \begin{bmatrix} 1 & \dfrac{b}{a} \\ 0 & 1 \end{bmatrix}$$

we have generated the permutation matrix $P$, that is,

$$\mathbf{P} = \mathbf{Q}_3 = \begin{bmatrix} 1 & \dfrac{b}{a} \\ 0 & 1 \end{bmatrix}$$

Furthermore, the required diagonal matrix simply contains the stored pivot elements:

$$\mathbf{D} = \begin{bmatrix} a & 0 \\ 0 & \dfrac{1}{a}(ac - b^2) \end{bmatrix}$$

One can readily verify the decomposition by performing the matrix multiplication, $\mathbf{P}^T \mathbf{D} \mathbf{P}$ to produce $\mathbf{Q}$.

With this construction, determining whether a symmetric matrix is positive or negative definite consists of merely performing Gaussian elimination on the matrix and saving the pivot elements. If all of the pivot elements are positive, the matrix is positive definite; if all are negative, the matrix is negative definite; otherwise it is indefinite.

### Example A.4.3

Given

$$\mathbf{Q} = \begin{bmatrix} 12 & -2 \\ -2 & 10 \end{bmatrix}$$

We investigate the definiteness of $\mathbf{Q}$ via the reduction process. $d_1 = 12$, and the first reduced array is

$$\begin{bmatrix} 1 & -\frac{1}{6} \\ -2 & 10 \end{bmatrix}$$

The final reduction gives

$$\begin{bmatrix} 1 & -\frac{1}{6} \\ 0 & \frac{29}{3} \end{bmatrix} \quad \text{or} \quad \begin{bmatrix} 1 & -\frac{1}{6} \\ 0 & 1 \end{bmatrix}$$

with $d_2 = \frac{29}{3}$. Accordingly, $\mathbf{Q}$ has been factored to

$$\mathbf{Q} = \begin{bmatrix} 1 & 0 \\ -\frac{1}{6} & 1 \end{bmatrix} \begin{bmatrix} 12 & 0 \\ 0 & \frac{29}{3} \end{bmatrix} \begin{bmatrix} 1 & -\frac{1}{6} \\ 0 & 1 \end{bmatrix}$$

Since the two diagonal elements are *positive*, $\mathbf{Q}$ is positive definite.

## A.5 CONVEX SETS

A set $S$ is said to be a convex set if for any two points in the set the line joining those two points is also in the set. Mathematically, $S$ is a convex set if for any two vectors $x^{(1)}$ and $x^{(2)}$ in $S$, the vector $x = \lambda x^{(1)} + (1 - \lambda)x^{(2)}$ is also in $S$ for any number $\lambda$ between 0 and 1.

### *Examples*

Figures A.1 and A.2 represent convex sets, while Figure A.3 is not a convex set.

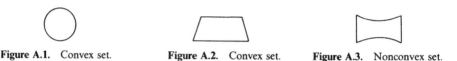

**Figure A.1.** Convex set.     **Figure A.2.** Convex set.     **Figure A.3.** Nonconvex set.

### *Theorem 1*

The set of all feasible solutions to a linear programming problem is a convex set.

### *Theorem 2*

The intersection of convex sets is a convex set (Figure A.4).

### *Theorem 3*

The union of convex sets is not necessarily a convex set (Figure A.4).

### *Definition*

A convex combination of vectors $x^{(1)}, x^{(2)}, \ldots, x^{(k)}$ is a vector $x$ such that

$$x = \lambda_1 x^{(1)} + \lambda_2 x^{(2)} + \cdots + \lambda_k x^{(k)}$$

$$\lambda_1 + \lambda_2 + \cdots + \lambda_k = 1$$

$$\lambda_i \geqslant 0 \qquad \text{for } i = 1, 2, \ldots, k$$

An *extreme point* or *vertex* of a convex set is a point in the set that cannot be expressed as the midpoint of any two points in the set. For example, consider the convex set $S = \{(x_1, x_2) | 0 \leqslant x_1 \leqslant 2, 0 \leqslant x_2 \leqslant 2\}$. This set has four extreme points given by $(0, 0)$, $(0, 2)$, $(2, 0)$, and $(2, 2)$.

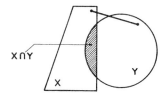

**Figure A.4.**   Intersection and union of convex sets.

A *hyperplane* is the set of all points $x$ satisfying $cx = z$ for a given vector $c \neq 0$ and scalar $z$. The vector $c$ is called the *normal* to the hyperplane. For example, $H = \{(x_1, x_2, x_3)|2x_1 - 3x_2 + x_3 = 5\}$ is a hyperplane.

A *half-space* is the set of all points $x$ satisfying $cx \leqslant z$ or $cx \geqslant z$ for a given vector $c \neq 0$ and scalar $z$.

### Theorem 4

A hyperplane is a convex set.

### Theorem 5

A half-space is a convex set.

# Appendix B

# Convex and Concave Functions

**Convex Function.** A function of $n$ variables $f(x)$ defined on a convex set $D$ is said to be a *convex function* if and only if for any two points $x^{(1)}$ and $x^{(2)} \in D$ and $0 \leqslant \lambda \leqslant 1$,

$$f\left[\lambda x^{(1)} + (1-\lambda)x^{(2)}\right] \leqslant \lambda f(x^{(1)}) + (1-\lambda)f(x^{(2)})$$

Figure A.5 illustrates the definition of a convex function of a single variable.

## Properties of Convex Functions

1. The chord joining any two points on the curve always falls entirely on or above the curve between those two points.
2. The slope or first derivative of $f(x)$ is *increasing* or at least *nondecreasing* as $x$ increases.
3. The second derivative of $f(x)$ is always *nonnegative* for all $x$ in the interval.
4. The linear approximation of $f(x)$ at any point in the interval always *underestimates* the true function value.
5. For a convex function, a local minimum is always a global minimum.

Figure A.6 illustrates property 4. The linear approximation of $f$ at the point $x^0$, denoted by $\tilde{f}(x; x^0)$, is obtained by ignoring the second and other higher order terms in the Taylor's series expansion

$$\tilde{f}(x; x^0) = f(x^0) + \nabla f(x^0)(x - x^0)$$

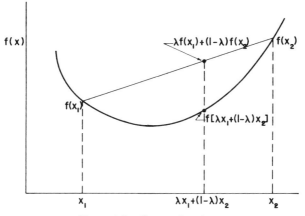

**Figure A.5.** Convex function.

For a convex function, property 4 implies that

$$f(x) \geqslant f(x^0) + \nabla f(x^0)(x - x^0) \qquad \text{for all } x$$

The *gradient* of a function $f(x_1, \ldots, x_n)$ is given by

$$\nabla f(x_1, \ldots, x_n) = \left[ \frac{\delta f}{\delta x_1}, \frac{\delta f}{\delta x_2}, \ldots, \frac{\delta f}{\delta x_n} \right]$$

The *Hessian matrix* of a function $f(x_1, \ldots, x_n)$ is an $n \times n$ symmetric matrix given by

$$\mathbf{H}_f(x_1, \ldots, x_n) = \left[ \frac{\delta^2 f}{\delta x_i \, \delta x_j} \right] = \nabla^2 f$$

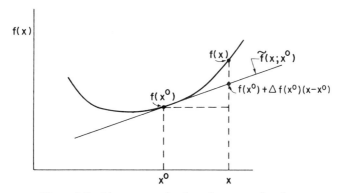

**Figure A.6.** Linear approximation of a convex function.

### Test for Convexity of a Function

A function $f$ is convex function if the Hessian matrix of $f$ is positive definite or positive semidefinite for all values of $x_1, \ldots, x_n$.

**Concave Function.**　A function $f(x)$ is a *concave function* in $D$ if and only if $-f(x)$ is a convex function in $D$.

### Test for Concavity of a Function

A function $f$ is concave if the Hessian matrix of $f$ is negative definite or negative semidefinite for all values of $x_1, \ldots, x_n$.

### Example B-1

$$f(x_1, x_2, x_3) = 3x_1^2 + 2x_2^2 + x_3^2 - 2x_1x_2 - 2x_1x_3$$

$$+ 2x_2x_3 - 6x_1 - 4x_2 - 2x_3$$

$$\nabla f(x_1, x_2, x_3) = \begin{pmatrix} 6x_1 - 2x_2 - 2x_3 - 6 \\ 4x_2 - 2x_1 + 2x_3 - 4 \\ 2x_3 - 2x_1 + 2x_2 - 2 \end{pmatrix}$$

$$\mathbf{H}_f(x_1, x_2, x_3) = \begin{bmatrix} 6 & -2 & -2 \\ -2 & 4 & 2 \\ -2 & 2 & 2 \end{bmatrix}$$

To show that $f$ is a convex function, we test for whether $\mathbf{H}$ is positive definite or positive semidefinite. Note that

1.　$\mathbf{H}$ is symmetric.
2.　All diagonal elements are positive.
3.　The leading principal determinants are

$$|6| > 0, \quad \begin{vmatrix} 6 & -2 \\ -2 & 4 \end{vmatrix} = 20 > 0, \quad |H_f| = 16 > 0$$

Hence $\mathbf{H}$ is a positive definite matrix, which implies $f$ is a convex function. (As a matter of fact, when $\mathbf{H}_f$ is positive definite, $f$ is said to be strictly convex with a unique minimum point.)

# Appendix C

# Gauss-Jordan Elimination Scheme

Systems of linear equations can be solved by the classical Gauss-Jordan elimination procedure. In this section, a brief review of this procedure is given through an example.

***Example C-1***

Consider the system of two equations in five unknowns denoted by

$$x_1 - 2x_2 + x_3 - 4x_4 + 2x_5 = 2 \qquad \text{(C.1)}$$

$(S_1)$

$$x_1 - x_2 - x_3 - 3x_4 - x_5 = 4 \qquad \text{(C.2)}$$

Since there are more unknowns than equations, this system will have more than one solution. The collection of all possible solutions to the system is called the *solution set*.

***Definition***

Two systems of equations are said to be *equivalent* if both systems have the same solution set. In other words, a solution to one system is automatically a solution to the other system, and vice versa.

The method of solving a system of equations is to get an equivalent system that is easy to solve. By solving the simple system, we simultaneously get the solutions to the original system.

There are two types of elementary row operations that can be used to obtain equivalent systems.

1. Multiply any equation in the system by a positive or negative number.
2. Add to any equation a constant multiple (positive, negative, or zero) of any other equation in the system.

An equivalent system to system $S_1$ can be obtained by multiplying Eq. (C.1) by $-1$, and adding the result to Eq. (C.2) as follows:

$$x_1 - 2x_2 + x_3 - 4x_4 + 2x_5 = 2 \qquad \text{(C.3)}$$

$(S_2)$

$$x_2 - 2x_3 + x_4 - 3x_5 = 2 \qquad \text{(C.4)}$$

From system $S_2$, another equivalent system can be obtained by multiplying Eq. (C.4) by 2, and adding it to Eq. (C.3). This gives system $S_3$:

$$x_1 - 3x_3 - 2x_4 - 4x_5 = 6 \qquad \text{(C.5)}$$

$(S_3)$

$$x_2 - 2x_3 + x_4 - 3x_5 = 2 \qquad \text{(C.6)}$$

Since systems $S_1$, $S_2$, and $S_3$ are equivalent, a solution to one automatically gives a solution to the other two. In this case, it is easy to write out all the possible solutions to system $S_3$. For example, setting $x_3 = x_4 = x_5 = 0$ gives $x_1 = 6$, $x_2 = 2$, which is a solution to all the three systems. Other solutions to system $S_3$ may be obtained by choosing arbitrary values for $x_3$, $x_4$, $x_5$ and finding the corresponding values of $x_1$ and $x_2$ from Eqs. (C.5) and (C.6). All these solutions are solutions to the original system. Systems like $S_3$ are called *canonical systems* or systems in *row echelon form*.

# Author Index

# Subject Index